POLAROGRAPHY

Second Edition

By I. M. Kolthoff and James J. Lingane

VOLUME I:

Theoretical Principles
Instrumentation and Technique

VOLUME II:

Inorganic Polarography
Organic Polarography
(with the collaboration of S. Wawzonek)
Biological Applications
Amperometric Titrations

POLAROGRAPHY

Second completely revised
and augmented edition
In Two Volumes

I. M. KOLTHOFF
*Professor and Head of Division
of Analytical Chemistry
University of Minnesota
Minneapolis, Minnesota*

JAMES J. LINGANE
*Professor of Chemistry
Harvard University
Cambridge, Massachusetts*

VOLUME II

**Inorganic Polarography
Organic Polarography
Biological Applications
Amperometric Titrations**

1952

INTERSCIENCE PUBLISHERS
NEW YORK • LONDON

Library of Congress Catalog Card Number 52-6638

INTERSCIENCE PUBLISHERS, Inc.
250 Fifth Avenue, New York 1, New York

For Great Britain and Northern Ireland:
INTERSCIENCE PUBLISHERS, Ltd.
2a Southampton Row, London W. C. 1, England

PRINTED IN THE UNITED STATES OF AMERICA

PREFACE TO VOLUME II

The general aims of this revision are stated in the Preface to the Second Edition at the front of Volume I. The first volume deals with the theory of polarography, with instrumentation, and with general methodology. The present Volume II is concerned with practical applications in inorganic, organic, and biological chemistry, and with amperometric titrations. Author and Subject Indexes for both volumes will be found at the end of this volume.

Part Three, which begins this volume, consists of a systematic review of the polarographic chemistry of the elements, and concludes with Chapter XXXV which presents selected typical procedures for the analysis of various types of alloys and other technical materials. Because Chapter XXXV is intended for bench use by the practicing analyst most of the procedures are given in sufficient detail to permit their application without recourse to the original literature. We are confident of the reliability of these procedures, within their stated limitations, and their successful application does not require one to be a specialist in polarography. However a modicum of fundamental knowledge is essential, and those not already familiar with polarography are urged to consider the advice given in lines 12 to 22 on page 15 of Volume I.

Part Four deals with the polarography of organic substances. This proved to be an especially difficult part of the book to prepare, not only because of our limited competence in the art of organic chemical interpretation, but because organic polarography is still in a chaotic state and is badly in need of systematic development. We are greatly indebted to Professor Stanley Wawzonek of the University of Iowa, who, as an organic chemist thoroughly conversant with polarography, rendered invaluable assistance in the major task of organizing these chapters.

We are also grateful to Dr. W. Stricks of the University of Minnesota for his help in gathering information for the enlarged Part Five on Biological Applications.

Minneapolis, Minnesota
Cambridge, Massachusetts
July 1, 1952

I. M. Kolthoff
James J. Lingane

Contents—Volume I

PART ONE: THEORETICAL PRINCIPLES

PART TWO: INSTRUMENTATION AND TECHNIQUE

Contents—Volume II

PART FOUR: ORGANIC POLAROGRAPHY

PART SIX: AMPEROMETRIC TITRATIONS

Part Three

INORGANIC POLAROGRAPHY

CHAPTER XX

Alkali Metals

The behavior of the alkali and alkaline earth metal ions was the subject of Heyrovsky's first polarographic investigation,[1] and one of the most frequently quoted favorable attributes of the polarographic method is its applicability to the determination of these elements. Although the alkali metal ions are characterized by very negative reduction potentials, well-developed waves are obtainable from neutral solutions without interference from hydrogen discharge by virtue of three facts: (*a*) the formation of stable compounds of mercury and the alkali metals, both in the solid state and in the liquid amalgams, decreases the reduction potentials by about 1 v. compared to the pure solid metals; (*b*) the reduction of hydrogen ion on mercury requires a potential about 0.8 v. more negative than on platinum (overvoltage); and (*c*) in neutral or alkaline solution the reduction potential of hydrogen ion is 0.5 v. to 0.8 v. more negative than in acid solutions—the pH(RT/F) factor in the Nernst equation. The cumulative result of these three effects is that the alkali metal ions are reduced more easily than hydrogen ion from neutral or alkaline solutions. In acid media, however, reduction waves of the alkali metals are precluded by hydrogen discharge because factors (*a*) and (*b*) are more than counteracted by the reversal of factor (*c*).

The choice of supporting electrolytes for the determination of the alkali metals is necessarily limited, and tetraalkylammonium halides or hydroxides are virtually the only usable salts. The half-wave potential of ammonium ion itself is -2.05 v. *vs.* S.C.E., but the tetraalkyl-substituted ammonium ions have reduction potentials several tenths of a volt more negative, and their employment extends the usable potential range to about -2.4 v. Because of their more negative reduction potentials the tetraethyl salts are preferable to the tetramethyl salts.

Unfortunately, high purity tetraalkylammonium salts are not commonly available, and commercial preparations may be contaminated with reducible impurities and alkali metals. Because of their great solubility the purification of these salts by recrystallization is not too easy, and the thermal instability of the free bases precludes purification of these by

[1] J. Heyrovsky, *Chem. Listy*, **16**, 256 (1922); *Phil. Mag.*, **45**, 303 (1923).

ordinary distillation methods. Tetramethyl- or tetraethylammonium bromide or iodide are the easiest salts to purify, and this is best achieved by recrystallization from water–ethanol solutions. The recrystallization is repeated until polarograms of the purified salt show the absence of reducible impurities. Solutions of the free bases are prepared by treating solutions of the purified halides with an excess of moist silver oxide and filtering off the silver halide and excess silver oxide. Since tetraalkylammonium hydroxides are strong bases their solutions should not be stored in unprotected glass containers for any length of time because this can lead to contamination by sodium and calcium from the glass. The storage vessel should be protected by a wax coating.

According to Weaver and Lykken,[2] whose paper is a valuable contribution to the polarographic determination of sodium and/or potassium in various materials, solutions of tetraethylammonium hydroxide deteriorate after 5 or 6 days as evidenced by a high residual current and poorly developed sodium wave. On the other hand, Zlotowski and Kolthoff[3] observed that precipitates were formed in freshly prepared aqueous tetraethylammonium hydroxide, and this was accompanied by a *decrease* in the residual current. Solutions which had stood in the dark for 8 days and then were filtered showed no further change in residual current after several weeks' standing. Alcoholic solutions are less stable and should be freshly prepared as needed. The suitability of the reagent solution should be verified by daily blank polarograms.

Peracchio and Meloche[4] observed that the alkali metal waves were developed better in mixtures of water and ethanol (and other alcohols) than in purely aqueous medium, and this has been verified by the systematic investigations of Zlotowski and Kolthoff.[3, 5] The polarographic characteristics of the alkali metal ions in 0.1 *M* tetraethylammonium hydroxide in 50 per cent ethanol are summarized in Table XX-1, which is based on the measurements of Zlotowski and Kolthoff, except for the half-wave potentials of rubidium and cesium which are derived from the report of Heyrovsky and Ilkovic.[6]

The half-wave potentials in pure water are approximately 0.05 v. more negative, and in 80 per cent ethanol about 0.04 v. more positive, than in 50 per cent ethanol. To some extent this includes changes in liquid-junc-

[2] J. R. Weaver and L. Lykken, *Ind. Eng. Chem., Anal. Ed.*, **19**, 372 (1947).
[3] I. Zlotowski and I. M. Kolthoff, *Ind. Eng. Chem., Anal. Ed.*, **14**, 473 (1942).
[4] E. S. Peracchio and V. W. Meloche, *J. Am. Chem. Soc.*, **60**, 1770 (1938).
[5] I. Zlotowski and I. M. Kolthoff, *J. Am. Chem. Soc.*, **64**, 1297 (1942).
[6] J. Heyrovsky and D. Ilkovic, *Collection Czechoslov. Chem. Communs.*, **7**, 198 (1935).

tion potential between the test solutions and the aqueous calomel reference electrode. The reduction proceeds reversibly as shown both by the wave slopes and the agreement of the observed half-wave potentials with the theoretical values computed from the standard potentials of the solid metals, their solubilities in mercury, and the free energies of formation of the amalgams as discussed in Chapter XI.

The values of $i_d/Cm^{2/3}t^{1/6}$ decrease with increasing ethanol concentration up to about 80 per cent and then increase somewhat with larger concentrations of ethanol. Matsuyama[7] demonstrated that this can be only approximately correlated with the change in viscosity and that solvation effects are also partly operative. The earlier measurements of Peracchio and Meloche[4] show that the wave heights decrease in the order: water, ethylene glycol, trimethylene glycol, and glycerol. The diffusion currents in 70 to 80 per cent dioxane are not as well developed as in the other mixed solvents.

TABLE XX-1

CHARACTERISTICS OF ALKALI METAL IONS IN 50 PER CENT ETHANOL AT 25°

0.1 *M* Tetraethylammonium hydroxide as supporting electrolyte. Half-wave potentials *vs.* saturated calomel electrode. Values of $i_d/Cm^{2/3}t^{1/6}$ at drop times greater than 1.5 sec.

Ion	$E_{1/2}$, volts	$\frac{i_d}{Cm^{2/3}t^{1/6}}$
Li^+	−2.31	1.16
Na^+	−2.07	1.40
K^+	−2.10	1.69
Rb^+	−1.99	—
Cs^+	−2.05	—

The close grouping of the half-wave potentials of Na^+, K^+, Rb^+, and Cs^+ makes impossible their polarographic discrimination, and mixtures of two or more of them produce only a single cumulative wave.[8] However, the half-wave potential of lithium is sufficiently removed from those of the other alkali metals that separate determinations of lithium and the sum of the others are quite feasible. Zlotowski and Kolthoff[3, 5] found that sodium and/or potassium can be determined accurately in the presence

[7] G. Matsuyama, Ph.D. Thesis, University of Minnesota, 1947.

[8] An indirect method of determining sodium and potassium from exaltation of the limiting current of the mixed chlorides without supporting electrolyte was proposed by Heyrovsky and Bures, *Collection Czechoslov. Chem. Communs.*, **8**, 446 (1936), and is discussed in Chapter VII.

of a 50-fold excess of lithium, and lithium can be determined in the presence of up to a 10-fold excess of sodium and/or potassium, in 80 per cent ethanol.

Because the residual current is relatively large at the very negative potentials at which the diffusion currents of the alkali metal ions are measured, proper correction for it is especially important. The correction is best made by running a blank polarogram with the supporting electrolyte alone and subtracting the residual current, measured at the potential at which the alkali metal diffusion current is measured, from the total current. The importance of the correction is exemplified by data presented by Zlotowski and Kolthoff which show that the residual current accounts for 28 per cent of the total current observed with 1 millimolar lithium ion at −2.5 v. and for 20 per cent of the total current produced by 1 millimolar potassium ion at −2.25 v.

Determinations of the alkali metal ions require the prior removal of most all other heavy metal ions. The classical methods customarily used for the isolation of the alkali metal chlorides in silicate analysis may of course be used, if one has the time and patience. However, the most expeditious procedure is electrolysis with the mercury cathode from acidic solutions. This convenient technique permits the removal of as much as 500 mg. of copper, lead, zinc, iron, chromium, cadmium, cobalt, nickel, etc., from as little as 1 mg. of sodium and/or potassium.[9] It does not remove such elements as aluminum, titanium, vanadium, or uranium. However, Weaver and Lykken (*loc. cit.*) have shown that even large amounts of aluminum do not interfere with the determination of sodium and/or potassium when tetraethylammonium hydroxide is used as supporting electrolyte, because the aluminate ion formed in such strongly alkaline medium is not reduced. Titanium is precipitated as the hydrous oxide in tetraethylammonium hydroxide and it should not interfere unless its amount is so large that coprecipitation of the alkali metal ions becomes significant.

Using aqueous 1 *M* tetraethylammonium hydroxide as supporting electrolyte, and a sodium ion concentration of 8 micrograms per milliliter, Weaver and Lykken found that as much as 300 micrograms of aluminum, manganese, copper, cadmium, or magnesium, 50 micrograms of calcium, 80 micrograms of mercuric mercury or lead, 20 micrograms of ammonium ion, 9 micrograms of zinc, and 1 microgram of iron, interfered only slightly with the sodium determination. Larger amounts of most of these elements interfered seriously, presumably due to coprecipitation of sodium

[9] H. O. Johnson, J. R. Weaver, and L. Lykken, *Ind. Eng. Chem., Anal. Ed.*, **19**, 481 (1947).

ion in those cases where precipitates are formed. Boron and phosphate do not interfere even in large amounts. In addition to electrolysis with the mercury cathode Weaver and Lykken recommended ether extraction for the removal of large quantities of iron, precipitation with ammonium carbonate for reducing calcium to a permissible concentration, and precipitation with hydrogen sulfide in acid solution for removal of large amounts of copper and presumably other heavy metals.

Majer[10] showed that sodium and potassium could be determined in the presence of moderate amounts of iron, aluminum, titanium, calcium, and magnesium, after precipitating the latter with phosphate and an excess of tetramethylammonium hydroxide. The precipitates need not be filtered off when their amount is small.

Separate Determinations of Sodium and Potassium. Abresch[11] developed the following procedure for the separate determination of potassium and sodium, which is based on the preliminary separation of the potassium by the perchlorate method.

A 1-ml. portion of a weakly acid solution, containing from 1 to 10 mg. each of sodium and potassium, is transferred to a small centrifuge tube and diluted with 3 ml. of absolute ethanol. The potassium is precipitated by the dropwise addition of 6 ml. of a solution composed of one volume of 70 per cent perchloric acid and two volumes of absolute ethanol,[12] and the mixture is centrifuged. The supernatant solution is drawn off, and the precipitate of potassium perchlorate is washed with another portion of the perchloric acid, and again centrifuged. The supernatant solution and washings are combined for the determination of sodium. The potassium perchlorate is dissolved in 5 ml. of water, treated with two drops of phosphoric acid, 2 ml. of 0.5 *N* tetramethylammonium hydroxide and diluted to 10 ml. The polarogram of the potassium in this solution is then recorded. Sodium is determined in the solution and washings from the precipitation of potassium perchlorate by diluting with an equal volume of water adding 1 ml. of 10 per cent sulfuric acid, and evaporating to dryness (caution!). The residue is taken up in water, treated with tetramethylammonium hydroxide, and the polarogram of the sodium obtained.

Lohuis, Meloche, and Juday[13] analyzed mixtures of sodium and potassium salts, by first determining the sum of sodium and potassium by the polarographic method, and then determining the sodium by the ordinary zinc uranyl acetate method. They applied this procedure to the determination of sodium and potassium in lake waters.

[10] V. Majer, *Z. anal. Chem.*, **92,** 321, 401 (1933).

[11] K. Albresch, *Angew. Chem.*, **48,** 683 (1935); *Chem. Fabrik*, **8,** 350 (1935).

[12] *Caution* must be observed with such solutions because they are explosive when warmed.

[13] D. Lohuis, V. W. Meloche, and C. Juday, *Trans. Wisconsin Acad. Sci.*, **31,** 285 (1938).

Methods for the determination of the alkali metals in natural waters have also been described by Reznikov and Starik-Smagina,[14] and by Komar.[15]

Carruthers[16] developed an ultramicro method for sodium based on the precipitation of sodium zinc uranyl acetate and polarographic measurement of the uranium in the precipitate.

Portnov and Afanas'ev[17] utilized precipitation of potassium by the magnesium salt of hexanitrodiphenylamine to separate it from sodium in connection with the polarographic determinations of potassium and sodium in carnallite and sylvanite.

Sodium and Potassium in Minerals, Ceramics, etc. Majer[18] developed the following simple procedure for the determination of the sum of potassium and sodium in silicates of high aluminum and silica content.

A 5- to 20-mg. sample of the silicate is evaporated to dryness with 0.3 ml. of concentrated sulfuric acid and 0.5 ml. of 40 per cent hydrofluoric acid in a platinum crucible. The residue is treated with one drop of 85 per cent phosphoric acid, 0.5 ml. of water, and 1 ml. of 1 *N* tetramethylammonium hydroxide. The solution is transferred to a micro cell and the polarogram of sodium plus potassium is recorded. Calibration of the dropping electrode is effected by the internal standard method (Chap. XVIII), by adding a known quantity of potassium ion and recording a second polarogram. According to Majer this procedure is accurate to about ±5 per cent with silicates that contain up to 2 per cent sodium and potassium, and the determination requires only about 40 minutes. This procedure cannot be expected to give accurate results when the ratio of sodium to potassium varies considerably in different samples, because it is based on the assumption that the diffusion current constants of potassium and sodium ions are equal and this is not true (*vide supra*, Table XX-1).

Mainzhausen[19] and Okac[20] adapted Majer's method to the determination of potassium and sodium in soils.

Weaver and Lykken[2] recommended the following procedure for determining as little as 0.1 per cent of sodium and/or potassium in alumina-silica samples, and it should also be applicable to silicate minerals of low iron content.

---

[14] A. A. Reznikov and A. S. Starik-Smagina, *Trans. All-Union Conf. Anal. Chem.*, **2,** 559 (1943).

[15] N. V. Komar, *Zavodskaya Lab.*, **6,** 1074 (1937).

[16] C. Carruthers, *Ind. Eng. Chem., Anal. Ed.*, **15,** 70 (1943).

[17] M. A. Portnov and S. K. Afanas'ev, *Zavodskaya Lab.*, **7,** 421 (1938).

[18] V. Majer, *Chimie & industrie*, **29** (No. 6), 211 (1933).

[19] L. Mainzhausen, *Bodenkunde u. Pflanzenernähr.*, **14,** 290 (1939).

[20] A. Okac, *Bodenkunde u. Pflanzenernähr.*, **36,** 37 (1945).

A 0.5 g.-sample in a platinum dish is moistened with water and treated with three separate 5-ml. portions of 48 per cent hydrofluoric acid, with evaporation to near dryness after each addition. 10 ml. of 9 *M* (1:1) sulfuric acid is then added and the mixture is evaporated to sulfur trioxide fumes without prolonged heating. (Radiant infrared heating aids in avoiding spattering!) The residue is taken up in 10 to 15 ml. of water, transferred to a 250-ml. quartz or Vycor flask (glass would be attacked by the residual fluoride with resultant sodium contamination), and the solution is evaporated and heated until all the sulfuric acid has been removed. A slow stream of air delivered into the flask by a silica tube aids in this removal. The residue is taken up in about 10 ml. of hot water, transferred to a 25-ml. volumetric flask, cooled, and diluted to the mark.

An aliquot of this sample solution is mixed with exactly five times its volume of 1 *M* tetraethylammonium hydroxide in a polarographic cell and the polarogram is recorded after removing dissolved air with nitrogen.

From their experience in the Shell Development Co. laboratory with several hundred samples of silica, alumina, iron oxide, and petroleum products, Weaver and Lykken state: "Experience with both the chemical and polarographic methods for determination of sodium and potassium has demonstrated the definite superiority of the latter method, especially at concentrations below 0.5%."

Sodium and Potassium in Organic Materials. The following procedure was recommended by Weaver and Lykken.[2]

The sample contained in a platinum dish is ashed in a muffle furnace at 550° C until all organic matter is destroyed. (Liquid samples are first evaporated on a hot plate, or if they are not very volatile are heated over a burner until they ignite, and the residue is then ashed.) The ash is taken up with about 10 ml. of water, and the solution is transferred to a 25 ml. volumetric flask and diluted to the mark. An aliquot of this sample solution is mixed with exactly 5 times its volume of 1 *M* tetraethylammonium hydroxide in a polarographic cell, air is removed with nitrogen, and the polarogram is recorded.

By this procedure Weaver and Lykken were able to determine 0.005 per cent sodium in sucrose with a relative accuracy of 1 to 2 per cent.

The polarographic determination of sodium and potassium in blood serum, after destruction of the organic matter by evaporation with sulfuric acid and washing, has been described by Schweitzer[21] and Prinsen Geerligs.[22]

Sodium in Aluminum Alloys. See Chapter XXXV.

Sodium in Glass. See Chapter XXXV.

---

[21] K. Schweitzer, *Acta Brevia Neerland. Physiol. Pharmacol. Microbiol.*, **6,** 110 (1936).

[22] G. F. Prinsen Geerligs, *Acta Brevia Neerland. Physiol. Pharmacol. Microbiol.*, **7,** 38 (1937).

CHAPTER XXI

Beryllium, Magnesium, Alkaline Earth Metals, and Radium

1. BERYLLIUM

Heyrovsky and Berezicky[1] concluded that no reduction wave for beryllium was obtained with pure solutions of beryllium salts, but that a hydrogen wave was produced due to the pronounced hydrolysis of the beryllium ion. However, Kemula and Michalski[2] later found that polarograms of solutions of beryllium sulfate and beryllium chloride showed two waves: the first at about -1.4 v. due to the discharge of hydrogen, and the second at -1.8 v. *vs.* the S.C.E. due to the reduction of beryllium ion. The beryllium wave was not well defined. Kemula and Michalski also reported that the beryllium wave coalesced with that of aluminum.

2. MAGNESIUM

Kimura[3] and Müller[4] found that magnesium produces a very poorly defined wave at about -2.2 v. *vs.* the S.C.E. in solutions of tetramethylammonium salts as supporting electrolyte. The wave shows a very pronounced maximum and the ill-defined diffusion current that follows it it abnormally large. Müller, and Heyrovsky and Berezicky,[1] attributed this abnormally large diffusion current to the rapid decomposition of water, and the consequent evolution of hydrogen, by the freshly formed magnesium amalgam. Müller found that the maximum was much less pronounced in heavy water (99.2 per cent D_2O) than in ordinary water.

An indirect method for the determination of 0.07 to 0.6 mg. of magnesium in biological material, based on the precipitation of magnesium hydroxyquinolate and measurement of the hydroxyquinoline wave in a solution of the precipitate in a phosphate buffer of pH = 7.1, was developed

---

[1] J. Heyrovsky and S. Berezicky, *Collection Czechoslov. Chem. Communs.*, **1**, 19 (1920).

[2] W. Kemula and M. Michalski, *Collection Czechoslov. Chem. Communs.*, **5**, 436 (1933).

[3] G. Kimura, *Collection Czechoslov. Chem. Communs.*, **4**, 492 (1932).

[4] O. H. Müller, *Collection Czechoslov. Chem. Communs.*, **7**, 321 (1935).

by Carruthers.[5] Stone and Furman[6] improved this method by measuring the wave of the excess 8-hydroxyquinoline left after precipitation of the magnesium salt. The latter authors recommended an ammonia-ammonium chloride buffer of pH = 10. In this medium the reduction of the 8-hydroxyquinoline produces a double wave with half-wave potentials of −1.39 v. and −1.61 v. *vs.* S.C.E. The second wave is poorly developed and only the first wave is measured. As in any indirect method maximal precision is attained when the amount of hydroxyquinoline added in excess is kept small compared to the amount precipitated. Most of the many interfering metals which form precipitates with 8-hydroxyquinoline are best removed by electrolyzing a moderately acid solution of the sample with the mercury cathode. Aluminum, calcium, and titanium are the only common interfering ions which remain in solution along with magnesium. Small amounts (0.5 mg. per 25 ml.) of calcium can be tolerated, and aluminum interference can be circumvented by complexing it with tartrate added to the ammoniacal buffer. Stone and Furman recommend removal of titanium by precipitation as the hydroxide.

3. ALKALINE EARTH METALS

Calcium. The polarography of calcium was originally investigated by Kimura,[3] and more recently Zlotowski and Kolthoff[7] have systematically studied its characteristics. In aqueous tetramethylammonium iodide as supporting electrolyte the calcium wave is distorted by a large, sharp maximum. The maximum is much less pronounced and the diffusion current much better defined when the supporting electrolyte contains 80 per cent ethanol. In water–ethanol media the maximum can be completely eliminated by adding a trace of barium ion, whose reduction wave precedes that of calcium ion by *ca.* 0.2 v. In the presence of barium ion in 50 to 80 per cent ethanol the calcium wave is excellently developed and reduction proceeds reversibly. In aqueous medium lanthanum ion serves better than barium ion to suppress the maximum.

In purely aqueous medium the half-wave potential of calcium is −2.22 v. *vs.* S.C.E., and in 80 per cent ethanol it is −2.13 v., the supporting electrolyte being tetramethylammonium chloride in both cases. The value of $i_d/Cm^{2/3}t^{1/6}$ decreases nearly linearly from 3.39 in purely aqueous medium at 25° to 2.47 in 80 per cent ethanol, and then increases to 2.6 in pure ethanol.

[5] C. Carruthers, *Ind. Eng. Chem., Anal. Ed.*, **15**, 412 (1943).

[6] K. G. Stone and N. H. Furman, *Ind. Eng. Chem., Anal. Ed.*, **16**, 596 (1944).

[7] I. Zlotowski and I. M. Kolthoff, *J. Phys. Chem.*, **49**, 386 (1945).

Because it produces a current (probably catalytic hydrogen discharge) between −2.0 and −2.3 v., magnesium ion interferes seriously with the calcium wave when present in concentrations greater than about 3×10^{-5} *M*.

The calcium wave in tetramethylammonium iodide in 50 per cent ethanol is well separated from the preceding waves of barium and strontium, and calcium can be determined accurately in the presence of these two elements provided the sum of their concentrations is not greater than about five times that of calcium. The controlled potential electrolysis technique with a mercury cathode[8] should serve most conveniently to separate interferingly large amounts of barium and/or strontium from calcium; this problem is being investigated in our laboratories.

An indirect method for the determination of calcium, based on its precipitation with excess picrolonic acid and polarographic measurement of the excess reagent, has been developed by Cohn and Kolthoff.[9] The precipitation of calcium picrolonate (solubility product = 5×10^{-11}) and the polarographic measurement of picrolonic acid is best made in an acetate buffer of pH 3.6 to 3.8. In this medium the reduction of the picrolonic acid produces a double wave at −0.3 and −0.7 v. *vs.* S.C.E. and the total diffusion current is well developed between −0.9 v. and −1.1 v. when gelatin is present as a maximum suppressor.

The neutral calcium solution in a 100-ml. volumetric flask is treated with 10 ml. of a buffer composed of 1 *M* acetic acid, 0.125 *M* lithium acetate, and 1 *M* lithium chloride,[10] and a measured volume of 0.01 *M* picrolonic acid[11] is added. Not more than a four-fold excess of picrolonic acid, and preferably less, should be added. Sufficient gelatin solution to make its final concentration 0.01 per cent is added and the solution is diluted to the mark. The mixture is allowed to stand overnight (preferably in a refrigerator when the amount of calcium is small), and the residual concentration of picrolonic acid in the supernatant solution is determined from the diffusion current at −1.1 v. *vs.* S.C.E.

Magnesium, potassium, ammonium, and sodium ions do not interfere provided their final concentrations do not exceed about 0.05 *M*. Phosphate up to about 0.05 *M* and sulfate up to about 0.02 *M* do not interfere, but larger concentrations may precipitate calcium depending on its concentration.

Strontium. An early study of the polarography of strontium was carried

[8] See Chapter XVIII, and J. J. Lingane, *Anal. Chim. Acta*, **2**, 584 (1948).

[9] G. Cohn and I. M. Kolthoff, *J. Biol. Chem.*, **147**, 705 (1943).

[10] Potassium or sodium salts are less desirable than lithium salts because large concentration of the former cations precipitate picrolonic acid.

[11] Standardized by titration with standard sodium hydroxide to a phenolphthalein end point.

out by Heyrovsky and Berezicky,[1] and Zlotowski and Kolthoff[12] more recently have systematically investigated its characteristics in tetramethylammonium iodide as supporting electrolyte in which the wave is very well defined.

The reduction of strontium ion proceeds reversibly both in aqueous medium and in media containing up to 80 per cent ethanol. The half-wave potential *vs.* S.C.E. is a linear function of the volume per cent of ethanol P according to:

$$E_{1/2} = -2.110 + 0.0010P$$

The value of $i_d/Cm^{2/3}t^{1/6}$ at 25° decreases nearly linearly from 3.44 in entirely aqueous medium to 2.50 in 80 per cent ethanol, and it increases to 2.67 in pure ethanol.

The strontium wave is so well separated from the preceding wave of barium ion that strontium can be determined in the presence of as much as a 20-fold excess of barium. Strontium may be determined in the presence of an even larger excess of calcium and magnesium because the waves of the latter two follow the strontium wave. The alkali metals (except lithium) and ammonium interfere with the strontium wave, as they do with calcium.

Barium. The reduction potential of barium ion is the most positive of the alkaline earth ions, and the barium wave is excellently developed in tetramethylammonium salt solutions. According to Zlotowski and Kolthoff[12] the half-wave potential *vs.* S.C.E. in ethanol–water mixtures decreases linearly with increasing volume per cent of ethanol P according to:

$$E_{1/2} = -1.940 + 0.0015P$$

The value of $i_d/Cm^{2/3}t^{1/6}$ at 25° decreases from 3.57 in aqueous medium to 2.67 in 80 per cent ethanol, and it increases to 2.82 in pure ethanol.

Small amounts of barium can be determined easily in the presence of relatively large amounts of the other alkaline earth ions, magnesium, and lithium. Amounts of sodium and potassium not exceeding the amount of barium should not interfere, but large concentrations of the former ions mask the full development of the barium diffusion current.

In 0.1 M lithium chloride as a supporting electrolyte the barium wave shows a rounded maximum which cannot be suppressed by gelatin, thymol, methyl red, or methylene blue, but in 0.05 M calcium chloride well-

[12] I. Zlotowski and I. M. Kolthoff, *J. Am. Chem. Soc.*, **66**, 1431 (1944).

developed normal waves are observed.[13] The diffusion current in magnesium chloride is about 1.5 times larger than in lithium or calcium chlorides and the diffusion current plateau is not as well developed as in calcium chloride. (Apparently Figures 2 and 4 in the paper of Kolthoff and Gregor are transposed.) Furthermore, in magnesium chloride i_d/C decreases significantly with increasing barium ion concentration, but in lithium or calcium chlorides it is constant.

Calcium chloride or tetramethylammonium chloride or iodide are the best supporting electrolytes for the determination of barium.

The polarographic behavior of calcium, strontium, and barium in anhydrous liquid ammonia at −36° has been studied by Nyman.[14] Well-developed reduction waves are observed in the presence of tetraethylammonium iodide, and the diffusion currents obey the Ilkovic equation. The relative half-wave potentials against a mercury pool anode are: Ca, −1.96 v.; Sr, −1.68 v.; and Ba, −1.54 v.

4. RADIUM

Heyrovsky and Berezicky reported a fairly well-developed wave for radium ion in potassium chloride medium; the half-wave potential is −1.84 v. *vs.* S.C.E. Since the wave of radium is slightly in advance of that of barium a fair separation of the two waves is obtained when the ratio of their concentrations is within the limits 0.1 to 10.

[13] I. M. Kolthoff and H. P. Gregor, *Anal. Chem.*, **20,** 541 (1948).
[14] C. J. Nyman, *J. Am. Chem. Soc.*, **71,** 3914 (1949).

CHAPTER XXII

Scandium, Yttrium, Lanthanum, and Rare Earth Elements

1. SCANDIUM

Leach and Terrey[1] reported fairly good waves for the reduction of Sc^{+++} in 0.1 *N* potassium chloride, barium chloride, or (better) lithium chloride, as supporting electrolytes. The half-wave potential is -1.80 v. *vs.* the S.C.E., in these supporting electrolytes.

With potassium chloride as supporting electrolyte, the scandium wave increased slowly at first and then rapidly until the diffusion current was reached, *i.e.*, there was a slight indication of a prewave before the main wave. Leach and Terrey showed that this prewave was due to the reduction of hydrogen ion formed by the hydrolysis of Sc^{+++}. When small concentrations of hydrochloric acid were added to the scandium solution, the hydrogen wave preceded, and was well separated from, the scandium wave. In such slightly acid solutions the diffusion current of scandium was directly proportional to its concentration, whereas in neutral solutions a linear relationship was not obtained. The best supporting electrolyte for analytical purposes is a 0.1 *N* solution of lithium chloride containing hydrochloric acid at such a concentration that the hydrogen wave is of the same order of magnitude as the scandium wave.

Noddack and Bruckl[2] studied the reduction of scandium sulfate solutions without any supporting electrolyte present, and they concluded that the reduction takes place in stages, first to the +2 state and then to the metal. However, in view of the experiments of Leach and Terrey this conclusion appears to be incorrect, and the first wave observed by Noddack and Bruckl was probably due to the discharge of hydrogen ion. From the evidence available it appears that the reduction of Sc^{+++} proceeds directly to the metallic state, under the conditions of a polarographic experiment.

In this connection it should be mentioned that Holleck and Noddack[3] studied the reduction of scandium sulfate solutions by strontium amalgams,

[1] R. H. Leach and H. Terrey, *Trans. Faraday Soc.*, **33,** 480 (1937).

[2] W. Noddack and A. Bruckl, *Angew. Chem.*, **50,** 362 (1937).

[3] L. Holleck and W. Noddack, *Angew. Chem.*, **50,** 819 (1937).

and they concluded that a small concentration of +2 scandium was formed. They based this conclusion on the fact that the reduced solutions reduced small amounts of permanganate, but their evidence is not conclusive.

2. YTTRIUM AND LANTHANUM

The only study of the polarography of these two elements is that of Noddack and Bruckl,[2] who employed 0.01 *M* solutions of the +3 sulfates without any supporting electrolyte. In both cases a double wave was observed, the half-wave potentials in the case of yttrium being −1.76 v. and −1.84 v. *vs.* S.C.E., and with lanthanum −1.90 v. and −2.00 v. Noddack and Bruckl concluded that reduction proceeded first to the +2 states and finally to the metals, but this is very doubtful. It is much more probable that the first wave results from reduction of hydrogen ion produced by the hydrolysis of the aquo Y^{+++} and La^{+++} ions, and that the second waves correspond to 3-electron reductions of the Y^{+++} and La^{+++} to the metals. It should be possible to decide this by studying polarograms in unbuffered tetramethylammonium chloride as supporting electrolyte with the pH adjusted to various values in the neighborhood of the "natural pH" of yttric and lanthanum chlorides.

3. RARE EARTH ELEMENTS

The first study of the polarography of the rare earth elements was made by Noddack and Bruckl,[2] who employed 0.01 *M* solutions of the sulfates without any supporting electrolyte. An incompletely separated double wave was observed in all cases, which Noddack and Bruckl attributed to stepwise reduction to the +2 states and to the metals. This is correct only in the cases of samarium, europium, and ytterbium, but in the other cases the first wave actually corresponds to reduction of hydrogen ion and the second to the direct reduction of the +3 ions to the metals.

Cerium. The half-wave potential of the second wave observed by Noddack and Bruckl with pure cerous sulfate is −1.97 v. *vs.* S.C.E., and this probably corresponds to reduction of cerous ion to the metal. It agrees closely with the standard potential (−1.924 v. *vs.* S.C.E.) of the cerous-cerium couple.[4]

The oxidation potential of the ceric-cerous couple is so large that the diffusion current of the reduction of ceric to cerous ion starts at zero applied e.m.f. Using 0.5 *N* sulfuric acid as supporting electrolyte Canneri

[4] G. C. Walters and T. De Vries, *J. Am. Chem. Soc.*, **65**, 119 (1943).

and Cozzi[5] utilized this diffusion current for the determination of cerium in the presence of the other rare earths, following oxidation to the +4 state by cobaltitriamminotrinitrate.

Praseodymium. From the systematic study of Swensen and Glockler[6] the half-wave potential for the reduction of Pr^{+++} to the metal without any supporting electrolyte decreases from −2.06 v. *vs.* S.C.E. at a concentration of 20 millimolar Pr^{+++} to −1.89 v. at a concentration of 2 millimolar (after proper correction for iR drop). With 0.1 M lithium chloride plus 0.01 per cent gelatin as supporting electrolyte the preceding hydrogen wave is well separated from the praseodymium wave and the latter was fairly well developed. The half-wave potential in this supporting electrolyte increased from −1.75 v. to −1.83 v. *vs.* S.C.E. when the concentration of $Pr_2(SO_4)_3$ was increased from 2.5 to 10 millimolar, but $i_d/Cm^{2/3}t^{1/6}$ remained constant at 3.59 ± 0.06 at 25°. Very nearly the same values were observed with 0.1 M tetramethylammonium iodide. When the pH of 6.75 millimolar $Pr_2(SO_4)_3$ in 0.1 M lithium chloride is decreased from its natural value of 5.88 to 2.40 by addition of sulfuric acid the half-wave potential increases from −1.85 v. to −1.93 v. *vs.* S.C.E. but the diffusion current constant is unchanged.

Neodymium. Contrary to the reports of earlier investigators Estee and Glockler[7] observed only a single drawn-out reduction wave with solutions of $Nd_2(SO_4)_3$ without supporting electrolyte. In 0.1 M lithium chloride plus 0.01 per cent gelatin the ratio i_d/C increases markedly with decreasing concentration of neodymium sulfate, apparently because of increasing hydrolysis with increasing dilution and coreduction of hydrogen ion with Nd^{+++} ion. When 0.002 M sulfuric acid is added to the above supporting electrolyte the preceding hydrogen wave is well separated from the Nd^{+++} wave, and the height of the latter in the weakly acid solution becomes directly proportional to the neodymium concentration. The half-wave potential of the neodymium wave under these conditions is −1.82 v. *vs.* S.C.E. independent of the neodymium concentration, and reduction proceeds directly to the metal. The same half-wave potential is observed in 0.1 M tetramethylammonium iodide–0.002 M sulfuric acid–0.01 per cent gelatin, and the value of $i_d/Cm^{2/3}t^{1/6}$ in this medium is 4.40 ± 0.07 at 25°.

Samarium. Samarium is one of the three rare earths whose +2 states are capable of existence in aqueous solution (the other two being ytterbium

[5] G. Canneri and D. Cozzi, *Gazz. chim. ital.*, **71**, 311 (1941).

[6] A. W. Swensen and G. Glockler, *J. Am. Chem. Soc.*, **71**, 1641 (1949).

[7] C. R. Estee and G. Glockler, *J. Am. Chem. Soc.*, **70**, 1344 (1948).

and europium), and under certain conditions a doublet wave corresponding to a stepwise reduction of Sm^{+++} to Sm^{++} and to the metal can be obtained.

From the thorough study by Timnick and Glockler[8] an optimum supporting electrolyte is 0.1 *M* tetramethylammonium iodide containing 0.001 *N* sulfuric acid and 0.01 per cent gelatin. The double samarium wave following the hydrogen wave is fairly well defined; the half-wave potentials are −1.80 v. and −1.96 v. *vs.* S.C.E. With 1 millimolar $Sm_2(SO_4)_3$ the separate $i_d/Cm^{2/3}t^{1/6}$ values of the first and second waves were found to be 3.85 and 8.2 at 25°, in agreement with successive 1- and 2-electron reductions. However, with 4 millimolar Sm^{+++} the separate values of $i_d/Cm^{2/3}t^{1/6}$ were 3.56 and 2.93, the second wave being relatively much smaller than with the smaller concentration of Sm^{+++}. The fact that the $i_d/Cm^{2/3}t^{1/6}$ value of the first wave is about twice as large as would be expected for a 1-electron reduction indicates that the samarous ion produced at the electrode surface is oxidized rapidly enough by hydrogen ion so that considerable cyclic reduction and oxidation of Sm^{+++} and Sm^{++} occurs at the electrode surface to increase the limiting current. Because the free hydrogen ion in the solution is already reduced at potentials corresponding to the first samarium wave the hydrogen ion which oxidizes samarous ion must be furnished by the dissociation of water, corresponding to the reaction:

$$2Sm^{++} + 2H_2O \rightleftharpoons 2Sm^{+++} + 2OH^-$$

Assuming that this is true, the decrease in the $i_d/Cm^{2/3}t^{1/6}$ value of the second wave with increasing $Sm_2(SO_4)_3$ concentration can then be attributed to increasing precipitation of hydrous samaric oxide (or samarous oxide, or basic salts) by the hydroxide ion produced in the diffusion layer.

Holleck[9] utilized the polarographic technique to study the conversion of samaric ion from one isoelectronic form to another, and concluded that three forms of samaric ion exist whose relative proportions depend on the age of the solution, its method of preparation, and the temperature.

Europium. Holleck[10] found that the reduction of europic to europous ion produces a well-developed wave which can be used for its determination in rare earth mixtures. This was confirmed by Laitinen and Taebel,[11] who demonstrated that the wave is excellently defined in 0.1 *M* ammonium chloride; reduction proceeds nearly reversibly and the half-wave potential

[8] A. Timnick and G. Glockler, *J. Am. Chem. Soc.*, **70**, 1347 (1948).
[9] L. Holleck, *Z. Elecktrochem.*, **45**, 249 (1939).
[10] L. Holleck, *Z. anal. Chem.*, **116**, 161 (1939); *Z. Elektrochem.*, **46**, 69 (1940).

is −0.671 v. *vs.* S.C.E. A well-defined wave is also obtainable in strongly acid medium, which distinguishes europium from all the other rare earths.

Since the wave of europium is 0.5 v. in advance of that of ytterbium, the next most easily reduced of the rare earth ions, europium can be determined easily in the presence of ytterbium and the other rare earths. For this purpose the oxide sample is dissolved in hydrochloric acid and neutralized with ammonia to about pH = 5.0 (methyl red), and the polarogram is recorded.

Europous ion does not show a reduction wave before the final current rise in ammonium chloride or alkali chloride as supporting electrolytes.

Holleck[12] polarographically investigated the complexes of europic ion with a variety of hydroxy-carboxylic acids. No complexation was detected with amino acids.

Gadolinium. Rabideau and Glockler[13] found that the 3-electron reduction of Gd^{+++} to the metal does not proceed reversibly but it does produce a well-formed wave in 0.1 *M* lithium or potassium chloride plus 0.01 per cent gelatin. The half-wave potential is −1.77 v. *vs.* S.C.E. and $i_d/Cm^{2/3}t^{1/6}$ is 3.7 at 25°. When increasing amounts of hydrochloric acid up to about 0.006 *M* are added to the lithium chloride supporting electrolyte the gadolinium half-wave potential increases from −1.77 v. to −1.93 v.

Terbium, Dysprosium, Holmium, Erbium, and Thulium. The only data available are those of Noddack and Bruckl,[2] which indicate that the +3 ions of all these elements are reduced to the metals at a potential in the neighborhood of −1.85 v. *vs.* S.C.E.

Ytterbium. Laitinen and Taebel[11] found that the reduction of ytterbic ion in 0.1 *M* ammonium chloride yields an excellent wave whose half-wave potential is −1.169 v. *vs.* S.C.E. Reduction proceeds only to the +2 state and ytterbous ion does not show a reduction wave in this medium. Laitinen[14] demonstrated that the ytterbic-ytterbous couple behaves reversibly at the dropping electrode and the anodic wave resulting from the oxidation of ytterbous to ytterbic ion is as well developed as the cathodic wave. Solutions used for measurement of the cathodic wave must be no more than slightly acid because the reduction of hydrogen ion masks the ytterbic wave. According to Laitinen and Taebel ytterbium can be determined very satisfactorily in rare earth oxide mixtures provided the amounts of cerium group elements present and of samarium are relatively

---

11 H. A. Laitinen and W. A. Taebel, *Ind. Eng. Chem., Anal. Ed.*, **13**, 825 (1941).

12 L. Holleck, *Z. Naturforsch.*, **2b**, 81 (1947).

13 S. W. Rabideau and G. Glockler, *J. Am. Chem. Soc.*, **70**, 1342 (1948).

14 H. A. Laitinen, *J. Am. Chem. Soc.*, **64**, 1133 (1942).

small. The determination of ytterbium in rare earth mixtures has also been discussed by Holleck.[15]

The dependence of the diffusion current of Yb^{+++} on the supporting electrolyte and pH has been studied by Barton and Kurbatov.[16] Using lithium chloride and ammonium chloride they found that the diffusion current is not affected by changes in pH between 3.5 and 6.0, but at higher pH values precipitation occurs. These authors also reported a wave for

TABLE XXII-1

HALF-WAVE POTENTIALS OF RARE EARTHS

In 0.1 *M* lithium chloride or tetramethylammonium chloride containing 0.005 per cent gelatin. For detailed specification of conditions see the text.

	$E_{1/2}$ *vs.* S.C.E.		
	$M^{+++} \rightarrow M^{++}$	$M^{++} \rightarrow M$	$M^{+++} \rightarrow M$
Ce	None	None	−1.97
Pr	None	None	−1.75 → −1.93[a, b]
Nd	None	None	−1.82
Sm	−1.80	−1.96	None
Eu	−0.67	None	None
Gd	None	None	−1.77 → −1.93[b]
Tb	None	None	−1.85 (?)
Dy	None	None	−1.85 (?)
Ho	None	None	−1.85 (?)
Er	None	None	−1.85 (?)
Tm	None	None	−1.85 (?)
Yb	−1.17	−2.05	None
Lu	None	None	−1.8 (?)

[a] Dependent on concentration of Pr^{+++}.
[b] Dependent on pH.

the reduction of ytterbous ion to the metal, with a half-wave potential between −2.00 v. and −2.07 v. *vs.* S.C.E. depending on the pH.

Lutecium. From the measurements of Noddack and Bruckl with 0.01 *M* $Lu_2(SO_4)_3$ without supporting electrolyte we guess that Lu^{+++} undergoes a 3-electron reduction to the metal at about −1.8 v. *vs.* S.C.E.

Summary. The half-wave potential data cited above for the +3 rare earth ions are summarized in Table XXII-1. Reduction to the +2 state occurs only with Sm, Eu, and Yb, and a wave for reduction from the +2 state to the metal is observed only with Sm and Yb. The single reduction waves of the other rare earths all correspond to direct 3-electron reductions

[15] L. Holleck, *Z. anal. Chem.*, **126,** 1 (1943).
[16] G. B. Barton and J. D. Kurbatov, *J. Phys. & Colloid Chem.*, **53,** 683 (1949).

to the metals, and the half-wave potentials are so nearly the same that there is little possibility of using these waves to distinguish these elements.

A small concentration of hydrogen ion, whose reduction precedes that of all the rare earth ions except Eu^{+++} and Yb^{+++}, appears to be conducive to optimum development of the waves, but in strongly acid medium all the waves except that of Eu^{+++} are masked by hydrogen discharge.

CHAPTER XXIII

Titanium, Zirconium, Hafnium, and Thorium

1. TITANIUM

Zeltzer[1] observed a good wave for the reduction of titanic salts in 0.1 *N* hydrochloric, sulfuric, or nitric acids. The reduction proceeds to the titanous state. From our experience the half-wave potential is −0.81 v. *vs.* S.C.E. in 0.1 *M* hydrochloric acid containing 0.005 per cent gelatin, and $i_d/Cm^{2/3}t^{1/6}$ is 1.56 at 25°. There is no reduction of +4 titanium (titanate ion) from strongly alkaline medium.

Zeltzer recommended the following simple procedure for the determination of titanium in titaniferous minerals, such as rutile, anatase, yttrotitanite, and brookite. The mineral sample is fused with potassium bisulfate in a platinum crucible. The cooled melt is then dissolved in 0.1 *N* sulfuric acid, and the polarogram is recorded. According to Zeltzer, aluminum or ferrous iron does not interfere with the titanic wave. He also reported that chromium does not interfere, but this is doubtful in view of the fact that the chromic wave occurs at about the same potential as the titanic wave.

Esin[2] claimed that the reduction of +4 titanium at a mercury jet electrode in dilute acid medium produces a double wave, and he ascribed the second wave to reduction of +3 titanium to the +2 state. Although the work of Forbes and Hall[3] demonstrates that +2 titanium can exist in dilute hydrochloric acid medium, polarograms of +4 titanium that we have obtained in 0.02 to 0.5 *M* hydrochloric acid at both 25° and 0° show only a single wave with no indication that reduction at the d.e. proceeds below the +3 state.

Strubl[4] found that the reduction of titanic chloride and the oxidation of titanous chloride are not reversible from dilute hydrochloric acid solutions. With a mixture of titanic and titanous chlorides in 0.01 *N* hydrochloric acid the composite cathodic-anodic wave has a cathodic half-wave potential

[1] S. Zeltzer, *Collection Czechoslov. Chem. Communs.*, **4,** 319 (1932).

[2] O. Esin, *Acta Physicochim. U.R.S.S.*, **13,** 429 (1940).

[3] G. S. Forbes and L. P. Hall, *J. Am. Chem. Soc.*, **46,** 385 (1924).

[4] R. Strubl, *Collection Czechoslov. Chem. Communs.*, **10,** 475 (1938). See also O. A. Esin, *J. Phys. Chem. U.S.S.R.*, **14,** 731 (1940).

of −0.81 v. and an anodic half-wave potential of −0.14 v. *vs.* the S.C.E. The anodic wave corresponds to the oxidation of titanous to titanic ion. However, when tartaric or citric acid is added to the solution the complex ions that are formed are reversibly reduced and oxidized, and the composite cathodic-anodic wave shows no inflection at zero current. In a saturated tartaric acid solution the cathodic and anodic half-wave potentials are identical and equal to −0.44 v. *vs.* the S.C.E. Both diffusion currents are well defined. Similar waves are obtained in acid solutions of citric acid, but when excess sodium hydroxide is added to a titanic-titanous solution in citric acid the composite wave acquires the irreversible form.

Cagliosi and Sartori[5] found that the reduction potential (not $E_{1/2}$) of titanic salts in alkaline solutions of sodium tartrate varied from −0.9 v. to −1.1 v. *vs.* the S.C.E. when the ratio of titanic salt to tartrate was decreased from $\frac{1}{3}$ to $\frac{1}{5}$. Hence acid solutions of tartaric or citric acid are preferable in practical work. In such solutions the cathodic and anodic diffusion currents are both directly proportional to the titanic and titanous concentrations, respectively.

The polarography of +4 titanium in oxalate, tartrate, and citrate supporting electrolytes, as a function of pH was systematically investigated by Vandenbosch.[6] In a supporting electrolyte comprising 0.1 *M* potassium oxalate, 1 *M* H_2SO_4, and 0.005 per cent gelatin the titanic wave is excellently developed; the half-wave potential is −0.173 v. *vs.* S.C.E. and $i_d/Cm^{2/3}t^{1/6}$ is 1.25 at 25°. The solution is stable for several hours but on longer standing the diffusion current decreases, apparently because of a slow hydrolytic precipitation of the titanium. Less strongly acid oxalate solutions are unsatisfactory because of the rapid hydrolytic precipitation of the titanium.

In tartrate medium the titanic wave is well developed only at pH values smaller than about 2, and between about pH 6 and 7. In a supporting electrolyte containing 0.4 *M* sodium tartrate and 0.005 per cent gelatin the half-wave potential of the cathodic titanic wave varies with pH according to:

pH	0.5	3.9	6.9	11.8
$E_{1/2}$ *vs.* S.C.E., v.	−0.28	−1.30	−1.32	−1.65

The titanic-titanous couple in tartrate media behaves reversibly only at pH values smaller than about 1, and at higher pH values the half-wave potential of the anodic titanous wave is more positive than that of the

[5] V. Cagliosi and G. Sartori, *Gazz. chim. ital.*, **66**, 741 (1936).
[6] V. Vandenbosch, *Bull. soc. chim. Belg.*, **58**, 532 (1949).

cathodic titanic wave. In a supporting electrolyte containing 0.4 *M* sodium tartrate and 0.005 per cent gelatin the value of $i_d/Cm^{2/3}t^{1/6}$ of the cathodic wave is 1.03 at a pH of about 0.5 and 1.22 at a pH of 6.5. Vandenbosch observed that the more strongly acid solutions were less stable with respect to hydrolytic precipitation of titanium than the solution of pH 6.5, which is stable for at least a day.

The titanic wave in 0.4 *M* citrate medium is very well developed at all pH values up to 11.5; above the latter pH the titanium precipitates. The half-wave potentials in 0.4 *M* sodium citrate–0.005 per cent gelatin at various pH values are:

pH	0	3.0	7.0	11.5
$E_{1/2}$ *vs.* S.C.E., v.	−0.28	−0.80	−0.95	−1.49

The plot of $E_{1/2}$ *vs.* pH consists of three linear sections over the pH ranges 0 to 3, 3 to 7, and 7 to 11.5, corresponding to three distinctly different complexes in these pH ranges. At pH 5.5 to 6.0 the value of $i_d/Cm^{2/3}t^{1/6}$ is 1.02, and it decreases significantly both at lower and higher pH values.

Vandenbosch concluded that a citrate solution of pH 5.5 to 6.0 was the most satisfactory of any of the media he investigated for the simultaneous determination of ferric iron and titanium. The ferric wave precedes the titanic wave by over 0.6 v., and both diffusion currents are excellently developed.

An acidic tartrate supporting electrolyte was recommended by Zan'ko, Geller, and Nikitin[7] for the determination of titanium in the presence of iron and aluminum. The wave of the ferric-tartrate complex precedes that of the titanyl-tartrate complex, and when the amount of iron greatly exceeds that of titanium the iron must be reduced to the noninterfering ferrous state. The same authors also developed a procedure for titanium in steel which is described in Chapter XXXV.

Titanium in Clay. Adams[8] developed the following procedure for the determination of a few per cent of titanium in clays, leached clay residues, alumina, and similar materials. A supporting electrolyte composed of 1 *N* sulfuric acid and 8 per cent urea and saturated with sodium oxalate is used. The wave of the ferric-oxalate complex precedes that of titanium but since the separation is good the determination can be carried out in the presence of up to a ten-fold excess of iron.

Adams claimed that the half-wave potential of the titanium wave in this medium shifted greatly to more negative potentials with increasing con-

[7] A. M. Zan'ko, V. A. Geller, and A. D. Nikitin, *Zavodskaya Lab.*, **9,** 976 (1940).
[8] D. F. Adams, *Anal. Chem.*, **20,** 891 (1948).

centrations of either titanium and/or iron, but Lingane and Vandenbosch[9] disproved this. Adams' erroneous conclusion resulted from the use of a very high resistance cell and failure to correct for the correspondingly very large iR drop. With 0.5 to 14 millimolar titanium, and up to 15 millimolar iron, the true half-wave potential in the above supporting electrolyte is constant at −0.285 v. *vs.* S.C.E.

A 0.2-g. sample (titania content about 7 mg.) is fused with 3 g. of potassium hydrogen sulfate, and the cooled melt is leached with dilute sulfuric acid, transferred to a 100-ml. volumetric flask, and diluted to the mark. A 10-ml. aliquot of this sample solution is placed in a 25-ml. volumetric flask, together with 8 ml. of 25 per cent urea solution, and 2.4 ml. of 10 N sulfuric acid (to bring the final acidity to approximately 1 N), and the solution is diluted to volume. A portion of the solution is placed in a polarographic cell and saturated with sodium oxalate. Air is removed with nitrogen and the polarogram is recorded.

Adams recommended the urea as a maximum suppressor, but Vandenbosch[6] found that pure urea does not suppress the titanium maximum and that 0.005 per cent gelatin should be used. From Vandenbosch's work a 0.4 M citrate supporting electrolyte of pH 5.5 to 6.0 is preferable to a strongly acidic oxalate solution for the simultaneous determination of iron and titanium.

Titanium in Steel. See Chapter XXXV.

Titanium in Minerals. Graham and Maxwell[10] developed a generally applicable procedure based on the removal of interfering elements by electrolysis of the sample solution with a mercury cathode and determination of the titanium in the residual solution in a tartaric–sulfuric acid supporting electrolyte.

With minerals containing 0.05 to 1 per cent titanium a 1-g. sample (ground to 100 mesh) is decomposed by heating with a sulfuric–hydrofluoric acid mixture in a platinum dish and the excess hydrofluoric acid is removed by evaporation to fuming. The solution is diluted to about 75 ml., transferred to a mercury cathode cell of the type described by Rabbitts,[11] and electrolyzed with a current density of 0.15 amp. cm.$^{-2}$ until spot tests with thiocyanate show that only trace amounts of iron remain (*ca.* 20 to 30 min.).

The residual solution is transferred to a 100-ml. volumetric flask, neutralized with concentrated ammonia, treated with 2.8 ml. of concentrated sulfuric acid and sufficient tartaric acid to make the final concentration of the latter 1.00 M, and diluted to volume. A portion of this solution is polarographed after removing air with nitrogen. The titanium wave is at −0.3 v. *vs.* S.C.E.

Of the elements not removed by the mercury cathode electrolysis only uranium interferes. The polarographic method yields results in close agreement with the

[9] J. J. Lingane and V. Vandenbosch, *Anal. Chem.*, **21,** 649 (1949).

[10] R. P. Graham and J. A. Maxwell, *Anal. Chem.*, **23,** 1123 (1951).

[11] F. T. Rabbitts, *Anal. Chem.*, **20,** 181 (1948). See also J. A. Maxwell and R. P. Graham, *Chem. Rev.*, **46,** 471 (1950).

hydrogen peroxide colorimetric method and is less subject to interferences than the latter method.

2. ZIRCONIUM

Laubengayer and Eaton[12] studied the polarographic behavior of solutions of zirconyl chloride, $ZrOCl_2$, in 0.1 *N* potassium chloride containing small additions of hydrochloric acid and sodium hydroxide. The pH of all the solutions investigated was between 2 and 5, and in all cases the discharge of hydrogen preceded the reduction of the zirconyl ion. However, with a 0.001 *M* solution of zirconyl chloride in 0.1 *N* potassium chloride (pH = 3) a reduction wave for the zirconium was noticeable after the hydrogen wave; the half-wave potential of this wave was −1.65 v. *vs.* the S.C.E. Laubengayer and Eaton concluded that the wave corresponded to the complete reduction of the zirconyl ion to the metallic state.

3. HAFNIUM AND THORIUM

Neither of these elements is determinable polarographically. They both exist only in the +4 oxidation state in aqueous solution and their estimated[13] reduction potentials are too negative (reducing), and the hydrolysis of their salts too great, to permit reduction waves before the reduction of hydrogen ion or other cations of supporting electrolytes.

[12] A. W. Laubengayer and R. B. Eaton, *J. Am. Chem. Soc.*, **62,** 2704 (1940).
[13] W. M. Latimer, *Oxidation Potentials*, Prentice-Hall, New York, 1938 (p. 254).

CHAPTER XXIV

Vanadium, Columbium, and Tantalum

1. VANADIUM

The ease with which relatively stable compounds of the several oxidation states of vanadium can be produced accounts for its versatile oxidation-reduction chemistry in aqueous solution, and thus for very useful polarographic characteristics under a variety of conditions. Early studies of the polarography of vanadium were carried out by Zeltzer,[1] von Stackelberg *et al.*,[2] Thanheiser and Williams,[3] and Voriskova.[4] More recently Lingane and Meites[5, 6] systematically investigated the behavior of all the oxidation states of vanadium in a variety of different supporting electrolytes.

+5 Vanadium. In dilute acid media +5 vanadium undergoes stepwise reduction to the +4 and +2 states. The first diffusion current starts from zero applied e.m.f. because the potential of the $V^{+5} \rightarrow V^{+4}$ couple (+0.76 *vs.* S.C.E.) is more positive than the oxidation potential of mercury. In 0.05 *M* sulfuric acid in the presence of 0.005 per cent gelatin the half-wave potential of the second $V^{+4} \rightarrow V^{+2}$ wave is −0.98 v. *vs.* S.C.E., and the $i_d/Cm^{2/3}t^{1/6}$ values of the first and total waves are 1.65 and 4.96 at 25°. In solutions of much higher hydrogen ion concentration the reduction of hydrogen ion interferes with the full development of the second wave, and even with 0.005 per cent gelatin present the second wave is distorted by a sharp maximum. That the second stage of reduction does not proceed reversibly is obvious from the fact that the reversible potentials of the $V^{+4} \rightarrow V^{+3}$ and $V^{+3} \rightarrow V^{+2}$ couples (+0.095 v. and −0.497 v. *vs.* S.C.E.)[7] are so well separated that the second wave should be a

[1] S. Zeltzer, *Collection Czechoslov. Chem. Communs.*, **4**, 319 (1932).

[2] M. von Stackelberg, P. Klinger, W. Koch, and E. Krath, *Tech. Mitt. Krupp Forschungsher.*, **2**, 59 (1939).

[3] G. Thanheiser and J. Williams, *Arch. Eisenhüttenw.*, **13**, 73 (1939).

[4] M. Voriskova, *Collection Czechoslov. Chem. Communs.*, **11**, 588 (1939).

[5] J. J. Lingane, *J. Am. Chem. Soc.*, **67**, 182 (1945).

[6] J. J. Lingane and L. Meites, *J. Am. Chem. Soc.*, **69**, 1021, 1882 (1947); **70**, 2525 (1948).

[7] Grinnell Jones and J. H. Colvin, *J. Am. Chem. Soc.*, **66**, 1563, 1573 (1944).

double wave if the reduction was reversible. The overvoltage amounts to nearly 1.1 v. Since the reduction of vanadic ion (V^{+++}) to vanadous ion (V^{++}) proceeds reversibly (*vide infra*), the great degree of irreversibility of this second wave evidently is associated with the vanadyl (VO^{++}) to vanadic stage.

The reduction of vanadate ion (VO_3^-) from ammoniacal medium also produces a doublet wave corresponding to reduction to the +4 and +2 states. The first wave is only slightly in advance of the second; in 1 *M* ammonia–1 *M* ammonium chloride–0.005 per cent gelatin the half-wave potentials are −0.97 v. and −1.26 v. *vs.* S.C.E., and the $i_d/Cm^{2/3}t^{1/6}$ value of the total wave is 4.72 at 25°. When the concentration of ammonia is increased from 1 to 6 *M* the half-wave potentials are −1.07 v. and −1.22 v., and $i_d/Cm^{2/3}t^{1/6}$ is 4.99 at 25°. The excellent development of the total diffusion current renders it very suitable for the practical determination of vanadium. With a large ratio of ammonium chloride to ammonia the wave becomes a poorly developed triplet, and hence in practical work the concentration of ammonia should be 1 *M* or greater when 1 *M* ammonium chloride is used. Page and Robinson[8] employed an ammoniacal supporting electrolyte for the determination of vanadium in medicinals.

When an originally acid solution of +5 vanadium is made ammoniacal the original yellow color may require 1 hr. or more to fade out, reflecting a slow conversion of the yellow species that predominate in acid medium (probably hydrated VO_2^+ and $H_2V_2O_7^{--}$) to the stable alkaline species probably VO_3^-, $V_2O_7^{----}$, and $V_3O_9^{---}$ depending on the pH). The conversion is greatly accelerated the higher the ammonia concentration, and also by increased temperature and by the presence of sulfite ion which may be added to reduce dissolved oxygen. To insure reproducibility the solutions should be allowed to decolorize completely before polarograms are recorded.

Zeltzer observed a poorly developed single wave at about −1.7 v. *vs.* S.C.E. with solutions of ammonium metavanadate in 0.1 *M* lithium hydroxide. The poor definition of the vanadate wave in strongly alkaline medium was confirmed by von Stackelberg *et al.* and by Thanheiser and Willems.

In 1 *M* potassium oxalate solutions of pH = 4.6 the reduction of +5 vanadium to the +4 state produces a diffusion current that starts at zero applied e.m.f., and a second wave at −1.33 v. *vs.* S.C.E. corresponding to reduction to the +2 state. The values of $i_d/Cm^{2/3}t^{1/6}$ are, respectively,

[8] J. E. Page and F. E. Robinson, *Analyst*, **68**, 269 (1943).

1.86 and 5.60. At a pH of 7.6 the diffusion currents become poorly developed.

+4 Vanadium. In 0.1 *M* H_2SO_4 vanadyl ion undergoes a 2-electron reduction to vanadous ion to produce a single wave with $E_{1/2} = -0.85$ v. *vs.* S.C.E. The diffusion current is not too well defined, and in stronger acid medium it is more or less masked by hydrogen discharge. As the acidity is reduced the wave shifts to more negative potentials, and in unbuffered 1 *M* potassium sulfate of pH = 6.1 a poorly defined double wave appears with half-wave potentials of approximately −1.0 v. and −1.29 v. *vs.* S.C.E. Lingane and Meites concluded that the first wave in neutral unbuffered medium results from the reduction of $VOOH^+$ formed by the hydrolysis of VO^{++}, and that the second wave is due to reduction of VO^{++} itself. They based this conclusion on the observation that only the second wave appears in an acetate buffer of the same pH (6.1).

Polarograms of +4 vanadium in ammoniacal medium display both a cathodic and an anodic wave. In 1 *M* ammonia–1 *M* ammonium chloride–0.08 *M* sodium sulfite–0.005 per cent gelatin the cathodic wave ($E_{1/2} = -1.28$ v. *vs.* S.C.E.) is well developed, but the anodic wave ($E_{1/2} = -0.32$ v.) is less well formed. The cathodic wave, for which $i_d/Cm^{2/3}t^{1/6} = 1.82$, results from reduction to the +2 state, and the anodic wave, for which $i_d/Cm^{2/3}t^{1/6} = 0.94$, corresponds to oxidation to the +5 state.

The addition of sulfite ion is essential when working with ammoniacal solutions of +4 vanadium. When it is not present a yellowish brown precipitate forms even though air has been displaced by nitrogen. Even with sulfite present the solutions are not very stable and gradually develop a turbidity.

In strongly alkaline medium, *e. g.*, 1 *M* NaOH, no cathodic wave is observed for the reduction of +4 vanadium (probably present as a polyvanadite ion such as $V_4O_9^{--}$) but an excellently developed anodic wave appears. In 1 *M* NaOH containing 0.08 *M* sodium sulfite the half-wave potential of this anodic wave is −0.432 v. *vs.* S.C.E., and oxidation to the +5 state proceeds reversibly. Addition of 0.01 per cent gelatin shifts the wave to −0.35 v., and since this shortens the diffusion current plateau the use of gelatin is undesirable. The value of $i_d/Cm^{2/3}t^{1/6}$ is 1.47 at 25° with a capillary whose $m^{2/3}t^{1/6}$ value is 2.43, and it is independent of the concentration of sodium hydroxide between 0.3 and 3 *M*. When the concentration of sodium hydroxide is smaller than about 0.25 *M* the wave divides into two waves, which apparently represent oxidation of two different species of +4 vanadium in sluggish equilibrium (probably $V_2O_5^{--}$ and $V_4O_9^{--}$). This anodic wave of +4 vanadium in strongly alkaline medium has been employed for the determination of vanadium in steel[9] (see Chap. XXXV).

Solutions of +4 vanadium in NaOH are so sensitive to air oxidation that it is essential that they be protected by using both sulfite ion and nitrogen, and composited by adding the sample of slightly acid vanadyl solution to the previously de-aired supporting electrolyte in the polarographic cell. Furthermore, the solutions are not stable when the vanadium concentration is larger than about 5 millimolar, but with 5 millimolar or less vanadium they are stable for at least 12 hours when properly protected from oxygen.

The anodic wave of +4 vanadium is also well developed in a buffer solution of pH = 9.4 composed of 0.05 *M* each of sodium carbonate and potassium bicarbonate. The half-wave potential is −0.342 v. *vs.* S.C.E. When the pH of the carbonate-bicarbonate buffer is decreased below about 9 the wave shifts to more positive potentials and the diffusion current plateau is incompletely developed because it begins to merge with the anodic dissolution current of mercury.

In 1 *M* sodium hydroxide the anodic wave remains well developed in the presence of large amounts of carbonate, phosphate, pyrophosphate, tetraborate, fluoride, oxalate, and citrate. In the presence of fluoride and tetraborate the value of $i_d/Cm^{2/3}t^{1/6}$ is the same as in 1 *M* sodium hydroxide alone, but significantly smaller diffusion current constants are observed with large concentrations of the other anions. Tartrate in strongly alkaline medium has a deleterious influence, and the same is true of thiocyanate. Large concentrations, *e. g.*, 1 *M*, of chloride shift the wave slightly to more positive potentials and thus tend to make the wave worse, but the effect is small enough so that the presence of considerable chloride is tolerable.

In oxalate media of pH between 5.5 and 6.1 a doublet cathodic wave is observed with +4 vanadium. The first wave is due to reduction of the vanadium to the +2 state, but the second results from the reduction of hydrogen oxalate ion (probably to CHOCOOH). This hydrogen oxalate wave decreases with increasing pH, and it disappears when the pH is greater than about 6.5, but the vanadium wave height decreases only slightly with increasing pH up to about 11.

+3 Vanadium. In 0.5 *M* sulfuric acid vanadic ion undergoes a reversible 1-electron reduction to vanadous ion to produce an excellently developed wave. With 0.005 per cent gelatin present the half-wave potential is −0.55 v., and without gelatin it is −0.508 v. The value of $i_d/Cm^{2/3}t^{1/6}$ in the presence of 0.005 per cent gelatin is 1.41 at 25°.

Because the wave of +3 vanadium is about 0.7 v. in advance of that of

[9] J. J. Lingane and L. Meites, *Ind. Eng. Chem., Anal. Ed.*, **19**, 159 (1947).

ferrous ion in 1 *M* sulfuric acid it should be easily possible to determine +3 vanadium in the presence of large amounts of iron provided the latter is converted to the ferrous state.

In weakly acid but noncomplexing media, such as acetate buffers of pH 4.1 to 6.3, the cathodic +3 vanadium wave divides into a doublet whose relative heights are a function of pH. This doublet apparently results from reduction of the hydrolytic species VOH^{++} and VO^{+} in sluggish equilibrium.

The intensely green oxalate complex of +3 vanadium produces a well-developed cathodic wave, corresponding to reduction to the +2 state, in 1 *M* potassium oxalate solutions of pH between 3.5 and 6.5. The half-wave potential is −1.136 v. *vs.* S.C.E. and $i_d/Cm^{2/3}t^{1/6}$ is 1.95 at 25°. An incompletely formed anodic wave also appears at −0.05 v., corresponding to oxidation to the +4 state.

Meites[10] found that +3 vanadium is oxidized directly but irreversibly to the +5 state and produces a very well-developed anodic wave in carbonate-bicarbonate-carbon dioxide buffers of pH between about 6.7 and 9.4. At pH = 9.4 the half-wave potential is −0.337 v. *vs.* S.C.E. and $i_d/Cm^{2/3}t^{1/6}$ is 2.80 at 25°.

The influence of various anions and pH on the anodic and cathodic waves of +3 vanadium was systematically studied by Meites and his thesis may be consulted for details.

+2 Vanadium. No reduction waves have been observed with +2 vanadium, but it yields well-developed anodic waves in various supporting electrolytes. In 0.5 *M* sulfuric acid in the absence of gelatin the anodic half-wave potential is identical with the cathodic half-wave potential of +3 vanadium (−0.508 v. *vs.* S.C.E.), showing that oxidation to the +3 state proceeds reversibly. The value of $i_d/Cm^{2/3}t^{1/6}$ is 1.74 at 25°. Although addition of 0.005 per cent gelatin causes slight irreversibility of the cathodic wave of +3 vanadium (*vide supra*) it has little or no influence on the anodic wave of +2 vanadium.

The reddish brown oxalate complex of +2 vanadium also yields a well-formed anodic wave due to its oxidation to the +3 state. In 1 *M* potassium oxalate of pH = 6.5 the oxidation proceeds reversibly; $E_{1/2} = -1.091$ v. *vs.* S.C.E., and $i_d/Cm^{2/3}t^{1/6} = 1.43$ at 25°. In 1 *M* potassium oxalate of pH = 4.5 the anodic half-wave potential is − 1.136 v. *vs.* S.C.E.; but at this pH the vanadous-oxalate complex is oxidized so rapidly by hydrogen ion that accurate measurement of the diffusion current is virtually impossible. At pH = 6.5, on the other hand, the complex is quite stable pro-

[10] L. Meites, Ph.D. Thesis, Harvard University, 1947. See also J. J. Lingane and L. Meites, *J. Am. Chem. Soc.*, **73**, 2165 (1951).

vided that it is protected from air. Although the vanadic-vanadous couple becomes irreversible in alkaline oxalate media the anodic wave remains fairly well developed up to a pH of about 10.

In a 1 *M* acetate buffer of pH = 5.4 the oxidation of +2 vanadium produces a double anodic wave, due to successive 1- and 3-electron oxidations to the +3 and +5 states. The half-wave potentials *vs.* S.C.E. are −0.89 v. and −0.11 v., and the $i_d/Cm^{2/3}t^{1/6}$ values of the separate waves are, respectively, 1.09 and 3.36, the total being 4.45, at 25°.

Meites[10] systematically investigated the anodic waves of +2 vanadium as a function of pH in supporting electrolytes containing phosphate, pyridine, carbonate, thiocyanate, cyanide, thiosulfate, pyrophosphate, tetraborate, benzoate, phthalate, salicylate, tartrate, and citrate, and his thesis may be consulted for details.

Vanadium in Steel. See Chapter XXXV.

2. COLUMBIUM (NIOBIUM)

Zeltzer[11] found that columbic acid produces a fairly well-defined wave from a 1 *N* nitric acid solution at about −0.8 v. *vs.* the S.C.E. In 0.1 *N* nitric acid the wave showed a rounded maximum. The reduction apparently proceeds to the +3 state. No reduction wave was observed with solutions of columbium pentoxide in hydrochloric acid, nor in strongly alkaline medium.

The reducibility of +5 columbium from nitric acid medium has been verified by Stromberg and Reinus[12] who found that the half-wave potential decreases from −0.84 v. *vs.* S.C.E. in 0.06 *N* nitric acid to −0.76 v. in 0.9 *N* nitric acid. Because the height of the wave was abnormally great compared to the expected value for a 2-electron reduction these authors concluded that the +5 columbium is reduced to the +3 state but that the +3 columbium is rapidly reoxidized to the +5 state by hydrogen ion at the electrode surface, the net result being a catalytic reduction of hydrogen ion. In spite of this the height of the wave was found to be directly proportional to the columbium concentration between 0.19 and 1.12 millimolar. Stromberg and Reinus found no indication of a reduction wave of +5 columbium in either sulfuric or hydrochloric acid media, nor in solutions of tartaric acid, salicyclic acid, or ammonium oxalate.

3. TANTALUM

According to Zeltzer[11] solutions of tantalum pentoxide, in either hydrochloric acid or strongly alkaline medium, do not show a reduction wave.

[11] S. Zeltzer, *Collection Czechoslov. Chem. Communs.*, **4**, 319 (1932).
[12] A. G. Stromberg and L. M. Reinus, *J. Phys. Chem. U. S. S. R.*, **20**, 693 (1946).

CHAPTER XXV

Chromium, Molybdenum, Tungsten, and Uranium

1. CHROMIUM

Prajzler[1] and Demassieux and Heyrovsky[2] found that chromic ion is reduced stepwise at the dropping electrode to produce a polarogram consisting of two well-defined waves. In 0.1 *N* potassium chloride or ammonium perchlorate as supporting electrolyte, and with 0.01 per cent gelatin present as a maximum suppressor, the half-wave potentials of the double wave are −0.91 v. and −1.47 v. *vs.* the S.C.E. The first wave corresponds to the reduction of chromic to chromous ion and the second to the reduction of chromous ion to the metal.

In a neutral unbuffered supporting electrolyte the height of the second wave is significantly greater than twice that of the first because hydrogen ion from the hydrolysis of chromic ion is coreduced and contributes to the second diffusion current. The discrepancy increases with decreasing concentration of the chromic salt and can be quantitatively correlated with the increased degree of hydrolysis resulting from increased dilution.[3] With 1 millimolar chromic chloride in 0.5 *M* sodium perchlorate or potassium chloride this "hydrolysis current" is of the same order of magnitude as the chromic to chromous diffusion current. Consequently only the first wave should be employed in practical determinations of chromium when the latter is present as a chromic salt in unbuffered supporting electrolytes, and the pH must be at least as small as the "natural" value resulting from the hydrolysis of chromic ion (3 to 4 depending on the chromic salt concentration) because at higher pH values hydrolytic precipitation of the chromium occurs.

In 0.5 *M* sodium perchlorate in the presence of 0.005 per cent gelatin the value of $i_d/Cm^{2/3}t^{1/6}$ for the first chromic to chromous wave is 1.46 at 25°, and in 0.25 *M* potassium chloride it is 1.50.

Since the first chromic wave is about 0.4 v. in advance of the ferrous wave, small amounts of chromic ion can be determined in the presence of a large amount of iron when care is taken to keep the latter in the ferrous state.

[1] J. Prajzler, *Collection Czechoslov. Chem. Communs.*, **3**, 406 (1931).
[2] M. Demassieux and J. Heyrovsky, *J. chim. phys.*, **26**, 219 (1929).
[3] J. J. Lingane and R. L. Pecsok, *J. Am. Chem. Soc.*, **71**, 425 (1949).

Willis[4] found that the reduction of the green species of chromic ion (probably $Cr(H_2O)_5SO_4^+$) is less reversible and requires larger overvoltage than the violet $Cr(H_2O)_6^{+++}$ ion. The half-wave potential of the green species is 0.22 v. more negative than that of the violet species in 1 *N* potassium sulfate as supporting electrolyte.

According to Voriskova[5] the chromic-ammonia complex produces a doublet wave, whose half-wave potentials in 1 *M* ammonia–1 *M* ammonium chloride–0.005 per cent gelatin are −1.42 v. and −1.70 v. *vs.* S.C.E. Spalenka[6] utilized an ammoniacal supporting electrolyte to determine chromium and zinc simultaneously in anodic zinc chromating solutions. Because the first chromic wave coincides with the zinc wave, the chromium content is first determined from the second chromium wave at −1.7 v., and the zinc content is obtained by subtracting the chromium diffusion current from the combined chromium-zinc wave.

The pyridine-chromic complex produces a well-developed reduction wave, whose half-wave potential in 0.1 *M* pyridine–0.1 *M* pyridinium chloride (pH *ca.* 5.4) is −0.95 v. *vs.* S.C.E.[7] Only the first reduction stage to the chromous state is observable because the final current rise occurs prior to the reduction potential of the chromous-pyridine complex.

In chloride supporting electrolytes +2 chromium produces a well-developed anodic wave when 0.005 per cent gelatin is present as a maximum suppressor.[8]

In 1 *M* potassium chloride the half-wave potential is −0.40 v. *vs.* S.C.E. and $i_d/Cm^{2/3}t^{1/6}$ is 1.54. In agreement with Kalousek,[9] Pecsok and Lingane found that the chromous-chromic couple behaves nearly reversibly in very concentrated chloride media. In saturated calcium chloride (*ca.* 10 *M* in chloride ion) a mixture of chromous and chromic ions yields a composite anodic-cathodic wave with no inflection at zero current. The half-wave potential in saturated calcium chloride is −0.51 v. *vs.* S.C.E. and the value of $i_d/Cm^{2/3}t^{1/6}$ for the anodic wave is 0.47. That this value is only about one-third as large as the value in 1 *M* potassium chloride appears to be chiefly due to the relatively great viscosity of saturated calcium chloride solution.

The polarographic characteristics of the anodic +2 chromium wave in a variety of supporting electrolytes, as summarized by Pecsok and Lingane,

---

[4] J. B. Willis, *J. Proc. Roy. Soc. N. S. Wales*, **78,** 239 (1946).
[5] M. Voriskova, *Collection Czechoslov. Chem. Communs.*, **11,** 588 (1939).
[6] M. Spalenka, *Metallwirtschaft*, **23,** 346 (1944).
[7] J. J. Lingane and H. Kerlinger, *Ind. Eng. Chem.*, *Anal. Ed.*, **13,** 79 (1941).
[8] R. L. Pecsok and J. J. Lingane, *J. Am. Chem. Soc.*, **72,** 189 (1950).
[9] M. Kalousek, *Collection Czechoslov. Chem. Communs.*, **11,** 592 (1939).

are given in Table XXV-1, and typical polarograms are shown in Fig. XXV-1.

The behavior of the yellow salicylate chromous complex (curve *g* in Fig. XXV-1) is especially remarkable. The oxidation of this complex proceeds reversibly, and its very negative half-wave potential (−1.23 v. *vs.* S.C.E.) shows that it is one of the most powerful reductants known in aqueous solution.

Satisfactory anodic waves are not obtained with +2 chromium in either tartrate, citrate, or oxalate supporting electrolytes.

The anodic chromous wave in hydrochloric or sulfuric acid was utilized by Zan'ko and Manusova[10] for the determination of chromium in steel, after reduction to the chromous state by liquid zinc amalgam.

Hume and Kolthoff[11] found that the chromicyanide-chromocyanide couple behaves reversibly at the d.e.; in 1 *M* potassium cyanide containing 0.01 per cent gelatin the half-wave potential (and formal standard potential) is −1.38 v. *vs.* S.C.E., and $i_d/Cm^{2/3}t^{1/6}$ for the cathodic wave is 1.55. The value of $i_d/Cm^{2/3}t^{1/6}$ for the anodic wave is nearly the same, but it is difficult to measure precisely because the chromocyanide complex is oxidized by water at an appreciable rate. The excellent development of the cathodic chromium wave in cyanide medium should make it useful for practical chromium determinations.

Chromate. The reduction of chromate ion produces a well-defined wave in 0.1 to 1 *N* sodium hydroxide as supporting electrolyte; the half-wave potential is −0.85 v. *vs.* the S.C.E.[12] The diffusion current is directly proportional to the concentration of chromate ion, and its magnitude is in good agreement with the value calculated from the Ilkovic equation for $n = 3$. Hence the reduction takes place according to:

$$CrO_4^{--} + 4H_2O + 3e \rightarrow Cr(OH)_3 + 5OH^-$$

or

$$CrO_4^{--} + 2H_2O + 3e \rightarrow CrO_2^- + 4OH^-$$

The latter reaction predominates in strongly alkaline medium in which chromic hydroxide is appreciably soluble. A sodium hydroxide supporting electrolyte is most suitable for the polarographic determination of chromate, *e. g.*, in steel (see Chap. XXXV).

Thanheiser and Willems[13] found that the diffusion current of chromate

[10] A. M. Zan'ko and F. A. Manusova, *Zavodskaya Lab.*, **10,** 565 (1941).
[11] D. N. Hume and I. M. Kolthoff, *J. Am. Chem. Soc.*, **65,** 1897 (1943).
[12] J. J. Lingane and I. M. Kolthoff, *J. Am. Chem. Soc.*, **62,** 852 (1940).
[13] G. Thanheiser and J. Willems, *Arch. Eisenhüttenw.*, **13,** 73 (1939).

TABLE XXV-1

POLAROGRAPHIC CHARACTERISTICS OF +2 CHROMIUM AT 25°

$m^{2/3}t^{1/6}$ = 1.610 mg.$^{2/3}$ sec.$^{-1/2}$; 0.005 per cent gelatin added to all solutions.

Supporting electrolyte	$E_{1/2}$, volts *vs.* S.C.E.	$\frac{i_d}{Cm^{2/3}t^{1/6}}$
0.1 *M* KCl	−0.34	−1.54
1.0 *M* KCl	−0.40	−1.54
Satd. KCl	−0.51	−1.55
Satd. $CaCl_2$	−0.51	−0.47
0.05 *M* KSCN	−0.71	−1.55
0.5 *M* KSCN	−0.78	−1.78
1.0 *M* KSCN	−0.80	−1.64
2.0 *M* KSCN	−0.82	−1.55
9.0 *M* KSCN	−0.85	−1.20
0.1 *M* NH_4OH–5 *M* NH_4Cl	−0.85	−1.14
0.5 *M* HCOOH	−0.30	−1.54
0.1 *M* $(CH_3)_4NBr$	−0.43	−1.54
0.1 *M* $NaC_7H_5O_3$–0.1 *M* NaOH	−1.23	−1.17

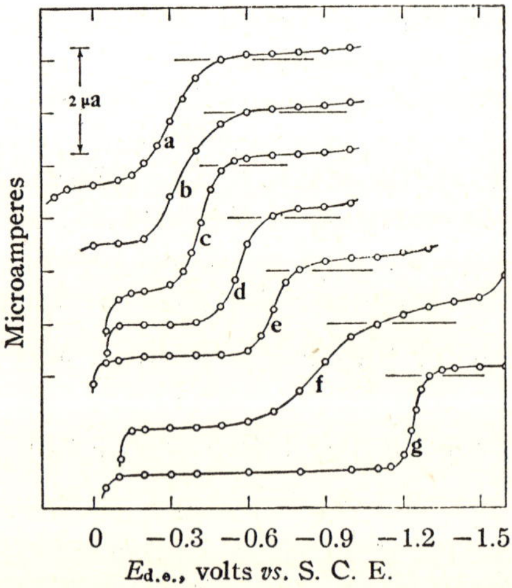

Fig. XXV-1. Polarograms of 1 millimolar chromium(II) in: (a) 0.5 *M* formic acid, pH 1.8; (b) 0.1 *M* potassium chloride; (c) 0.1 *M* tetraethylammonium bromide; (d) saturated potassium chloride; (e) 0.05 *M* potassium thiocyanate; (f) 5 *M* ammonium chloride–0.1 *M* ammonium hydroxide; (g) 0.5 *M* sodium salicylate–0.1 *M* sodium hydroxide.[8] All solutions contained 0.005 per cent gelatin. Zero current for each curve is indicated on ordinate axis; each curve has been shifted downward by one microampere.

decreased markedly when the concentration of sodium hydroxide was increased above 1 *N*.

From neutral and unbuffered solutions of potassium chloride as supporting electrolyte the polarogram of chromate consists of four waves at −0.3 v., −1.0 v., −1.5 v., and −1.7 v. *vs.* the S.C.E. Lingane and Kolthoff showed that the last three waves corresponded to the reduction of the chromate to the Cr^{+3}, Cr^{+2}, and metallic states, respectively. The height of the first wave at −0.3 v. was not directly proportional to the concentration of chromate ion, but its *relative* height increased markedly when the concentration of chromate was decreased. It was concluded that this wave was due to the reduction of chromate ion, but that its height does not correspond to $n = 3$ because of the formation of a film of either chromic hydroxide, or basic chromic chromate, at the surface of the mercury drops which interferes with the complete reduction.

In ammonium chloride-ammonia buffers of pH between 8 and 9, and in a diethylammonium chloride-diethylamine buffer of pH = 10, only two waves are observed at −0.35 v. and −1.7 v. These waves correspond to the reduction of chromate to Cr^{+++} and to the metal, respectively.

The chromate-dichromate equilibrium has been studied polarographically by Stevenson and Langer.[14]

Chromium in Steel. See Chapter XXXV.

2. MOLYBDENUM

Von Stackelberg *et al.*[15] found that molybdate does not produce a wave from neutral or alkaline solutions, but a well-defined wave is obtained in sulfuric acid solutions. The polarographic reduction of +6 molybdenum in sulfuric, hydrochloric, perchloric, and phosphoric acid media was studied by Höltje and Geyer.[16] In all they observed two to four waves, depending on the acidity, which they ascribed to reduction to +5 molybdenum, to the red +3 molybdenum, to the green +3 molybdenum, and to $M(OH)_3$. Carritt[17] investigated the polarography of +6 molybdenum and +3 molybdenum in various concentrations of sulfuric and hydrochloric acids, and

[14] B. T. Stevenson and A. Langer, Westinghouse Research Lab., Scientific Paper No. 1022 (1941).

[15] M. von Stackelberg, P. Klinger, W. Koch, and E. Krath, *Tech. Mitt. Krupp Forschungsber.*, **2,** 59 (1939).

[16] R. Höltje and R. Geyer, *Z. anorg. u. allgem. Chem.*, **246,** 258 (1941). See also Ya. P. Hokhstein, *J. Gen. Chem. U. S. S. R.*, **10,** 1725 (1940).

[17] D. E. Carritt, Ph.D. Thesis, Harvard University, 1947. See also G. Jongh, Doctoral Thesis, Amsterdam, 1950, who independently carried out a very similar study.

employed the coulometric analysis technique at controlled potential to determine reduction states.

In 0.3 *M* hydrochloric acid the polarogram of +6 molybdenum comprises two waves, which result from reduction to +5 molybdenum and +3 molybdenum, with half-wave potentials of −0.26 v. and −0.63 v. *vs.* S.C.E. The first diffusion current plateau is incompletely developed but the total diffusion current is well defined. In the same medium +5 molybdenum yields a single wave ($E_{1/2}$ = −0.60 v. *vs.* S.C.E.) corresponding to a 2-electron reduction to +3 molybdenum. Controlled potential reduction of hydrochloric acid solutions of +6 molybdenum, or +5 molybdenum at a large mercury cathode produces the green form of +3 molybdenum with virtually 100 per cent current efficiency. Solutions of +3 molybdenum in 0.3 *M* hydrochloric acid show no cathodic wave, but an incompletely developed anodic wave appears at −0.22 v. *vs.* S.C.E.

In 0.8 *M* hydrochloric acid three waves are observed on the polarogram of +6 molybdenum, the half-wave potentials being −0.12 v., −0.25 v., and −0.55 v. *vs.* S.C.E. The first two waves are not completely resolved. Controlled potential reduction of the +6 molybdenum solution at a mercury cathode at a potential (−0.15 v.) between the first and second waves produced a solution of +5 molybdenum which showed only the wave at −0.55 v. This led Carritt to conclude that the first two waves in the original +6 molybdenum polarogram correspond to the reduction to +5 molybdenum of two different molybdenum species in sluggish equilibrium. Controlled potential reduction of the +5 molybdenum solution at −0.75 v. produced a solution of green +4 molybdenum with nearly 100 per cent current efficiency; the calculated current consumption for a 1-electron reduction was 401 coulombs and the value observed after the current decreased virtually to zero was 403 coulombs. The solution of +4 molybdenum showed no cathodic wave but an incompletely developed anodic wave was observed at the same potential (−0.15 v.) as the anodic wave of +3 molybdenum. Carritt concluded that +4 molybdenum is produced only by the reduction of +5 molybdenum, and that controlled potential reduction of solutions of +6 molybdenum at the same potential proceeds directly all the way to the +3 state. This latter condition is the one that prevails in reduction of +6 molybdenum solution at the d.e. at potentials corresponding to the total diffusion current.

The polarograms of +6 molybdenum in sulfuric acid medium are less complex and more satisfactory from a practical analytical viewpoint than those in hydrochloric acid solutions. In 10 to 14 *M* sulfuric acid two waves are observed: the first corresponding to reduction to the +5 state starts at zero applied e.m.f., and the second which results from reduction to the +3 state is at −0.13 v. *vs.* S.C.E. In more dilute sulfuric acid solutions the first wave divides into an incompletely resolved doublet wave. The value of $I = i_d/Cm^{2/3}t^{1/6}$ for the total diffusion current decreases linearly with increasing sulfuric acid concentration, and obeys the relation:

$$I = 4.24 - 0.25M$$

where M is the molarity of sulfuric acid between 0.2 and 14.4.

Carritt found that gelatin and agar exert a marked distortive and suppressive effect on the molybdenum waves in both sulfuric and hydrochloric acid media.

Boltz, De Vries, and Mellon[18] reported that molybdate present in a supporting electrolyte of pH = 2.8 composed of *ca.* 0.03 M disodium hydrogen phosphate, 0.1 M citric acid, and 0.1 M potassium chloride produces a well-developed double wave with half-wave potentials of −0.23 v. and −0.58 v. *vs.* S.C.E. The height of the first wave is one-half that of the second, indicating reduction to the +5 and +3 states.

According to Boltz, De Vries, and Mellon the molybdiphosphoric acid complex present in an acetic acid-sodium acetate buffer of pH = 3.5 (composed of 1 M acetic acid, 0.1 M sodium acetate, and 0.1 M potassium chloride) produces a double wave with half-wave potentials of about −0.45 v. and −0.65 v. *vs.* S.C.E. The heights of both waves are directly proportional to the molybdenum concentration but the diffusion currents are not very well developed. When the concentration of sodium acetate was increased from 0.1 to 1 M to increase the pH to 4.6, only a single well-formed wave resulted with $E_{1/2}$ = −0.57 v. *vs.* S.C.E. According to Boltz, De Vries, and Mellon the height of this wave is not strictly proportional to the concentration of molybdenum. When the concentration of potassium chloride in the acetate buffer of pH = 4.6 was increased from 0.1 to 0.5 M the total wave height increased markedly and a small wave appeared in advance of the main wave.

Molybdisilicic acid in the above acetate buffer of pH = 3.5 produces a single, double, and triple wave with increasing concentration with half-wave potentials in the neighborhood of −0.2 v., −0.5 v., and −0.55 v. *vs.* S.C.E. The total diffusion current is clearly developed and directly proportional to the molybdenum concentration.

Indirect Determination of Phosphorus. Uhl[19] reported that molybdate in a nitric acid-ammonium nitrate solution containing lactic acid produces

[18] D. F. Boltz, T. De Vries, and M. G. Mellon, *Anal. Chem.*, **21,** 563 (1949).
[19] F. A. Uhl, *Z. anal. Chem.*, **110,** 202 (1937).

a double wave with half-wave potentials of about -0.35 and -0.5 v. *vs.* S.C.E. He applied this wave to the indirect determination of phosphate in fertilizers, by precipitation as ammonium phosphomolybdate with a measured excess of molybdate and determination of the excess molybdate. Stern,[20] who applied the Uhl method to the determination of phosphorus in biological material, found that the addition of lactic acid is unnecessary, and with a supporting electrolyte containing approximately 0.25 *M* nitric acid and 3 *M* ammonium nitrate he obtained only a single well-formed wave with a half-wave potential of -0.41 v. *vs.* S.C.E. A high concentration of ammonium nitrate and a relatively small concentration of nitric acid are essential for optimum development of the molybdate wave. With a large concentration of nitric acid the wave is distorted by a rounded maximum. This can be suppressed by 0.3 per cent methylcellulose, but when this is done the wave becomes drawn out and is less satisfactory than that obtained with a small nitric acid concentration.

Molybdenum in Plants and Soils. Nichols and Rogers[21] developed a method for the determination of 1 to 100 micrograms of molybdenum in the ash from plants and soils based on its preliminary separation by α-benzoinoxime from acid solution, and ultimate measurement of the $+6$ molybdenum wave in a nitric acid-ammonium nitrate supporting electrolyte. Since this procedure requires preliminary separation of the molybdenum it has no particular advantage over the colorimetric method.

Molybdenum in Ores. Kanevskii and Shvartsburd[22] obtained a fairly well-defined double wave for the reduction of molybdate from a phosphoric acid solution, but they did not report the half-wave potential. According to Höltje and Geyer[16] the half-wave potentials of the doublet wave in 1 *M* phosphoric acid are -0.33 v. and -0.90 v. *vs.* S.C.E., but in 10 *M* phosphoric acid a triplet wave is obtained with half-wave potentials of -0.01 v., -0.44 v., and -0.99 v. *vs.* S.C.E. Kanevskii and Shvartsburd described the following method for the determination of molybdenum in ores.

The sample of molybdenum ore is fused with a mixture of potassium hydroxide and sodium peroxide, leached with hot water, boiled with hydrogen peroxide to remove manganese, and filtered. A 5-ml. portion of the filtrate is then added to 3 ml. of 85 per cent phosphoric acid, and the polarogram of the molybdate is recorded. They obtained results that were accurate to about ± 2 to ± 5 per cent with ores that contained from 0.005 to over 50 per cent molybdenum.

Molybdenum in Steel. See Chapter XXXV.

[20] A. Stern, *Ind. Eng. Chem., Anal. Ed.*, **14,** 74 (1942).

[21] M. L. Nichols and L. H. Rogers, *Ind. Eng. Chem., Anal. Ed.*, **16,** 137 (1944).

[22] E. A. Kanevskii and L. A. Shvartsburd, *Zavodskaya Lab.*, **9,** 283 (1940); *Chem. Abstracts*, **34,** 5782 (1940).

3. TUNGSTEN (WOLFRAM)

Von Stackelberg, Klinger, Koch, and Krath[23] studied the polarography of +6 tungsten in concentrated hydrochloric acid and reported only a single reduction wave, which they utilized for the determination of tungsten in steel (see Chap. XXXV). In neutral or alkaline solutions +6 tungsten does not produce a reduction wave.

Because hydrous WO_3 is insoluble in noncomplexing acids its polarography is necessarily limited to complexing acids, and concentrated halogen acids in particular. Lingane and Small[24] investigated the polarography of the various oxidation states of tungsten in hydrochloric acid media and employed the controlled potential coulometric electrolysis technique to identify the oxidation states corresponding to the observed waves. In 12 *M* hydrochloric acid +6 tungsten undergoes stepwise reduction to +5 tungsten and the red +3 state. The first diffusion current starts from zero applied e.m.f. and the half-wave potential of the second wave is −0.56 v. *vs.* S.C.E. In 10, 8, 6, and 4 *M* hydrochloric acid the half-wave potentials of the second wave are −0.62 v., −0.62 v., −0.64 v., and −0.66 v. *vs.* S.C.E. In 12, 10, and 8 *M* hydrochloric acid at 25° the values of $i_d/Cm^{2/3}t^{1/6}$ for the total double wave are 3.82, 4.33, and 4.70.

+5 Tungsten in 12 *M* HCl produces a single well-formed wave corresponding to a reversible 2-electron reduction, whose half-wave potential is −0.56 v. *vs.* S.C.E. and the value of $i_d/Cm^{2/3}t^{1/6}$ is 2.53 at 25°. The wave remains well formed with decreasing hydrochloric acid concentration down to 6 *M*, but with more dilute acid the original blue solution becomes progressively more yellow and the wave disappears.

A mixture of +5 tungsten and the red form of +3 tungsten in 12 *M* hydrochloric acid produces a composite cathodic-anodic wave at −0.55 v. with no inflection at zero current, which usually is taken as a criterion of reversible behavior. However, in this case the cathodic and anodic reactions are different—the former being a 2-electron reduction of +5 tungsten to the +3 state but the latter being only a 1-electron oxidation of the red +3 tungsten to the +4 rather than the +5 state.

The red form of +3 tungsten in concentrated hydrochloric acid transforms spontaneously to a yellow modification which displays an anodic wave at −0.53 v. *vs.* S.C.E. The oxidation of this yellow +3 tungsten, like that of the red form, proceeds only to the +4 state in concentrated hydrochloric acid, but in 3 *M* hydrochloric acid a double anodic wave is

[23] M. von Stackelberg, P. Klinger, W. Koch, and E. Krath, *Tech. Mitt. Krupp Forschungsber.*, **2**, 59 (1939).

[24] J. J. Lingane and L. A. Small, *J. Am. Chem. Soc.*, **71**, 973 (1949).

obtained resulting from stepwise oxidation to the +4 and +5 states. The familiar green form of +3 tungsten (probably $W_2Cl_9^{---}$) shows no anodic wave in either concentrated or dilute hydrochloric acid.

Solutions of +4 tungsten in 12 *M* hydrochloric acid (prepared by electrolytic oxidation of yellow +3 tungsten) disproportionate to red +3 tungsten and the +5 state, and the +5 tungsten combines with +4 tungsten to form a species with an apparent oxidation state of 4.5. The polarogram of this species shows a single cathodic wave at −0.56 v. *vs.* S.C.E., corresponding to reduction to the +3 state.

Souchay[25] employed the polarograph as a tool in his systematic investigations of tungstates and iso- and heteropoly acids of tungsten.

Tungsten in Steel. See Chapter XXXV.

4. URANIUM

The first study of the polarography of uranium was carried out by Herasymenko,[26] who observed that in neutral or weakly acid medium uranyl ion undergoes stepwise reduction to produce a polarogram comprising three waves. Herasymenko concluded that these result from successive reduction to the +5, +4, and +3 states, and as evidence in support of this conclusion he demonstrated that the polarogram of a uranyl solution which had been photochemically reduced to the +4 state showed only the third wave. Herasymenko's interpretation has been verified by Harris and Kolthoff[27] who systematically investigated the polarography of uranyl ion as a function of pH and other variables.

In moderately acid solution (0.01 to 0.02 *M* hydrochloric acid) the reduction of uranyl ion produces a doublet wave, whose half-wave potentials are −0.18 v. and −0.92 v. *vs.* S.C.E. independent of the acidity, the concentration of uranyl ion, or whether potassium chloride is present in concentrations up to about 0.5 *M*. The first wave results from the reduction of +6 uranium to the +5 state and the second, which is twice as large as the first, corresponds to a further 2-electron reduction to the +3 state. In the same supporting electrolyte +4 uranium yields a single reduction wave whose half-wave potential is the same as that of the second wave observed with uranyl ion, *i. e.*, +4 and +5 uranium are reduced at virtually the same potential.

In a supporting electrolyte composed of 0.1 *M* potassium chloride, 0.01

[25] P. Souchay, *Ann. chim.*, **18,** 61, 73, 169 (1943); **19,** 102 (1944); **20,** 73, 96 (1945); **21,** 232 (1946).

[26] P. Herasymenko, *Chem. Listy*, **19,** 172 (1925); *Trans. Faraday Soc.*, **24,** 272 (1928).

[27] W. E. Harris and I. M. Kolthoff, *J. Am. Chem. Soc.*, **67,** 1484 (1945).

M hydrochloric acid, and 2×10^{-4} per cent thymol as a maximum suppressor, the value of $i_d/Cm^{2/3}t^{1/6}$ for the first uranyl wave is 1.51 at 25°. This value, which coincides with the values observed in many other known 1-electron reductions, constitutes conclusive proof that the first wave results from reduction of +6 uranium to the +5 state. Since the slope of the first wave corresponds to reversible reduction, and because the half-wave potential is virtually independent of the hydrogen ion concentration, Harris and Kolthoff concluded that the reduction reaction is:

$$UO_2^{++} + e \rightleftharpoons UO_2^{+}$$

The fact that the log plot of the second wave does not produce a straight line led Harris and Kolthoff to conclude that the reduction of +5 uranium to the +3 state does not proceed reversibly.

The value of $i_d/Cm^{2/3}t^{1/6}$ of the second uranyl wave in 0.1 M potassium chloride, 0.01 M hydrochloric acid, and 2×10^{-4} per cent thymol is 3.20 at 25°, which is 2.12 times larger than that of the first wave rather than exactly twice as large, as would be expected for successive 1- and 2-electron reductions. No explanation was offered for this anomaly. It may reflect a small amount of hydrogen ion reduction catalyzed by the simultaneous reduction of the uranium.

In strongly acid medium, *i. e.*, hydrogen ion concentration greater than about 0.2 M, the first uranyl wave increases with increasing acidity, due to increasing instability of +5 uranium with respect to disproportionation into UO^{++} and UO_2^{++}, but the total height of the doublet wave remains constant.[28] Because the rate constant of the disproportionation of +5 uranium increases with increasing uranyl concentration, the limiting current of the first wave in solutions of high acidity, *e. g.*, 2 M hydrochloric acid, is not proportional to the uranyl ion concentration, but is relatively larger at higher concentrations.

A micromethod for the determination of 0.05 to 0.5 mg. of sodium, based on precipitation as sodium zinc uranyl acetate and measurement of the uranyl diffusion current in 0.5 M hydrochloric acid, has been described by Carruthers.[29] In view of the above findings of Kolthoff and Harris pertinent to strongly acid media, it is evident that the total height of the doublet wave should be measured to avoid error due to disproportionation of +5 uranium when such a strongly acid solution is used to dissolve the sodium "triple salt," or better a less strongly acid-supporting electrolyte should be employed.

[28] I. M. Kolthoff and W. E. Harris, *J. Am. Chem. Soc.*, **68**, 1175 (1946).
[29] C. Carruthers, *Ind. Eng. Chem., Anal. Ed.*, **15**, 70 (1943).

Harris and Kolthoff recommended the first uranyl wave in 0.1 M potassium chloride, 0.01 M hydrochloric acid, and 2×10^{-4} per cent thymol for the practical determination of uranium. The interference of even large amounts of ferric iron is eliminated, as originally recommended by Strubl,[30] by reducing it to the ferrous state by warming the solution containing 2 M hydroxylamine hydrochloride at about 50° for ten minutes. This permits the determination of uranium in the presence of at least a 200-fold excess of iron. The diffusion current of the first wave is not affected by up to 2 M hydroxylamine hydrochloride, but it decreases significantly when the concentration of thymol maximum suppressor is increased from 2×10^{-4} to 10^{-3} per cent.

The fact that the reduction of either uranyl ion or uranous ion catalyzes the reduction of nitrate ion[31] was utilized by Harris and Kolthoff[27] for the determination of traces of uranium down to about 3×10^{-7} M. The catalyzed nitrate limiting current is not directly proportional to the uranium concentration, but the procedure can be empirically calibrated with various known uranium concentrations, using a supporting electrolyte comprising 0.005 M potassium nitrate, 0.5 M potassium chloride, and 0.05 M hydrochloric acid.

Kern and Orlemann[32] prepared solutions of +5 uranium in sodium perchlorate-perchloric acid solutions by controlled potential reduction of uranyl ion at −0.55 v. *vs.* S.C.E. at a mercury cathode. They found that mixtures of +5 and +6 uranium produce a reversible anodic-cathodic wave and they utilized the well-defined anodic diffusion current of +5 uranium to study the kinetics of its disproportionation into UO_2^{++} and UO^{++}. The rate law at 25° in solutions composed of mixtures of sodium perchlorate and perchloric acid of total ionic strength equal to 0.4 is second order with respect to +5 uranium concentration and is given by:

$$-[dUO_2^+]/dt = ka_{H^+}[UO_2^+]^2$$

where $k = 130$ (moles liter)$^{-1}$ sec.$^{-1}$. Kern and Orlemann concluded that the standard potential of the $UO_2^{++} + e \rightleftharpoons UO_2^+$ couple is $+0.062 \pm 0.002$ v. *vs.* N.H.E. (or −0.180 v. *vs.* S.C.E.), which agrees exactly with the value observed by Harris and Kolthoff in chloride media.

The polarography of +5 uranium has also been studied by Kraus,

[30] R. Strubl, *Collection Czechoslov. Chem. Communs.*, **10**, 466 (1938).

[31] I. M. Kolthoff, W. E. Harris, and G. Matsuyama, *J. Am. Chem. Soc.*, **66**, 1782 (1944).

[32] D. M. H. Kern and E. F. Orlemann, *J. Am. Chem. Soc.*, **71**, 2102 (1949). See also H. G. Heal, *Nature*, **157**, 225 (1946).

Nelson and Johnson,[33] whose results further confirm the reversibility of the +6 to +5 couple at the d.e., and the standard potential quoted above. These authors also concluded that the optimum stability range of +5 uranium is from about pH 2 to 4, and their results are in accord with those of Kern and Orlemann with respect to the rate law for the disproportionation of +5 uranium into UO^{++} and UO_2^{++}. Kraus, Nelson, and Johnson demonstrated that the $i_d/Cm^{2/3}t^{1/6}$ values of the cathodic uranyl wave and the anodic wave of +5 uranium are equal to within 2 per cent.

Another excellent study of the polarography of uranium was reported by Kritchevsky and Hindman[34] simultaneously with the reports of Kern and Orlemann and Kraus, Nelson, and Johnson. In addition to supplying additional confirmation of the reversibility of the $UO_2^{++} + e \rightleftharpoons UO_2^{+}$ couple and its standard potential, Kritchevsky and Hindman demonstrated that the $U^4 - U^3$ couple also behaves reversibly at the d.e. An acidic solution of +3 and +4 uranium produces a reversible composite anodic-cathodic polarogram and the cathodic and anodic diffusion currents both correspond to 1-electron reactions. With increasing acidity the half-wave potential of the $U^4 - U^3$ couple shifts to more positive values; in 0.01 *M* perchloric acid plus 0.09 *M* sodium perchlorate $E_{1/2}$ is −0.921 v. *vs.* S.C.E. and 0.1 *M* perchloric acid it is −0.862 v. This shift is in accord with the conclusions of Kraus and Nelson,[35] who demonstrated that +4 uranium exists partly as UOH^{+++} ion in weakly acid medium, and that the hydrolysis constant of the reaction:

$$U^{++++} + H_2O \rightleftharpoons UOH^{+++} + H^+$$

is 7.1×10^{-2} in chloride and perchlorate solutions of ionic strength equal to 0.1. From this constant it follows that the predominant species of +4 uranium is UOH^{+++} up to about 0.1 *M* hydrogen ion, and if this is reduced to U^{+++} according to:

$$UOH^{+++} + H^+ + e \rightleftharpoons U^{+++} + H_2O$$

the half-wave potential should shift to a more positive value by 59 mv. per 10-fold increase in the hydrogen ion concentration. This is exactly the shift shown by the data of Kritchevsky and Hindman, although these authors did not interpret the shift as a hydrogen ion dependence. Data presented by the latter investigators also show that $E_{1/2}$ becomes virtually constant at −0.87 v. in perchloric acid solutions of concentration above

[33] K. A. Kraus, F. Nelson, and G. L. Johnson, *J. Am. Chem. Soc.*, **71,** 2510 (1949).
[34] E. S. Kritchevsky and J. C. Hindman, *J. Am. Chem. Soc.*, **71,** 2096 (1949).
[35] K. A. Kraus and F. Nelson, *J. Am. Chem. Soc.*, **71,** 2517 (1949).

about 0.1 *M*, which is to be expected from the hydrolysis constant of the +4 uranium.

Kritchevsky and Hindman also showed that addition of gelatin or thymol to an acidic uranyl solution shifts the half-wave potential of the second wave to a more negative value, and with 0.02 per cent gelatin the second wave is virtually obliterated. This may explain why Harris and Kolthoff, who employed thymol, observed the more negative value of −0.92 v. for the half-wave potential.

In a very weakly acid medium containing +4 uranium (0.001 *M* hydrochloric acid plus 0.1 *M* potassium chloride) Kritchevsky and Hindman observed an incompletely developed anodic wave at −0.19 v. *vs.* S.C.E., which probably results from the oxidation of UOH^{+++} to UO_2^+. The wave does not appear in more strongly acid solutions.

Harris and Kolthoff[36] observed that in an unbuffered very weakly acid medium (pH between 2.7 and 3.2) the second reduction wave of uranyl divides into two waves. At pH values between about 3.3 and 4, the third wave is no longer observed and the height of the remaining second wave is the same as that of the first wave. In neutral 0.1 *M* potassium chloride two waves are also obtained but the second at −0.6 v. *vs.* S.C.E. closely follows the first at −0.2 v. Harris and Kolthoff attributed these phenomena to hydrolysis of the uranyl ion:

$$UO_2^{++} + H_2O \rightleftharpoons UO_2OH^+ + H^+$$

and they determined the hydrolysis constant of this reaction to be 8.1×10^{-5} at 25°. They concluded that the first wave in neutral 0.1 *M* potassium chloride results from reduction of only UO_2^{++}, that the hydrolyzed species UO_2OH^+ is not reduced until the potential of the second wave is reached, and that the second wave results from the simultaneous reduction of +5 uranium to the +4 state (solid UO_2) and the UO_2OH^+ to only the +5 state (solid UO_2OH). The fact that the ratio $i_d/Cm^{2/3}t^{1/6}$ for the first wave in 0.1 *M* potassium chloride decreases from 1.22 to 1.02 when the total concentration of +5 uranium is decreased from 2.6 to 0.2 millimolar lends support to this interpretation, because the degree of hydrolysis of UO_2^{++} increases and its *relative* concentration decreases with decreasing concentration. However, if the hydrolyzed +6 uranium is reduced only to the +5 rather than to the +4 state to which UO_2^{++} is reduced, the value of $i_d/Cm^{2/3}t^{1/6}$ of the *total* double wave should decrease with decreasing uranium concentration because the *total* electron transfer

[36] W. E. Harris and I. M. Kolthoff, *J. Am. Chem. Soc.*, **69**, 446 (1947). See also E. S. Kritchevsky and J. C. Hindman, **71**, 2096 (1949).

should decrease from 2 to a value between 1 and 2. Actually the data of Harris and Kolthoff show that this ratio increases from 2.46 to 2.65 (rather than decreases) as the uranium concentration is decreased from 2.6 to 0.2 millimolar, contrary to the expected behavior if the hydrolyzed +6 uranium is reduced only to the +5 state.

Addition of small quantities of sodium hydroxide to a uranyl solution precipitates the uranium (perhaps as a mixture of sodium and/or potassium diuranate and $UO_2(OH)_2$) and the height of the doublet wave diminishes. The first wave decreases more rapidly than the second; for example with 1 millimolar uranyl ion in 0.1 *M* potassium chloride with 0.31, 0.94, and 1.24 millimolar sodium hydroxide the ratio of the *total* doublet wave height to that of the first wave was 2.42, 4.06, and 5.7.

Strubl[37] observed that +6 uranium in an ammonium carbonate supporting electrolyte undergoes stepwise reduction to produce a doublet wave, which he ascribed to reduction of the uranyl carbonate complex to the +4 and +2 states. However Harris and Kolthoff[36] showed conclusively from the magnitudes of the diffusion currents that the doublet wave corresponds to successive 1-electron reductions to the +5 and +4 states, rather than to the +4 and +2 states. A solution of +4 uranium in ammonium carbonate medium shows no reduction wave, nor does it show any oxidation wave. Harris and Kolthoff also reported that the total limiting current in 1 *M* ammonium carbonate solution is not strictly proportional to the uranyl concentration, and for this reason they advised against the use of an ammonium carbonate supporting electrolyte for the practical determination of uranium.

In 0.1 *M* sodium carbonate +6 uranium produces only a single reduction wave, which is well formed, but, according to Harris and Kolthoff, the limiting current is not proportional to the uranium concentration.

A method for the polarographic determination of uranium in atmospheric dust has been developed by Ballenger.[38]

[37] R. Strubl, *Collection Czechoslov. Chem. Communs.*, **10**, 466 (1938).

[38] H. F. Ballenger, U. S. Government Pub. AECD-2065.

CHAPTER XXVI

Manganese and Rhenium

1. MANGANESE

Manganous ion produces a good wave and well-defined diffusion current in neutral alkali chloride or alkaline earth chloride supporting electrolytes, when a trace of gelatin is present as a maximum suppressor. In 1 *N* potassium chloride the half-wave potential is −1.51 v. *vs.* the S.C.E. Verdier[1] found that $E_{1/2}$ changed slightly with the drop time in 1 *N* potassium chloride; with a drop time of 2 sec. it was −1.513 v., and with a drop time of 6 sec. it was −1.499 v. *vs.* the S.C.E.

Hamamoto[2] claimed that the diffusion current in an alkali chloride supporting electrolyte was not a strict linear function of the manganous ion concentration at very small concentrations. However, according to our experience, a strict linear relation is obtained when proper correction is made for the residual current.

The presence of free acid interferes with the manganous wave, because the hydrogen wave precedes that of manganous ion. Consequently, in strongly acid solutions no wave for manganese is observed before hydrogen discharge.

Prajzler[3] has shown that a barium chloride supporting electrolyte is suitable for the simultaneous determination of zinc and manganese, nickel and manganese, or zinc, cobalt, and manganese.

A fairly well-defined wave for manganese is obtained in ammoniacal medium.[4] In a supporting electrolyte composed of 1 *M* ammonia, 1 *M* ammonium chloride, and about 0.005 per cent gelatin, the half-wave potential is −1.65 v. *vs.* S.C.E.[5] Heyrovsky[6] demonstrated that copper, cadmium, nickel, zinc, and manganese, whose waves appear in the order listed, can be determined simultaneously in a supporting electrolyte composed of 1 *M* ammonia, 1 *M* ammonium chloride, and 0.005 per cent gelatin,

[1] E. T. Verdier, *Collection Czechoslov. Chem. Communs.*, **11,** 216 (1939).
[2] E. Hamamoto, *Collection Czechoslov. Chem. Communs.*, **6,** 325 (1934).
[3] J. Prajzler, *Collection Czechoslov. Chem. Communs.*, **3,** 406 (1931).
[4] J. Brezina, *Rec. trav. chim.*, **44,** 520 (1925).
[5] M. Voriskova, *Collection Czechoslov. Chem. Communs.*, **11,** 588 (1939).
[6] J. Heyrovsky, *Polarographie,* Springer, Vienna, 1941 (p. 249).

provided that the various metals are present in approximately equal concentrations.

A well-formed manganese wave is obtained in 1 *M* potassium thiocyanate in the presence of 0.01 per cent gelatin; the half-wave potential is −1.55 v. *vs.* the S.C.E. and the diffusion current is well defined. Verdier[1] found that the half-wave potential in 1 *N* potassium thiocyanate shifted from −1.553 v. to −1.533 v. *vs.* the S.C.E. when the drop time was increased from 2 to 6 sec., but it was independent of the concentration of manganese. Lyalikov[7] recommended an ammoniacal thiocyanate supporting electrolyte for the determination of manganese in steels and slags.

The reduction of manganese from cyanide medium has been studied in some detail by Verdier.[1] From 1.5 *N* potassium cyanide the manganous-cyanide complex produces a single well-defined wave with $E_{1/2} = -1.33$ v. *vs.* the S.C.E. In 1 *N* cyanide the height of the wave corresponds to complete reduction of the manganous-cyanide complex to the metal. However, with concentrations of cyanide between about 0.5 and 1 *N* a double wave was observed; with 1 *N* cyanide the half-wave potentials were −1.33 v. and −1.8 v. *vs.* the S.C.E. Since the relative heights of the two waves in buffered solutions containing 1 *N* potassium cyanide varied with the pH between 8 and 12, Verdier concluded that the first wave was due to the reduction of a basic manganous cyanide (Verdier postulated MnOHCN). For analytical purposes the optimum cyanide concentration is 1.5 *N*.

Verdier[8] recommended the use of a cyanide supporting electrolyte for the determination of manganese in the presence of copper, chromium, zinc, and *ferrous* iron, since none of these metals produces a wave in cyanide medium. He also developed the following procedure for the determination of manganese in steel.

The steel sample is dissolved in hydrochloric acid, evaporated nearly to dryness, and, after being dissolved in a few milliliters of water, the iron is reduced to the ferrous state by addition of sodium sulfite and boiling for a few minutes. Sufficient solid potassium cyanide is then added to the slightly acid solution (under a hood!) to make its final concentration 1.5 *N*, the solution is diluted to 25 ml. in a volumetric flask, and the manganese wave in an aliquot portion is recorded. Cobalt interferes because its wave coincides with the manganese wave in cyanide medium, but none of the other common metals in steel interferes.

Spalenka[9] employed a cyanide supporting electrolyte for the determination of manganese in aluminum and zinc alloys (see Chap. XXXV for procedure). To avoid air oxidation of the manganous-cyanide complex

[7] Yu. S. Lyalikov, *Trans. All-Union Conf. Anal. Chem.*, **2,** 439 (1943).
[8] E. T. Verdier, *Collection Czechoslov. Chem. Communs.*, **11,** 238 (1939).
[9] M. Spalenka, *Z. anal. Chem.*, **128,** 42 (1947).

he recommended that Metol (monomethylparamidophenol sulfate) be added to the solution.

Oscilloscopic studies of the reduction of the ammonia and cyanide complexes of +2 manganese by Heyrovsky[10] and Bourrelier[11] show that reduction does not proceed reversibly in either case. Bourrelier's results also show that the ammonia complex is formed very rapidly, but the formation of the cyanide complex is rather slow.

Verdier[12] found that the manganous-tartrate complex formed in alkaline tartrate solutions is oxidized at the dropping electrode, and produces a well-defined anodic wave at −0.4 v. *vs.* the S.C.E. He employed a 5% sodium potassium tartrate solution that was about 2 *N* with respect to sodium hydroxide. The manganous-tartrate complex is apparently oxidized to the +3 state. The anodic diffusion current is directly proportional to the manganese concentration. The curves obtained by Verdier also show an ill-defined cathodic wave at −1.7 v., that is apparently due to the reduction of the manganous-tartrate complex to the metal. The manganous-tartrate complex is easily air oxidized, and hence care must be taken to exclude air from the solutions. When air was passed through an alkaline manganous-tartrate solution the anodic wave disappeared, and an ill-defined triple cathodic wave was obtained. Verdier concluded that the manganous-tartrate complex was oxidized by air to the +4 state, and that the triple wave corresponded to the stepwise reduction of the +4 manganese complex to the metal. Verdier recommended the use of the anodic manganous-tartrate wave for determining small amounts of Mn in the presence of Ni and Co. He also applied the anodic wave for determining small amounts of manganese in the presence of iron. The iron must first be converted to the ferric state, because ferrous iron forms a complex in alkaline tartrate solution that is oxidized and produces an extended anodic wave between −0.6 and −0.9 v. which masks the anodic wave of small amounts of manganese.

Kolthoff and Watters[13] discovered that the pyrophosphate complex of +3 manganese (probably $Mn(H_2P_2O_7)_3^{---}$) is reduced at the dropping electrode to produce a diffusion current that starts at zero applied e.m.f., and they utilized this for the practical determination of manganese. The preliminary oxidation of the manganese to the +3 state is effected by shaking the pyrophosphate solution of the +2 manganese with excess lead dioxide at a pH between 2 and 4; the excess lead dioxide is filtered off and

[10] J. Heyrovsky, *Österr. Chem. Ztg.*, **48,** 24 (1947).

[11] M. Bourrelier, *Compt. rend.*, **228,** 395 (1949).

[12] E. T. Verdier, *Collection Czechoslov. Chem. Communs.*, **11,** 233 (1939).

[13] I. M. Kolthoff and J. I. Watters, *Ind. Eng. Chem., Anal. Ed.*, **15,** 8 (1943).

the filtrate is polarographed. The plumbous lead resulting from the reduction of the PbO_2 produces a wave at −0.4 v., but since the manganic pyrosphosphate diffusion current is fully developed at potentials well in advance of this the lead wave does not interfere. The optimum concentration of pyrophosphate is about 0.4 *M*. The pH must be between about 2 and 4; at lower pH values iron interferes and at higher pH values the violet manganic-pyrophosphate complex tends to disproportionate into MnO_2 and +2 manganese.

The sample solution equivalent to 1 to 100 mg. of manganese is transferred to a 100-ml. volumetric flask and enough 1:1 nitric acid (or sulfuric acid) is added to make the total amount of mineral acid present equivalent to 12 ml. of 1:1 nitric acid (or 4.5 ml. of 1:1 sulfuric acid). Then 20 ml. of 2 *M* potassium pyrophosphate solution is added slowly with thorough swirling. Thymol blue indicator is added, the pH is adjusted to 2.0 to 2.4 by addition of either acid or ammonia, 5 ml. of 2 per cent gum arabic (or 5 ml. of 2 per cent peptone) is added as a maximum suppressor, and the solution is diluted to the mark.

A portion of this solution is transferred to a polarographic cell and the residual current is measured at +0.1 v. *vs.* S.C.E. Another portion of the solution is treated with 1 g. of lead dioxide for every 25 to 50 mg. of manganese present, and after shaking for 5 to 10 minutes the mixture is filtered through a dry filter or centrifuged. A portion of the clear oxidized solution is transferred to a polarographic cell and the diffusion current is measured at +0.1 v. *vs.* S.C.E. The previously measured residual current is subtracted from the total current to obtain the corrected diffusion current of +3 manganese. The value of $i_d/Cm^{2/3}t^{1/6}$ under these conditions at 25° is 1.17.

Because the diffusion current of the manganic-pyrophosphate complex is measured at a potential more positive than the reduction potential of oxygen, it is not necessary to remove dissolved air from the solutions. The manganic complex is slowly reduced by contact with mercury and therefore an H-cell with separate anode should be used and the d.e. should not be placed in the test solution until all is in readiness for the measurement of the diffusion current. A mercury-mercurous sulfate-saturated potassium sulfate anode in the H-cell is preferable to a saturated calomel anode.

The test solution should not contain more than traces of chloride ion, because the latter depolarizes the d.e. and produces an anodic current in the neighborhood of +0.1 v. and it also is more or less oxidized by the lead dioxide to produce reducible chlorine and hypochlorite. Iron in amounts up to 0.2 g. is innocuous, provided the pH is between 2 and 4, but chromium, vanadium, and cerium interfere by contributing to the diffusion current at +0.1 v. Most other metals are harmless.

Permanganate. Von Stackelberg, *et al.*[14] reported that neutral solutions of permanganate in a barium chloride supporting electrolyte produced a poorly defined wave that started at zero applied e.m.f., corresponding to reduction to manganous ion, and a second well-defined wave in the neighborhood of −1.5 v., resulting from reduction of manganous ion to the

[14] M. von Stackelberg, P. Klinger, W. Koch, and E. Krath, *Tech. Mitt. Krupp Forschungsber.*, **2**, 59 (1939).

metal. They concluded that the waves were not suitable for analytical purposes.

Manganese in Biological Material. Hamamoto[15] recommended the following procedure for the determination of *traces of manganese in biological material.*

A 1- to 2-g. sample of tissue that has been dried at 100° is ignited at 600° to 800° C. in an electric furnace. The ash is dissolved in a few milliliters of 12 *N* hydrochloric acid, diluted with three or four times its volume of water, and most of the acid is neutralized with sodium carbonate. Iron is removed by addition of an excess of sodium acetate, and the solution is filtered. The manganese in the filtrate is oxidized and precipitated as manganese dioxide. The latter is taken up with 1 ml. of 3 *N* hydrochloric acid, and evaporated to dryness to remove excess acid. The residue of manganous chloride is then dissolved in 1 to 3 ml. of 0.01 *N* lithium chloride, the solution is transferred to a micro cell and the polarogram is obtained in the usual way. Hamamoto obtained good results with amounts of manganese from 0.1 to 30 γ (10^{-4} to 0.03 mg.).

Manganese in Alloys. See Chapter XXXV.

2. RHENIUM

Heyrovsky[16] found that rhenium catalyzes the evolution of hydrogen from an acetic acid-acetate buffer after treatment with hydrogen sulfide, and he recommended this catalytic wave for the detection of traces of rhenium in manganese salts. A solution of the manganous salt is prepared in an acetic acid-acetate buffer (pH = 4.7), and saturated with hydrogen sulfide. The excess hydrogen sulfide is removed by boiling, and the solution is filtered. The polarogram of the filtrate shows the catalytic wave due to perthiorhenate ion, ReS_4^-, at − 1.2 v. *vs.* the S.C.E. Heyrovsky claimed that 10^{-6} *M* rhenium could be detected in this way, and that none of the common metals interfered.

In a phosphate buffer of pH = 7 perrhenate ion also produces a catalytic wave at −1.6 v. *vs.* S.C.E.[17] The wave height is several times greater than corresponds to the maximal possible 8-electron reduction to the −1 state (rhenide ion), and it is increased still more by traces of gelatin. The magnitude of this wave is greatly increased by increased temperature; the temperature coefficient being about +4.3 per cent per degree at 25°. The *net* reaction responsible for the catalytic wave very probably is reduc-

---

[15] E. Hamamoto, *Collection Czechoslov. Chem. Communs.*, **6,** 325 (1934).
[16] J. Heyrovsky, *Nature*, **135,** 870 (1935).
[17] J. J. Lingane, *J. Am. Chem. Soc.*, **64,** 1005 (1942).

tion of hydrogen ion, and Lingane postulated the following mechanism:

$$Re^{+7} + 8e \rightleftharpoons Re^{-1} \quad (1)$$

$$Re^{-1} + 2H^+ \text{ (or HA)} \rightleftharpoons Re^{+1} + H_2 \quad (2)$$

$$Re^{+1} + 2e \rightleftharpoons Re^{-1} \quad (3)$$

The small fraction of the total current due to the 8-electron reduction to rhenide ion ought to be diffusion controlled, and the remainder is due to the rapid cyclic recurrence of reactions 2 and 3. Reaction 2 probably is the rate-controlling step.

The reduction of perrhenate ion from 0.1 *M* hydrochloric acid produces only a slight indication of a wave, but with sufficiently large concentrations of hydrochloric or perchloric acid a fairly well-developed wave is obtained, corresponding to a 3-electron reduction to the +4 state[18]. The reduction does not proceed reversibly and the wave is nearly obliterated by even as little as 0.01% gelatin. In 2 *M* hydrochloric acid the half-wave potential is −0.45 v. *vs.* S.C.E., and in 4.2 *M* hydrochloric acid it is −0.31 v. In 4 *M* perchloric acid the definition of the wave is better than in hydrochloric acid and the diffusion current is well developed and directly proportional to the concentration of perrhenate ion from 0.1 to 1 millimolar. The value of $i_d/Cm^{2/3}t^{1/6}$ in this medium at 25° is 6.69. The half-wave potential in 4 *M* perchloric acid shifts from −0.44 v. *vs.* S.C.E. to −0.39 v. when the concentration of perrhenate ion is increased from 0.1 to 1 millimolar.

From unbuffered 2 *M* potassium chloride solution polarograms of perrhenate ion show two waves. The first at −1.41 v. *vs.* S.C.E. corresponds to an irreversible 8-electron reduction to the −1 state (rhenide ion):

$$ReO_4^- + 4H_2O + 8e \rightleftharpoons Re^- + 8OH^-$$

and the second at 1.7 v. is a catalytic wave of the type described above. Although the height of the first wave corresponds quite closely to the theoretical value for $n = 8$, it is not strictly proportional to the perrhenate ion concentration; $i_d/Cm^{2/3}t^{1/6}$ decreases from 22.4 with 0.02 millimolar perrhenate ion to 17.9 with 0.3 millimolar perrhenate (theory = 17.9). This probably results in part from the fact that the wave is distorted by a rounded maximum which renders exact measurement uncertain. Addition of as little as 0.01 per cent gelatin obliterates the first wave and increases the remaining catalytic wave at −1.7 v.

Rulfs and Elving[19a] confirmed the earlier conclusion[18] that according to the Ilkovic equation the reduction of perrhenate in 4 *M* perchloric acid

[18] J. J. Lingane, *J. Am. Chem. Soc.*, **64,** 1001 (1942).

corresponds to a 3-electron process. However when they attempted to verify this by controlled potential coulometric analysis (see p. 392) they found that hydrogen gas was the chief product. Although Rulfs and Elving offered no explanation for this result, it seems very probable that +4 rhenium is the primary product, but that during the extended time of the coulometric analysis it reduces hydrogen ion in the very strongly acid solution and is thus reoxidized to a higher state which is again reduced and reoxidized cyclically. Apparently this does not occur to any appreciable extent during the brief life of each drop at the d.e.

Solutions of rhenide ion, produced by reducing cold solutions of perrhenate ion in dilute sulfuric or perchloric acid with amalgamated zinc, produce a complicated anodic polarogram.[19] In 1 *N* sulfuric acid at 0° there are three anodic waves with half-wave potentials of −0.54 v., −0.34 v., and −0.07 v. *vs.* S.C.E. In 1 *N* perchloric acid the second wave divides incompletely into two waves at −0.42 v. and −0.26 v. Lingane tentatively concluded that the wave at −0.54 v. corresponds to a 3-electron oxidation of the rhenide ion to the +2 state, that the wave at −0.34 v. results from oxidation to the +5 state (or the doublet at −0.42 v. and −0.26 v. to the +3 and +5 states), and that the wave at −0.07 v. reflects complete oxidation to the +7 state.

Rulfs and Elving[19a] disagree with the foregoing interpretation of the anodic waves of rhenide ion. They ascribe the wave at −0.50 v. to oxidation to the +1 state, that at −0.27 v. to oxidation to the +2 state, and that at −0.02 v. to oxidation to the +3 state. These different interpretations are both based on very indirect evidence, and neither can be regarded as conclusive.

Epple[20] observed that chlororhenite ion, $ReCl_6^{--}$, undergoes a 1-electron reduction from dilute hydrochloric acid medium. The wave is well formed, and its half-wave potential in 1 *M* hydrochloric acid is −0.51 v. *vs.* S.C.E. The slope of the wave corresponds closely to the expected value for a reversible 1-electron reduction, but the fact that solutions of +3 rhenium in the same medium do not show an oxidation wave demonstrates that the reduction of $ReCl_6^{--}$ is not perfectly reversible. Rhenium tribromide in 2 *M* perchloric acid undergoes irreversible stepwise reduction to the +2 state and to the metal; the half-wave potentials of the two waves are −0.28 v. and −0.46 v. *vs.* S.C.E.

Epple also found that perrhenate ion is catalytically reduced to the +3 state by the correduction of uranyl ion.

[19] J. J. Lingane, *J. Am. Chem. Soc.*, **64,** 2182 (1942).
[19a] C. L. Rulfs and P. J. Elving, *J. Am. Chem. Soc.*, **73,** 3284, 3287 (1951).
[20] R. Epple, Ph.D. Thesis, Massachusetts Institute of Technology, Sept. 1946.

CHAPTER XXVII

Iron, Cobalt, Nickel, and Platinum Metals

1. IRON

Because the standard potential of the ferric ion-ferrous ion couple (+0.51 v. *vs.* S.C.E.) is somewhat more positive than the anodic dissolution potential of mercury (*ca.* +0.4 v. in the absence of complexing or precipitating agents) the diffusion current resulting from the reduction of uncomplexed ferric ion to ferrous ion starts at the potential of the anodic mercury current. This is followed by the wave due to the reduction of ferrous ion to the metal, whole half-wave potential varies from −1.3 v. *vs.* S.C.E. in alkali or alkaline earth chloride supporting electrolytes to −1.5 v. in perchlorate media. In the presence of 0.005 per cent gelatin the diffusion current of ferric ion is well developed and suitable for the determination of iron, except that it cannot be distinguished from that due to any other similarly strong oxidant.

The ferrous ion wave is well formed. The fact that the half-wave potential is 0.6 to 0.8 v. more negative than the standard ferrous ion-iron potential shows that a large overvoltage is involved in the reduction. This seems to be quite typical of metals that are insoluble in mercury. Prajzler[1] found that the half-wave potential is −1.3 v. *vs.* S.C.E. in 0.05 *N* barium chloride, and the same is true in neutral alkali chloride supporting electrolytes. In 1 *M* ammonium perchlorate as supporting electrolyte, where complexation of ferrous ion is minimal, we observed that the half-wave potential is more negative and it shifts from −1.48 to −1.44 v. *vs.* S.C.E. when the ferrous ion concentration is increased from 0.1 to 10 millimolar. Such a shift is expected in the reduction of the ion of a metal that is insoluble in mercury.

Since the ferrous ion wave is virtually coincident with that of hydrogen ion it is easily masked by hydrogen discharge, and hence the pH of the supporting electrolyte must be above about 5 and yet below the value (pH *ca.* 7) at which precipitation of ferrous hydroxide occurs. The pH cannot be regulated with buffers, because the acid constituents of buffers can show a hydrogen wave.

---

[1] J. Prajzler, *Collection Czechoslov. Chem. Communs.*, **3,** 406 (1931).

Verdier[2] reported that the ferric-tartrate complex produces a cathodic wave and that the ferrous-tartrate complex yields an anodic wave, and he recommended the anodic wave for the determination of iron in manganiferous materials. The polarography of ferric and ferrous iron in tartrate, citrate, and oxalate media has been studied systematically by Lingane.[3] In all three media the reduction of the ferric complexes to the ferrous complexes produces very well-formed waves at pH values smaller than about 6, and the diffusion currents are well developed in the presence of 0.005 per cent gelatin. The second stage of reduction of the ferrous complexes to the metal is more or less masked by hydrogen ion or alkali metal ion discharge, and is clearly developed only in strongly alkaline citrate and ammoniacal tartrate media.

In all three media the ferric-ferrous complex couples behave reversibly at pH values less than about 7, and under this condition the half-wave potentials of the anodic waves of the ferrous complexes are identical with the cathodic half-wave potentials of the ferric complexes. In alkaline citrate and tartrate solutions the ferric-ferrous couples behave irreversibly, the anodic ferrous waves being at a much more positive potential than the cathodic ferric waves. In oxalate media of pH greater than 7.9 hydrous ferric oxide precipitates.

In solutions containing 0.5 *M* sodium citrate and 0.005 per cent gelatin the half-wave potential of the ferric to ferrous wave is a linear function of pH between values of 4 and 12, and its value *vs.* S.C.E. at 25° follows the equation:

$$E_{1/2} = 0.426 - 0.108\text{pH}$$

In the same supporting electrolyte between pH 5 and 12 the value of $i_d/Cm^{2/3}t^{1/6}$ is constant at 0.93 ± 0.01 at 25°. The characteristics of the anodic ferrous wave in citrate medium are optimum between pH 6 and 7, and the value of $i_d/Cm^{2/3}t^{1/6}$ in this pH range in 0.5 *M* sodium citrate is 0.90.

In 0.5 *M* sodium tartrate containing 0.005 per cent gelatin the half-wave potentials of the cathodic ferric to ferrous wave at pH values below 6 are very nearly the same as in citrate medium. At pH values between 6.2 and 8.0 the cathodic ferric wave divides into two parts, which apparently result from the reductions of different ferric-tartrate complexes in sluggish equilibrium. The half-wave potentials *vs.* S.C.E. in 0.5 *M* sodium tartrate

[2] E. T. Verdier, *Collection Czechoslov. Chem. Communs.*, **11**, 140 (1939). See also V. F. Toropova, *J. Gen. Chem. U. S. S. R.*, **15**, 603 (1945).

[3] J. J. Lingane, *J. Am. Chem. Soc.*, **68**, 2448 (1946).

containing 0.005 per cent gelatin are:

pH	5.8	6.2	6.8	7.4	8.2	9.4
$E_{1/2}$	−0.17	−0.21	−0.29	−0.33	−1.19	−1.20
	−1.50	−1.53	−1.15	−1.18	−1.72	−1.73
			−1.68	−1.71		

The largest negative value at each pH pertains to the wave of the reduction of the ferrous complex. Between pH 4 and 6 in this supporting electrolyte the $i_d/Cm^{2/3}t^{1/6}$ value of the cathodic ferric wave is 1.11 ± 0.03 at 25°, and the anodic ferrous wave has virtually the same value.

The half-wave potentials *vs.* S.C.E. of the ferric wave in 0.2 *M* sodium oxalate containing 0.005 per cent gelatin are:

pH	2.3	3.7	5.25	7.9	8.3
$E_{1/2}$	−0.140	−0.240	−0.245	−0.270	−0.275

The half-wave potentials of the anodic ferrous wave are the same at pH values less than about 7.9. The considerably more positive half-wave potentials in oxalate medium as compared with citrate and tartrate media show that the oxalate complex of ferric ion is weaker, or that the oxalate complex of ferrous ion is stronger, or both, than the citrate and tartrate complexes. This doubtless is connected with the fact that complexation with tartrate or citrate directly involves replacement by ferric or ferrous ion of the hydrogens of the hydroxyl groups, which are lacking in oxalate ion. Over the pH range from 2 to 6 in the above supporting electrolyte $i_d/Cm^{2/3}t^{1/6}$ for the cathodic ferric wave is 1.50 ± 0.03 at 25°. This value is considerably larger than for the citrate or tartrate complexes, indicating a much smaller size for the oxalate complexes.

In oxalate media of pH greater than about 7.9 hydrous ferric oxide precipitates. In 0.2 *M* sodium oxalate of pH = 8.3 the observed ferric diffusion current corresponds to a solubility of 0.39 millimolar at 25°, and in 0.2 *M* oxalate containing 0.1 *M* excess hydroxide ion the solubility is so small that no wave is observed.

In tartrate or citrate media of pH in the neighborhood of 6 the ferric complexes slowly undergo photochemical reduction to the ferrous complexes when the solutions are exposed to white light. Because the cathodic and anodic diffusion current constants are virtually identical the increase in the anodic wave height equals the decrease in the cathodic wave height and the total composite wave height remains constant. When acidic oxalate solutions of ferric iron are exposed to light for several days a bright yellow precipitate forms, and a polarogram of the solution shows no wave for either ferric or ferrous iron.

Citrate, tartrate, or oxalate supporting electrolytes are very useful for determining a small amount of ferric iron in the presence of a large amount of ferrous, or *vice versa*. To avoid error from photochemical reduction the solutions should be polarographed without delay after they are prepared. The ferrous complexes are easily air oxidized and it is, therefore, important to exclude atmospheric oxygen carefully when the solutions are composited. A good technique consists of adding a relatively small measured volume of sample solution to a known volume of the previously de-aerated supporting electrolyte in the polarographic cell.

An oxalate supporting electrolyte was used by Knanishu and Rice[4] for the simultaneous determination of iron and zinc in phosphate coatings on ferrous surfaces. Tartrate supporting electrolytes have been used for the determination of iron in ores and slags[5] and in aluminate solutions.[6]

The ammonia complex of ferrous iron produces a well-defined reduction wave, whose half-wave potential is −1.48 v. *vs.* S.C.E. in a supporting electrolyte containing 1 *M* each of ammonia and ammonium chloride, 0.005 per cent gelatin, and sulfite ion to remove oxygen.[7]

Strubl[8] observed two waves for ferric iron in an approximately 1 *M* ammonium carbonate solution. The half-wave potentials were −0.44 v. and −1.52 v. *vs.* the S.C.E. The first wave apparently is due to the reduction of a carbonato ferric complex to the ferrous complex, and the second to the reduction of the ferrous complex to the metal.

Komarek[9] found that ferric iron produces two waves in 0.1 to 5 *N* solutions of potassium hydroxide that contain 8 per cent mannitol, with which ferric and ferrous ions form soluble complexes. The first wave, with a half-wave potential of about −0.9 v. *vs.* S.C.E., is due to the reduction of the ferri-mannitol complex to the ferrous state, and the second wave, with a half-wave potential of about −1.5 v. *vs.* S.C.E., corresponds to the reduction of the ferro-mannitol complex to the metal. Most of the polarograms given by Komarek show the height of the second wave just twice that of the first wave, as is to be expected, but in some instances the second wave height was equal to that of the first wave. No explanation was offered for this anomaly. Komarek also found a small double wave with saturated solutions of hydrous ferric oxide in concentrated alkali solutions, due to a small concentration of ferrite ion, FeO_2^-. The first wave cor-

[4] J. Knanishu and T. Rice, *Ind. Eng. Chem.*, **17**, 444 (1945).
[5] Yu. S. Lyalikov, *Zavodskaya Lab.*, **7**, 395 (1938).
[6] N. A. Suvoroyskaya, *J. Applied Chem. U. S. S. R.*, **17**, 156 (1944).
[7] M. Voriskova, *Collection Czechoslov. Chem. Communs.*, **11**, 588 (1939).
[8] R. Strubl, *Collection Czechoslov. Chem. Communs.*, **10**, 466 (1938).
[9] K. Komarek, *Collection Czechoslov. Chem. Communs.*, **7**, 404 (1935).

responds to the reduction of ferrite ion to hypoferrite ion, FeO_2^{--}, and the second to the reduction of the hypoferrite ion to the metal. The half-wave potentials of these ions are practically identical with those of the corresponding mannitol complexes.

Schrager[10] found a good wave with saturated solutions of ferrous hydroxide in 1 *N* sodium hydroxide, with a half-wave potential of −1.46 v. *vs.* the S.C.E. This wave is due to the reduction of the hypoferrite ion. From the magnitude of the diffusion current, Schrager calculated that the solubility of ferrous hydroxide in 1 *N* sodium hydroxide is 3×10^{-5} *M*.

Von Stackelberg and von Freyhold[11] reported that ferric iron produces a reduction wave in fluoride medium (pH not specified), and that the half-wave potential is constant at −1.36 v. *vs.* S.C.E. when the concentration of potassium fluoride is increased from 0.04 to 0.8 *M*. Heyrovsky[12] also shows a polarogram of the ferric-fluoride complex in 0.1 *M* KHF_2 as supporting electrolyte. According to Heyrovsky ferrous iron in 0.1 *M* KHF_2 produces an anodic wave corresponding to oxidation to the ferric-fluoride complex. The anodic ferrous wave is at a considerably more positive potential than the cathodic ferric wave, showing that the ferric-ferrous couple does not behave reversibly in fluoride medium.

On the contrary, West and Dean,[13] using a 1 *M* sodium fluoride supporting electrolyte of pH between about 4 and 7 containing 0.004 per cent gelatin, claimed that no reduction wave for the ferric-fluoride complex is observable below the discharge potential of sodium ion. The reason for these contradictory conclusions is not evident, and a more thorough systematic investigation of the polarography of iron in fluoride media over a wide range of pH and fluoride concentrations is highly desirable. It is quite possible that the supporting electrolytes used by von Stackelberg and von Freyhold and by Heyrovsky were more acidic than pH 4, and that reduction of the ferric-fluoride complex only occurs in fairly acidic fluoride solutions, or that gelatin obliterates the wave.

Furness, Crawshaw, and Davies[14] reported that solutions containing the trisodium salt of ethylenediamine tetraacetate (Trilon B), excess ferric iron, 0.4 *M* potassium nitrate as supporting electrolyte, and 0.01 per cent gelatin, yield the usual diffusion current due to the excess ferric ion and this is followed by a wave at 0.0 v. *vs.* S.C.E. which results from the re-

---

10 B. Schrager, *Collection Czechoslov. Chem. Communs.*, **1,** 275 (1929); *Chem. News*, **138,** 354 (1929).

11 M. von Stackelberg and H. von Freyhold, *Z. Elektrochem.*, **46,** 120 (1940).

12 J. Heyrovsky, "*Polarographie*," Springer, Vienna, 1941 (p. 106).

13 P. W. West and James F. Dean, *Ind. Eng. Chem., Anal. Ed.*, **17,** 686 (1945).

14 W. Furness, P. Crawshaw, and W. C. Davies, *Analyst*, **74,** 629 (1949).

duction of the complexed iron. The fact that a separate wave of the complexed iron is observed shows that the rate of dissociation of the complex is relatively small.

For determining small amounts of iron in the presence of large amounts of copper, *e. g.*, in brass, Souchay and Faucherre[15] recommended a supporting electrolyte containing 0.1 *M* ethylenediamine tetraacetate (Trilon B, Versene, etc.) and 2 *M* sodium acetate, the pH presumably being adjusted to the neighborhood of 6 to 7. In this medium the reduction of the ferric complex produces a well-formed wave at about −0.13 v., which is nearly 0.3 v. in advance of the wave of the cupric complex.

Ferric iron is readily reduced to the ferrous state by hydroxylamine in warm dilute hydrochloric acid solution. Strubl[16] made use of this fact to eliminate the interference of ferric ion in the polarographic determination of various metal ions whose reduction potentials are smaller than that of ferrous ion. The solution to be analyzed is acidified with hydrochloric acid, treated with an excess of hydroxylamine hydrochloride, and heated for a few minutes to effect complete reduction of the ferric iron. The polarogram of the cooled solution is then obtained in the usual way. Hydroxylamine is not reducible in acid medium, and its presence is not harmful. In this way Strubl claimed that small amounts of copper, bismuth, antimony, tin, lead, cadmium, indium, chromium, titanium, and zinc could be determined in the presence of an excess of iron.

Ferrocyanide ion is not reducible at the dropping electrode.[17]

Ferricyanide ion is easily reduced to ferrocyanide at the dropping electrode, and a fairly well-defined diffusion current, which is directly proportional to the ferricyanide ion concentration, is obtained in neutral 0.1 *N* potassium chloride.[18] The potential of an equimolar mixture of ferricyanide and ferrocyanide in 0.1 *N* potassium chloride is about +0.23 v. *vs.* the S.C.E., and hence the diffusion current of ferricyanide is obtained at the very start of the polarogram. In 0.1 *N* potassium chloride solutions the ferricyanide wave has a pronounced maximum, but this is easily eliminated by a small amount of methyl red.

Iron in Alloys. See Chapter XXXV.

2. COBALT

In noncomplexing supporting electrolytes, the half-wave potential of the hexaquo cobaltous ion is about −1.4 v. *vs.* the S.C.E. The diffusion

[15] P. Souchay and J. Faucherre, *Anal. Chim. Acta*, **3**, 259 (1949).
[16] R. Strubl, *Collection Czechoslov. Chem. Communs.*, **10**, 466 (1938).
[17] E. B. Sanigar, *Rec. trav. chim.*, **44**, 549 (1925).
[18] J. J. Lingane and I. M. Kolthoff, *J. Am. Chem. Soc.*, **61**, 825 (1939).

current is well defined when a trace of gelatin is present as a maximum suppressor.

Brdicka[19] found that pink solutions of cobaltous choride, without any other supporting electrolyte, are reduced irreversibly. The cobalt is present in these solutions chiefly as the pink $Co(H_2O)_6^{++}$ ion, and the reduction potential is several tenths of a volt more negative than one would predict from the reversible potential of the Co^{++}–Co system. When very large concentrations of chloride were added to form the blue chloro complex ions, *e. g.*, $Co(H_2O)_2Cl_4^{--}$, the reduction potential was shifted markedly to more positive potentials, and the slopes of the waves increased. Brdicka also observed that the temperature coefficient of the reduction potential was much smaller in the presence of large amounts of chloride. Evidently the reduction of the blue chloro complex ion is more reversible than the reduction of the hexaquo cobaltous ion.

The polarography of solutions of cobaltous chloride in methanol and methanol–benzene mixtures has been studied by Sartori,[20] who reported that the pink alcoholated $Co(MeOH)_n^{++}$ ion is reversibly reduced, in contradistinction to the irreversible reduction of the aquocobaltous ion.

In neutral solutions of cobaltous chloride Brdicka observed a small "prewave" about 0.2 v. before the main cobalt wave. This prewave increased when very small concentrations of potassium hydroxide were added to the solution, but it disappeared when the solution was acidified. Brdicka concluded that the prewave was due to the $Co(H_2O)_5OH^+$ ion formed by hydrolysis. The fact that two waves are obtained indicates that the equilibrium:

$$Co(H_2O)_6^{++} \rightleftharpoons Co(H_2O)_5OH^+ + H^+$$

is established quite slowly.

Prajzler[21] pointed out that the cobaltous wave is fairly well defined between the waves of zinc and manganous ions in a barium chloride supporting electrolyte, and the simultaneous determination of all three metals is possible when their concentrations are of about the same order of magnitude. Large amounts of zinc interfere with the cobaltous wave, but it should be possible to determine small amounts of cobalt in the presence of a considerable excess of manganous ion, since the half-wave potential of the latter is about 0.3 v. beyond that of cobaltous ion.

The hexamminocobaltous ion, $Co(NH_3)_6^{++}$, produces a well-defined wave

---

[19] R. Brdicka, *Collection Czechoslov. Chem. Communs.*, **2,** 489, 545 (1930); **3,** 396 (1931).

[20] G. Sartori, *Gazz. chim. ital.*, **71,** 263 (1941).

[21] J. Prajzler, *Collection Czechoslov. Chem. Communs.*, **3,** 406 (1931).

in the presence of gelatin as a maximum suppressor. In a solution containing 1 N ammonium chloride and 1 M ammonia the half-wave potential is −1.30 v. *vs.* the S.C.E. The hexamminocobaltous ion is oxidized by dissolved oxygen in ammoniacal solution, and the resulting hexamminocobaltic ion, $Co(NH_3)_6^{+++}$, produces a double wave with half-wave potentials of about −0.5 v. and −1.3 v. *vs.* the S.C.E.[22] The height of the first wave is one-half that of the second wave, corresponding to:

$$Co(NH_3)_6^{+++} + e \rightleftharpoons Co(NH_3)_6^{++}$$

The second wave corresponds to

$$Co(NH_3)_6^{++} + 2e + Hg \rightleftharpoons Co(Hg) + 6NH_3$$

Willis, Friend, and Mellor[23] systematically investigated the polarography of cobalt ammines, and of aquo, chloro, nitro, and other substituted ammonia and ethylenediamine complexes of +3 cobalt. All the ammine complexes, substituted ammine complexes, and ethylenediamine complexes, produced doublet waves corresponding to successive 1- and 2-electron reductions. The half-wave potentials of the first cobaltic to cobaltous waves varied with the nature of the substituents, from −0.03 v. *vs.* S.C.E. for $Co(NH_3)_3(NO_2)_3$ to −0.47 v. for $Co(NH_3)_5H_2O^{+++}$. The half-wave potential of the first wave was different for the *cis* and *trans* isomers of $Co(NH_3)_4(NO_2)_2^{+}$, being −0.04 v. for the *cis* form and −0.21 v. for the *trans*, but it was exactly the same for the *cis* and *trans* forms of $Co(NH_3)_4Cl_2^{+}$ (−0.42 v.). The half-wave potential of the second wave is the same (−1.28 v. *vs.* S.C.E.) for the various complexes, except for $Co(NO_2)_6^{---}$, in which case it is −1.43 v. Willis, Friend, and Mellor interpreted this to mean that "the first step corresponds to the reduction and disruption of the complex ion to form $Co(H_2O)_6^{++}$; the second step corresponding to the reduction of $Co(H_2O)_6^{++}$ to cobalt." However, this does not reconcile with the fact that the half-wave potential of hexaquocobaltous ion (−1.425 v. according to W., F., and M.) is 0.14 v. more negative than that of the second wave of the complexes.

The effect of various capillary-active substances and supporting electrolytes on the doublet wave of the hexamminocobaltic ion without excess ammonia present has been systematically investigated by Laitinen, Bailar, Holtzclaw, and Quagliano.[24] Gelatin and octyl alcohol shift the half-wave

[22] R. Brdicka, *Collection Czechoslov. Chem. Communs.*, **5**, 112 (1933).

[23] J. B. Willis, J. A. Friend, and D. P. Mellor, *J. Am. Chem. Soc.*, **67**, 1680 (1945).

[24] H. A. Laitinen, J. C. Bailar, Jr., H. F. Holtzclaw, and J. V. Quagliano, *J. Am. Chem. Soc.*, **70**, 2999 (1948).

potential of the first wave by several tenths of a volt to more negative values, increase the steepness of the wave (an advantage), and decrease the diffusion current somewhat. Similar but much less pronounced effects were observed with methyl red. Sulfate, citrate, and tartrate shift the first wave by 0.1 to 0.2 v. to more negative values, and the latter two also cause a negative shift of the second wave. These shifts are all accompanied by somewhat decreased diffusion currents. Laitinen *et al.* interpreted these effects by postulating the formation of "super-complexes" in which the hexamminocobaltic ion acts as the central ion.

Watters and Kolthoff[25] developed the following procedure for the determination of cobalt based on oxidation to the +3 ammonia complex by perborate ion, the excess of which is easily removed (reduced) by boiling.

The sample equivalent to 1 to 120 mg. of cobalt as the nitrate, sulfate, chloride, or perchlorate in about 25 ml. of solution in a 250-ml. Erlenmeyer flask is treated with 10 ml. of 1 *M* ammonium chloride, 40 ml. of 6.25 *M* ammonia, and about 1 g. of sodium perborate ($NaBO_3 \cdot 4H_2O$). The flask is fitted with a small reflux condenser and the solution is boiled gently in a boiling water bath for 10 min. to decompose the excess perborate. The solution is then cooled, transferred to a 100-ml. volumetric flask, treated with 2 ml. of fresh 1 per cent gelatin, and diluted to the mark. A portion of this solution is transferred to a polarographic cell and the polarogram is recorded. The cobalt content is computed from the height of the first wave, whose half-wave potential under the specified conditions is −0.53 v. *vs.* S.C.E. The value of $i_d/Cm^{2/3}t^{1/6}$ is 1.68 at 25°, and is independent of the concentration of excess ammonia between about 0.5 and 4 *M* and of the concentration of ammonium chloride between 0.1 and 0.5 *M*.

This method is particularly valuable for determining cobalt in the presence of as much as a 1000-fold excess of nickel. Even large amounts of zinc do not interfere. Large amounts of iron and manganese interfere due to coprecipitation of the cobalt with the hydrous ferric oxide and manganese dioxide which precipitate in the procedure, but amounts of manganese and iron not larger than the amount of cobalt can be tolerated. Large amounts of copper interfere because the copper wave precedes that of the cobaltic ammine. Chromium interferes because it is oxidized to chromate ion, which produces a reduction wave before the cobalt wave. Large amounts of iron can be removed by isopropyl ether extraction, copper can be removed by reduction to the metal with zinc amalgam, and chromium can be removed after the oxidation by precipitation as barium chromate. Manganese (and copper and ferric iron) can be separated from the cobalt by precipitation with 1-nitroso-2-naphthol in 50 per cent acetic acid solution.

The polarographic characteristics of the +3 cobalt complex of ethylenediaminetetraacetate (Trilon B, Versene, etc.) have been studied by Souchay and Faucherre,[26] who prepared the complexes *in situ* by oxidizing

[25] J. I. Watters and I. M. Kolthoff, *Anal. Chem.*, **21**, 1466 (1949).
[26] P. Souchay and J. Faucherre, *Anal. Chim. Acta*, **3**, 252 (1949).

solutions of +2 cobalt containing ethylenediaminetetraacetate with lead dioxide. Although specific half-wave potential data are not given by Souchay and Faucherre, curves which they present indicate that the +3 cobalt complex is reduced to the +2 state at a potential ranging from about −0.1 to −0.4 v. in solutions of pH about 6 to 9.5 containing 0.1 *M* Trilon B. These curves also indicate that the cobaltic wave begins at the potential of the anodic dissolution of mercury, and hence the true reduction potential of the cobaltic complex appears to be more positive (more oxidizing) than the oxidation potential of mercury. The cobaltous complex does not show a reduction wave. Souchay and Faucherre recommended a Trilon-pyridine supporting-electrolyte for the determination of small amounts of cobalt in the presence of much iron; the iron is precipitated by the pyridine and the cobalt in the filtrate is oxidized to the +3 state with lead dioxide in the presence of tripotassium ethylenediaminetetraacetate. A more definitive study of this method is desirable because it appears to have considerable promise in steel analysis, and it also permits the determination of cobalt in the presence of much nickel.

Cobaltous cobalt produces a well-defined wave in 0.1 *M* pyridine containing 0.1 *M* pyridinium chloride; the half-wave potential is −1.07 v. *vs.* the S.C.E.[27] The wave shows a pronounced maximum, but this is easily suppressed by gelatin and the diffusion current is then well defined. A pyridine-pyridinium chloride supporting electrolyte is well suited for the simultaneous determination of nickel and cobalt; the nickel wave precedes the cobalt wave by about 0.3 v.

Lingane and Kerlinger reported that the half-wave potential of +2 cobalt in 1 *M* potassium thiocyanate is −1.03 v. *vs.* the S.C.E. In neutral thiocyanate solutions the diffusion current is well defined when gelatin is present. In slightly acid thiocyanate solutions the diffusion current shows a peculiar minimum, even in the presence of gelatin, which makes it unsuitable for analytical purposes.

Early studies of the polarography of cobalt in cyanide media were carried out by Emelianova,[28] and Sartori,[29] and Hume and Kolthoff[30] have recently investigated this subject more thoroughly. Addition of excess cyanide ion to a cobaltous salt solution forms the aquopentacyanocobaltate (II) ion, $Co(CN)_5H_2O^{---}$, which is reduced only to a complex of +1 cobalt to produce a rather extended wave (irreversible reduction). The

[27] J. J. Lingane and H. Kerlinger, *Ind. Eng. Chem., Anal. Ed.*, **13,** 77 (1941).
[28] N. V. Emelianova, *Rec. trav. chim.*, **44,** 529 (1925).
[29] G. Sartori, *Gazz. chim. ital.*, **66,** 688 (1936).
[30] D. N. Hume and I. M. Kolthoff, *J. Am. Chem. Soc.*, **71,** 867 (1949).

half-wave potential is −1.3 v. *vs.* S.C.E. in 1 *M* potassium cyanide, and the diffusion current is developed very satisfactorily. The +1 cobalt complex is neither reducible nor oxidizable at the d.e. Experiments by Hume and Kolthoff indicate that the hexacyanocobaltate (II) ion, $Co(CN)_6^{----}$, is unstable in aqueous solution and rapidly undergoes aquation to form the aquopentacyano complex.

The aquopentacyanocobaltate (III) complex, formed by warming solutions of the cobaltous complex, in 1 *M* potassium cyanide:

$$Co(CN)_5H_2O^{---} + H_2O \rightleftharpoons Co(CN)_5H_2O^{--} + OH^- + \frac{1}{2}H_2$$

is irreversibly reduced to a complex of +1 cobalt, to produce an extended wave starting at −1.2 v. *vs.* S.C.E., and the diffusion current plateau is not very well formed. The hexactyanocobaltate (III) ion, $Co(CN)_6^{---}$, formed by prolonged treatment of the aquopentacyano complex with excess cyanide, is not reducible at the d.e. The use of a cyanide supporting electrode for the determination of cobalt does not appear to be very promising.

Watters[31] found that the oxalate complex of cobaltous ion is so stable that no reduction wave is observed below the final current rise in solutions containing excess oxalate and buffered to a pH of 5. However, the green oxalate complex of +3 cobalt—probably $Co(C_2O_4)_3^{---}$—under the same conditions is reduced to the +2 state. The wave starts from zero applied e.m.f. and the diffusion current is well developed between +0.2 and −0.5 v. *vs.* S.C.E. Watters utilized this diffusion current for the determination of cobalt in the presence of iron, copper, nickel, and most other common elements. The cobalt is oxidized to the +3 state by shaking the solution with excess lead dioxide.

A 25-ml. sample solution containing 2 to 60 mg. of cobalt as the nitrate, perchlorate, or sulfate is neutralized to a permanent turbidity with sodium hydroxide solution. Then 2 ml. of glacial acetic acid, 10 ml. of 2 *M* ammonium acetate, and 50 ml. of 2 *M* potassium oxalate are added and the solution is diluted to exactly 100 ml. The pH should be close to 5. About 75 ml. of this solution is treated with about 1 g. of lead dioxide (manganese free!), shaken for 5 min., and filtered or centrifuged. A measured 25-ml. portion of the clear oxidized solution is treated with 1 ml. of 0.5 per cent gelatin and polarographed after removing dissolved air with nitrogen. The diffusion current is measured between 0 and −0.1 v. *vs.* S.C.E. The residual current is determined from a polarogram of a portion of the unoxidized solution, and subtracted from the total current to obtain the true diffusion current of the +3 cobalt.

Iron, nickel, and zinc do not interfere, but chloride ion and manganese do interfere.

---

[31] J. I. Watters, Ph.D. Thesis, University of Minnesota, 1943.

The polarography of the cobaltous dimethylglyoxime complex has been studied by Komarek.[32] Although it is claimed that dimethylglyoxime itself is not reducible in alkaline medium[33] its addition in excess to a strongly alkaline cobaltous solution increases the cobalt wave 4- to 5-fold. Komarek recommended this phenomenon for the determination of cobalt in the presence of nickel, ferric iron, manganese, aluminum, and cadmium. Ferrous iron, zinc, and lead interfere because their complexes with dimethylglyoxime produce waves similar to that of the cobaltous-dimethylglyoxime complex. Although the bright yellow cobaltous complex is easily air oxidized this produces no effect on the cobalt wave.

3. NICKEL

In supporting electrolytes that do not form complexes with nickel, such as alkali perchlorates or nitrates, the half-wave potential of the hexaquonickel ion is about −1.1 v. *vs.* the S.C.E. In the presence of gelatin, or methyl red in slightly acid solutions, the diffusion current is well defined and directly proportional to the concentration of nickelous ion. The reduction of the hexaquonickel ion from non-complex-forming supporting electrolytes is irreversible, as shown by the fact that the half-wave potential is more than 0.5 v. more negative than one would predict from the reversible potential of the Ni^{++}–Ni couple (about −0.50 v. *vs.* the S.C.E.). Furthermore, the slope of the nickel wave is abnormally small especially in the presence of gelatin.

With increasing temperature the reduction potential of the hexaquonickel ion is shifted markedly to a more *positive* value,[34] indicating a more reversible reduction at higher temperatures.

Pavlick[34] observed that the addition of large concentrations of chloride to a nickel salt solution shifted the reduction potential markedly to a more positive value. In the presence of a high concentration of chloride the wave also had a more nearly normal slope, and the temperature coefficient of the reduction potential was considerably smaller. These effects appear to be due to the formation of complex ions between the chloride and nickelous ions, which apparently are reduced more reversibly than the hexaquonickel ion.

Lingane and Kerlinger[35] found that the thiocyanate and pyridine com-

[32] K. Komarek, *Collection Czechoslov. Chem. Communs.* **12,** 399 (1947).

[33] See, however, I. M. Kolfhoff and A. Langer, *J. Am. Chem. Soc.*, **62,** 211 (1940), who showed that d.m.g. is reduced in ammoniacal medium.

[34] M. Pavlick, *Collection Czechoslov. Chem. Communs.*, **3,** 223 (1931).

[35] J. J. Lingane and H. Kerlinger, *Ind. Eng. Chem., Anal. Ed.*, **13,** 77 (1941).

plexes of nickelous ion are reduced at a much more positive potential than the hexaquonickel ion. In 1 *N* potassium thiocyanate the half-wave potential of nickel is −0.70 v., and in 1 *N* potassium chloride containing 0.5 *M* pyridine it is −0.78 v., *vs.* the S.C.E. These media are ideally suited to the determination of small amounts of nickel in the presence of a large excess of cobalt, since the waves of the nickel complexes are about 0.3 v. in advance of that of cobalt. A pyridine supporting electrolyte permits the rapid and accurate determination of nickel present as an impurity in cobalt and cobalt salts and the simultaneous determination of nickel and cobalt in steels and other ferrous materials. The hydrous ferric oxide precipitate formed in a pyridine-pyridinium ion buffer of pH *ca.* 5.5 does not coprecipitate nickel and cobalt.

A thiocyanate supporting electrolyte was recommended by Stout and Levy[36] for simultaneous determination of nickel and zinc, and for the determination of nickel in the presence of excess zinc. These authors also developed procedures for the polarographic determination of nickel in *plant materials.*

Nickel produces an excellent wave in ammoniacal medium in the presence of 0.005 per cent gelatin. With 1 *M* ammonia and 0.2, 1, and 3 *M* ammonium chloride the half-wave potentials *vs.* S.C.E. and $i_d/Cm^{2/3}t^{1/6}$ values at 25° are:[37] −1.06 v., 3.54; −1.09, 3.56; and −1.12, 3.33. An ammoniacal supporting electrolyte is well suited to the simultaneous determination of nickel and zinc;[37, 38] the nickel wave precedes that of the zinc by about 0.3 v. (see Chap. XXXV). Ammoniacal supporting electrolytes have also been recommended for the simultaneous determination of nickel and cobalt.[39, 40] However, the separation of the nickel and cobalt waves in ammoniacal medium is not nearly as good as in pyridine solution.

The polarography of nickel in cyanide medium has been systematically investigated by Hume and Kolthoff[41] using solutions prepared both from pure $Na_2Ni(CN)_4 \cdot 2H_2O$, and by mixing nickelous ion in solution with excess cyanide ion. Solutions of $Ni(CN)_4^{--}$ in either 0.1 *M* potassium chloride or sodium hydroxide produce an extended wave starting at −1.35

[36] P. R. Stout and J. Levy, *Collection Czechoslov. Chem. Communs.*, **10**, 136 (1938).

[37] J. J. Lingane, *Ind. Eng. Chem., Anal. Ed.*, **18**, 430 (1946).

[38] P. Pavolov and I. Pavlenko, *J. Gen. Chem. U. S. S. R.*, **7**, 2246, 2259 (1937).

[39] G. Thanheiser and G. Maassen, *Mitt. Kaiser-Wilhelm Inst. Eisenforsch. Düsseldorf*, **19**, 27 (1937); *Arch. Eisenhüttenw.*, **10**, 441 (1937). G. Maassen, *Angew. Chem.*, **50**, 375 (1937).

[40] M. von Stackelberg, P. Klinger, W. Koch, and E. Krath, *Tech. Mitt. Krupp Forschungsber.*, **2**, 59 (1939).

[41] D. N. Hume and I. M. Kolthoff, *J. Am. Chem. Soc.*, **72**, 4423 (1950).

v. *vs.* S.C.E. with a well-defined diffusion current. Addition of excess cyanide ion shifts the wave to *less* negative potentials, increases the wave slope, and greatly improves the wave form. The half-wave potentials in 0.01, 0.1, and 1 *M* potassium cyanide containing 0.01 per cent gelatin are −1.47, −1.42, and −1.36 v. *vs.* S.C.E. Addition of 0.1 per cent gelatin to the solution in 1 *M* potassium cyanide virtually obliterates the wave. Hume and Kolthoff confirmed our previous[42] conclusion that reduction proceeds directly to the metal rather than to the +1 state as Emelianova[43] originally, and more recently Caglioti, Sartori, and Silvestroni,[44] erroneously concluded.

The results of Hume and Kolthoff also contradict Hochstein's claims[45] that the $Ni(CN)_4^{--}$ ion is not reduced from alkaline medium, and that the half-wave potential varies with drop time and galvanometer sensitivity.

Solutions of the cyanide complex of +1 nickel (presumably $Ni(CN)_3^{--}$) in 1 *M* potassium cyanide show a well-developed anodic wave ($E_{1/2}$ = −0.80 v. *vs.* S.C.E.) resulting from the 1-electron oxidation to the +2 complex. The fact that the half-wave potential of this anodic wave is 0.55 v. more positive than the cathodic half-wave potential for the 2-electron reduction of $Ni(CN)_4^{--}$ ion demonstrates that neither reaction proceeds reversibly at the d.e.

Hume and Kolthoff also found that saturated solutions of nickelous cyanide in 0.1 *M* sodium chloride produce two waves, the first at −0.8 v. corresponding to the reduction of free nickelous ion and the second at −1.4 v. to the reduction of $Ni(CN)_4^{--}$. They interpreted this as evidence that the solid nickelous cyanide is nickelous tetracyanonickelate $Ni[Ni(CN)_4]$. Since separate waves of Ni^{++} and $Ni(CN)_4^{--}$ are obtained the rate of dissociation of $Ni(CN)_4^{--}$ must be very small.

A cyanide supporting electrolyte is useful for the determination of nickel in the presence of excess copper and zinc, since the latter two metals do not produce waves in the presence of excess cyanide.

From few experiments we have made it appears that the tartrate complex of nickel is so stable that no reduction wave is observed with nickel in neutral or alkaline tartrate media.

According to West and Dean[46] nickel yields a very good wave in a fluo-

---

[42] *Polarography*, 1st ed., 1941 (p. 283).

[43] N. V. Emelianova, *Rec. trav. chim.*, **44,** 529 (1925).

[44] V. Caglioti, G. Sartori, and P. Silvestroni, *Atti acad. nazl. Lincei, Classe sci. fis. mat. e nat.*, **3,** 448 (1947).

[45] J. P. Hochstein, *J. Gen. Chem. U. S. S. R.*, **7,** 2488 (1937).

[46] P. W. West and J. F. Dean, *Ind. Eng. Chem., Anal. Ed.*, **17,** 686 (1945). See also P. W. West, J. F. Dean, and E. J. Breda, *Collection Czechoslov. Chem. Communs.*, **13,** 1 (1948).

ride supporting electrolyte; with 0.5 *M* sodium fluoride containing 0.01 per cent gelatin, and at pH values between 4.3 and 6.4, the half-wave potential is −1.12 v. *vs.* S.C.E., and $i_d/Cm^{2/3}t^{1/6}$ is 2.29 at 25°. Cobalt in this same medium produces a wave only slightly beyond that of nickel, and large amounts of cobalt tend to mask the full development of the nickel wave but small amounts can be tolerated. West and Dean found that large amounts of ferric iron are precipitated, and small amounts are sequestered, in fluoride medium, and they utilized this fact to eliminate iron interference in the determination of nickel in steel and in nickel ores (see Chap. XXXV).

Nickel in Alloys. See Chapter XXXV.

4. PLATINUM METALS

Most of the platinum group metals catalyze the reduction of hydrogen ion at the dropping electrode, and the characteristics of such catalytic hydrogen waves are discussed in Chapter XIII.

The oxidation potentials of the various couples involving the platinum group metals are all rather high (oxidizing), so that from most solutions the platinum metals tend to be reduced to the metallic states by contact with mercury. To obtain characteristic reduction waves it is generally necessary to select complexing agents which lower the potentials below the oxidation potential of mercury, but which do not complex the particular platinum metal so strongly as to prevent its reduction before the reduction of the cations of the supporting electrolyte.

Ruthenium. Willis,[47] who systematically investigated the polarography of the platinum metals, found that no reduction waves for either +4 or +3 ruthenium are observed with solutions of the chloro complexes, nor from supporting electrolytes containing cyanide, thiocyanate, ammonia, or pyridine.

Neidrach and Tevebaugh[48] studied the polarography of +4 ruthenium in perchloric acid media scrupulously freed from chloride. In 0.1 to 5 *M* perchloric acid three well-formed waves are obtained. In 1 *M* perchloric acid the first diffusion current is observed at zero applied e.m.f., and the half-wave potentials of the second and third waves are +0.21 v. and −0.34 v. *vs.* S.C.E. The sum of the first two diffusion currents corresponds to an irreversible 1-electron reduction to the +3 state of two different hydrolytic species of +4 ruthenium in such sluggish equilibrium that each produces a separate wave. The third wave results from a nearly reversible

[47] J. B. Willis, *J. Am. Chem. Soc.*, **67**, 547 (1945).

[48] L. W. Niedrach and A. D. Tevebaugh, private communication.

reduction of the +3 ruthenium to the +2 state. In 1 *M* perchloric acid the value of $i_d/Cm^{2/3}t^{1/6}$ for the sum of the first two waves is 1.53, and for the sum of all three waves it is 2.91 at 25°. The +2 ruthenium formed by the total reduction is slowly oxidized by perchlorate ion all the way back to the +4 state.

Rhodium. Willis[49] observed that several complexes of +3 rhodium produce more or less well-developed reduction waves. A fairly well-defined wave is obtained from 1 *M* potassium cyanide ($E_{1/2}$ = −1.47 v. *vs.* S.C.E.), corresponding to a 1-electron reduction of $Rh(CN)_6^{---}$ to a cyanide complex of the +2 state. The reduction is not reversible.

Chloropentamminorhodic ion, $Rh(NH_3)_5Cl^{++}$, in 1 *M* ammonium chloride undergoes a reversible 2-electron reduction, the half-wave potential being −0.93 v. *vs.* S.C.E. In a supporting electrolyte containing 1 *M* each of ammonia and ammonium chloride the half-wave potential is the same as in the absence of ammonia but the diffusion current is decreased by about 25 per cent, which Willis attributed to formation of aquopentamminorhodic ion. In 1 *M* potassium nitrate and in 0.5 *M* potassium sulfate in the presence of 0.01 per cent gelatin the half-wave potential of the choropentamminorhodic ion is −0.96 v. *vs.* S.C.E., and reduction proceeds reversibly to the +1 state.

In 0.9 *M* potassium thiocyanate +3 rhodium (probably $Rh(CNS)_6^{---}$) produces a wave whose half-wave potential is −0.39 v. *vs.* S.C.E., and thus much more positive than in the other supporting electrolytes. Application of the Ilkovic equation indicates that n = 2. Willis concluded that this does not correspond to reduction to the +1 state, but rather that the +3 rhodium is reduced to the +2 state at a potential more positive than that at which the anodic dissolution current occurs, and that the wave at −0.39 v. results from reduction of the +2 rhodium complex to the metal.

A very well-formed wave corresponding to a 1-electron reduction of the +3 rhodium pyridine complex is obtained from supporting electrolytes containing 1 *M* pyridine and 1 *M* potassium chloride or potassium bromide. The half-wave potential is −0.41 v. *vs.* S.C.E. Willis concluded that this supporting electrolyte is the most promising of those investigated for the determination of rhodium.

No reduction waves are observable with +3 rhodium in supporting electrolytes containing nitrite, oxalate, tartrate, or ethylenediamine. The chloro complex of +3 rhodium is reduced to the metal by contact with mercury in chloride supporting electrolytes.

---

[49] J. B. Willis, *J. Am. Chem. Soc.*, **66**, 1067 (1944).

The polarography of +3 rhodium has also been studied by Repin,[50] who agrees with Willis that a pyridine supporting electrolyte is the most suitable for the determination of rhodium. According to Repin rhodium can be determined in the presence of an excess of platinum, by employing a pyridinium hydrochloride supporting electrolyte and reducing the platinum to the +2 state with sulfite. Iridium does not interfere.

Palladium. Willis[49] observed that +2 palladium (probably $Pd(CN)_4^{--}$) in 1 *M* potassium cyanide is reduced reversibly to the metal. The half-wave potential (−1.77 v. *vs.* S.C.E.) is so close to the final current rise that the diffusion current plateau is not developed fully.

The tetramminopalladous ion $Pd(NH_3)_4^{++}$ produces a well-formed reduction wave from a supporting electrolyte containing 1 *M* each of ammonia and ammonium chloride and 0.005 per cent methyl red as a maximum suppressor. The reduction does not proceed reversibly, and the half-wave potential shifts from −0.72 v. *vs.* S.C.E. with 0.2 millimolar palladium to −0.80 v. with 5 millimolar. However, the diffusion current is directly proportional to the palladium concentration, so that the wave is suitable for practical analytical purposes.

The pyridine complex of +2 palladium (probably $Pd(py)_4^{++}$) undergoes reversible reduction to the metal and the diffusion current is directly proportional to concentration. In a solution containing 1 *M* pyridine and 1 *M* potassium chloride Willis observed that the half-wave potential shifted from −0.31 v. *vs.* S.C.E. to −0.36 v. when the palladium concentration was increased from 0.2 to 5 millimolar.

From a supporting electrolyte containing 0.1 *M* ethylenediamine, 1 *M* potassium chloride, and 0.005 per cent methyl red as a maximum suppressor, a well-defined, 2-electron reduction wave is obtained with concentrations of +2 palladium smaller than about 2 or 3 millimolar, but not with much larger concentrations. The half-wave potential shifted from −0.63 v. to −0.65 v. *vs.* S.C.E. when the palladium concentration was increased from 0.2 to 1 millimolar. A large concentration (1 *M*) of ethylenediamine produces an unsatisfactory wave, and the half-wave potential is shifted to −0.76 v.

Osmium. The polarography of osmium tetroxide and of +6 osmium in the form of osmate ion OsO_4^{--} has been investigated by Crowell, Heyrovsky, and Engelkemeir.[51] Acid solutions could not be used because of direct reduction by contact with mercury, and polarograms obtained from

---

[50] S. A. Repin, *J. Applied. Chem. U. S. S. R.*, **20,** 46, 55 (1947).

[51] W. R. Crowell, J. Heyrovsky, and D. W. Engelkemeir, *J. Am. Chem. Soc.*, **63,** 2888 (1941).

solutions containing sodium, potassium, lithium, or tetramethylammonium hydroxides were unsatisfactory.

In saturated calcium hydroxide as supporting electrolyte osmium tetroxide produces three excellently developed waves, corresponding to stepwise reduction to the +6, +5, and +3 states. The first wave starts at zero applied e.m.f., and the half-wave potentials of the second and third waves are −0.41 and −1.16 v. *vs.* S.C.E. Two very well-formed waves at −0.41 v. and −1.16 v. are obtained with osmate ion in the same supporting electrolyte, corresponding to reduction from the +6 to the +4 and +3 states. In all cases the diffusion currents are directly proportional to the osmium concentration, and the excellent wave characteristics render them well suited to the practical polarographic determination of osmium.

Willis[49] found that the cyanide complex of +2 osmium, $Os(CN)_6^{----}$, does not show a reduction wave in either 1 *M* potassium cyanide or potassium sulfate.

Iridium. According to Willis neither +4 nor +3 iridium shows a reduction wave in 1 *M* potassium cyanide, ammonia-ammonium chloride, potassium thiocyanate, ethylenediamine, or pyridine solutions. Repin[50] reported that $IrCl_6^{--}$ is reduced to the metal by contact with mercury or by merely passing hydrogen through the solution.

Platinum. Willis reported that +2 platinum does not produce a reduction wave from potassium cyanide, ammonia-ammonium chloride, pyridine, or ethylenediamine solutions, and this is supported by Repin's observations. Solutions containing either +4 or +2 platinum tend to be reduced to the metal by contact with mercury; this occurs especially easily with the chloro complexes.

CHAPTER XXVIII

Copper, Silver, and Gold

1. COPPER

Half-wave potentials and values of $i_d/Cm^{2/3}t^{1/6}$ for cupric copper in a number of common supporting electrolytes are listed in Table XXVIII-1.[1]

From the following standard potentials of the couples involving simple cupric ion, simple cuprous ion, and metallic copper it follows that reduction of simple cupric ion to the metal requires less energy than its reduction to simple cuprous ion:

$$Cu^+ + e \rightleftharpoons Cu; \qquad E^\circ = +0.280 \text{ v. } vs. \text{ S.C.E.}$$

$$C^{++} + 2e \rightleftharpoons Cu; \qquad E^\circ = +0.103 \text{ v. } vs. \text{ S.C.E.}$$

$$Cu^{++} + e \rightleftharpoons Cu^+; \qquad E^\circ = -0.074 \text{ v. } vs. \text{ S.C.E.}$$

In other words, *simple* cuprous ion is incapable of existing in any appreciable concentration and disproportionates into cupric ion and metallic copper:

$$2Cu^+ \rightleftharpoons Cu + Cu^{++}; \qquad K = 1 \times 10^6$$

Consequently in the absence of complexing agents, *e. g.*, in nitrate, perchlorate or sulfate supporting electrolytes, the polarogram of cupric ion comprises only a single wave corresponding to direct reduction to the metal. The same is true in the presence of substances, such as tartrate or citrate ion, which form more stable complexes with cupric than with cuprous ion.

In the presence of substances, like ammonia or chloride ion in high concentration, which form more stable complexes with cuprous than with cupric ion the cuprous state is stabilized sufficiently so that stepwise reduction of the cupric complexes occurs. The polarogram in such cases consists of a doublet wave whose two parts are of equal height.[2, 3] Solutions of cuprous copper in such supporting electrolytes can show, in addition to the cathodic wave for the reduction of the cuprous complex, an anodic wave resulting from oxidation of the cuprous complex to the cupric state.

---

[1] J. J. Lingane, *Ind. Chem., Anal. Ed.*, **15,** 583 (1943).

[2] G. Thanheiser and G. Maassen, *Mitt. Kaiser-Wilhelm Inst. Eisenforsch. Düsseldorf*, **19,** 27 (1937). G. Maassen, *Angew. Chem.*, **50,** 375 (1937).

[3] M. Voriskova, *Collection Czechslov. Chem. Communs.*, **11,** 588 (1939).

In addition to ammonia and concentrated chloride media, stepwise reduction of cupric copper also occurs in supporting electrolytes containing thiocyanate or pyridine. In 0.1 *M* potassium thiocyanate the half-wave potentials are −0.02 v. and −0.39 v., and in 0.1 *M* pyridine–0.1 *M* pyridinium chloride[4] they are +0.05 and −0.25 v. *vs.* S.C.E. Because the first wave starts so close to zero applied e.m.f. correction for the residual current usually requires a blank experiment with the supporting electrolyte alone, and to avoid this inconvenience it is preferable to measure only the second wave in practical analytical work.

TABLE XXVIII-1

POLAROGRAPHIC CHARACTERISTICS OF COPPER

All solutions contained 0.01 per cent gelatin and the data pertain to a temperature of 25° C. Half-wave potentials are referred to the saturated calomel electrode, and values of $i_d/Cm^{2/3}t^{1/6}$ are based on i_d in microamp., *C* in millimoles per liter, *m* in mg. sec.$^{-1}$ and *t* in sec.

Supporting electrolyte	$E_{1/2}$, v.	$\frac{i_d}{Cm^{2/3}t^{1/6}}$
0.1 *M* KCl (HCl)	+0.04	3.23
1 *M* HCl (KCl)	+0.04	
	−0.22	3.39[a]
0.5 *M* H_2SO_4	0.00	2.12
1 *M* HNO_3	−0.01	3.24
1 *M* NaOH	−0.42	2.91
0.5 *M* Na tartrate (pH = 4.5)	−0.09	2.37
0.5 *M* Na tartrate (pH = 9)	−0.12	2.24
0.5 *M* Na tartrate plus 0.1 *M* NaOH	−0.4	
	−1.6	[b]
1 *M* NH_3 plus 1 *M* NH_4Cl	−0.24	
	−0.50	3.75[a]

[a] Total double wave.
[b] Wave distorted by presence of gelatin.

Thanheiser and Maassen[2] reported that the linear relation between the diffusion current and concentration of copper in ammoniacal medium failed at small concentrations. However, we have obtained data under identically the same conditions employed by Thanheiser and Maassen which show that i_d/C is truly constant from 4×10^{-5} to 0.01 *M* copper when proper correction is made for the residual current. (See p. 378.)

Copper does not show a reduction wave in cyanide medium because cupric ion is reduced by cyanide ion to a cyanide complex of the cuprous state:

$$2Cu^{++} + 8CN^{-} \rightleftharpoons 2Cu(CN)_3^{--} + C_2N_2$$

[4] J. J. Lingane and H. Kerlinger, *Ind. Eng. Chem., Anal. Ed.*, **13**, 77 (1941).

and the cuprous-cyanide complex is so stable that its reduction potential is more negative than the discharge potential of the potassium or sodium ions of the supporting electrolyte. This fact is utilized to prevent copper interference in the determination of other metals in copper-base alloys (see Chap. XXXV).

The polarography of copper in tartrate medium was investigated originally by Suchy[5] who recommended a tartrate supporting electrolyte for the simultaneous determination of copper, bismuth, lead, and cadmium, whose waves appear in the order listed. Lingane[6] emphasized the pronounced effect of pH on the half-wave potentials of the tartrate complexes and showed that the optimum pH range was 4 to 5 for maximal separation of the waves of copper, bismuth, lead, and cadmium. The same author developed a systematic procedure for the analysis of mixtures of the copper group metals in any proportion, based on controlled potential electrolytic separations. This procedure is described in Chapter XVIII.

Most recently Meites[7] has studied thoroughly the influence of pH and total tartrate concentration, and the effect of gelatin on the polarographic characteristics of copper in tartrate media. At pH values below about 6 only a single wave is observed whose half-wave potential varies linearly with pH. In the pH range from about 6 to 7 the wave divides into two parts, accompanied by an abrupt negative shift of the half-wave potentials; the doublet wave apparently results from the reduction of separate tartrate complexes in sluggish equilibrium. The doublet wave persists up to a pH of 13.5. Meites found that gelatin has a profound influence on the wave character, and it causes distortion of the waves when present in concentrations greater than about 0.001 per cent. This effect is most marked in alkaline tartrate solutions; addition of 0.01 per cent gelatin to a cupric solution in 0.5 *M* tartrate plus 1 *M* potassium hydroxide virtually obliterates the wave.

Meites concluded that the predominant complex species is $Cu(II\ Tart)_4^{--}$ in solutions of pH less than 7, that at pH values between 7 and 11 "the copper is present mainly as a tartrate complex whose formula cannot be determined from polarographic measurements," and that the chief species in very strongly alkaline solutions is $Cu(OH)_2(Tart)_2^{----}$. On the basis of polarographic and other evidence obtained in unpublished investigations we are inclined to a different interpretation. When pure solutions of cupric sulfate and sodium tartrate are mixed the pH of the mixture is

[5] K. Suchy, *Collection Czechslov. Chem. Communs.*, **3,** 354 (1931).
[6] J. J. Lingane, *Ind. Eng. Chem., Anal. Ed.*, **16,** 147 (1944).
[7] L. Meites, *J. Am. Chem. Soc.*, **71,** 3269 (1949).

significantly *smaller* than that of the original cupric sulfate solution. This shows that hydrogen ion is a product of the complexation reaction between cupric ion and tartrate ion. The complexation must involve the replacement of the protons of the hydroxyl groups of tartrate ion by cupric ion rather than mere coordination of hydrogen tartrate or tartrate ion. This replacement is favored by increasing pH, and in strongly alkaline tartrate medium (corresponding to Fehling solution) the complexation may be represented by:

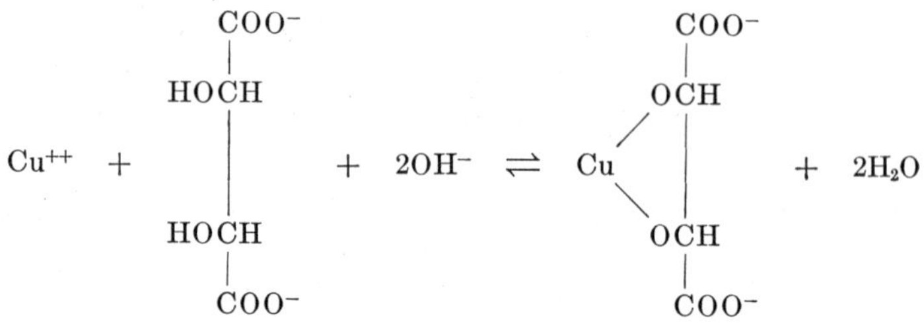

The pH titration curve, obtained by mixing cupric sulfate solution with excess sodium tartrate solution and titrating the mixture with sodium hydroxide, shows two distinct inflection points at pH values of about 7.5 and 10, corresponding to the successive replacement of the two hydroxyl hydrogens of the tartrate ion by cupric ion. At low and intermediate pH values polynuclear complexes, containing two or more cupric ions attached to two or more tartrate ions, doubtless exist. It seems highly probable that complexation of copper and other metal ions with citrate ion must also involve replacement of the hydroxyl hydrogen as it does with tartrate ion.

As Meites[8] has shown the polarography of copper in citrate medium is analogous to that in tartrate medium, and a similar great effect of pH on the half-wave potential is observed. The half-wave potential of the single wave in 0.5 M potassium citrate shifts from -0.02 v. at pH = 3 to -0.22 v. *vs.* S.C.E. at pH = 7.5. Between pH 7.5 and 8.5 in 0.5 M citrate the half-wave potential shifts abruptly to -0.38 v. *vs.* S.C.E. and remains constant at this value up to a pH of about 12, and then increases up to -0.50 v. at pH = 14.

A single well-defined wave is observed with copper in oxalate media, and Meites[9] has studied the effect of pH on the half-wave potential. In the presence of 0.1 M potassium oxalate the half-wave potential shifts from

[8] L. Meites, *J. Am. Chem. Soc.*, **72,** 180 (1950).
[9] L. Meites, *J. Am. Chem. Soc.*, **72,** 184 (1950).

-0.03 v. *vs.* S.C.E. at pH = 1 to -0.20 v. at pH = 5.7, and it remains constant at the latter value up to a pH of about 10. In 1 *M* potassium oxalate between a pH of 5.7 and 10 the half-wave potential is -0.27 v. *vs.* S.C.E. Meites concluded that the complex species is $Cu(C_2O_4)_2^{--}$ at pH values above about 5.7. Above a pH of about 10 the single wave divides into a doublet wave, and this is accompanied by the precipitation of what is apparently a basic cupric oxalate.

A well-developed doublet wave was observed by Meites[9] in carbonate media of pH 9.5 to 11; the half-wave potentials in 0.25 *M* carbonate are -0.17 and -0.24 v. *vs.* S.C.E. Since the two parts of the incompletely resolved doublet are of approximately equal height, and their relative heights are independent of the total carbonate concentration, stepwise reduction *via* the cuprous state may be responsible for the doublet wave. In carbonate media of pH above about 11 cupric hydroxide precipitates.

Meites found that addition of ammonia to alkaline tartrate, citrate, or oxalate supporting electrolytes containing copper produced the doublet wave characteristic of the ammonia-cupric complex.

The polarography of the pyrophosphate complexes of copper has been studied briefly by Rogers and Reynolds,[10] and more systematically by Laitinen and Onstott.[11] The former investigators observed two waves in 0.1 *M* sodium pyrophosphate solution (pH apparently about 9) containing 0.002 per cent gelatin, which they attributed to stepwise reduction of the cupric complex to the cuprous state and to the metal. However, Laitinen and Onstott, who systematically studied the influence of pH over a wide range, demonstrated that the double wave does not result from stepwise reduction, but rather from reduction to the metal of two different cupric complex species in sluggish equilibrium. Solutions containing excess pyrophosphate and having a pH smaller than about 5 or greater than about 12.5 produce only a single normal wave, but the double wave appears over the pH range from 5 to 12.5. Addition of gum ghatti as a maximum suppressor greatly diminished the height of the first wave in this critical pH range; an effect reminiscent of the suppressive influence of gelatin in similar cases. Without any maximum suppressor the height of the first wave was constant between pH 7.5 and 12.0, and increased by about 10 per cent between pH 12.0 and 12.6. With 0.1 *M* sodium pyrophosphate and 0.2 *M* sodium acetate the half-wave potential of the single wave is -0.085 v. *vs.* S.C.E. at pH = 4.5, and the slope of the log plot of the wave corresponds to a reversible 2-electron reduction. At pH = 12.5 with 0.05 *M* pyrophosphate

[10] L. B. Rogers and C. A. Reynolds, *J. Am. Chem. Soc.*, **71,** 2084 (1949).
[11] H. A. Laitinen and E. I. Onstott, *J. Am. Chem. Soc.*, **72,** 4729 (1950).

the half-wave potential is −0.3 v. At pH = 4.5 the half-wave potential shifts in the negative direction by 0.052 v. per 10-fold increase in the pyrophosphate concentration, indicating that at this pH two monohydrogen pyrophosphate ions are coordinated per cupric ion. Additional detailed data will be found in the paper of Laitinen and Onstott.

West, Dean, and Breda[12] reported that cupric copper in 0.5 *M* sodium fluoride, containing 0.01 per cent gelatin and adjusted to pH = 5.1, produces a single well-formed wave corresponding to direct reduction to the metal. The half-wave potential is −0.003 v., indicating only slight complexation with fluoride ion, and $i_d/Cm^{2/3}t^{1/6}$ is 3.48 at 25°. The same authors give data for many other metal ions in fluoride medium.

The polarography of cupric glycinate and cupric alaninate complexes has been investigated by Keefer.[13] Calvin and Bailes[14] studied the polarography of a large number of chelated copper complexes of salicylaldehyde derivatives in 50 per cent aqueous pyridine as a solvent.

The precipitation of copper by salicylaldoxime was utilized by Neuberger for the amperometric titration of copper (see Chap. XLVII). The decrease in the diffusion current of salicylaldoxime caused by addition of copper was employed by Carruthers[15] for the determination of microgram quantities of copper in biological materials. In a phosphate buffer of pH = 5.4 the half-wave potential for the reduction of salicylaldoxime is −0.98 v. *vs.* S.C.E. and the diffusion current is directly proportional to the salicylaldoxime concentration. Using a 10^{-4} *M* solution of salicylaldoxime in an 0.08 *M* phosphate buffer quantities of copper from about 3 to 75 micrograms per 10 ml. can be determined with an accuracy of 1 to 3 per cent.

The reduction of the complexes of copper with ethylenediamine, propylenediamine, diethylenetriamine, and glycine produces a single wave corresponding to direct reduction to the metal, and from data presented by Laitinen, Onstott, Bailar, and Swann[16] the reduction appears to proceed reversibly in all cases. In supporting electrolytes containing 0.1 *M* potassium nitrate, 0.01 per cent gelatin and 1 *M* amine the half-wave potentials *vs.* S.C.E. and $i_d/Cm^{2/3}t^{1/6}$ values at 25° are, respectively: ethylenediamine, −0.568 v., 3.19; propylenediamine, −0.583 v., 2.90; diethylenetriamine, −0.602 v., 2.48. In a supporting electrolyte composed of 0.05 *M* potas-

[12] P. W. West, J. Dean, and E. J. Breda, *Collection Czechoslov. Chem. Communs.*, **13,** 1 (1948).

[13] R. M. Keefer, *J. Am. Chem. Soc.*, **68,** 2329 (1946).

[14] M. Calvin and R. H. Bailes, *J. Am. Chem. Soc.*, **68,** 949 (1946).

[15] C. Carruthers, *Ind. Eng. Chem., Anal. Ed.*, **17,** 398 (1945).

[16] H. A. Laitinen, E. I. Onstott, J. C. Bailar, and S. Swann, *J. Am. Chem. Soc.*, **71,** 1550 (1949).

sium dihydrogen phosphate and 1 *M* glycine the half-wave potential is a linear function of pH between 6 and 8 (−0.238 v. *vs.* S.C.E. at pH = 6.3 and −0.325 v. at pH = 7.9) but the value of $i_d/Cm^{2/3}t^{1/6}$ is constant at 3.13. Laitinen *et al.* concluded that in all cases two molecules of the amine or of glycine are coordinated to the cupric ion when a large excess of the complexing agent is present.

According to Furness, Crawshaw, and Davies[17] the cupric complex of ethylenediamine tetraacetate (Trilon, Versene, etc.) in ammoniacal medium produces a single 2-electron reduction wave at −0.6 v. *vs.* S.C.E. which is separate from the preceding doublet wave due to reduction of the tetraminocupric ion. When a solution containing excess cupric ion and sodium ethylenediamine tetraacetate in 0.2 *M* potassium nitrate is treated with excess magnesium oxide the free cupric ion is precipitated and a polarogram of the remaining solution shows only a single well-formed wave of the complexed copper at −0.44 v. *vs.* S.C.E. Furness, Crawshaw, and Davies utilized this fact for the determination of sodium ethylenediamine tetraacetate in commercial preparations of this sequestering agent.

Copper in Presence of Iron. The diffusion current of ferric iron overlaps that of copper in many supporting electrolytes, but ferrous iron is reduced at a much more negative potential. Hence the interference of iron can be eliminated by reducing it to the ferrous state with hydroxylamine or sulfur dioxide (excess removed by boiling) in dilute hydrochloric or sulfuric acid media and recording the copper polarogram in the same medium. Practical application of this principle is made in determining copper in magnesium and aluminum alloys by procedures described in Chapter XXXV.

Davies and Key[18] recommended fluoride complexation of ferric iron to eliminate its interference in the determination of copper in pyrites and in various alloys.

It should be possible to utilize an alkaline tartrate supporting electrolyte to determine copper in the presence of much ferric iron; a solution composed of 0.5 *M* sodium tartrate and adjusted to pH = 12 by addition of sodium hydroxide should be suitable. In this solution the reduction of the ferric-tartrate complex does not begin below about −1.0 v. *vs.* S.C.E.,[19] and the cupric-tartrate complex produces a good wave at −0.4 v.[20] The copper wave does not show a maximum and no maximum suppressor is required.

---

[17] W. Furness, P. Crawshaw, and W. C. Davies, *Analyst*, **74**, 629 (1949).
[18] W. C. Davies and C. Key, *Ind. Chemist*, **19**, 555 (1943).
[19] J. J. Lingane, *J. Am. Chem. Soc.*, **68**, 2448 (1946).
[20] L. Meites, *J. Am. Chem. Soc.*, **71**, 3270 (1949).

Gelatin must not be added because it obliterates the copper wave in alkaline tartrate media.

Copper in Alloys. See Chapter XXXV.

Copper in Ores. Methods for the determination of copper in zinc sulfide ores have been described by Shakhov,[21] Sinyakova,[22] and Aref'eva,[23] and Portnow and Suschencorva[24] developed a method for its determination in pyrolusite.

Malyuga[25] advocated precipitation by rubeanic acid for the preliminary separation of copper prior to its polarographic determination in minerals, soils, and plant material.

Preliminary Concentration of Copper by an Ion-Exchange Resin. In connection with the determination of copper in milk products Cranston and Thompson[26] found that microgram quantities of copper could be concentrated conveniently from solutions of pH 3.5 to 6.5 by passage through a column of Amberlite IR-100. The copper is subsequently eluted from the resin with dilute hydrochloric acid and determined polarographically.

2. SILVER

The standard potential of the silver ion-silver couple (+0.7995 v. *vs.* N.H.E.) is virtually coincident with that of the mercurous ion-mercury couple (+0.7986 v.). Consequently the reduction of silver ion from noncomplexing supporting electrolytes, *e. g.*, nitric or perchloric acid, or nitrate or perchlorate salts, produces a diffusion current that starts from zero applied e.m.f., and the apparent half-wave potential corresponds to the anodic dissolution potential of mercury rather than the true reduction potential of silver ion. Because the types and degree of stability of various complexes of mercury and silver are so very similar, this same effect is usually observed in most supporting electrolytes. Consequently there are very few data in the literature on the polarography of silver.

With alkali nitrates or perchlorates as supporting electrolytes the diffusion current of silver is well defined and can be used for its determination provided it is known that other oxidants which produce a similar diffusion current are absent. We also observed a good diffusion current for the reduction of the silver thiosulfate complex from 0.01 to 1 *M* solutions of sodium thiosulfate.

[21] A. S. Shakhov, *Zavodskaya Lab.*, **9,** 293 (1940).
[22] S. I. Sinyakova, *Zhur. Anal. Khim.*, **1,** 241 (1946).
[23] T. C. Aref'eva, *Trans. All-Union Conf. Anal. Chem.*, **2,** 463 (1943).
[24] M. A. Portnow and G. I. Suschencorva, *Zavodskaya Lab.*, **10,** 462 (1941).
[25] D. P. Malyuga, *Proc. Biochem. Lab. Acad. Sci.* (*U. S. S. R.*), **7,** 86 (1944).
[26] H. A. Cranston and J. B. Thompson, *Ind. Eng. Chem., Anal. Ed.*, **8,** 323 (1946).

Sanigar[27] found that fairly concentrated solutions of potassium argentocyanide, without excess cyanide present, produce a wave that starts at zero applied e.m.f. The wave has a pronounced maximum, but the diffusion current following it is well defined. With a 10^{-4} *M* solution of $KAg(CN)_2$, without excess cyanide, Sanigar obtained a well-defined wave, whose half-wave potential was about −0.3 v. *vs.* the S.C.E.

3. GOLD

Since the reduction potential of auric ion is more positive than the potential of anodic dissolution of mercury, it must be transformed into a complex ion that is more stable than the corresponding complex with mercuric ion in order to obtain a true wave.

Herman[28] found that solutions of auric chloride in 2 *N* potassium hydroxide produced a polarogram consisting of three waves at about −0.2 v., −0.4 v., and about −1.1 v. *vs.* the S.C.E. The last wave was very flat. On standing, or when the solution was heated, the wave at −0.4 v. decreased and that at −1.1 v. increased. After very long standing the third wave also decreased. These effects appear to be due to a slow decomposition of the auric-hydroxide complex first to the aurous complex and finally to the metal.

In the presence of excess cyanide Herman found that auric gold produced a double wave. The first wave at zero applied e.m.f. is probably due to the reduction of the auric-cyanide complex to the aurous complex, and the second wave at about −1.4 v. *vs.* the S.C.E. to the reduction of the aurous complex to the metal. Solutions of aurous gold in 0.1 *N* potassium cyanide showed a fairly well-defined wave at about −1.4 v. *vs.* the S.C.E. The character of the wave was greatly improved when the solutions were filtered, due to extraction of capillary-active substances from the filter paper. The aurous cyanide complex gives the best wave for analytical purposes.

For the polarographic determination of gold Herman recommended the following procedure. The solution of auric gold obtained from ores, bullion, etc., is treated with excess potassium iodide to reduce the gold to the aurous state, and the liberated iodine is removed with excess sodium sulfite. The solution is then made about 0.1 *N* with respect to potassium cyanide and 2 *N* with respect to potassium hydroxide, and, after filtering, the polarogram of the aurous-cyanide complex is recorded.

Herman obtained evidence that the conversion of gold hydroxide com-

---

[27] E. B. Sanigar, *Rec. trav. chim.*, **44**, 549 (1925).

[28] J. Herman, *Collection Czechoslov. Chem. Communs.*, **6**, 37 (1934).

plexes into the cyanide complexes is relatively slow. When excess cyanide was added to an alkaline gold solution the characteristic wave of the auric-hydroxide complex persisted and when excess sodium hydroxide was added to a solution containing the auric-cyanide complex the characteristic wave of the cyanide complex remained unchanged. The latter result is not surprising, because the cyanide complex is the more stable of the two, but the first experiment shows that the conversion of the hydroxide complex into the cyanide complex is very slow.

CHAPTER XXIX

Zinc, Cadmium, and Mercury

1. ZINC

Zinc produces well-developed waves from many different supporting electrolytes and thus is readily amenable to polarographic determination. This is of special practical importance because zinc is not one of the easiest elements to determine by classical methods. The half-wave potentials and $i_d/Cm^{2/3}t^{1/6}$ values of zinc in a number of common supporting electrolytes are assembled in Table XXIX-1.[1]

The reduction wave of hydrogen ion follows that of zinc ion so closely that in strongly acid medium the diffusion current of zinc is not fully developed before reduction of hydrogen ion begins. The same is true in tartrate media of pH less than about 5 or 6.

The fact that the half-wave potential of zinc is more negative than those of such other common metals as copper, cadmium, lead, etc., usually requires the prior removal of such metals before zinc can be determined unless the total concentration of the interfering metals does not exceed that of the zinc. Complexation can be utilized only to a very limited extent to eliminate interferences of other metals in zinc determinations because with most complexing agents zinc itself forms complexes that usually are as stable as those of the interfering metals. Controlled potential electrolytic separations have proven to be very useful for removing large amounts of such elements as copper, cadmium, bismuth, lead, antimony, and tin prior to zinc determinations. (See p. 387.)

Prajzler[2] has shown that small amounts of zinc can be determined in the presence of nickel, cobalt, and manganese by using a nearly saturated solution of ammonium oxalate as supporting electrolyte. Presumably the pH should be adjusted to about 7. The complex oxalates of nickel, cobalt, and manganese are so stable that no wave for them is observed below the discharge potential of ammonium ion. The complex zinc oxalate ion

[1] J. J. Lingane, *Ind. Eng. Chem., Anal. Ed.*, **15,** 583 (1943).

[2] J. Prajzler, *Collection Czechoslov. Chem. Communs.*, **3,** 406 (1931). See also A. G. Hybbinetti, *Svensk. Kem. Tid.*, **57,** 6 (1945).

produces a well-defined wave, with a half-wave potential of −1.3 v. *vs.* the S.C.E. in a nearly saturated ammonium oxalate solution.

The polarographic behavior of zinc in cyanide medium has been studied in some detail by Pines.[3] He found that polarograms of saturated solutions of zinc cyanide in 0.001 to 0.01 N potassium cyanide consist of three distinct waves, with $E_{1/2}$ values of −1.0 v., −1.2 v., and −1.4 v. *vs.* the S.C.E. These waves obviously cannot be due to stepwise reduction to intermediate oxidation states since none exist, and Pines concluded that they were due to the reduction of three different species of zinc-cyanide complexes whose rate of interconversion is so slow that each one produces its own wave. According to Pines the first wave at −1.0 v. is due to the reduction of the simple zinc ion, the second at −1.2 v. to the $Zn(CN)_4^{--}$ ion, and the third at −1.4 v. to the $Zn(CN)_5^{---}$ ion. When large excess of potassium cyanide is added to a zinc solution no wave at all is observed, which Pines interpreted as due to the formation of $Zn(CN)_6^{----}$ whose reduction potential apparently is more negative than that of potassium ion.

TABLE XXIX-1

POLAROGRAPHIC CHARACTERISTICS OF ZINC AND CADMIUM IN VARIOUS SUPPORTING ELECTROLYTES

Half-wave potentials in volts *vs.* the saturated calomel electrode and $i_d/Cm^{2/3}t^{1/6}$ values in the units microamp. per millimole per liter per mg.$^{2/3}$ sec.$^{-1/2}$ at 25°. In all cases 0.01 per cent gelatin was present as a maximum suppressor.

Supporting electrolyte	Cd		Zn	
	$E_{1/2}$	$i_d/Cm^{2/3}t^{1/6}$	$E_{1/2}$	$i_d/Cm^{2/3}t^{1/6}$
0.1 M KCl	−0.599	3.51	−0.995	3.42
1 M KCl	.642		−1.022	
1 M HCl	.64	3.58	[a]	[a]
0.5 M H_2SO_4	.59	2.6[b]	[a]	[a]
1 M HNO_3	.59	3.06	[a]	[a]
1 M NaOH	Insol.		−1.53	3.14
0.5 M tartrate (pH = 4.5)	−0.64	2.34	−1.23	[a]
0.5 M tartrate (pH = 9)	.64	2.34	−1.15	2.30
0.5 M tartrate + 0.1 M NaOH	.79	[c]	−1.42	2.65
1 M NH_3 + 1 M NH_4Cl	.81	3.68	−1.33	3.82

[a] Diffusion current masked by reduction of hydrogen ion.
[b] Diffusion current not well developed.
[c] Limiting current abnormally small and not proportional to cadmium concentration.

The wave of the tetramminozinc ion is so excellently defined that the

[3] I. Pines, *Collection Czechoslov. Chem. Communs.* **1,** 429 (1929).

use of an ammoniacal supporting is advantageous whenever possible. An ammoniacal medium is especially useful for the simultaneous determination of nickel and zinc.[4, 5] The nickel wave precedes that of the zinc, and the separation of the two is best in a solution containing a relatively large concentration of ammonia and a relatively small concentration of ammonium ion. In solutions containing 1 *M* ammonia, 0.005 per cent gelatin, and 0.2, 1, and 3 *M* ammonium chloride the separation of the nickel and zinc half-wave potentials is *seriatim* 0.27, 0.22, and 0.24 v. The values of $i_d/Cm^{2/3}t^{1/6}$ for zinc in solutions containing 1 *M* ammonia, 0.005 per cent gelatin and 0.2, 1, and 3 *M* ammonium chloride are 3.78, 3.82, and 3.94.

A well-defined wave for zinc is obtained in thiocyanate medium; in 0.1 *N* potassium thiocyanate the half-wave potential is −1.01 v. *vs.* the S.C.E. Since this value is practically identical with that of the simple zinc ion, there appears to be little or no complex formation. A thiocyanate supporting electrolyte has been recommended by Stout and Levy[6] for the simultaneous determination of zinc and nickel; the nickel wave precedes that of the zinc by about 0.2 v.

Zinc in Alloys. Procedures for the determination of zinc in various types of alloys are described in Chapter XXXV.

Zinc in Cadmium Compounds. The determination of small amounts of zinc present as an impurity in cadmium and its compounds is of special interest. Because the cadmium wave precedes that of zinc in all supporting electrolytes the prior removal of most of the cadmium is necessary before small amounts of zinc can be determined. We have attempted to precipitate large amounts of cadmium as cadmium hydroxide in sodium hydroxide medium and to measure the zincate wave in the filtrate or supernatant solution. Unfortunately zinc coprecipitates to such an extent that this method fails when the ratio of cadmium to zinc is large. A hydrogen sulfide separation may be employed,[7] but difficulties due to coprecipitation and postprecipitation of zinc with the cadmium sulfide may be anticipated when only traces of zinc are present.

Lingane[8] found that as much as 4 g. of cadmium can be separated from a few tenths of a milligram of zinc by controlled potential electrolytic deposition of the cadmium into a mercury cathode from an ammoniacal solution. A mercury cathode is used rather than platinum, because when

---

[4] P. Pavlov and I. Pavlenko, *Zhur. Obshcheĭ Khim.*, **7**, 2246, 2259 (1937).

[5] J. J. Lingane, *Ind. Eng. Chem., Anal. Ed.*, **18**, 429 (1946).

[6] P. R. Stout and J. Levy, *Collection Czechoslov. Chem. Communs.*, **10**, 136 (1938).

[7] D. Cozzi, *Ann. chim. applicata*, **31**, 227 (1941).

[8] J. J. Lingane, *Anal. Chim. Acta*, **2**, 595 (1948).

attempts are made to remove large amounts of cadmium on a platinum cathode the deposited metal has a pronounced tendency to form "trees" which grow rapidly to such an extent that short-circuiting of the anode and cathode results.

A sample of up to about 4 g. of cadmium metal (or equivalent sample of cadmium salt) is dissolved in a minimal amount of hydrochloric acid. Most of the excess acid is removed by evaporation to near, but not complete, dryness. The residue is dissolved in about 150 ml. of water, the solution is neutralized with concentrated (15 *M*) ammonia, and 50 ml. of ammonia is added in excess. The solution is electrolyzed in a cell containing a well-stirred mercury pool cathode and a platinum wire anode, with the cathode potential controlled at -1.00 ± 0.02 v. *vs.* S.C.E. (A cell comprising a mercury pool cathode on the bottom of a 250-ml. beaker and a platinum wire cathode in the form of a flat spiral, with provision for mechanical stirring of the cathode mercury, is most convenient. The general technique of controlled potential electrolysis is described in Chapter XVIII.) The complete deposition of the cadmium requires about one hour, during which time the current decreases from an initial value of several amperes to a final value of about 0.01 amp. The residual solution is transferred to a 250-ml. volumetric flask, 2 g. of sodium sulfite (to remove oxygen) and 2.5 cc. of 0.1 per cent gelatin solution are added, and the solution is diluted to the mark. A portion of this solution is transferred to a polarographic cell and the polarogram of zinc is recorded.

Tests of this procedure with known added amounts of zinc showed that 1 mg. of zinc could be determined in 4 g. of cadmium with a relative accuracy of a few per cent, and 0.1 mg. of zinc could be detected unambiguously.

Kovalenko and Dmitrieva[9] recommended the use of aluminum electrodes for the electrolytic removal of cadmium prior to the determination of zinc, but this does not appear to be as advantageous as a mercury cathode separation.

Zosimovick, Tsimmergakl, and Khaimovick[10] determined zinc in cadmium alloys by dissolving the alloy in a small amount of mercury in contact with a potassium hydroxide solution. It was claimed that all the zinc enters the solution, and the cadmium remains insoluble (apparently as cadmium hydroxide), so that after filtration the zinc can be determined from a polarogram of the filtrate.

Zinc in Ferrous Materials. Zinc can be determined in the presence of as much as a 2000-fold excess of iron if the bulk of the latter is reduced to the ferrous state with hydroxylamine and the remainder with ferrum reductum (finely divided pure iron powder) in dilute hydrochloric acid me-

[9] P. N. Kovalenko and V. L. Dmitrieva, *Zhur. Anal. Khim.*, **2,** 85, 334 (1947).

[10] D. P. Zosimovick, V. A. Tsimmergakl, and R. S. Khaimovick, *Zavodskaya Lab.* **14,** 149 (1948).

dium. Ringbom and Törn[11] applied this principle to the determination of small quantities of zinc in iron ores, roasted pyrites, and sinters.

A solution of the sample is prepared in about 0.5 *M* hydrochloric acid and the bulk of the iron is reduced to the ferrous state by adding hydroxylamine hydrochloride and boiling. The solution is filtered to remove silica, etc., and made up to a known volume. An aliquot is warmed with ferrum reductum to complete the reduction of the ferric ion, and is finally polarographed for zinc in a sodium acetate-acetic acid buffer.

Titanium, manganese, and cobalt do not interfere. Copper is removed by the ferrum reductum and does not interfere. The nickel wave coincides with that of zinc in the acetate supporting electrolyte, but if the amount of nickel does not greatly exceed the amount of zinc, the former can be separately determined in a thiocyanate supporting electrolyte and the appropriate correction is then subtracted from the apparent zinc wave.

Zinc in Plant and Biological Materials. Stout, Levy, and Williams[12] developed the following procedure for the determination of small amounts of zinc in the presence of copper, nickel, cobalt, cadmium, lead, and bismuth, with particular reference to the determination of traces of *zinc in plants.*

A 1- to 2-g. sample of the plant material is ashed below 500° C. The ash is dissolved in a small volume of 1 *N* hydrochloric acid, transferred to a 250-ml. separatory funnel with about 100 ml. of water, and treated with 5 ml. of 1 *N* ammonium citrate. The solution is then made slightly alkaline with ammonium hydroxide, and shaken with 10 ml. of a solution of 1 to 3 mg. of diphenylthiocarbazone ("dithizone") in chloroform. The zinc, nickel, cadmium, and copper enter the chloroform layer as the dithizone complexes, and iron, manganese, etc., remain in the aqueous layer. The chloroform layer is removed and shaken with 10 ml. of 0.5 *N* hydrochloric acid which extracts the zinc, lead, and cadmium into the aqueous layer. To insure complete transfer of the zinc the extraction with hydrochloric acid is repeated. The combined hydrochloric acid extracts are evaporated to dryness, and the zinc in the residual salt is taken up in a solution of 0.1 *N* ammonium acetate and 0.025 *N* potassium thiocyanate. The polarogram of this solution shows well-separated waves of lead, cadmium, nickel, and zinc in the order named. It is evident that the determination of zinc by this procedure will only be feasible when its concentration is at least as large as or larger than the concentrations of lead, cadmium, and nickel.

A procedure very similar to the above has also been recommended by Walkley.[13] In connection with their thorough study of the determination

[11] A. Ringbom and L. Törn, *Finska Kemistsamfundets Medd.*, **56,** 12 (1947).

[12] P. R. Stout, J. Levy, and Lee C. Williams, *Collection Czechoslov. Chem. Communs.*, **10,** 129 (1938).

[13] A. Walkley, *Australian J. Exptl. Biol. Med. Sci.*, **20,** 139 (1942).

of traces of zinc in biological materials Cholak, Hubbard, and Burkey[14] advocated the use of di-β-naphthylthiocarbazone in place of dithizone for the extraction of zinc, which leads to more complete extraction over a wider range of pH.

Takagawa and Sherman[15] utilized procedures essentially the same as those mentioned above for the determination of zinc in soils, following an alkali carbonate fusion of the sample and dissolution of the melt in dilute hydrochloric acid.

Reed and Cummings[16] simplified the polarographic determination of zinc in plants, peat, and soil by omitting the dithizone separations, and removing iron, aluminum, etc., by ammonia precipitation at a controlled pH.

A 1-g. sample of the plant material is ashed below 450° C., and the ash is treated with 1 to 2 ml. of concentrated hydrochloric acid and evaporated to dryness. The residue is taken up in 2 to 3 ml. of 1 N hydrochloric acid, diluted to 20 to 30 ml., and the pH is adjusted to between 4 and 5 with dilute ammonia (glass electrode control) to precipitate iron and aluminum. The solution is filtered, and the filtrate is evaporated to dryness on the steam bath. The residue is dissolved in exactly 25 ml. of a solution composed of 0.1 N ammonium acetate and 0.025 N potassium thiocyanate buffered with acetic acid to a pH of 4.6, and the polarogram of the zinc is then recorded. This procedure is applicable to samples of plant material that contain as little as 0.0005 per cent zinc, and according to its authors the amounts of manganese, nickel, lead, cadmium, or copper ordinarily present in plants do not interfere. Cobalt interferes when its concentration is greater than twice that of the zinc, because the zinc and cobalt waves tend to coalesce.

Zinc Oxide in Lithopone. Knoke[17] recommends the following procedure for determining zinc oxide impurity in lithopone (barium sulfate and zinc sulfide).

A 0.5-g. sample of the lithopone is shaken for fifteen minutes with 20 ml. of a solution composed of 2 N ammonium chloride, 1.5 M ammonia, and 0.4 per cent tylose. The residue is allowed to settle and the polarogram of zinc in the supernatant solution is then recorded in the usual way. Knoke recommended that the polarogram be recorded without removing air from the solution, but it would probably be better to remove dissolved oxygen by adding a few tenths of a gram of sodium sulfite.

Zinc in Paints. See Chapter XXXV.

---

[14] J. Cholak, D. M. Hubbard, and R. E. Burkey, *Ind. Eng. Chem., Anal. Ed.*, **15**, 754 (1943).

[15] F. Takagawa and G. D. Sherman, *J. Assoc. Offic. Agr. Chemists*, **30**, 182 (1947).

[16] J. F. Reed and R. W. Cummings, *Ind. Eng. Chem., Anal. Ed.*, **12**, 489 (1940).

[17] S. Knoke, *Angew. Chem.*, **50**, 728 (1937); *Paint Varnish Production Mgr.*, **19**, 188 (1939).

2. CADMIUM

Cadmium possesses very favorable polarographic characteristics and can be determined in a variety of supporting electrolytes. The wave form is excellent in neutral alkali metal chloride, nitrate, and sulfate supporting electrolytes, and in nitric, hydrochloric, or perchloric acid media. Half-wave potential and $i_d/Cm^{2/3}t^{1/6}$ data in various supporting electrolytes are listed above in Table XXIX-1.

The behavior of cadmium in 0.5 *M* sulfuric acid containing 0.01 per cent gelatin is unsatisfactory because the diffusion current does not attain a perfectly constant value but continues to increase gradually up to the potential of hydrogen discharge.

W found that the half-wave potential of the iodide-cadmium complex (probably CdI_4^{--}) in 1 *M* potassium iodide solution is −0.74 v. *vs.* S.C.E. The wave shows a pronounced maximum which is difficult to suppress.

The reduction of the complex present in acidic or neutral (up to pH = 9) tartrate supporting electrolytes produces a normal well-developed wave, but a strongly alkaline tartrate medium (0.5 *M* tartrate plus 0.1 *M* excess hydroxide ion) is not suitable for the determination of cadmium. In the latter medium the wave itself is well enough defined but the limiting current is abnormally small and it is not directly proportional to the cadmium concentration. In a typical series of experiments[18] the quantity $i_d/Cm^{2/3}t^{1/6}$ decreased from 1.36 to 0.33 when the cadmium concentration was increased from 1 to 18.6 millimolar in a supporting electrolyte composed of 0.5 *M* sodium tartrate, 0.1 *M* sodium hydroxide, and 0.01 per cent gelatin. These data indicate that the limiting current is not diffusion controlled and hence it has little analytical value. This anomalous behavior may be due in part to the presence of gelatin, and the influence of the concentration of gelatin deserves further study.

The tetramminocadmium ion present in ammoniacal media produces an excellently defined wave.[19]

Sartori[20] has made a polarographic study of the oxalate and pyrophosphate complexes of cadmium, but his data cannot be interpreted exactly because he measured only the 45° tangent reduction potential and did not report complete polarograms.

Pines[21] and Demassieux and Heyrovsky[22] found that cadmium gives a well-defined wave from cyanide medium. These authors studied only the

18 J. J. Lingane, *Ind. Eng. Chem., Anal., Ed.*, **15**, 587 (1943).

19 M. Dobryszycki, *Collection Czechoslov. Chem. Communs.*, **2**, 134 (1930).

20 G. Sartori, *Gazz. chim. ital.*, **64**, 3 (1934).

21 I. Pines, *Collection Czechslov. Chem. Communs.*, **1**, 387 (1929).

22 M. Demassieux and J. Heyrovsky, *Bull. soc. chim.*, **45**, 30 (1929).

45° tangent reduction potential, and they concluded that the reduction was reversible. However, we found that the slope of the wave is considerably smaller than it would be if the reduction were perfectly reversible. According to our experience the half-wave potential is -1.18 v. *vs.* the S.C.E. in 1 *N* potassium cyanide. Von Euler[23] reported 1.4×10^{-17} for the dissociation constant of the $Cd(CN)_4^{--}$ ion. From this value, and the observed half-wave potential of the simple cadmium ion in 1 *N* potassium nitrate (-0.586 v.), we compute that the half-wave potential of the cyanide complex in 1 *N* cyanide should be -1.09 v. *vs.* the S.C.E. The fact that the observed value (-1.18 v.) is only 0.09 v. more negative than this theoretical value indicates that the overvoltage is not very great in the reduction of the $Cd(CN)_4^{--}$ ion.

Spalenka[24] employed a cyanide supporting electrolyte for the determination of traces of cadmium (and lead) in zincic materials.

In most supporting electrolytes the cadmium wave is so far in advance of that of zinc that the polarographic method is ideally suited to the determination of even traces of cadmium in zinc and zinc compounds.[24-28] Large amounts of such metals as copper, bismuth, and lead whose waves precede that of cadmium and hence interfere can be removed conveniently by controlled potential electrolysis with a mercury cathode.[29] Copper may also be rendered innocuous by employing an alkaline cyanide supporting electrolyte (Spalenka[24]).

Cadmium in Alloys. See Chapter XXXV.

Cadmium in Biological Materials. Cholak and Hubbard[30] developed a procedure which is very similar to that mentioned in a preceding section under Zinc in Biological Materials.

Cadmium in Atmospheric Dusts. The polarographic method has been applied advantageously in industrial hygiene studies of cadmium dust and fumes in air.[31, 32]

---

[23] H. von Euler, *Ber.*, **36**, 3400 (1903).

[24] M. Spalenka, *Z. anal. Chem.*, **126**, 49 (1943).

[25] S. I. Siniakova, *Comp. rend. acad. sci. (U. R. S. S.)*, **29**, 376 (1940).

[26] S. A. Pletenev and T. V. Aref'eva, *Trans. All-Union Conf. Anal. Chem.*, **2**, 451 (1943).

[27] D. Cozzi, *Mikrochemie ver. Mikrochim. Acta*, **31**, 37 (1943); *Ann. chim. applicata* **31**, 227 (1941).

[28] R. C. Hawkings and H. G. Thode, *Ind. Eng. Chem., Anal. Ed.*, **16**, 71 (1944).

[29] J. J. Lingane, *Ind. Eng. Chem., Anal. Ed.*, **16**, 147 (1944).

[30] J. Cholak and D. M. Hubbard, *Ind. Eng. Chem., Anal. Ed.*, **16**, 333 (1944).

[31] F. L. Feicht and E. H. Shrenk, *U. S. Bur. Mines Repts. Investigations*, No. 3639 (1942).

[32] L. Silverman, *Chem. Analyst*, **35**, 53 (1946).

3. MERCURY

Both mercurous and mercuric ions yield well-defined diffusion currents that are directly proportional to the concentration of the mercurous or mercuric salt. When an internal mercury anode is used, the waves start immediately from an applied e.m.f. of zero. The waves of simple mercurous and mercuric ions show very prominent maxima, but the latter are easily suppressed by various dyes or gelatin.[33]

The characteristics of the waves of mercurous and mercuric salt solutions have been studied in some detail by Kolthoff and Miller.[34] In the reduction of mercurous ion the electrode reaction is:

$$Hg_2^{++} + 2e \rightleftharpoons 2Hg \tag{1}$$

On the assumption that the concentration of mercurous ion is equal to its activity, and since the activity of the deposited mercury is constant, we have:

$$E_{\text{d.e.}} = E^0_{\text{Hg}_2^{++}} + (RT/nF_y) \ln C^0_{\text{Hg}_2^{++}} \tag{2}$$

where $C^0_{\text{Hg}_2^{++}}$ is the concentration of mercurous ion at the electrode surface, and $E^0_{\text{Hg}_2^{++}}$ is the standard potential of reaction 1 (+0.553 v. *vs.* the S.C.E.). It follows, therefore, that the equation of the wave should be:

$$E_{\text{d.e.}} = E^0_{\text{Hg}_2^{++}} - 0.0296 \log k + 0.0296 \log (i_d - i) \tag{3}$$

Hence a plot of $E_{\text{d.e.}}$ *vs.* log $(i_d - i)$ should produce a straight line with a slope of 0.0296 v. at 25° C. Kolthoff and Miller studied the waves of mercurous nitrate in 0.1 *N* nitric acid, and they found that eq. 3 was valid. They obtained a slope for the log plot of 0.039 v. in fair agreement with the theoretical value. The fact that the observed slope is somewhat larger than the theoretical value may have been due to the presence of gelatin, which was added to suppress the maximum.

Since the curve of eq. 3 is unsymmetrical about the midpoint, the half-wave potential depends on the concentration of mercurous ion, and should shift by 0.030 v. to a more positive value with a 10-fold increase in the concentration of mercurous salt. It is evident from eq. 3 that the half-wave potential is given by:

$$E_{1/2} = E^0_{\text{Hg}_2^{++}} + 0.0296 \log (C/2) \tag{4}$$

where C is the concentration of mercurous ion in the body of the solution.

---

[33] P. Herasymenko, J. Heyrovsky, and K. Tancakivsky, *Trans. Faraday Soc.*, **25**, 152 (1929).

[34] I. M. Kolthoff and C. S. Miller, *J. Am. Chem. Soc.*, **63**, 1013 (1941).

wrong, should be 2732

Kolthoff and Miller found that solutions of completely dissociated mercuric salts, *e. g.*, mercuric perchlorate or mercuric nitrate, produce waves that are identical with those obtained with mercurous salts, except that the diffusion current of mercurous ion is slightly greater than that of mercuric ion at the same concentration. Kolthoff and Miller found that the apparent diffusion coefficient of mercurous ion is 0.92×10^{-5} cm.^{2} sec.$^{-1}$, while that of mercuric ion is 0.82×10^{-5} cm.^{2} sec.$^{-1}$, with 0.1 N nitric acid as supporting electrolyte at 25° C.

Since the diffusion currents of completely dissociated mercury salts are reached before the reduction of oxygen begins, it is not necessary to remove air from the solutions in order to determine the mercury. This is no longer true, however, when mercury is determined in solutions in which it forms stable complexes.

The reduction of mercuric cyanide has been studied by Kemula,[35] Tomes,[36] and Dillinger.[37] The characteristics of the wave are discussed in Chapter XII.

Poma[38] studied the polarography of the iodide, cyanide, and thiocyanate complexes of mercury, but his report contains too little information on important experimental conditions to permit reliable judgment concerning the validity of his conclusions.

---

[35] W. Kemula, *Collection Czechoslov. Chem. Communs.*, **2,** 347, 502 (1930).
[36] J. Tomes, *Collection Czechoslov. Chem. Communs.*, **9,** 81 (1937).
[37] M. Dillinger, *Collection Czechoslov. Chem. Communs.*, **1,** 638 (1929).
[38] K. Poma, *Bull. soc. chim. Belg.*, **54,** 277 (1945).

CHAPTER XXX

Aluminum, Gallium, Indium, and Thallium

1. ALUMINUM

The first study of the polarographic behavior of aluminum ion was made by Prajzler,[1] who observed a fairly well-developed reduction wave in 0.05 *N* barium chloride as supporting electrolyte. The half-wave potential is −1.75 v. *vs.* S.C.E., which is so close to the discharge potentials of barium, sodium, and potassium ions that the wave of aluminum is not fully developed when very large concentrations of barium, sodium, or potassium ions are present. It is better, therefore, to employ lithium salts or tetraalkyl ammonium salts as supporting electrolytes. Aluminate ion, present in strongly alkaline solutions, does not show a reduction wave.

The reduction wave of hydrogen ion precedes that of aluminum ion, but when the concentration of hydrogen ion does not exceed that of aluminum ion the waves are well separated. However, large concentrations of hydrogen ion completely mask the aluminum wave. On the other hand, the pH must not be larger than the natural pH resulting from hydrolysis of the aluminum ion as otherwise hydrous aluminum oxide will precipitate. It is evident, therefore, that the successful polarographic determination of aluminum is very critically dependent on proper control of pH.

According to Semerano and Ronchi[2] the pH of pure aluminum chloride solutions depends on the molar concentration C according to pH = 2.65 − 0.54 log C, from which the natural pH of 0.01 M aluminum chloride solution is 3.73. Since the wave of hydrogen ion precedes that of aluminum ion the pH must not be smaller than about 3, as otherwise the aluminum wave will be masked by the hydrogen wave. Gull[3] employed a pH of 3.8 for the determination of aluminum in magnesium alloys, while Semerano and Ronchi recommend a pH of 3.25. The solution must not contain any weak acids, because—even though the pH is correctly adjusted to the neighborhood of 3.5—weak acids can produce a large hydrogen wave which will completely obscure the aluminum wave. Unfortunately, therefore,

[1] J. Prajzler, *Collection Czechoslov. Chem. Communs.*, **3,** 407 (1931).

[2] G. Semerano and I. Ronchi, *Atti reale ist. Veneto sci., Pt. 2.*, **100,** 401 (1942).

[3] H. C. Gull, *J. Soc. Chem. Ind.*, **56,** 177 (1937).

buffer solutions cannot be used for pH regulation. Ordinarily one starts with a solution acidified only slightly (pH 1 to 3) with hydrochloric, sulfuric, or perchloric acids. If necessary large amounts of free acid are preliminarily removed by evaporation to dryness. The residue is then taken up in very dilute, *e. g.*, not over 0.1 *M*, hydrochloric acid. Final pH adjustment is made with lithium, tetramethylammonium, barium, or sodium hydroxides. Lithium hydroxide is preferable because the aluminum wave is indistinct in the presence of much sodium ion. Although the proper pH can be attained with the aid of indicators or indicator test papers, the fact that the pH adjustment is so critical makes advisable the use of a glass electrode pH meter.

Because the waves of ferric ion and chromic ion precede and are separated from the aluminum wave, it is possible to determine aluminum in the presence of iron and chromium provided that the concentrations of the latter do not exceed the concentration of aluminum. Large amounts of iron and/or chromium, or any other substance which is reduced at a potential more positive than about −1.5 v., obscure the aluminum wave. Electrolysis with a mercury cathode serves excellently to remove large amounts of such elements as iron, copper, chromium, nickel, cobalt, zinc, and most other heavy metals, prior to the polarographic determination of aluminum.[4]

Aluminum in Clay. The simultaneous determination of aluminum and iron in clays has been investigated by Hövker,[5] whose procedure as quoted by Heyrovsky[6] is as follows.

A solution of the clay (method of preparation not specified) containing about 50 mg. of aluminum is prepared in a volume of 100 ml. A 5-ml. aliquot is treated with about 3 ml. of concentrated sulfuric acid and evaporated to dryness. The residue is treated with a small amount of water and again evaporated to dryness. The residue is treated with 8 ml. of 0.1 *N* sulfuric acid, diluted to 30 ml., heated for a few minutes to 80° to 90°, filtered into a 100-ml. volumetric flask, cooled, and diluted to the mark. The pH of this solution should be 1.7 to 1.8. Exactly 10 ml. of the solution is mixed with exactly 30 ml. of a supporting electrolyte containing 1.5 g. of lithium oxide, 2.0 g. of lithium sulfate, 5.0 g. of tylose (methyl cellulose maximum suppressor), and 1 g. of hydrazine sulfate per liter, and the polarogram is recorded without removing air. (It would seem advisable to check the final pH to be sure it is between 3.5 and 3.8.)

The wave of iron occurs prior to that of aluminum. Because the diffusion current of hydrogen ion adds to the apparent height of the iron wave, Heyrovsky recommends

[4] H. O. Johnson, J. R. Weaver, and L. Lykken, *Ind. Eng. Chem., Anal. Ed.*, **19,** 481 (1947).

[5] G. Hövker, Dissertation, University of Hamburg, 1938.

[6] J. Heyrovsky, *Polarographie*, Springer, Vienna, 1941 (p. 304).

that the polarogram of a blank, prepared exactly like the sample solution, be recorded to determine the correction to be subtracted from the apparent diffusion current of the iron. To avoid such an uncertain correction it would be better practice to determine iron in a separate aliquot of the sample solution using a supporting electrolyte, such as 0.5 *M* sodium tartrate of pH between about 4 and 6, in which ferric iron produces a well-developed normal wave.

Aluminum Oxide in Portland Cement. The following procedure is based on the recommendations of Ford and LeMar.[7]

Silica is removed from a 0.5-g. sample by the ammonium chloride method (A. S. T. M. Methods C114-46, 1946). The silica is volatized by hydrofluoric acid treatment in platinum. The ammonium hydroxide group is separated by double precipitation from the filtrate of the silica, and placed in the platinum vessel containing the residue of the silica volatilization; it is then dried and ignited. The ignited precipitate is fused with a small amount of sodium carbonate, dissolved in a slight excess of 1 *M* (1:10) hydrochloric acid, and diluted to about 60 ml. The solution is neutralized carefully with 0.1 *N* potassium hydroxide to a pH of 3.60 ± 0.03 (glass electrode pH meter). Then 30 ml. of 0.5 *M* barium chloride solution is added, and the solution is transferred to a 200-ml. volumetric flask and diluted to volume. The final pH should be between 3.55 and 3.70. A portion of the solution is transferred to a polarographic cell, dissolved air is removed with nitrogen, and the polarogram is recorded.

Ford and LeMar recommend that the procedure be calibrated with a Bureau of Standards sample of argillaceous limestone or with a cement sample of known aluminum oxide content. With 27 cement samples containing from about 3 to 6 per cent aluminum oxide the polarographic analyses agreed very well with values determined by standard chemical procedure and by spectrographic analysis.

Aluminum in Wine. To determine aluminum that enters port wine as the result of storage in aluminum alloy containers Almeida[8] recommended the following procedure.

A 50-ml. sample is evaporated and ashed in a platinum crucible. The ash is transferred to a 10-ml. test tube, treated with 5 ml. of calcium chloride solution, bromphenol blue indicator, and neutralized to pH = 3.8 with 0.05 *N* hydrochloric acid. After standing for 1 hour to insure complete solution of hydrous aluminum oxide, the solution is diluted to exactly 10 ml. and polarographed. Almeida found aluminum contents from 5 to 13 mg. per liter of wine.

Willard and Dean Method. A new type of method for the polarographic determination of aluminum has been developed by Willard and Dean,[9, 10] based on the complexation of aluminum ion with di-*o*-hydroxyazo dyes and the consequent change in the character of the polarogram of the dye.

---

[7] C. L. Ford and L. LeMar, *A. S. T. M. Bull.*, **157,** 66 (1949).
[8] H. de Almeida, *Anais. inst. vinho Porto,* **No. 7,** 11 (1946).
[9] H. H. Willard and J. A. Dean, *Anal. Chem.*, **22,** 1264 (1950).
[10] J. A. Dean, Ph.D. Thesis, University of Michigan, 1949.

They employed Pontochrome Violet SW (Color Index 169) which is the sodium salt of 5-sulfo-2-hydroxy-α-benzene-azo-β-naphthol with the *cis* and *trans* structures:

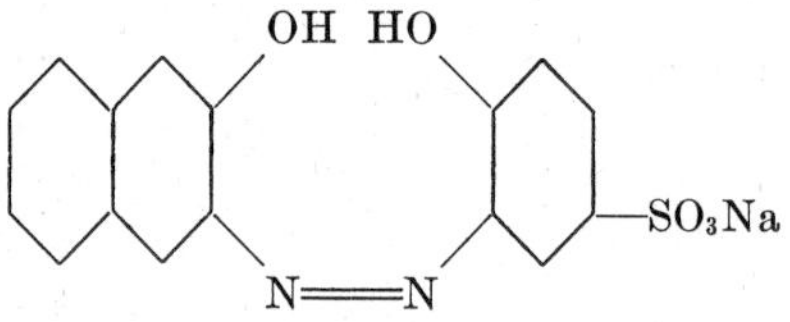

(I)

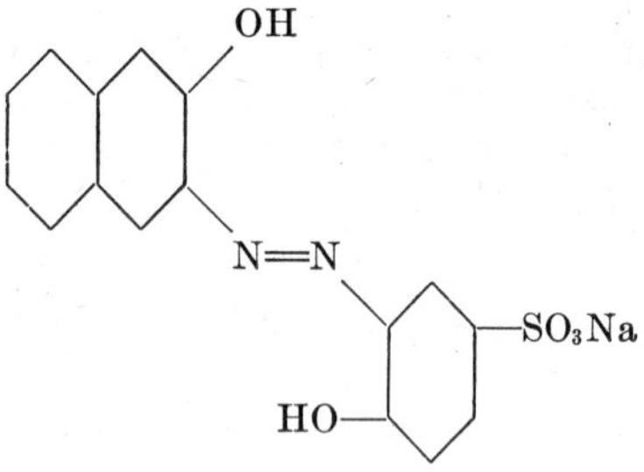

(II)

In buffered solutions of pH between about 3 and 6 this dye produces a single well-developed reduction wave whose half-wave potential *vs.* S.C.E. varies linearly with pH according to:

$$E_{1/2} = 0.021 - 0.069\text{pH}$$

When a small amount of aluminum ion is added to a solution containing an excess of the dye a double wave is obtained, the first part corresponding to the reduction of the excess unreacted dye and the second to the reduction of the portion of the dye that is complexed with aluminum. The second wave is displaced by about 0.2 v. from the first, its half-wave potential *vs.* S.C.E. between a pH of about 3 to 6 being given by:

$$E_{1/2} = -0.258 - 0.058\text{pH}$$

Provided that an excess of dye is present the diffusion current of the *second* wave is directly proportional to the concentration of aluminum, and can be used for the determination of aluminum. This second wave results from the reduction of the complexed dye rather than the aluminum ion. Because the wave of the aluminum-dye complex occurs at a relatively positive potential (−0.52 v. at pH = 4.5) this ingenious method circumvents the nuisance of exact pH adjustment of unbuffered solutions required in the usual method for aluminum and it permits the use of buffers for pH control.

The height of the second wave is greatly dependent on pH; below pH 4 it decreases rapidly with decreasing pH and above pH 5 it increases rapidly with increasing pH. The optimum pH range, within which the rate of change of wave height with pH is relatively small, is between 4 and 5. Willard and Dean recommend a pH of 4.6 ± 0.1 which is obtained conveniently with an acetate buffer.

Because of the limited solubility of the dye (apparently about 2 millimolar or about 0.07 per cent) the maximal concentration of aluminum should not exceed about 0.2 millimolar. With a 0.2 *M* sodium acetate supporting electrolyte adjusted to pH = 4.7 with perchloric acid, and presumably containing about 0.4 millimolar (0.01 per cent) of the Violet SW, Willard and Dean report that $i_d/Cm^{2/3}t^{1/6}$ for the second wave is 10.2 ±0.3 at 25° with concentrations of aluminum between 0.005 and 0.13 millimolar. For maximal precision in measuring the second wave the concentration of excess dye presumably should not be much greater than two or three times that of the aluminum-dye complex.

Willard and Dean found that the rate of formation of the aluminum complex was fairly slow at room temperature, but that equilibrium is attained within about 5 minutes if the solution is warmed to 60° C. after adding the dye and then cooled to room temperature before recording the polarogram.

The fact that the dye in the aluminum complex is reduced at a more negative potential than from the free state was interpreted by Willard and Dean by assuming that the complex involves a rearranged *cis* form of the dye stabilized by complexation with aluminum ion and that the free dye exists in a more easily reducible *trans* form. They concluded that the complex contains two dye molecules per aluminum ion but the evidence is not perfectly conclusive. Whatever the structure of the complex may be the production of two waves, whose relative heights apparently are not dependent on the concentration of total dye provided that it is present in moderate excess, must depend on the very slow attainment of equilibrium between the free and complexed dye.

With regard to interfering ions, chloride, sulfate, calcium, and potassium should be avoided because they cause precipitation of the dye. Only perchloric acid or sodium hydroxide should be used in preliminary steps requiring an acid or base. Fluoride and citrate interfere because they complex more strongly with aluminum than does the Violet SW. Phosphate does not interfere provided that its concentration does not exceed ten times that of the aluminum. Titanium and vanadium interfere by complexing the dye. The presence of manganese, magnesium, chromium, and zinc in amounts not exceeding the aluminum concentration is permissible. Even very large amounts of beryllium do not interfere. Copper and iron both interfere seriously. Interfering heavy metals are best removed by electrolysis with a mercury cathode from a perchloric acid solution, which, however, will not remove titanium or vanadium.

Willard and Dean recommend the following procedure for determining aluminum in limestones, iron ores, and other minerals.

The usual scheme of analysis is followed up through the precipitation of the R_2O_3 group by ammonia. The washed (but not ignited) R_2O_3 precipitate is dissolved in 20 ml. of hot 1 *M* perchloric acid, and the solution is diluted to 50 ml. and electrolyzed in a mercury cathode cell. The residual solution is transferred to a 100-ml. volumetric flask and diluted to the mark. An aliquot corresponding to 0.01 to 0.3 mg. of aluminum is transferred to a 50-ml. volumetric flask and neutralized to methyl red with 2.5 *M* sodium hydroxide solution. The solution is then acidified with exactly 1 ml. of 5 *M* perchloric acid, and 5 ml. of 2 *M* sodium acetate followed by 20 ml. of a 0.05 per cent (1.4 millimolar) aqueous solution of Pontochrome Violet SW are added, and the solution is diluted to the mark. The flask is immersed in a beaker of water at 55° to 70° C. for 5 minutes, and then cooled back to room temperature. A portion of the solution is transferred to a polarographic cell, dissolved air is removed with nitrogen, and the polarogram is recorded. The aluminum content is computed from the height of the second wave i_2 and the measured m and t values of the dropping electrode by the relation: $C = i_2/Im^{2/3}t^{1/6}$, where I is the diffusion current constant evaluated with known concentrations of aluminum carried through the same procedure.

Aluminum in Alloys. See Chapter XXXV.

Aluminum in Glass and Ceramics. See Chapter XXXV.

2. GALLIUM

The fairly negative standard potential of the gallic ion-gallium couple (−0.82 v. *vs.* S.C.E.[11]) coupled with the facts that the gallic ion hydrolyzes to a considerable degree and the reduction wave of hydrogen ion is in the neighborhood of −1.3 v. *vs.* S.C.E., renders its polarographic determination rather difficult.

Zeltzer[12] observed a fairly well-developed wave for the 3-electron reduction of gallic ion to the metal at −1.1 v. *vs.* S.C.E. from solutions containing not more than 10^{-3} *M* hydrochloric acid. In the presence of greater concentrations of hydrogen ion the gallium wave is masked by the hydrogen wave which immediately follows it.

Challenger observed a single well-developed wave for gallic ion from potassium chloride supporting electrolytes having the natural pH due to hydrolysis of the gallic ion; from the relative magnitudes of the gallium and hydrogen waves he was able to compute a value for the first hy-

[11] G. Challenger, Ph.D. Thesis, Harvard University, 1945.

[12] S. Zeltzer, *Collection Czechoslov. Chem. Communs.*, **4,** 319 (1032).

drolysis constant of gallic ion which agreed fairly well with the value determined by other methods. In 0.1 M potassium chloride containing 0.005 per cent gelatin Challenger found that the half-wave potential shifted from -1.03 to -1.15 v. *vs.* S.C.E. when the concentration of gallic ion was increased from 0.1 to 5 millimolar. The slope of the plot of log $i_d/(i_d - i)$ *vs.* E in 0.1 M potassium chloride is 0.083 compared to the value 0.019 expected for a reversible 3-electron reduction. It is conceivable that the excessive wave slope may be due to incompletely resolved stepwise reduction to the $+2$ state and to the metal, although the fact that the half-wave potential is several tenths of a volt more negative than the reversible potential of the gallic ion-gallium couple leaves little doubt that the reduction proceeds irreversibly at the d.e.

Some indication of a reduction wave was observed by both Zeltzer and Challenger in ammoniacal media, but the wave is too indistinct to be of practical use. Gallate ion present in strongly alkaline solutions does not show a reduction wave, and the same is true in tartrate supporting electrolytes.

3. INDIUM

The polarography of indium has been studied by Heyrovsky,[13] Takagi,[14] Tomes,[15] and Lingane.[16] The latter observed that the aquoindic ion present in perchloric acid solutions is irreversibly reduced to produce an extended wave whose half-wave potential is -0.95 v. *vs.* S.C.E. The wave shows a pronounced maximum which is easily suppressed by methyl red or gelatin. When chloride or iodide is added to the solution the wave shifts to a more positive value and becomes steeper and well defined, apparently because of the formation of chloro complex ions of the $+3$ indium which are reduced more reversibly than the aquoindic ion. In 0.1 and 1 M potassium chloride the half-wave potentials are, respectively, -0.561 and -0.597 v. *vs.* S.C.E., and the wave slope corresponds closely to a reversible 3-electron reduction.

In 1 N potassium chloride the diffusion current is well defined for about 0.3 v. beyond the half-wave potential but it then decreases markedly, reaches a minimum at about -1.4 v., and then rises to its initial value just before the final current rise. When methyl red is added to the solution

[13] J. Heyrovsky, *Chem. Listy*, **19,** 168 (1925).

[14] S. Takagi, *J. Chem. Soc. London*, **1928,** 301.

[15] J. Tomes, *Collection Czechoslov. Chem. Communs.*, **9,** 12 (1937).

[16] J. J. Lingane, Ph.D. Thesis, University of Minnesota, (1938); *J. Am. Chem. Soc.*, **61,** 2099 (1939).

the minimum is first decreased, but with larger amounts of the dye it develops into a maximum. This peculiar behavior has not been explained. We have also found that large concentrations of sulfate ion greatly suppress the diffusion current. The best diffusion current is obtained in 0.1 *N* potassium chloride in the presence of a small amount of gelatin.

Since the wave of indium is about 0.4 v. in advance of that of zinc, traces of indium can be determined easily in zinc and zinc salts.

According to Takagi[14] it is also possible to determine small amounts of indium in gallium and its salts, since the half-wave potential of the latter is about 0.5 v. more negative than that of indium.

The determination of indium in zinc ores has been described by Siniakova,[17] and Sinigoglia[18] has discussed its determination in grain zinc.

Mukhina[19] described a procedure for the determination of indium in lead bearing metal. The bulk of the lead is removed by precipitation as lead sulfate, and the filtrate is polarographed for indium after addition of ammonium acetate and adjustment of the pH to 3.9 to 4.2. The indium wave follows the small wave due to unprecipitated lead. When iron is present hydroxylamine hydrochloride is added to reduce it to the noninterfering ferrous state.

4. THALLIUM

Thallous ion produces well-defined waves in practically all the common supporting electrolytes, and since it displays little tendency toward complex formation its half-wave potential does not change very much in different supporting electrolytes.

In 1 *M* solutions of potassium nitrate, potassium chloride, potassium sulfate, ammonium chloride, ammonia, potassium hydroxide, and the common mineral acids, the half-wave potential of thallous ion is -0.475 ± 5 v. *vs.* the S.C.E. In 0.1 *M* solutions of these electrolytes the half-wave potential is -0.460 ± 5 v.[20-22]

Heyrovsky and Ilkovic[20] reported that the half-wave potential of thallous ion in cyanide solutions was the same as in other common supporting

[17] S. I. Siniakova, *Comp. rend. acad. sci. (U. R. S. S.)*, **29,** 376 (1940); *Zhur. Anal. Khim.*, **1,** 241 (1946); *Chem. Abstracts*, **43,** 5330[1] (1949).

[18] S. Sinigoglia, *Metallwaren-ind. u. Galvano-Tech.*, **40,** 308 (1942).

[19] A. S. Mukhina, *Zavodskaya Lab.*, **14,** 354 (1948).

[20] J. Heyrovsky and D. Ilkovic, *Collection Czechoslov. Chem. Communs.*, **7,** 198 (1935).

[21] J. J. Lingane, *J. Am. Chem. Soc.*, **61,** 2099 (1939).

[22] J. Klumpar, *Collection Czechoslov. Chem. Communs.*, **11,** 459 (1939).

electrolytes. However, Kolthoff and Miller[23] have pointed out that the *apparent* half-wave potential of thallous ion in cyanide solutions corresponds to the potential at which cyanide ion depolarizes the dropping electrode, and not to the true half-wave potential of thallous ion. 0.1 *M* cyanide ion depolarizes the dropping electrode (formation of mercuric cyanide) at about −0.6 v. *vs.* the S.C.E., and hence at the potential at which thallous ion is actually reduced (−0.475 v.) the large anodic wave of the cyanide obscures the thallous wave. When the concentration of cyanide is about the same as that of the thallous salt, the thallous wave is actually observable on the *anodic* part of the c.v. curve. At potentials between −0.48 v. and −0.6 v. the reduction of thallous ion to produce a cathodic current, and the anodic dissolution of mercury to produce a negative current take place simultaneously, and the net current corresponds to the difference between the two currents of these opposing reactions. The cathodic diffusion current of thallous ion in cyanide medium does not appear until the potential is more negative than about −0.6 v. Similar phenomena are observed in sulfide solutions.

Terui[24] employed the polarographic method for determining the solubility of thallous bromide in potassium bromide solutions, and the solubility of thallous iodide in potassium iodide solutions.

The determination of thallium in the presence of such metals as cadmium,[25] indium, gallium, and zinc is relatively easy. In supporting electrolytes which do not complex lead ion the thallium and lead waves coincide, but by employing a sodium or potassium hydroxide supporting electrolyte to shift the half-wave potential of lead to −0.76 v. thallium can easily be determined in the presence of lead.

In 0.1 to 1 *M* perchloric acid +3 thallium produces a diffusion current that starts a zero applied e.m.f. corresponding to the 2-electron reduction to thallous ion, and this is followed by the thallous ion wave.[26]

---

[23] I. M. Kolthoff and C. S. Miller, *J. Am. Chem. Soc.*, **62,** 2171 (1940).
[24] Y. Terui, *Bull. Inst. Phys. Chem. Research Tokyo*, **17,** 595 (1938).
[25] F. Emsslin, H. Dreyer, and K. Abraham, *Metall. u. Erz*, **39,** 184 (1942).
[26] G. K. Hughes and N. S. Hush, *Australian J. Sci.*, **10,** 184 (1948).

CHAPTER XXXI

Germanium, Tin, and Lead

1. GERMANIUM

Germanium forms compounds of the −4 (GeH_4), +2 and +4 oxidation states but practically no quantitative thermodynamic data are available. Latimer[1] makes the following very approximate estimates of the standard potentials of some germanium half-reactions (*vs.* N.H.E.).

$$GeO_2 + 4H^+ + 2e \rightleftharpoons Ge^{++} + 2H_2O \qquad E^\circ \sim -0.2 \text{ v.}$$

$$HGeO_3^- + H_2O + 2e \rightleftharpoons HGeO_2^- + 2OH^- \qquad E^\circ \sim -1.4 \text{ v.}$$

$$GeO_2 + 4H^+ + 4e \rightleftharpoons Ge + 2H_2O \qquad E^\circ \sim -0.3 \text{ v.}$$

$$HGeO_3^- + 2H_2O + 4e \rightleftharpoons Ge + 2OH^- \qquad E^\circ \sim -1.2 \text{ v.}$$

$$Ge^{++} + 2e \rightleftharpoons Ge \qquad E^\circ \sim -0.4 \text{ v.}$$

These data indicate that in acid medium +4 germanium should undergo stepwise reduction first to the +2 state and then to the metal, but in alkaline medium reduction should proceed directly to the metal.

Österud and Prytz[2] concluded from a polarographic and oscillographic study of solutions of $GeCl_4$ in dilute perchloric acid that reduction of +4 germanium from acid solutions does occur in two steps. On the contrary, Alimarin and Ivanov-Emin[3] observed no reduction waves with the dropping electrode for +4 germanium from either acid or alkaline solutions, nor from solutions containing fluoride or oxalate.

Alimarin and Ivanov-Emin observed a well-defined wave for the reduction of +2 germanium to the metal in 6 *N* HCl. With 10^{-4} *M* germanium the half-wave potential was approximately −0.45 v. (apparently *vs.* the S.C.E.), and it shifted to more negative values with decreasing concentration of +2 germanium. According to these authors practical polarographic determinations of germanium are feasible provided that the element is first reduced to the +2 state with hypophosphite in hydrochloric

[1] W. M. Latimer, *Oxidation Potentials*, Prentice-Hall, New York, 1938 (p. 135).
[2] T. Österud and M. Prytz, *Arch. Math. Naturvidenskab*, **47**, 73 (1943).
[3] I. P. Alimarin and B. N. Ivanov-Emin, *J. Applied Chem.* (*U. S. S. R.*), **17**, 204 (1944).

acid medium. Elements, such as arsenic, lead, and tin, which produce waves in the vicinity of −0.4 v. interfere with the germanium wave.

2. TIN

Stannous. The reduction of stannous tin to the metal produces excellently defined waves from 1 *N* hydrochloric acid ($E_{1/2} = -0.47$ v. *vs.* S.C.E.), 1 *N* nitric acid ($E_{1/2} = -0.44$ v. *vs.* S.C.E.), and 1 *N* sulfuric acid ($E_{1/2} = -0.46$ v. *vs.* S.C.E.), when 0.005 to 0.01 per cent gelatin is present

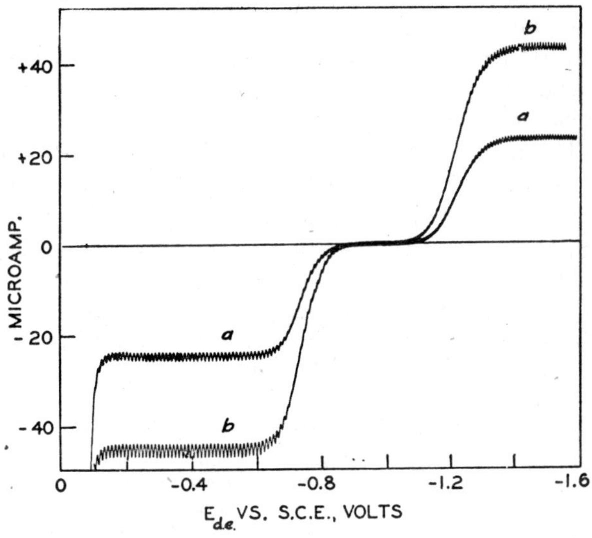

Fig. XXXI-1. Polarograms of stannite ion in 1 *N* sodium hydroxide in the presence of 0.01 per cent gelatin.[5] Concentrations of tin: (*a*) 3.04, (*b*) 5.59 millimolar.

as a maximum suppressor.[4, 5] The values of $i_d/Cm^{2/3}t^{1/6}$ in these three supporting electrolytes are *seriatim*[5] 4.07, 4.02, and 3.54 at 25°. In 1 *N* hydrochloric acid an indication of an anodic wave, corresponding to the oxidation of the chlorostannite ion to the chlorostannate ion, is observed[5] at about −0.1 v. *vs.* S.C.E., but the wave is too poorly developed to be of analytical use.

Lingane[5] observed that stannite ion in 1 *N* sodium hydroxide containing 0.01 per cent gelatin produces both cathodic and anodic waves as shown in Fig. XXXI-1. The cathodic wave ($E_{1/2} = -1.22$ v. *vs.* S.C.E.) results from the reduction of stannite ion to the metal, and the anodic wave

[4] J. Smrz, *Rec. trav. chim.*, **44**, 580 (1925).
[5] J. J. Lingane, *Ind. Eng. Chem., Anal. Ed.*, **15**, 586 (1943).

($E_{1/2}$ = −0.73 v. *vs.* S.C.E.) corresponds to the oxidation of stannite ion to stannate ion. The cathodic half-wave potential is several tenths of a volt more negative than the reversible potential of the stannite-tin couple, and the anodic half-wave potential is 0.5 v. more positive than the reversible stannite-stannate couple. The value of $i_d/Cm^{2/3}t^{1/6}$ is 3.45 at 25° for both the cathodic and anodic diffusion currents.

Stannite ion in 1 *N* sodium hydroxide is unstable to the extent of about 7.7 kcal. per mole with respect to disproportionation into stannate ion and metallic tin, and therefore when it is used as a determining form of tin the solution must be freshly prepared and protected very carefully from air oxidation. Lingane[5] observed no changes in the characteristics of the stannite polarogram after a solution stood for one hour in the absence of oxygen, and hence with pure solutions the disproportionation is evidently very slow. This may not be true in the presence of other metal ions.

Polarograms of stannous tin in tartrate supporting electrolytes comprise a cathodic wave, resulting from reduction of the stannous-tartrate complex to the metal, and an anodic wave due to the oxidation of the stannous-tartrate complex to a stannic-tartrate complex.[6, 7] Both waves show sharp maxima but these are easily suppressed by 0.005 to 0.01 per cent gelatin. With 0.5 *M* total tartrate and 0.01 per cent gelatin the anodic and cathodic half-wave potentials *vs.* S.C.E. at various pH values with 5 millimolar tin are:[7]

pH	2.3	3.4	4.3	9.0	9.8	13.0
Anodic	−0.14	−0.20	−0.28	−0.33	−0.34	−0.71
Cathodic	−0.49	−0.54	−0.59	−0.92	−0.95	−1.16

This pronounced effect reflects a change in the nature of the stannous complex at different pH values, which is typical of all tartrate and citrate complexes. At pH = 4.3, and presumably also at other pH values below about 9, the anodic half-wave potential shifts positively from −0.35 to −0.27 v. when the concentration of stannous tin is increased from 1.9 to 7 millimolar, but the cathodic half-wave potential is independent of the concentration of the stannous-tartrate complex.

The anodic wave in tartrate media develops a minimum when the concentration of stannous tin is greater than about 2 millimolar, as shown by the typical polarograms in Fig. XXXI-2. When the concentration of the stannous-tartrate complex is greater than about 10 millimolar the entire anodic wave nearly disappears (curve *d* in Fig. XXXI-2). This

[6] R. Strubl, *Collection Czechoslov. Chem. Communs.*, **10**, 490 (1938).
[7] J. J. Lingane, *J. Am. Chem. Soc.*, **65**, 866 (1943).

suppression of the anodic wave is not caused by the gelatin present as a maximum suppressor, and it was equally evident in the absence of gelatin. This peculiar phenomenon is only observed when the pH is smaller than about 9, and in strongly alkaline tartrate media the anodic wave is normal in all respects. In solutions containing 0.5 *M* total tartrate at a pH of 4.5, and with 0.01 per cent gelatin present, the anodic limiting current at

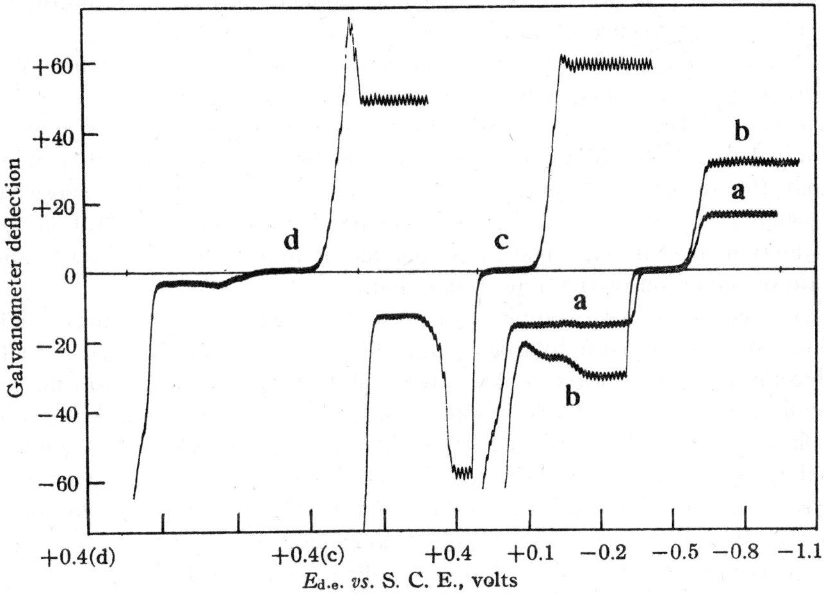

Fig. XXXI-2. Various concentrations of stannous tin in 0.38 *M* disodium tartrate, 0.12 *M* sodium hydrogen tartrate, and 0.01 per cent gelatin.[7] Concentrations of tin: (a) 1.88, (b) 3.68, (c) 7.08, (d) 14.6 millimolar. Galvanometer sensitivity was 0.670 microamp. per mm. for curves a, b and c; 1.675 microamp. per mm. for curve d.

−0.2 v. just preceding the minimum is strictly proportional to the concentration of the stannous-tartrate complex between 0.1 and about 7 millimolar, in spite of the existence of the minimum. The value of $i_d/Cm^{2/3}t^{1/6}$ for the anodic wave (2.43) is identical with that (2.42) for the cathodic wave. In an alkaline tartrate supporting electrolyte (0.5 *M* sodium tartrate, 0.1 *M* sodium hydroxide, and 0.01 per cent gelatin) the value of $i_d/Cm^{2/3}t^{1/6}$ for both waves is 2.86.

Shakhov[8] reported that the reduction potential of tin (apparently in the stannous state) is −0.8 v. *vs.* S.C.E. in an alkaline citrate medium and

[8] A. S. Shakhov, *J. Applied Chem.* (*U. S. S. R.*), **12,** 1555 (1939).

he employed this supporting electrolyte for the determination of tin in potassium fluocolumbate and in metallic columbium (niobium).

Stannic. Stannic tin does not produce a reduction wave from sodium hydroxide solutions, nor from tartrate media of any pH, nor from acidic oxalate solutions. Although disappointing from the standpoint of determining tin itself, this can be turned to advantage in connection with the determination of other metals, such as copper, antimony, bismuth, or lead, in the presence of tin.

In sulfuric or nitric acids the aquostannic ion is too greatly hydrolyzed, with the precipitation of hydrous stannic oxide or basic salts, to permit the use of these media as supporting electrolytes. Solutions of stannic tin in 1 to 2 *M* perchloric acid show no reduction wave, in spite of the fact that the reversible potential of the stannic-stannous couple is positive enough (−0.10 v. *vs.* S.C.E.) so that the thermodynamic condition for reduction is satisfied. The inhibiting factor appears to be a very slow rate of reduction of the aquostannic ion.

The chlorostannate complex ion present in solutions containing a high concentration of both hydrogen and chloride ions does produce a well-developed doublet wave.[9] The great influence of the concentration of hydrochloric acid on the development of this wave is demonstrated by the polarograms in Fig. XXXI-3. With increasing concentrations of hydrochloric acid the doublet wave gradually develops and its total height becomes constant when the chloride ion concentration is greater than about 4 *M*. With increasing chloride and hydrogen ion concentrations increasing proportions of higher order chloro complexes are formed until finally the predominant complex is $SnCl_6^{--}$, which is reduced rapidly enough so that the limiting current becomes diffusion controlled. The first wave results from the reduction of the chlorostannate ion to chlorostannite ion, and the second to complete reduction to the metal.

Lingane[9] found that an optimum supporting electrolyte was 4 *M* ammonium chloride and 1 *M* hydrochloric acid containing 0.005 per cent gelatin. This provides the high chloride ion concentration essential to full development of the chlorostannate wave, and a small enough hydrogen ion concentration so that the diffusion current plateau is well developed before the final current rise due to the reduction of hydrogen ion. The half-wave potential of the first wave is −0.25 v. and that of the second is −0.52 v. *vs.* S.C.E. At 25° in this supporting electrolyte $i_d/Cm^{2/3}t^{1/6}$ is 2.84 for the first wave and 3.49 for the second (6.33 total), showing that the first diffusion current is not fully developed before the second stage of reduction begins. This creates no difficulty because the relative heights

[9] J. J. Lingane, *J. Am. Chem. Soc.*, **67**, 919 (1945).

of the two waves are constant regardless of the tin concentration. Measurement of the second wave is recommended in practical analyses.

Tin in Copper-Base Alloys. The chlorostannate wave has been employed for the determination of tin in copper-base alloys.[10] The complete procedure is described in Chapter XXXV. Copper, bismuth, antimony, and lead interfere. In the analysis of copper-base alloys the first three of these are removed by controlled potential electrolysis from dilute hydrochloric acid solution, and correction for the lead (separately determined) is sub-

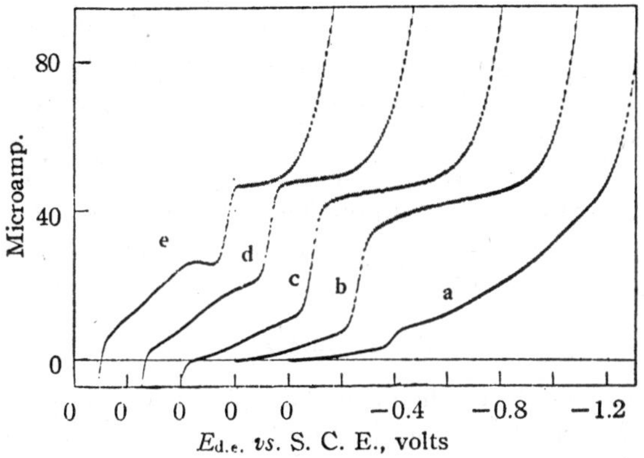

Fig. XXXI-3. Influence of chloride ion concentration on the polarogram of stannic tin.[9] Concentration of stannic chloride was 2.60 millimolar in each case, and 0.005 per cent gelatin was present as a maximum suppressor. Concentrations of hydrochloric acid: (a) 0.2, (b) 1.0, (c) 2.0, (d) 4.0, (e) 6.0 *N*.[9]

tracted from the tin wave. When lead exceeds tin it can be removed by controlled potential deposition on a platinum cathode at −0.5 v. *vs.* S.C.E. from a 0.5 *M* tartrate solution of pH 4 to 5, and the tin can then be determined in the remaining solution after acidifying with hydrochloric acid (1 *M*) and addition of ammonium chloride (4 *M*).

Tin in Steel. Allsopp and Damerell[11] employed the chlorostannate wave in 1 *M* hydrochloric acid–4 *M* ammonium chloride for the determination of tin in steel. The complete procedure is described in Chapter XXXV.

Tin Ores. Alimarin, Ivanov-Emin, and Pevzner[12] developed a method

---

[10] J. J. Lingane, *Ind. Eng. Chem., Anal. Ed.*, **18,** 429 (1946).

[11] W. E. Allsopp and V. R. Damerall, *Anal. Chem.*, **21,** 677 (1949).

[12] I. P. Alimarin, B. N. Ivanov-Emin, and S. M. Pevzner, *Trudy Vsesoyuz. Konferentsii Anal. Khim.*, **2,** 471 (1943).

for the determination of tin in ores based on sodium peroxide fusion of the sample, and hydrogen sulfide precipitation of tin (and other Hydrogen Sulfide Group metals) from a dilute sulfuric acid solution containing tartrate. Stannic and arsenic sulfides are extracted with Na_2S_2 solution, reprecipitated by acidifying the extract with acetic acid, dissolved in a mixture of nitric and sulfuric acids, and evaporated to remove the excess nitric acid, etc. The residue is finally made up to a known volume in 6 N hydrochloric acid and the chlorostannate wave is recorded.

Tin in Foods and Biological Materials. The chlorostannate wave in hydrochloric acid–ammonium chloride medium was utilized by Godar and Alexander[13] for the determination of tin in foods and biological materials. The sample is decomposed by wet digestion with sulfuric and nitric acids followed by perchloric acid to complete the destruction of organic material, the tin is precipitated as the hydrous oxide on hydrous aluminum oxide as a carrier, and is finally polarographed in a hydrochloric acid–ammonium chloride supporting electrolyte. The method is applicable to amounts of tin as small as 0.5 parts per million.

3. LEAD

The chemical and electrochemical characteristics of lead are very favorable for its polarographic determination from a variety of supporting electrolytes. The half-wave potentials and values of $i_d/Cm^{2/3}t^{1/6}$ in several different supporting electrolytes in the presence of 0.01 per cent gelatin as summarized by Lingane[14] are listed in Table XXXI-1.

The lead waves are very well developed in all these supporting electrolytes, and, except for the strongly alkaline tartrate medium, the wave slopes correspond closely to reversible reduction. In strongly alkaline tartrate solution the voltage span of the wave is abnormally large but the diffusion current is easily measurable. The fact that $i_d/Cm^{2/3}t^{1/6}$ is much smaller in alkaline tartrate than in sodium hydroxide demonstrates that a tartrate complex is present, but the nearly identical half-wave potentials in the two media show that the tartrate complex is not much more stable than the hydrogen plumbite ion.

In 0.01 to 0.5 M sodium hydroxide the half-wave potential of the hydrogen plumbite ion increases linearly with the concentration of hydroxide ion according to:[15]

$$E_{1/2} = -0.765 - 0.083 \log C_{OH^-}$$

[13] E. M. Godar and O. R. Alexander, *Ind. Eng. Chem., Anal. Ed.*, **18**, 681 (1946).
[14] J. J. Lingane, *Ind. Eng. Chem., Anal. Ed.*, **15**, 583 (1943).
[15] J. J. Lingane, *Chem. Revs.*, **29**, 1 (1941).

A 1 *M* sodium hydroxide supporting electrolyte is especially useful for the determination of lead in the presence of tin, antimony, and arsenic. Stannate ion and arsenite ion do not show reduction waves in this medium and the wave of antimonite ion is 0.5 v. more negative than that of plumbite ion.

Suchy[16] recommended a tartrate supporting electrolyte for the simultaneous determination of copper, bismuth, lead, and cadmium. The waves appear in the order listed and are best separated[17] at a pH between 4 and 5.

Lead produces a good wave in cyanide medium; in 1 *M* potassium cyanide the half-wave potential is −0.72 v. *vs.* S.C.E. A cyanide supporting electrolyte is useful for determining lead in the presence of much cadmium,[18]

TABLE XXXI-1

POLAROGRAPHIC CHARACTERISTICS OF LEAD IN VARIOUS SUPPORTING ELECTROLYTES AT 25°

Supporting electrolyte	$E_{1/2}$ *vs.* S.C.E., v.	$i_d/Cm^{2/3}t^{1/6}$
0.1 *M* KCl (or HCl)	−0.396	3.80
1 *M* HCl (or KCl)	.435	3.86
1 *M* HNO_3	.405	3.67
1 *M* NaOH	.755	3.39
0.5 *M* tartrate, pH = 4.5	.48	2.37
0.5 *M* tartrate, pH = 9	.50	2.30
0.5 *M* tartrate + 0.1 *M* NaOH	.75	2.39

because the wave of the lead-cyanide complex precedes that of the cadmium-cyanide complex by about 0.4 v.

Lead in Alloys. Procedures for the determination of lead in various types of alloys are described in Chapter XXXV.

Lead in Smelter Products. Hered and Hered[19] described a rapid procedure for the determination of lead in flue dusts and slags without any preliminary separations.

The sample is fused with sodium peroxide, and the melt is decomposed with water, and made up to known volume in dilute hydrochloric acid. A portion of this sample solution is then polarographed in either 1 *M* hydrochloric acid or in an alkaline tartrate supporting electrolyte depending on the other metals present. The authors

[16] K. Suchy, *Collection Czechoslov. Chem. Communs.*, **3,** 354 (1931).
[17] J. J. Lingane, *Ind. Eng. Chem., Anal. Ed.*, **16,** 147 (1944).
[18] M. Spalenka, *Z. anal. Chem.*, **126,** 49 (1943).
[19] W. Hered and G. R. Hered, *Ind. Eng. Chem., Anal. Ed.*, **17,** 780 (1945).

do not clearly specify which of these two supporting electrolytes should be employed for a given type of sample. Presumably the alkaline tartrate supporting electrolyte should be used when elements such as iron, copper, antimony, tin, or bismuth, which produce waves in hydrochloric acid medium which precede or overlap the lead wave, are present.

The determination of lead (and indium, cadmium, and copper) in sphalerites and other minerals has been described by Sinyakova.[20]

Lead in Natural Waters. Methods for the polarographic determination of lead in natural waters have been developed by Kuroda[21] and Reznikov.[22]

Lead Arsenate Insecticides. Caldwell and Reznek[23] devised the following rapid procedure.

A 0.5-g. sample is dissolved in 50 cc. of 1.25 *M* sodium hydroxide, with the aid of heat if necessary, and the solution is diluted to 500 cc. A 5-cc. aliquot of this solution is then made up to exactly 50 cc. in 0.25 *M* sodium hydroxide, 0.05 per cent Methocel (methyl cellulose maximum suppressor), and 0.5 per cent sodium sulfite (to remove dissolved oxygen), and the diffusion current of the plumbite ion in this solution is measured and compared with standards.

The efficiency of sulfite ion in removing oxygen from alkaline lead solutions is questionable.[24] However, by calibrating the procedure empirically Caldwell and Reznek obtained results with lead arsenate and basic lead arsenate insecticides which agreed to 1 to 2 per cent with the lead contents determined by the gravimetric chromate method.

Tetraethyllead in Gasoline. Methods for the determination of tetraethyllead in gasoline have been described by Frediani and Bass[25] and by Borup and Levin.[26] The following simple procedure was recommended by the latter authors.

Measure 10 ml. of 12 *M* hydrochloric acid and 10 ml. of the gasoline sample into a 50-ml. volumetric flask and shake for 5 min. Add a glass bead and digest on a steam bath for 30 min. Withdraw the gasoline layer, add 0.5 ml. of 0.05 per cent gelatin solution, and dilute to exactly 50 ml. Transfer a portion of this solution to a polarographic cell, remove dissolved air with nitrogen, and record the polarogram of lead. The procedure may be standardized with samples of known lead content. According to Borup and Levin thirty-five determinations can be completed per man day, with a precision equal to that of the standard A. S. T. M. chemical method.

---

[20] S. I. Sinyakova, *Zhur. Anal. Khim.*, **1,** 241 (1946).
[21] K. Kuroda, *Bull. Chem. Soc. Japan*, **15,** 153 (1940).
[22] A. A. Reznikov, *Trudy Vsesoyuz. Konferentsii Anal. Khim.*, **2,** 573 (1943).
[23] B. P. Caldwell and S. Reznek, *Ind. Eng. Chem., Anal. Ed.*, **14,** 187 (1942).
[24] J. J. Lingane, *Ind. Eng. Chem., Anal. Ed.*, **18,** 431 (1946).
[25] H. A. Frediani and L. A. Bass, *Oil Gas J.*, **39,** 51 (1940).
[26] R. Borup and H. Levin, *Proc. Am. Soc. Testing Materials*, **47,** 1010 (1947).

Lead in Paints. The polarographic determination of lead and zinc in paints has been described by Abraham and Huffman.[27] Their procedures are described in Chapter XXXV.

Dravnieks[28] developed a method for the determination of lead (and cobalt and manganese) in varnish and lacquers.

Lead in Foodstuffs and Plants. Jones and Brasher[29] described a method for the determination of 5 to 40 p.p.m. of lead in a 5-g. sample of such foodstuffs as canned meat and fish based on destruction of the organic matter by heating with sulfuric and nitric acids followed by perchloric acid. The solution is diluted to a known volume, treated with hydroxylamine to reduce iron to the noninterfering ferrous state, neutralized carefully to a pH of 2.5 to 3 (to eliminate interference by tin), and the polarogram of lead is recorded after removing dissolved air with hydrogen. The original paper should be consulted for the detailed procedure.

The determination of traces of lead in plants has been studied by Hagen[30] and Prát,[31] and in soils and plants by Siniakova.[32]

Lead in Physiological Materials. Yosida[33] developed a procedure for the polarographic determination of traces of lead in physiological materials, based on the preliminary electrolytic separation of the lead as PbO_2 on a platinum anode.

Procedures for the polarographic determination of traces of lead in blood have been described by Teisinger,[34] Forche,[35] and Harmand.[36]

Teisinger recommended the following simple procedure. A 3-ml. sample of blood serum is treated with two drops of concentrated hydrochloric acid, and transferred to a small cell. Air is removed from the solution with pure hydrogen and the polarogram is then recorded. The lead wave is at about −0.4 v. *vs.* the S.C.E. Since the amount of lead will usually be extremely small it is advisable to compensate the residual current when the polarogram is recorded. Apparently none of the organic substances present in blood serum interferes.

In order to increase the accuracy of the lead determination, Forche employed a

[27] B. M. Abraham and R. S. Huffman, *Ind. Eng. Chem., Anal. Ed.*, **12,** 656 (1940).

[28] A. Dravnieks, *Z. anal. Chem.*, **126,** 300 (1943).

[29] F. R. Jones and D. M. Brasher, *Analyst*, **72,** 423 (1947).

[30] S. K. Hagen, *Dansk Tids. Farm.*, **18,** 165 (1944).

[31] S. Prát, *Preslia. Bull. Soc. Chim. Botan. Tchecoslov.*, **6,** 72 (1928).

[32] S. I. Siniakova, *Compt. rend. acad. sci.*, **48,** 414, 648 (1945).

[33] S. Yosida, *Oriental J. Diseases Infants*, **23,** 15 (1938).

[34] J. Teisinger, *Z. ges. exptl. Med.*, **98,** 520 (1936); *Z. Biochem.*, **277,** 178 (1935); *Presse med.*, **46,** 676 (1938).

[35] E. Forche, *Polarographische Studien*, University of Leipzig, 1938.

[36] J. Harmand, *Contribution to the Study of Lead Poisoning*, Berger-Levrault, Paris, 1941 (pp. 161–204).

50-ml. sample of fresh whole blood and decomposed the organic matter by wet digestion with 50 ml. of concentrated nitric acid and 20 ml. of concentrated sulfuric acid. After the easily decomposable material has been destroyed, the solution is evaporated nearly to dryness and the digestion is completed by adding 20 ml. of 70 per cent perchloric acid, and evaporating slowly to dryness.[37] The residue is taken up in 5 ml. of 1 *M* sodium potassium tartrate, and the solution is made slightly alkaline by addition of sodium hydroxide. The residue should be stirred with the tartrate solution for about two hours to insure complete extraction of the lead. The solution is finally filtered into a 10-ml. volumetric flask, made up to the mark, and the polarogram obtained in the usual way.

The determination of lead in bone has been described by Wenig,[38] and in industrial hygiene samples by Feicht, Schrenk, and Brown.[39]

Application of the polarographic lead determination to the clinical diagnosis of plumbic encephalitis has been made by Hamamoto,[40] and the same author[41] studied the polarographic determination of lead in toilet articles. The polarographic determination of lead in atmospheric samples has been described by Landry[42] and Levine.[43] The microdetermination of lead in toxicological investigations has been discussed by Wheeler[44] and Heusghem, Ponsart, and Lenaers.[45]

[37] *Caution* must be observed in such procedures not to add the perchloric acid until after the easily oxidizable material has been destroyed; otherwise dangerous explosions may result.

[38] E. Wenig, *Z. physiol. Chem.*, **273,** 158 (1942); **275,** 72 (1942).

[39] F. L. Feicht, H. H. Schrenk, and C. E. Brown, *U. S. Bur. Mines Repts. Investigations*, No. 3639 (1942).

[40] E. Hamamoto, *Oriental J. Diseases Infants*, **19,** 18 (1936).

[41] E. Hamamoto, *Oriental J. Diseases Infants*, **19,** 19 (1936).

[42] A. S. Landry, *J. Ind. Hyg. Toxicology*, **29,** 168 (1947).

[43] L. Levine, *J. Ind. Hyg. Toxicology*, **27,** 171 (1945).

[44] H. J. Wheeler, *J. Ind. Hyg. Toxicology*, **29,** 158 (1947).

[45] C. Heusghem, L. Ponsart, and R. Lenaers, *J. pharm. Belg.*, **4,** 27 (1949).

CHAPTER XXXII

Nitrogen, Phosphorus, Arsenic, Antimony, and Bismuth

1. NITROGEN

Nitrate and Nitrite. The reduction of nitrate and nitrite at the dropping electrode was originally studied by Tokuoka[1] and Tokuoka and Ruzicka.[2] No wave is observable for either nitrate or nitrite in neutral or alkaline solutions of potassium or sodium salts as supporting electrolytes, but an ill-defined wave is obtained when lithium chloride or tetramethylammonium chloride is used. The reduction potentials of NO_3^- and NO_2^- are shifted markedly to a more positive value, and well-defined waves are obtained, when polyvalent cations are added to the solutions. The reduction potentials of NO_3^- and NO_2^- are identical and both reductions are irreversible and require a large overvoltage.

According to Schwarz[3] nitrite produces an extended wave between −0.6 and −1.6 v. in a lithium chloride solution containing acetic acid and a trace of acetate. The fact that nitrous acid produces a reduction wave at −1.0 v. *vs.* S.C.E. in acid media, without polyvalent cations or catalysts present, has been confirmed by Keilin and Otvos,[4] but the reduction product under these conditions has not been established.

Tokuoka and Ruzicka reported the following values of the reduction potentials (45° tangent) of either nitrate or nitrite in 0.1 *N* solutions of the indicated supporting electrolytes: $LaCl_3$, −1.22 v.; $CeCl_3$, −1.23 v.; $MgCl_2$, −1.74 v.; $CaCl_2$, −1.78 v.; $SrCl_2$, −1.79 v.; $(CH_3)_4NCl$, −2.15 v.; and LiCl, −2.17 v. *vs.* the N.C.E. The reduction potentials fall into three groups, according to the charge on the cation of the supporting electrolyte, but in each group they are practically independent of the particular cation that is used. In very concentrated supporting electrolytes the reduction potentials approach the same value of about −1.5 v., regardless of the charge of the cation of the supporting electrolyte. Tokuoka and

[1] M. Tokuoka, *Collection Czechoslov. Chem. Communs.*, **4,** 444 (1932); *Mem. Faculty Sci. Agr., Taihoku Imp. Univ.*, **9,** (No. 6), 197 (1936).

[2] M. Tokuoka and J. Ruzicka, *Collection Czechoslov. Chem. Communs.*, **6,** 339 (1934).

[3] K. Schwarz, *Z. anal. Chem.*, **115,** 161 (1939).

[4] B. Keilin and J. W. Otvos, *J. Am. Chem. Soc.*, **68,** 2665 (1946).

Ruzicka attributed this effect of polyvalent cations to the formation of "ion pairs," or loose complexes, in which the nitrate or nitrite ion is electrostatically polarized and hence in a more easily reducible state than in the normal nitrate or nitrite ion. They also assume that the ion pairs have a resultant positive charge which facilitates their access to the negatively charged electrode.

One of the most striking characteristics of the nitrate or nitrite wave is

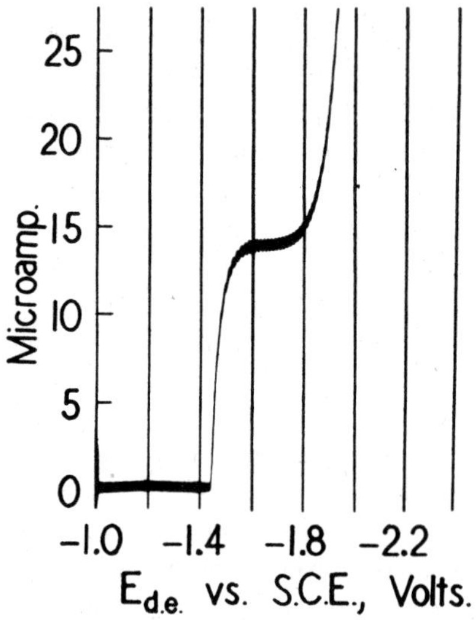

Fig. XXXII-1. Typical polarogram of nitrate ion, obtained with 5×10^{-4} *M* lithium nitrate in 0.017 *M* lanthanum chloride.

the unusual abruptness with which the current increases. This is demonstrated by the typical polarogram of 5×10^{-4} *M* lithium nitrate in 0.017 *M* lanthanum chloride shown in Fig. XXXII-1. Instead of the usual exponential increase, the current remains perfectly constant until the reduction potential is reached and then increases very abruptly. We found that an increase in the applied potential of only three or four millivolts in the neighborhood of the reduction potential causes the current to increase suddenly from practically zero to a value well above the half-wave current. When a trace of sodium hydroxide is added to the solution the abrupt increase in current disappears, and the foot of the wave assumes the usual exponential shape.

Kolthoff, Harris, and Matsuyama[5] observed that the potential at which the nitrate wave occurs in 0.01 *M* lanthanum chloride is *ca.* 0.09 more positive with decreasing applied potential than when the polarogram is recorded in the usual way with increasing applied potential. With decreasing applied potential the intermediate reduction products are already present at the electrode surface, and they apparently catalyze the slow step in the over-all reduction and thus reduce the overvoltage.

Because they observed that the diffusion current of nitrate was approximately four times greater than that of an equimolar concentration of cadmium ion, Tokuoka and Ruzicka concluded that 8 electrons are involved in the reduction of nitrate ion and that ammonia or ammonium ion is the reduction product. They also concluded, more from analogy than definite evidence, that nitrite ion is also reduced to ammonia. Kolthoff, Harris, and Matsuyama disagree, and from the value of $i_d/Cm^{2/3}t^{1/6}$ in 0.01 *M* lanthanum chloride (15.3 at 25°) and the infinite dilution diffusion coefficient of nitrate ion (1.92×10^{-5} cm.^{2} sec.$^{-1}$), they concluded that $n = 6$, and that the reduction proceeds only to hydroxylamine:

$$NO_3^- + 7H^+ + 6e \rightarrow NH_2OH + 2H_2O$$

The fact that hydroxylamine itself is not reduced at the d.e. adds to the plausibility of this conclusion. However, if a lanthanum nitrate complex is the actual diffusing species its diffusion coefficient must be much smaller than that of nitrate ion. Consequently the n-value computed by Kolthoff, Harris, and Matsuyama based on the diffusion coefficient of nitrate ion is probably smaller than the true value.

Tokuoka and Ruzicka observed that the limiting current of nitrate increases with increasing concentration of lanthanum chloride or magnesium chloride until the equivalent concentrations of these salts are about fifty times greater than the nitrate ion concentration. They also observed that the quotient i_d/C_{NO_3} decreases appreciably with increasing nitrate concentration in 0.03 *M* lanthanum chloride, but with 0.05 *M* magnesium chloride the quotient was constant. Kolthoff, Harris, and Matsuyama confirmed this; with 0.01 *M* lanthanum chloride they found that i_d/C_{NO_3} was constant to ±3 per cent between 0.02 and 0.12 millimolar nitrate but it decreased by about 20 per cent when the nitrate ion concentration was increased to 0.48 millimolar.

The presence of sulfate ion decreases the limiting currents of both nitrate and nitrite ions in a lanthanum chloride supporting electrolyte; with a

[5] I. M. Kolthoff, W. E. Harris, and G. Matsuyama, *J. Am. Chem. Soc.*, **66**, 1782 (1944).

large concentration of sulfate ion the nitrate wave is almost completely obliterated. Tokuoka and Ruzicka concluded that sulfate ion replaces nitrate ion in the complex with lanthanum ion.

According to Tokuoka and Ruzicka the limiting current of nitrate ion is about one and one-half times greater in 0.033 *M* lanthanum chloride than in 0.05 *M* magnesium chloride. These investigators also observed that the nitrate limiting current was decreased by the simultaneous reduction of zinc ion and other metal ions.[6]

Kolthoff, Harris, and Matsuyama[5] discovered that the simultaneous reduction of uranyl ion catalyzes the reduction of nitrate ion in dilute hydrochloric acid medium, as shown by the typical polarograms in Fig. XXXII-2. The first uranium wave (curve *2*) results from the reduction of uranyl ion to the +5 state, and the second corresponds to reduction of +5 uranium to the +3 state (see Chap. XXV). The catalyzed nitrate wave (curve *3* in Fig. XXXII-2) occurs at the same potential as the second uranium wave. The catalytic mechanism has not been identified, but the fact that nitrate ion is not reduced until the potential is negative enough so that +3 uranium is being formed at the electrode surface strongly suggests that +3 uranium is the actual nitrate-reducing agent.

With 0.2 millimolar uranyl ion as "activator" in a supporting electrolyte composed of 0.1 *M* potassium chloride and 0.01 *M* hydrochloric acid (same conditions as in Fig. XXXII-2) the limiting current of the nitrate ion is directly proportional to its concentration up to 0.5 millimolar. With larger nitrate ion concentrations the ratio i_d/C_{NO_3} decreases when 0.2 millimolar uranyl ion is present. The minimal uranyl ion concentration required to maintain a linear relation between i_d and C_{NO_3} is 0.2 millimolar with up to 0.5 millimolar nitrate, 0.1 millimolar with up to 0.1 millimolar nitrate, and 0.02 millimolar with up to 0.01 millimolar nitrate. Because the nitrate limiting current is measured as the difference between the total current at −1.2 v. *vs.* S.C.E. and the diffusion current of the uranyl ion determined in a blank experiment, an unnecessarily large uranyl ion concentration obviously is undesirable. Best results are obtained when the uranyl ion concentration is kept within the limits indicated above.

With 0.2 millimolar uranyl ion in 0.1 *M* potassium chloride–0.01 *M* hydrochloric acid the half-wave potential of the nitrate wave increases from −0.91 v. *vs.* S.C.E. with 0.04 millimolar nitrate ion to −0.98 v. with 0.4 millimolar nitrate ion. The value of $i_d/Cm^{2/3}t^{1/6}$ under these conditions is 13.8 at 25°. From this value, and the diffusion coefficient of nitrate ion, Kolthoff, Harris, and Matsuyama concluded that the nitrate

[6] See also G. Cinquina, *Chimica (Milan)*, **3**, 363 (1948).

ion undergoes a 5-electron reduction to elemental nitrogen:

$$NO_3^- + 6H^+ + 5e \rightarrow \frac{1}{2}N_2 + 3H_2O$$

in contrast to the reduction to hydroxylamine or ammonia in neutral lanthanum chloride solutions.

Kolthoff, Harris, and Matsuyama recommended the following procedures for the determination of 0.05 to 0.4 millimolar nitrate.

Prepare a stock uranyl chloride solution which is 1 millimolar in respect to uranyl chloride, 0.5 *M* in potassium chloride, and 0.05 *M* in hydrochloric acid. Dilute 5.00

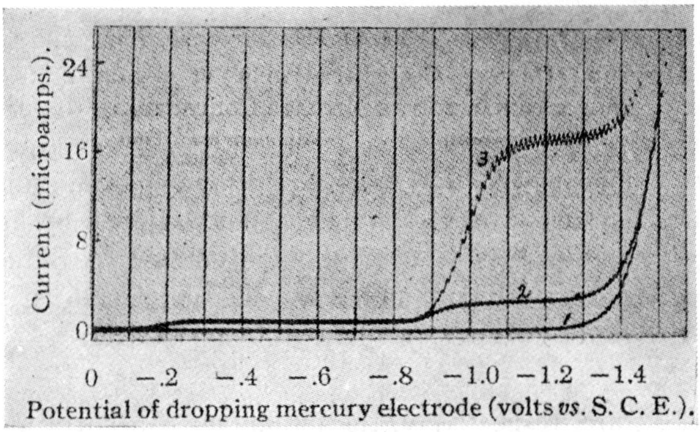

Fig. XXXII-2. Polarograms in 0.01 *M* hydrochloric acid, 0.1 *M* potassium chloride: (*1*) 4 × 10^{-4} *M* potassium nitrate; (*2*) 2 × 10^{-4} *M* uranyl chloride; (*3*) 4 × 10^{-4} *M* potassium nitrate, 2 × 10^{-4} *M* uranyl chloride.[5]

ml. of this stock solution to 25.00 ml., place the solution in a polarographic cell, remove dissolved air with nitrogen, and measure the uranium diffusion current at −1.2 v. *vs.* S.C.E. This is the "blank" to be subtracted from the total current when nitrate is determined.

Measure a suitable volume of the unknown nitrate solution (such that the final nitrate ion concentration will be between 0.05 and 0.4 millimolar) into a 25-ml. volumetric flask, add 5.00 ml. of the stock uranyl chloride solution, and dilute to volume. Place the solution (or an unmeasured portion of it) in the polarographic cell, remove dissolved air with nitrogen, and measure the total current at −1.2 v. *vs.* S.C.E. Subtract the previously determined uranium blank from the total current to obtain the nitrate diffusion current. The procedure may either be standardized with known nitrate concentrations, or the concentration of nitrate may be computed from the relation $C = i_d/13.8m^{2/3}t^{1/6}$ when the measurements are made at 25°.

Basic or strongly acidic nitrate solutions must first be neutralized to a pH of about 5 (methyl red) with hydrochloric acid or sodium hydroxide to avoid significant alteration of the acidity of the uranyl supporting electrolyte.

Very large amounts of sulfate ion interfere by decreasing the nitrate diffusion current, but amounts up to about 50 times the nitrate concentration cause no significant error. Chlorate does not interfere, and the foregoing procedure may be used to determine traces of nitrate in chlorate salts.

Oxalate interferes because it yields a reduction wave at the same potential as the nitrate wave.

Phosphate precipitates uranyl ion and must be removed.

Other substances which produce reduction waves before the nitrate wave interfere.

Keilin and Otvos[4] found that nitrite (nitrous acid) yields the same wave as nitrate in the presence of uranyl ion. The $i_d/Cm^{2/3}t^{1/6}$ value for nitrite is 7.45 at 25°, corresponding to a 3-electron reduction of nitrous acid to elemental nitrogen:

$$HNO_2 + 3H^+ + 3e \rightarrow \frac{1}{2}N_2 + 2H_2O$$

Therefore the sum of nitrite and nitrate is determined by the foregoing procedure of Kolthoff *et al.*

Keilin and Otvos showed that nitrite and nitrate in mixtures in which their ratio is not too disproportionate can be determined by first applying the foregoing procedure of Kolthoff *et al.* The cumulative diffusion current (corrected for the uranyl blank) obeys the relation (at 25°):

$$i_d = m^{2/3}t^{1/6}\ (7.45C_1 + 13.8C_2)$$

where C_1 and C_2 are, respectively, the millimolar concentrations of nitrite and nitrate. The nitrite in a second sample of the unknown solution is then oxidized quantitatively to nitrate with hydrogen peroxide in acid solution:

$$HNO_2 + H_2O_2 \rightarrow NO_3^- + H^+ + H_2O$$

The excess hydrogen peroxide is removed by making the solution basic and adding a small amount of manganese dioxide to catalyze the disproportionation of the hydrogen peroxide. After oxygen evolution ceases the solution is neutralized and carried through the procedure of Kolthoff *et al.* to obtain a nitrate diffusion current which obeys the relation:

$$i_d = 13.8m^{2/3}t^{1/6}(C_1 + C_2)$$

The equations for the two diffusion currents are solved simultaneously to compute C_1 and C_2. Although Keilin and Otvos do not supply any performance data on this ingenious method, it is evident that maximal precision will be obtained when $C_1 = C_2$, and the method will fail when the ratio C_1/C_2 is much outside the limits 0.1 to 10.

For the determination of traces of nitrate in nitrite salts Haslam and

Cross[7] recommend the destruction of the nitrite with excess sodium azide in boiling, dilute hydrochloric acid medium:

$$HNO_2 + 3HN_3 \rightarrow 5N_2 + 2H_2O$$

The solution is then neutralized and polarographed for nitrate in a lanthanum chloride supporting electrolyte.

A method for the determination of nitrate in blood and urine has been described by Scott and Bambach.[8]

Nitric Oxide. Heyrovsky and Nejedly[9] reported that acidified solutions of nitrite produce a wave starting at −0.77 v. *vs.* S.C.E. The wave disappears when hydrogen or nitrogen is passed through the solution for some time. Furthermore, when the effluent gas is passed through a sodium hydroxide solution, and the latter is then acidified, the wave at −0.77 v. appears. Heyrovsky and Nejedly concluded that the wave is due to the reduction of nitric oxide, formed by the disproportionation of the nitrous acid produced when a nitrite solution is acidified:

$$3HNO_2 \rightleftharpoons 2NO + NO_3^- + H^+ + H_2O$$

From a comparison of the nitric oxide wave with that due to an equimolar concentration of thallous ion, Heyrovsky and Nejedly concluded that 5 electrons were required and that the reduction product was ammonium ion:

$$NO + 5e + 6H^+ \rightarrow NH^+ + H_2O$$

Tokuoka[10] employed the nitric oxide wave for the determination of nitrite impurity in nitrate salts. The neutral or slightly alkaline solution of the nitrate salt is first freed from air by passage of hydrogen or nitrogen, then acidified to 0.1 *M* in hydrochloric acid, and the nitric oxide wave is recorded without further passage of gas through the solution.

From the experiments of Keilin and Otvos[4] it appears that this wave attributed to nitric oxide actually results from the reduction of nitrous acid. According to these authors the decomposition of nitrous acid is first order in the presence of air at room temperature, and at the concentrations of interest in polarography about 6 per cent of the nitrous acid decomposes in one-half hour. If the nitrite solution is polarographed immediately after acidification the error arising from nitrous acid decomposition can be kept below 3 per cent.

---

[7] J. Haslam and S. H. Cross, *J. Soc. Chem. Ind.*, **64,** 259 (1945).

[8] E. W. Scott and K. Bambach, *Ind. Eng. Chem., Anal. Ed.*, **14,** 136 (1942).

[9] J. Heyrovsky and V. Nejedly, *Collection Czechoslov. Chem. Communs.*, **3,** 126 (1931); *Chem. News*, **142,** 193 (1931).

[10] M. Tokuoka, *Collection Czechoslov. Chem. Communs.*, **4,** 444 (1932).

According to Heyrovsky and Nejedly, small amounts of *nitrite in nitrocellulose* can be determined in the following way. A 1 per cent suspension of the nitrocellulose is prepared in 0.1 N sodium hydroxide, and freed from air in the polarographic cell. An excess of air-free hydrochloric acid is then added, and the nitric oxide (nitrous acid) wave is recorded without further passage of hydrogen or nitrogen.

The foregoing technique cannot be applied to the determination of *nitrite in gunpowders that contain nitroglycerin*, because nitroglycerin is easily reducible and its wave interferes with that of the nitric oxide. Heyrovsky and Nejedly recommended the following procedure. A suspension of the gunpowder is prepared in dilute hydrochloric acid in a gas-washing bottle, and hydrogen or nitrogen is passed through the mixture to remove the nitric oxide. The effluent gas is passed through a 0.1 N solution of sodium hydroxide to absorb the nitric oxide. The alkaline absorbing solution may conveniently be in the polarographic cell. The absorbing solution is finally acidified with air-free hydrochloric acid and the nitric oxide wave obtained. Heyrovsky and Nejedly did not give any detailed results, nor any information about the time required to remove the nitric oxide completely from the acid suspension of the gunpowder.

Cyanogen and Oxamic Acid. Brezina and Heyrovsky[11] found that cyanogen is reduced at the dropping electrode from neutral or slightly acid or slightly alkaline media, but not from strongly acid or strongly alkaline solutions, nor from potassium cyanide solutions. They prepared the cyanogen by mixing solutions of cupric sulfate and potassium cyanide in a gas-washing bottle, and swept the gas into a polarographic cell with nitrogen. With solutions of cyanogen in 0.1 N sodium acetate they observed a well-defined double wave, with half-wave potentials of about -1.2 v. and -1.55 v. *vs.* S.C.E. When the solution was allowed to stand the height of the first wave decreased while that of the second wave increased. Brezina and Heyrovsky concluded that the first wave was due to the reduction of C_2N_2 according to:

$$C_2N_2 + 2e \rightarrow 2CN^-$$

They concluded that the second wave was due to the reduction of a hydrolysis product of cyanogen, probably oxamic acid, $CO(NH_2)(COOH)$, or its salt. This conclusion was substantiated by the fact that solutions of oxamic acid in 0.1 N sodium acetate produced a well-defined wave at -1.5 v., which is the position of the second wave in the reduction of cyanogen.

The fact that no wave was observed in strongly alkaline or strongly acid

[11] J. Brezina and J. Heyrovsky, *Collection Czechoslov. Chem. Communs.*, **8**, 114 (1936).

solutions is due to the fact that cyanogen is unstable in both acid and alkaline solutions:

$$C_2N_2 + 2H_2O + H^+ \rightarrow HCN + NH_4 + CO_2$$

$$C_2N_2 + 2OH^- \rightarrow CN^- + CNO^- + H_2O$$

and cyanate ion is not reducible at the dropping electrode in alkaline medium.

Cyanide. Cyanide ion or hydrogen cyanide depolarizes the d.e. to produce an anodic wave:

$$Hg + 2CN^- \rightleftharpoons Hg(CN)_2 + 2e$$

Kolthoff and Miller[12] demonstrated that the depolarization proceeds reversibly, and the equation of the wave has the form:

$$E = \text{const.} + \frac{0.0591}{2} \log \frac{i}{(i_d - i)^2}$$

the squared term arising from the fact that two cyanide ions are involved per molar unit of the reaction. Correspondingly the half-wave potential shifts by 0.030 v. in the positive direction for a ten-fold decrease in the cyanide concentration.

Cyanide is best determined in air-free solutions that contain sodium hydroxide to prevent loss of hydrocyanic acid by volatilization. With 5 millimolar cyanide in 0.1 *M* sodium hydroxide the wave starts at -0.45 v. *vs.* S.C.E. and the anodic diffusion current is well developed.

Thiocyanate. Thiocyanate ion depolarizes the d.e. reversibly like cyanide ion to yield a well-formed anodic wave:

$$Hg + 2CNS^- \rightleftharpoons Hg(CNS)_2 + 2e$$

In neutral or only weakly alkaline medium the half-wave potential of 1 millimolar thiocyanate ion is $+0.18$ v. *vs.* S.C.E.[12]

2. PHOSPHORUS

No direct electrochemical reactions of phosphorus compounds at the d.e. have been reported.

Phosphate. Indirect methods have been developed for the determination of phosphate based on the decrease of the diffusion current of bismuth, uranyl, and other heavy metal ions due to their precipitation by phosphate (see Chap. XLVII under Amperometric Titrations), and on the precipitation of ammonium phosphomolybdate (see Chap. XXV).

[12] I. M. Kolthoff and C. S. Miller, *J. Am. Chem. Soc.*, **63**, 1405 (1941).

Pyrophosphate. A method for the determination of pyrophosphate based on precipitation as $Cd_2P_2O_7$ from an acetate buffer of pH = 3.6, solution of the washed precipitate in dilute hydrochloric acid, and measurement of the cadmium diffusion current has been developed by Cohn and Kolthoff.[13] By precipitating the cadmium pyrophosphate from a weakly acid medium precipitation of cadmium orthophosphate is prevented. Pyrophosphate can be determined in the presence of a 15-fold excess of orthophosphate by a single precipitation, and in the presence of a 40-fold excess of orthophosphate if a double precipitation is used. Calcium was found to have a beneficial effect when orthophosphate is present.

To 25 ml. of the neutral or weakly alkaline solution, which must be 0.002 to 0.01 *M* in pyrophosphate (corresponding to 7 to 35 mg. of phosphorus pentoxide), sufficient 0.2 *M* orthophosphate solution is added, if necessary, to make the final orthophosphate concentration approximately 0.03 *M*. 5 ml. of acetate buffer of pH 3.6 (8 *M* acetic acid, 1 *M* ammonium acetate) is added and 2 ml. of 1 *M* calcium chloride (or less, if calcium is present). Then 15 to 20 ml. of 0.4 *M* cadmium acetate solution is added, and the mixture is gently shaken for about 5 to 6 hours. When the originally voluminous and slimy precipitate has changed into well-developed crystals, the precipitate is filtered or centrifuged and washed with water. The cadmium in the precipitate is determined polarographically after being dissolved in a few milliliters of hydrochloric acid (1 to 4), adding 25 ml. of 2 *M* potassium chloride solution, and diluting to 100 ml. in a volumetric flask.

For the determination of smaller amounts of pyrophosphate correspondingly reduced volumes are used.

If the orthophosphate content is larger than stated above, the precipitate, after being shaken for 5 hours and separated from the supernatant liquid, is dissolved in a few milliliters of hydrochloric acid (1 to 4) at room temperature. The solution is neutralized immediately with sodium hydroxide (6 *M*), using Tropeolin OO as indicator. After neutralization the volume is made about the same as in the first precipitation and the same amounts of buffer solution and of cadmium acetate are added as before.

3. ARSENIC

When present in the +5 state arsenic is not reduced at the dropping electrode from any of the various supporting electrolytes so far investigated, but +3 arsenic does produce reduction waves under limited conditions.[14-19]

[13] G. Cohn and I. M. Kolthoff, *Ind. Eng. Chem., Anal. Ed.*, **14,** 886 (1942).
[14] V. Bayerle, *Rec. trav. chim.*, **44,** 514 (1925).
[15] K. Kacirkova, *Collection Czechoslov. Chem. Communs.*, **1,** 477 (1929).
[16] K. Bambach, *Ind. Eng. Chem., Anal. Ed.*, **14,** 265 (1942).
[17] J. J. Lingane, *Ind. Eng. Chem., Anal. Ed.*, **15,** 583 (1943).
[18] N. Ya. Khlopin, *Zhur. Obshcheĭ Khim.*, **18,** 364, 1019 (1948).
[19] N. Ya. Khlopin, N. A. Rafalovich, and G. P. Aksenova, *Zhur. Anal. Khim.*, **3,** 16 (1948).

In 1 N hydrochloric acid containing 0.01 per cent gelatin the complicated reduction wave of +3 arsenic starts at −0.4 v. *vs.* S.C.E. as shown in Fig. XXXII-3. Some semblance of a limiting current is observed between −0.8 and −0.9 v. and this is followed by another ill-developed reduction wave at about −1.0 v. Lingane[17] found that $i_d/Cm^{2/3}t^{1/6}$ for the limiting current at −0.8 to −0.9 v. was 8.6, which is large enough to correspond to a complete 6-electron reduction to arsine. However the fact that this limiting current is followed by another reduction wave indicates that the

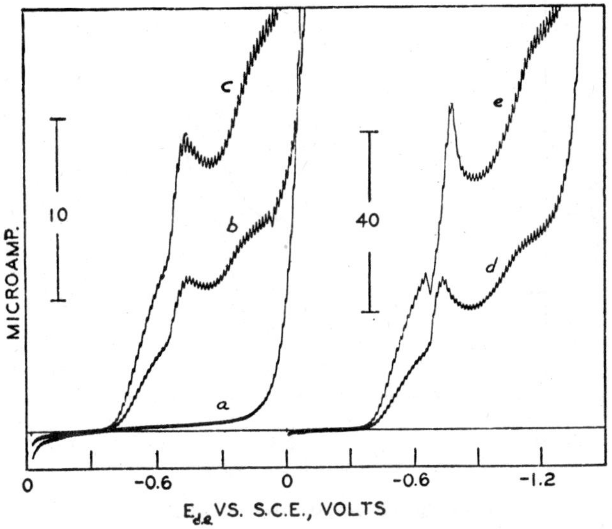

Fig. XXXII-3. Polarograms of arsenious acid in 1 N hydrochloric acid in the presence of 0.01 per cent gelatin.[17] As_2O_3 concentrations: (*a*) 0, (*b*) 0.196, (*c*) 0.385, (*d*) 0.742, (*e*) 1.53 millimolar.

current at −0.8 to −0.9 v. can correspond only to reduction to the elemental state. The final wave probably represents reduction to arsine.

As shown in Fig. XXXII-4 the wave form in 1 N sulfuric or nitric acid is much better than in hydrochloric acid; two distinct waves are observed whose half-wave potentials in the presence of 0.01 per cent gelatin are[17] −0.7 and −1.0 v. *vs.* S.C.E., and the $i_d/Cm^{2/3}t^{1/6}$ value for the first wave is 8.4 in 1 N sulfuric acid and 8.8 in 1 N nitric acid. The first wave in sulfuric acid is well enough developed to be of analytical value; the height of this wave is directly proportional to the concentration of +3 arsenic.

Arsenite ion is not reduced at the dropping electrode from its solutions in sodium hydroxide or from alkaline tartrate media. Reduction does occur from acidic tartrate media (pH = 4.5) but the wave merges with the final current rise (curve *b* in Fig. XXXII-4).

Bambach[16] developed a procedure for the determination of arsenic in biological material. The arsenic in the prepared sample is evolved as arsine which is collected in mercuric chloride solution. The resulting mercuric arsenide is decomposed by heating to arsenious oxide, the mercury is removed by reduction with hydroxylamine, and the resulting solution in

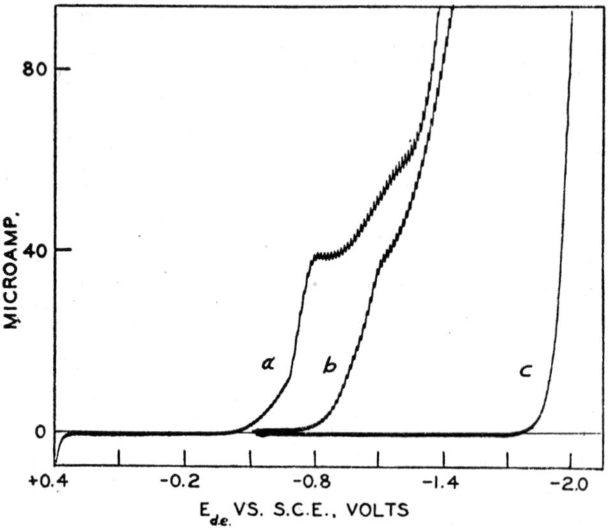

Fig. XXXII-4. Polarograms of arsenic in: (a) 1 *N* sulfuric acid; (b) 0.4 *M* sodium tartrate plus 0.1 *M* sodium hydrogen tartrate (pH = 4.5); (c) 1 *N* sodium hydroxide. Concentration of +3 arsenic was 2.00 millimolar in each case, and 0.01 per cent gel was present as a maximum suppressor.[17]

hydrochloric acid is polarographed. (Dilute sulfuric acid should be a better supporting electrolyte.) The method is applicable to the determination of 1 to 1000 microgram quantities of arsenic in various types of biological material using sample weights up to about 10 g. (or 25 ml. of liquids), and Bambach concluded that it has several advantages over more classical procedures.

Kryukova[20] described procedures for the polarographic determination of arsenic in commercial phosphoric and sulfuric acids.

Kolthoff and Probst[21] discovered that +3 arsenic in strongly alkaline

[20] T. A. Kryukova, *Zavodskaya Lab.*, **6,** 1385 (1937); **7,** 273 (1938); **9,** 950 (1940).
[21] I. M. Kolthoff and R. L. Probst, *Anal. Chem.*, **21,** 753 (1949).

medium is oxidized to the +5 state at the dropping electrode. In 0.5 *M* potassium hydroxide containing 0.025 per cent gelatin the anodic half-wave potential is −0.26 v. *vs.* S.C.E. and the diffusion current is directly proportional to the concentration of arsenite ion, the value of $i_d/Cm^{2/3}t^{1/6}$ being 3.82 at 25°. In view of the difficulty of obtaining satisfactory reduction waves with arsenic compounds this anodic wave should be of considerable utility in practical analysis.

4. ANTIMONY

When present in the +3 state antimony produces very well-developed polarograms from hydrochloric acid, nitric acid, sulfuric acid, and sodium hydroxide solutions in the presence of gelatin as a maximum suppressor.[22-24] Typical polarograms[24] are shown in Figs. XXXII-5 and XXXII-6. The peculiar discontinuity near the top of the wave in 1 *N* hydrochloric acid disappears when the concentration of gelatin is increased from 0.01 to 0.1 per cent.[23, 24] In all these supporting electrolytes reduction proceeds to the metal. The second incompletely developed wave at −1.0 v. in sulfuric acid medium probably corresponds to an incomplete 6-electron reduction to stibine.[24]

The polarograms of +3 antimony in tartrate supporting electrolytes of various pH values shown in Fig. XXXII-7 demonstrate that only in a strongly alkaline tartrate medium (curve *d*) is the diffusion current well enough developed to be used for exact quantitative purposes. The wave (curve *a* in Fig. XXXII-7) in "neutral" tartrate solution (pH = 9) in the absence of a maximum suppressor comprises three distinct parts. Because reduction to oxidation states between +3 and the metal is highly improbable, this triplet wave apparently corresponds to the reduction of three different molecular species (tartrate complexes) of +3 antimony.

Half-wave potentials for +3 antimony (and arsenic and bismuth) in various supporting electrolytes reported by Lingane[24] are assembled in Table XXXII-1.

The slope of the wave in 1 *N* hydrochloric acid corresponds closely to a reversible 3-electron reduction (probably of $SbCl_4^-$) but the somewhat smaller wave slopes in nitric and sulfuric acid media indicate that the reduction of the SbO^+ (the probable predominant species) does not proceed reversibly.

A hydrochloric acid supporting electrolyte is particularly useful for de-

---

22 K. Kacirkova, *Collection Czechoslov. Chem. Communs.*, **1,** 477 (1929).
23 J. E. Page and F. A. Robinson, *J. Soc. Chem. Ind.*, **61,** 93 (1942).
24 J. J. Lingane, *Ind. Eng. Chem., Anal. Ed.*, **15,** 585 (1943).

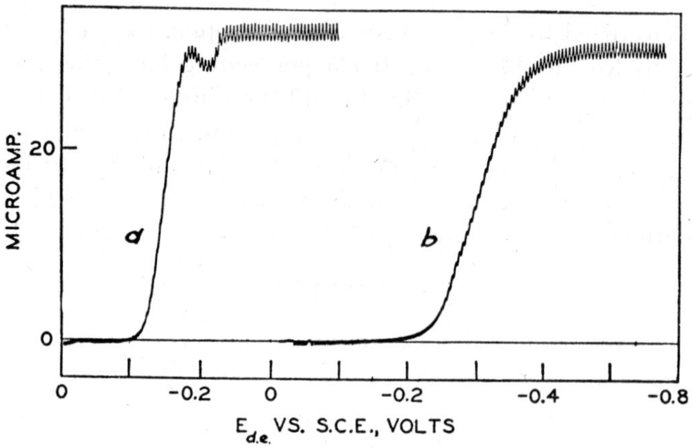

Fig. XXXII-5. Polarograms of 2.50 millimolar +3 antimony in 1 *N* hydrochloric acid (a) and 1 *N* nitric acid (b), in the presence of 0.01 per cent gelatin.[24]

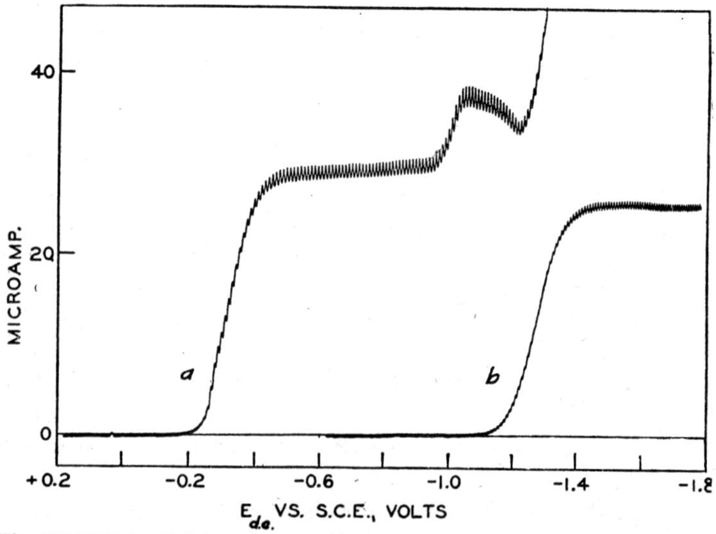

Fig. XXXII-6. Polarograms of 2.50 millimolar +3 antimony in 1 *N* sulfuric acid (a) and 1 *N* sodium hydroxide (b), in the presence of 0.01 per cent gelatin.[24]

termining antimony in the presence of lead, *e.g.*, in antimonial alloys such as battery plate. Page and Robinson[23] demonstrated the utility of a hydrochloric acid supporting electrolyte for determining antimony in various organic antimonial medicinals.

When present in the +5 state antimony is not easily reduced at the

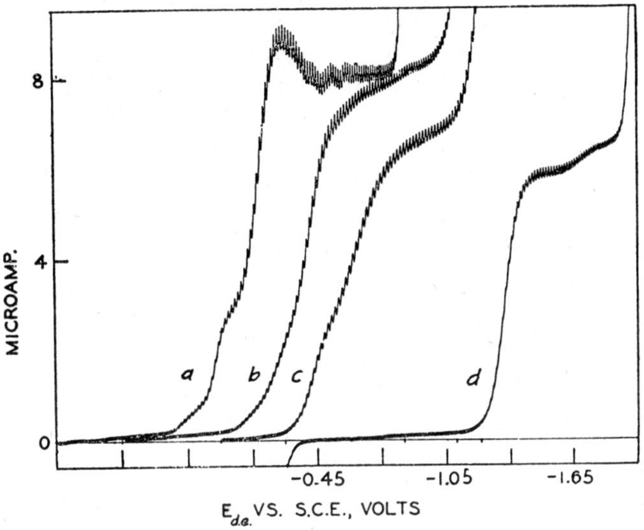

Fig. XXXII-7. Antimony in tartrate media: (a) 0.91 millimolar potassium antimonyl tartrate in 0.5 *M* sodium tartrate; (b) like (a) except 0.01 per cent gelatin present; (c) 0.86 millimolar potassium antimonyl tartrate in 0.4 *M* sodium tartrate plus 0.1 *M* sodium hydrogen tartrate plus 0.01 per cent gelatin; (d) 0.71 millimolar potassium antimonyl tartrate in 0.5 *M* sodium tartrate plus 0.1 *N* sodium hydroxide plus 0.01 per cent gelatin. Each curve starts at −0.3 volt.

TABLE XXXII-1

HALF-WAVE POTENTIALS OF THE +3 OXIDATION STATES OF ARSENIC, ANTIMONY, AND BISMUTH

Half-wave potentials in volts *vs.* S.C.E. at 25°. 0.01 per cent gelatin present in all cases. N.R. signifies "not reduced."

Supporting electrolyte	$E_{1/2}$		
	As	Sb	Bi
1 *N* HNO_3	−0.7	−0.30	−0.01
1 *N* HCl	−0.4 −0.9	−0.15	−0.09
1 *N* H_2SO_4	−0.7 −1.0	−0.32	−0.04
1 *N* NaOH	N.R.	−1.26	Insol.
0.5 *M* tartrate (pH = 4.5)	−1.0	−0.8	−0.29
0.5 *M* tartrate (pH = 9)	N.R.	−1.0 1.2	−0.7
0.5 *M* tartrate + 0.1 *M* NaOH	N.R.	−1.32	−1.0

d.e. Solutions of +5 antimony in perchloric acid media show no indication of a reduction wave before reduction of hydrogen ion. However, Lingane and Nishida[25] demonstrated that the chloro complex ion present in solutions that contain both chloride ion and hydrogen ion at high concentrations does produce a well-developed doublet wave as shown in Fig. XXXII-8. This has also been observed by Kraus and Novak.[26] The first wave in 4 to 6 *N* hydrochloric acid results from the 2-electron reduction of the chloro complex of +5 antimony (probably $SbCl_6^-$) to a chloro

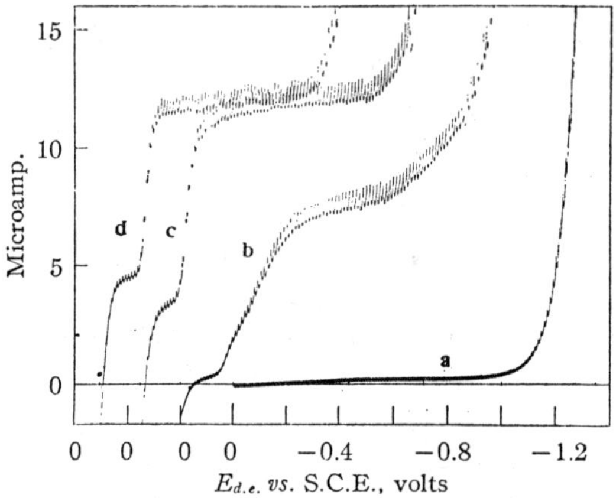

Fig. XXXII-8. Polarograms of 0.975 millimolar +5 antimony in (a) 0.2, (b) 2, (c) 4, and (d) 6 *N* hydrochloric acid, with 0.005 per cent gelatin as a maximum suppressor.[25] The starting voltage of each curve *vs.* the saturated calomel electrode is indicated by 0, and each voltage increment corresponds to 0.2 v. In the absence of gelatin the second wave in 6 *N* hydrochloric acid displayed a sharp maximum.[25]

complex of +3 antimony and the second wave corresponds to complete reduction to the metal. In 6 *N* hydrochloric acid the half-wave potential of the second wave is −0.257 v. *vs.* S.C.E. The potential at which the first wave starts corresponds to the anodic dissolution potential of mercury:

$$Hg + 4Cl^- \rightleftharpoons HgCl_4^{--} + 2e$$

and is much more negative (more reducing) than the true potential of the $Sb^{+5} \rightarrow Sb^{+3}$ couple.

Kraus and Novak[26] employed the wave of +5 antimony in hydrochloric acid for the determination of antimony in hard lead. A 0.2-g. sample is

[25] J. J. Lingane and F. Nishida, *J. Am. Chem. Soc.*, **69**, 530 (1947).
[26] R. Kraus and J. V. A. Novak, *Die Chemie*, **56**, 302 (1943).

dissolved in 15 ml. of hot 12 *M* hydrochloric acid with the aid of a few drops of bromine. After boiling off the excess bromine the solution is diluted to 50 ml. in 2 *N* hydrochloric acid containing 0.01 per cent gelatin and polarographed. In this relatively small concentration of hydrochloric acid the first wave is not fully developed and the analysis is based on the second wave at −0.2 v. Arsenic does not interfere in this procedure but the possibility of interference by copper and iron must be borne in mind, since both of these elements yield diffusion currents in the region of the antimony wave.

Kolthoff and Probst[21] observed that +5 antimony yields a single wave and well-defined diffusion current (corresponding to a 5-electron reduction to the metal) in a supporting electrolyte containing 1 *M* hydrochloric acid and 4 *M* (nearly saturated) potassium bromide. Only one wave is observed because the true reduction potential of the bromo complex of +5 antimony is more positive than the anodic dissolution potential of mercury in this medium, and the apparent half-wave potential of −0.35 v. *vs.* S.C.E. actually corresponds to this mercury potential. This supporting electrolyte is especially useful for determining small amounts of antimony in the presence of large amounts of arsenic.

Kolthoff and Probst also discovered that +3 antimony (antimonite ion) produces an anodic wave (in addition to the familiar cathodic wave) in a strongly alkaline supporting electrolyte. The anodic half-wave potential is −0.34 v. *vs.* S.C.E. in 0.1 *M*, and −0.45 v. in 1 *M*, potassium hydroxide, and the anodic reaction is the oxidation to +5 antimony. The limiting anodic current is not well developed because a pronounced minimum occurs just prior to the anodic dissolution current of the mercury. This effect, which is also observed in the anodic waves of hydroquinone,[27] rhenide ion,[28] and stannous tin,[29] appears to result from an inhibition of the oxidation of other substances at the dropping electrode when the potential is positive enough so that the mercury itself is oxidized.

The polarographic determination of antimony was utilized by Goodwin and Page[30] in a study of the physiological behavior of antimonial medicinals.

Volpi[31] described procedures for the determination of antimony (and lead, zinc, and manganese) in magnesium-base alloys.

---

27 O. H. Müller and J. P. Baumberger, *Trans. Am. Electrochem. Soc.*, **71**, 169, 181 (1937).

28 J. J. Lingane, *J. Am. Chem. Soc.*, **64**, 2182 (1942).

29 J. J. Lingane, *J. Am. Chem. Soc.*, **65**, 868 (1943).

30 L. G. Goodwin and J. E. Page, *Biochem. J.*, **37**, 198, 482 (1943).

31 A. Volpi, *Ricerca sci. e ricostruz.*, **17**, 1105 (1947).

Hourigan[32] has discussed the determination of antimony in lead cable sheathing.

5. BISMUTH

The polarographic characteristics of bismuth are very favorable for practical analytical purposes. Very well-developed waves for the reduction of +3 bismuth to the metal are obtained in hydrochloric, nitric, or sulfuric acid media containing gelatin as a maximum suppressor.[33-36] The half-wave potentials in various supporting electrolytes are listed in Table XXXII-1.

From hydrochloric and nitric acid solutions the observed[36] slope of the bismuth wave is close to that expected for a reversible 3-electron reduction. In sulfuric acid media the reduction proceeds somewhat less reversibly but the wave form is nonetheless very satisfactory. In 1 *N* sulfuric or nitric acids the half-wave potential of bismuth is about 0.28 v. more positive than that of +3 antimony, but in 1 *N* hydrochloric acid the difference is much less (0.06 v.). Therefore, 1 *M* sulfuric or nitric acid is a much better supporting electrolyte than hydrochloric acid for the determination of bismuth in the presence of antimony.

Because of its relatively positive half-wave potential in strongly acid media the determination of bismuth in the presence of elements with which it is commonly associated (such as arsenic, antimony, lead, tin, etc.) is usually easily accomplished. Ferric iron interferes but this can easily be circumvented by reducing the iron to the ferrous state. Copper also interferes in strongly acid medium but this can be eliminated when the concentration of copper does not greatly exceed that of bismuth by employing a tartrate supporting electrolyte of pH 4–5.

The use of a tartrate supporting electrolyte for the determination of bismuth was first recommended by Suchy.[37] The pronounced influence of pH on the polarographic behavior of bismuth in tartrate media has been studied by Lingane[38] and Swinehart, Garrett, and MacNevin.[39] A well-

---

[32] H. F. Hourigan, *Analyst*, **71,** 524 (1946).

[33] V. Bayerle, *Rec. trav. chim.*, **44,** 514 (1925).

[34] K. Kacirkova, *Collection Czechoslov. Chem. Communs.*, **1,** 477 (1929).

[35] J. E. Page and F. A. Robinson, *J. Soc. Chem. Ind.*, **61,** 93 (1942).

[36] J. J. Lingane, *Ind. Eng. Chem., Anal. Ed.*, **15,** 585 (1943).

[37] K. Suchy, *Collection Czechoslov. Chem. Communs.*, **3,** 354 (1931); *Chem. News*, **143,** 213 (1929).

[38] J. J. Lingane, *Ind. Eng. Chem., Anal. Ed.*, **15,** 585 (1943); **16,** 147 (1944).

[39] D. F. Swinehart, A. B. Garrett, and W. M. MacNevin, *Ind. Eng. Chem., Anal. Ed.*, **15,** 729 (1943).

developed wave is obtained only when the pH of the tartrate solution is smaller than about 7. A similar pH effect doubtless would be observed in citrate supporting electrolytes.

Lingane[38] found that the optimum pH range for the simultaneous determination of copper, bismuth, lead, and cadmium in a 0.5 *M* tartrate supporting electrolyte is 4–5; at pH values outside these limits the waves more or less coalesce.

Swinehart, Garrett, and MacNevin[39] found that the diffusion current of bismuth in a supporting electrolyte composed of 0.1 *M* tartrate, 0.05 *M* acetic acid, and 0.05 *M* sodium acetate of pH = 4.75 was strictly proportional to the concentration of bismuth up to 0.5 millimolar. At higher concentrations the ratio i_d/C decreased slightly.

Cozzi[40] developed polarographic procedures for the determination of traces of bismuth (and Cu, Pb, Cd, and Tl) in highly refined zinc. Pletenev, Aref'eva, Tal, and Dubovitskaya[41] described the determination of bismuth in lead and tin alloys.

The polarography of bismuth (and a variety of other metals) in a fluoride supporting electrolyte has been studied by West, Dean, and Breda.[42]

[40] D. Cozzi, *Mikrochem. Acta,* **31,** 37 (1943).

[41] S. A. Pletenev, T. V. Aref'eva, E. M. Tal, and E. I. Dubovitskaya, *Zavodskaya Lab.,* **12,** 38 (1946); *Chem. Abstracts,* **40,** 7046[8].

[42] P. W. West, J. Dean, and E. J. Breda, *Collection Czechoslov. Chem. Communs.,* **13,** 1 (1948).

CHAPTER XXXIII

Oxygen, Sulfur, Selenium, and Tellurium

1. OXYGEN

In 1924 Heyrovsky[1] first reported that oxygen dissolved in electrolyte solutions is reduced at the dropping electrode to produce two waves of equal height. The good development of the diffusion currents, and their linear dependence on oxygen concentration, makes possible the accurate polarographic determination of oxygen. This is of considerable practical importance because there are few other satisfactory methods for the determination of dissolved oxygen. The polarographic method excels most other methods both in convenience and accuracy.

Typical polarograms of *ca.* 0.25 millimolar oxygen, obtained with an air-saturated 0.05 *M* potassium chloride solutions, are shown in Fig. XXXIII-1. The first wave results from reduction to hydrogen peroxide:

$$O_2 + 2H^+ + 2e \rightarrow H_2O_2 \qquad \text{(acid medium)} \qquad (1)$$

$$O_2 + 2H_2O + 2e \rightarrow H_2O_2 + 2OH^- \qquad \text{(neutral or alkaline medium)} \qquad (2)$$

and the second corresponds to the reduction of the hydrogen peroxide either to water or hydroxyl ion depending on pH:

$$H_2O_2 + 2H^+ + 2e \rightarrow 2H_2O \qquad \text{(acid medium)} \qquad (3)$$

$$H_2O_2 + 2e \rightarrow 2OH^- \qquad \text{(alkaline medium)} \qquad (4)$$

These reactions proceed very irreversibly at the dropping electrode, and the observed reduction potentials include large overvoltages.

The first oxygen wave displays a very pronounced maximum, but this is easily suppressed by capillary-active substances, such as gelatin, dyestuffs, tylose, alkaloids, thymol, etc.[2] (see Chap. X). The suppressive action of 0.0005 per cent methyl red is demonstrated by curve *2* in Fig. XXXIII-1.

Kolthoff and Miller[3] found that the half-wave potential of the first wave

---

[1] J. Heyrovsky, *Trans. Faraday Soc.*, **19,** 785 (1924).

[2] J. Rasch, *Collection Czechoslov. Chem. Communs.*, **1,** 560 (1929). E. Varasova, *ibid.*, **2,** 8 (1930). B. Rayman, *ibid.*, **3,** 314 (1931). E. Hamamoto, *ibid.*, **5,** 427 (1933).

[3] I. M. Kolthoff and C. S. Miller, *J. Am. Chem. Soc.*, **63,** 1013 (1941).

was practically independent of pH, and is −0.05 v. *vs.* S.C.E. in most buffer solutions of pH 1 to 10. In biphthalate buffers, however, the half-wave potential is shifted to −0.1 to −0.15 v. The second wave is very broad, and it extends over the potential range from about −0.5 v. to −1.3 v. *vs.* S.C.E. The value of $i_d/Cm^{2/3}t^{1/6}$ for the total wave is 12.3 at 25° in 0.1 *M* potassium nitrate, corresponding to a diffusion coefficient of 2.6×10^{-5} cm.^{2} sec.$^{-1}$.

Because the waves of oxygen mask the waves of most other substances its removal is generally necessary when other substances are to be de-

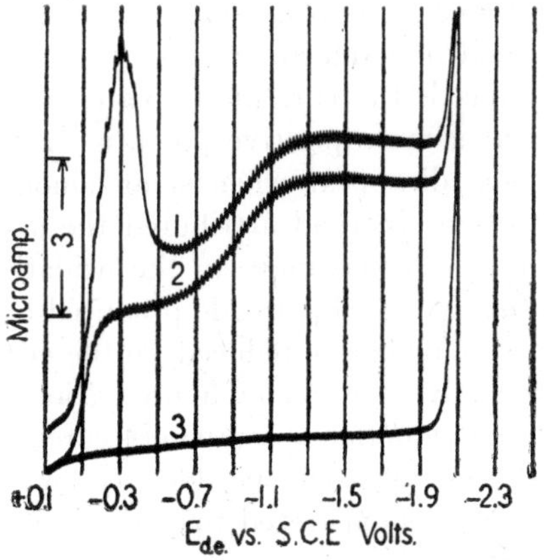

Fig. XXXIII-1. Polarograms of oxygen. (*1*) 0.05 *N* potassium chloride solution saturated with air. (*2*) After addition of a trace of methyl red. (*3*) Residual current after removal of air by nitrogen.

termined. Aside from this masking effect oxygen can influence polarograms of other substances in more subtle ways because of the hydrogen peroxide and hydroxide ion produced (or hydrogen ion consumed) by its reduction. For example, the hydroxide ion produced by the reduction of oxygen in neutral, unbuffered media can precipitate heavy metal ions in the diffusion layer and thus diminish their diffusion currents. As discussed in detail in Chapter VI, the diffusion current of hydrogen ion is diminished when oxygen is present, because of its neutralization in the diffusion layer by the hydroxide ion from the reduction of oxygen.[3] The waves of substances whose reduction involves hydrogen ion obviously can be profoundly influenced by the concomitant reduction of oxygen from unbuffered solutions.

The hydroxide ion produced by reaction of oxygen with mercury and chloride ion in an unbuffered medium:

$$O_2 + 2Hg + 2Cl^- + 2H_2O \rightleftharpoons H_2O_2 + Hg_2Cl_2 + 2OH^-$$

may also cause precipitation of hydrous metal oxides or hydroxides, and therefore neutral unbuffered metal salt solutions containing oxygen should not be allowed to stand in contact with mercury.[4]

Strnad[5] observed that small concentrations of heavy metal ions exert a a pronounced influence on the oxygen wave; the effect of lead ion is especially remarkable. When lead ion is added to air-saturated solutions of potassium nitrate, potassium sulfate, or calcium nitrate, the oxygen maximum is greatly increased and the limiting current following it is much greater than the normal first diffusion current of oxygen. At −0.45 v., where lead ion is reduced, the current decreases abruptly. The second oxygen wave begins at a more positive potential in the presence of lead ion. These phenomena disappear when small amounts of acid or base are added to the solution. Strnad attributed the abnormally large current following the enhanced maximum to the oxidation of plumbous ion to the +4 state by the hydrogen peroxide produced by the oxygen reduction. The +4 lead is immediately reduced, so the net result is the reduction of oxygen all the way to water or hydroxyl ion at a potential more positive than the normal reduction potential of hydrogen peroxide. Strnad claimed that 10^{-7} *M* lead ion could be detected by this catalytic effect.

Kolthoff and Laitinen[6] demonstrated that a *rotating* platinum electrode has an advantage over the d.e. when very small concentrations of oxygen are to be determined, because the diffusion current is much larger with the former electrode. The technique is described in Chapter XIX.

Oxygen in Gases. The oxygen content of electrochemically inert gases can be determined by equilibrating them with an electrolyte solution, *e. g.*, 0.01 *M* potassium chloride containing a trace of methyl red or other maximum suppressor.[7] It is not difficult to devise continuous recording techniques.[8]

The solubility of oxygen is a linear function of its partial pressure, and at 1 atmosphere and 25° it is 1.33 millimolar in pure water. In electrolyte solutions the solubility is smaller to an extent dependent on the ionic strength.

---

[4] E. R. Smith, J. K. Taylor, and R. E. Smith, *J. Research Natl. Bur. Standards*, **37,** 151 (1946).

[5] F. Strnad, *Collection Czechoslov. Chem. Communs.*, **11,** 391 (1939).

[6] I. M. Kolthoff and H. A. Laitinen, *Science*, **92,** 152 (1940).

[7] V. Vitek, *Collection Czechoslov. Chem. Communs.*, **7,** 537 (1935).

[8] W. S. Wise, *Chemistry and Industry*, No. 3, 37 (1948).

Calibration is most easily accomplished by saturating the same supporting electrolyte with air at measured atmospheric pressure, and the partial pressure of oxygen in the test gas is then computed from the ratio of the two diffusion currents (corrected for residual current) and the known partial pressure of oxygen in water-saturated air. By this procedure partial pressures of oxygen from a few millimeters up to atmospheric pressure can be determined with an accuracy of ±2 per cent or better.

Vitek[7] pointed out that the diffusion current of oxygen in anhydrous ethanol or methanol is several times larger than in water, chiefly because of the greater solubility of oxygen in these alcohols. The use of these solvents is advantageous when very small concentrations of oxygen in a gas are to be determined.

Polarographic Measurement of Respiration Rates. The polarographic determination of oxygen was employed by Petering and Daniels[9] to study the photosynthetic and respiration rates of green algae (*Chlorella pyrenoidosa*), and the respiration rates of yeast, blood cells, and animal tissue. These investigators concluded that the polarographic method is superior to the classical manometric and Winkler methods for investigations of this kind, especially when the rate of oxygen uptake is very small.

Cellular respiration in plant roots was studied polarographically by Wanner,[10] by measuring the rate of oxygen diffusion through the liquid layer in contact with the roots.

Blinks and Skow[11] used both the dropping electrode and the platinum micro wire electrode for the determination of oxygen in the course of photosynthesis.

Baumberger[12] studied the relation between the oxygen consumption rate of yeast cell suspensions and the redox potential of the medium.

Oxygen in Physiological Media. Beecher, Follansbee, Murphy, and Craig[13] developed a technique for polarographic measurement of the oxygen content of 1-ml. samples of lymph and other biological fluids. The chief problem was transfer of the sample from the animal to the cell without bringing it in contact with air or another gas, which would alter the oxygen content. This was solved by use of the cell shown in Fig. XVII-9, on p. 366. The sample, collected under mineral oil, is transferred to the cell

[9] H. G. Petering and F. J. Daniels, *J. Am. Chem. Soc.*, **60,** 2796 (1938).

[10] H. Wanner, *Helv. Chim. Acta*, **28,** 23 (1945).

[11] L. R. Blinks and R. K. Skow, *Proc. Natl. Acad. Sci. U. S.*, **24,** 413, 420 (1938). L. R. Blinks, M. L. Darsie, and R. K. Skow, *J. Gen. Physiol.*, **22,** 255 (1938).

[12] J. P. Baumberger, *Cold Spring Harbor Symposia Quant. Biol.*, **7,** 195 (1939).

[13] H. K. Beecher, R. Follansbee, A. J. Murphy, and F. N. Craig, *J. Biol. Chem.*, **146,** 197 (1942).

by starting with the latter completely filled with mercury, and then withdrawing the mercury to draw in the sample.

Methods for the determination of oxygen in blood plasma have been described by Berggren[14] and Wiesinger.[15]

Hill[16] described the continuous recording of oxygen concentrations in physiological fluids using a platinum microelectrode.

Oxygen in Natural Waters. According to Giguere and Lauzier[17] dissolved oxygen in sea water can be determined with either the d.e. or the platinum microelectrode down to a concentration of 0.02 cc. per liter. Manning[18] developed a rugged dropping electrode assembly which can be immersed in a lake or stream and connected with a recorder on shore to provide a continuous record of changes in the concentration of dissolved oxygen.

A technique for the continuous measurement of oxygen in water flowing through a polarographic cell has been developed by Spoor.[19] Zan'ko, Manusova, and Nikitin[20] compared polarographic and colorimetric (oxidation of $Cu(NH_3)_4^{++}$) methods for determining oxygen in water.

Oxygen in Sewage. The polarographic determination of oxygen and biological oxygen demand in sewage and sewage-polluted streams is of special practical importance. This application has been pioneered by Ingols,[21] who developed continuous recording apparatus suitable for use in sewage disposal plants. The advantages of the polarographic method for determining oxygen in sewage have also been exploited by Moore, Morris, and Okum,[22] who utilized simple manual instrumentation.

Miscellaneous. The polarographic determination of dissolved oxygen in dairy products has been described by Hartman and Garrett,[23] in soils by Karsten,[24] and in orange juice by Lewis and McKenzie.[25]

Ultrasensitive Laitinen-Higuchi-Czuha Technique. An isolated dropping mercury electrode which is completely free of reducible or oxidizable

[14] S. M. Berggren, *Acta Physiol. Scandinavica*, **4**, Suppl. XI, 92 (1942).
[15] K. Wiesinger, *Helv. Physiol. Pharmacol. Acta*, **6**, 13 (1948).
[16] D. K. Hill, *J. Physiol.*, **105**, 24 (1946).
[17] P. A. Giguere and L. Lauzier, *Can. J. Research*, **B23**, 76 (1945).
[18] W. M. Manning, *Ecology*, **21**, 509 (1940).
[19] W. A. Spoor, *Science*, **108**, 421 (1948).
[20] A. N. Zan'ko, F. A. Manusova, and A. D. Nikitin, *Zavodskaya Lab.*, **8**, 937 (1939).
[21] R. S. Ingols, *Sewage Works J.*, **13**, 1097 (1941); *Ind. Eng. Chem., Anal. Ed.*, **14**, 256 (1942).
[22] E. W. Moore, J. C. Morris, and D. A. Okum, *Sewage Works J.*, **20**, 1041 (1948).
[23] G. H. Hartman and O. F. Garrett, *Ind. Eng. Chem., Anal. Ed.*, **14**, 641 (1942).
[24] S. Karsten, *Am. J. Botany*, **25**, 14 (1938).
[25] V. H. Lewis and H. A. McKenzie, *Anal. Chem.*, **19**, 643 (1947).

substances acquires a potential corresponding to the maximum on the electrocapillary curve of mercury (−0.6 v. *vs.* the saturated calomel electrode in chloride solutions). If the potential is forced by an applied e.m.f. to assume some other value, a continuous current is observed to flow due to the establishment of the electrical double layer on the continuously forming surface. This is the familiar "charging current" or "condenser current" which is zero at the potential of the electrocapillary maximum, which becomes increasing anodic when the potential is made more positive than the electrocapillary zero, and increasingly cathodic when the potential is shifted in the negative direction. The slope, di/dE, of the charging current curve is so small that an exceedingly slight cathodic current resulting from a mere trace of a substance which is reduced at a potential more positive than the electrocapillary zero produces a relatively large positive shift in the potential at which the observed current is zero. Conversely, a trace of a substance which is oxidizable at a potential more negative than the electrocapillary maximum will produce a large shift of the zero current potential in the negative direction.

Laitinen, Higuchi, and Czuha[26] utilized this principle to develop a sensitive method for the determination of traces of oxygen in gases or in solution. These authors showed that the shift, $E - E_{\text{max.}}$, of the zero current potential increases with time during the life of the mercury drop according to (see also Chap. IX):

$$E - E_{\text{max.}} = 17.2Ct^{1/2}$$

where C is the concentration of oxygen in millimoles per liter, t is expressed in seconds, and the constant includes the capacity of the double layer, the diffusion coefficient of oxygen, and certain geometrical constants. A high-capacity condenser was inserted in the measuring circuit, so that the observed value of $E - E_{\text{max.}}$ corresponded to a maximal value during the drop life. A special flow-type cell (see Chap. IX) was designed to provide rapid equilibration of the supporting electrolyte (0.1 or 1 M potassium chloride) with the gas being tested. When a small amount of a maximum suppressor, such as methyl red, was present the observed shifts of the zero current potential were linear functions of oxygen concentration up to about 1 per cent by volume in the gas mixture, and amounts of oxygen as small as 0.01 per cent could be detected.

Hydrogen Peroxide. Although the wave of hydrogen peroxide is very extended (see Fig. XXXIII-1) the diffusion current is clearly developed and can be used for the determination of hydrogen peroxide. Kolthoff

[26] H. A. Laitinen, T. Higuchi, and M. Czuha, *J. Am. Chem. Soc.*, **70**, 561 (1948).

and Miller[3] found that the half-wave potential was constant at −0.94 v. *vs.* S.C.E. in Clark and Lubs buffer solutions from pH 2 to 10, but it was influenced slightly by the nature of the anions in the solution.

Pellequer[27] demonstrated that the polarographic method is capable of detecting as little as 1 microgram of hydrogen peroxide in a 2-ml. sample. When the concentration of hydrogen peroxide is small its disproportionation according to:

$$2H_2O_2 \rightleftharpoons O_2 + 2H_2O$$

does not significantly change the total current, because the oxygen formed is reduced to produce a current virtually equivalent to the hydrogen peroxide decomposed. On the other hand, Giguere and Jaillet[28] observed that the linear relation between i_d and C failed above about 45 millimolar hydrogen peroxide, apparently because at this very high concentration the oxygen formed by the disproportionation partially escapes from the solution.

Shaw[29] describes the interesting observation that a 0.000225 *M* hydrogen peroxide solution in 0.1 *M* sodium phenobarbital (pH 8.2 to 8.7) gives a sharply defined wave between about −1.15 and −1.25 v. (S.C.E.). However, with increasing peroxide concentration the wave flattens out and ultimately assumes the same shape as in potassium nitrate as supporting electrolyte. No explanation is offered for the peculiar behavior. Apparently, the phenobarbital effect is caused by its surface activity at the mercury–water interface. Semerano[30] reports that a small amount of the strongly capillary-active detergent Aerosol OT makes the hydrogen peroxide wave much better defined and gives the appearance of a reversible wave. Dr. E. P. Parry, at the University of Minnesota, found the effect of Aerosol OT on the peroxide wave truly remarkable, especially at higher peroxide concentrations. At these concentrations, a small hydrogen peroxide current is noticeable even at 0 v. (S.C.E.), the current increasing slowly with increasing negative potential until the sharp rise occurs at around −0.9 v. In the presence of Aerosol OT the long drawn-out wave disappears.

2. SULFUR

Sulfate. Sulfate ion is not reducible at the dropping electrode, but it can be determined indirectly by measuring the decrease in the diffusion

27 H. Pellequer, *Compt. rend.*, **222,** 1220 (1946); **225,** 116 (1947).
28 P. A. Giguere and J. B. Jaillet, *Can. J. Research*, **B26,** 767 (1948).
29 E. H. Shaw, Jr., *Proc. S. Dakota Acad. Sci.*, **27,** 92 (1948).
30 G. Semerano, *Rend. Accad. Nazl. Lincei*, **6,** 721 (1949).

current of barium ion caused by precipitation of barium sulfate,[31, 32] or by amperometric titration with barium ion or lead ion as described in Chapter XLVII.

Sulfur Dioxide and Sulfite Ion. In neutral or alkaline solutions sulfite ion is not reduced at the dropping electrode, but a well-defined wave is obtained from the reduction of sulfurous acid (sulfur dioxide) from acid solutions. In 1 *N* hydrochloric acid the wave starts at about −0.2 v. *vs.* the S.C.E. Gosman[33] assumed that sulfur dioxide was reduced to hyposulfite ion according to:

$$2SO_2 + 2e \rightarrow S_2O_4^{--} \quad (5)$$

However, Kolthoff and Miller[34] could not confirm this. According to these authors two electrons are required for the reduction of one molecule of sulfurous acid or sulfur dioxide in acid solution, and they concluded that the reduction product is sulfoxylic acid:

$$SO_2 + 2H^+ + 2e \rightarrow H_2SO_2 \quad (6)$$

The height of the wave was found to decrease markedly with increasing pH between 1 and 6 in buffered solutions, as shown in Fig. XXXIII-2. At a pH of 6 two small waves are observed with half-wave potentials of −0.67 v. and −1.23 v. *vs.* the S.C.E. At pH values greater than 6 no reduction occurs.

Kolthoff and Miller postulated that the first wave at −0.67 v. at a pH of 6 is due to the reduction of bisulfite ion to HSO_2, which then polymerizes to hyposulfurous acid, which in turn ionizes to form hyposulfite ion:

$$HSO_3^- + 2H^+ + e \rightarrow HSO_2 + H_2O \quad (7)$$

$$2HSO_2 \rightarrow H_2S_2O_4 \rightleftharpoons 2H^+ + S_2O_4^{--} \quad (\text{at pH} = 6) \quad (8)$$

They concluded that the second wave was due to the further reduction of the hyposulfite ion to thiosulfate ion:

$$S_2O_4^{--} + 2H^+ + 2e \rightarrow S_2O_3^{--} + H_2O \quad (9)$$

The fact that only one wave is obtained in strongly acid solution (pH less than 4) was attributed to the rapid disproportionation of the hyposulfurous acid first formed:

$$2H_2S_2O_4 + H_2O \rightarrow H_2S_2O_5 + 2H_2SO_2 \quad (10)$$

$$H_2S_2O_5 + H_2O \rightarrow 2H_2SO_3 \quad (11)$$

---

[31] J. Heyrovsky and S. Berezicky, *Collection Czechoslov. Chem. Communs.*, **1,** 19 (1929).

[32] Ya. P. Hokhstein, *Zavodskaya Lab.*, **5,** 1444 (1936).

[33] B. Gosman, *Collection Czechoslov. Chem. Communs.*, **2,** 185 (1930).

[34] I. M. Kolthoff and C. S. Miller, *J. Am. Chem. Soc.*, **63,** 2818 (1944).

For analytical purposes the best supporting electrolyte for the determination of sulfurous acid, or sulfite, is a 0.1 to 1 N solution of a strong acid. In such solutions the apparent diffusion coefficient of sulfur dioxide is 2.0×10^{-5} cm.^{2} sec.$^{-1}$ at 25° C. In 0.1 N nitric acid Kolthoff and Miller found that the half-wave potential was -0.38 v. *vs.* the S.C.E., and it was independent of the concentration of sulfurous acid. In buffered solutions

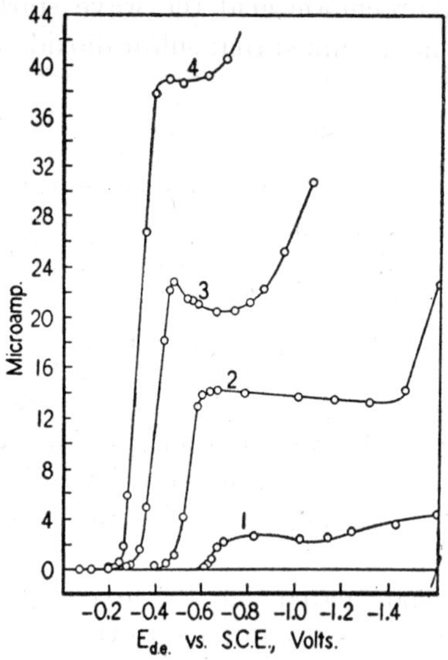

Fig. XXXIII-2. Current–voltage curves of sulfurous acid. (*1*) 1.93×10^{-3} M sulfite in a phosphate buffer of pH = 6.0. (*2*) 1.98×10^{-3} M sulfite in an acetate buffer of pH = 3.6. (*3*) 1.54×10^{-3} M sulfite in 0.1 N nitric acid. (*4*) 2.75×10^{-3} M sulfite in 1 N nitric acid.

of pH less than 4 the reduction of the sulfurous acid is reversible, and the equation of the wave agrees well with the theoretical equation:

$$E = E_{1/2} - \frac{RT}{2F_y} \ln \frac{i}{i_d - i} \qquad (\text{pH} = \text{constant}) \qquad (12)$$

Kolthoff and Miller found that a plot of E against log $i/(i_d - i)$ produced a straight line with a slope of 0.030 v., in good agreement with the theoretical value 0.0296 v. for $n = 2$ (reduction to sulfoxylic acid).

Prater, Johnson, and Pool[35] applied the sulfurous acid wave to the determination of sulfur dioxide in dehydrated foodstuffs.

---

[35] A. N. Prater, C. H. Johnson, and M. F. Pool, *Ind. Eng. Chem., Anal. Ed.*, **16,** 153 (1944).

Sulfite ion depolarizes the d.e. to yield an anodic wave, corresponding to the reaction:

$$2SO_3^{--} + Hg \rightleftharpoons Hg(SO_3)_2^{--} + 2e \tag{13}$$

This reaction proceeds reversibly, and Kolthoff and Miller[36] showed that the equation of the wave is:

$$E = E_{1/2} + \frac{0.0591}{2} \log \frac{i}{(i_d - i)^2} \tag{14}$$

at pH values above about 6. Because of the asymmetrical nature of the reaction, reflected by the squared $(i_d - i)$ term in eq. 14, the half-wave potential depends on the sulfite ion concentration; it shifts 0.0296 v. in the negative direction for each ten-fold increase in sulfite ion concentration. With 1 millimolar sulfite ion $E_{1/2}$ is -0.01 v. *vs.* S.C.E.

Hyposulfite Ion. Furness[37] found that hyposulfite ion in mildly alkaline medium yields an anodic wave corresponding to oxidation to sulfite ion:

$$S_2O_4^{--} + 4OH^- \rightleftharpoons 2SO_3^{--} + 2H_2O + 2e \tag{14a}$$

In a supporting electrolyte composed of 0.5 *M* diammonium hydrogen phosphate, 1 *M* ammonia, and 0.01 per cent gelatin the anodic half-wave potential is -0.43 v. *vs.* S.C.E., and $i_d/Cm^{2/3}t^{1/6}$ is 4.09 at 25°. A small well-developed cathodic wave was also observed at -1.03 v. *vs.* S.C.E., whose height was only about 3 per cent that of the anodic wave. Furness ascribed this cathodic wave to a different hyposulfite species in equilibrium with $S_2O_4^{--}$.

According to Furness the polarographic determination of hyposulfite in commercial hyposulfites is more specific than commonly used chemical procedures. Thiosulfate, sulfite, and aldehyde-sulfoxalates do not interfere with the anodic hyposulfite wave. Sulfide does interfere because its anodic wave (*vide infra*) occurs at a more negative potential than the hyposulfite wave, but this can be circumvented by adding zinc ion to the solution to precipitate the sulfide.

Thiosulfate Ion. Thiosulfate ion is not reduced at the d.e., but because it forms a fairly stable complex with mercuric ion it produces an anodic wave corresponding to the reaction:

$$2S_2O_3^{--} + Hg \rightleftharpoons Hg(S_2O_3)_2^{--} + 2e \tag{15}$$

The equation of this anodic wave has the same form as that of sulfite ion (eq. 14 above), and $E_{1/2}$ is a function of the thiosulfate ion concentration.

---

36 I. M. Kolthoff and C. S. Miller, *J. Am. Chem. Soc.*, **63,** 1405 (1941).

37 W. Furness, *J. Soc. Dyers and Colourists*, **66,** 270 (1950).

With 1 millimolar thiosulfate ion Kolthoff and Miller[36] found that $E_{1/2} = -0.15$ v. *vs.* S.C.E. The diffusion current is well formed.

Kolthoff and Miller noted a second, incompletely developed wave at +0.15 v. which they attributed to the irreversible oxidation of thiosulfate ion to sulfite ion.

Furness[37] showed that small amounts (down to 0.1 per cent) of thiosulfate in commercial hyposulfite preparations can be determined from its anodic wave. Because hyposulfite ion produces an anodic wave which masks the anodic thiosulfate wave the sample solution is first treated with excess formaldehyde to convert the hyposulfite to nonreactive formaldehyde-sulfoxalate ion and formaldehyde-bisulfite ion:

$$S_2O_4^{--} + 2HCHO + H_2O \rightarrow HSO_3 \cdot CH_2O^- + CH_2(OH)SO_3^-$$

Large amounts of chloride interfere because the large anodic chloride wave masks the thiosulfate wave. Sulfite does not interfere because it is converted to nonreactive formaldehyde-bisulfite ion.

Elemental Sulfur. According to Proske[38] elemental sulfur produces a cathodic wave at −0.7 v. *vs.* S.C.E. in a medium composed mainly of pyridine with sodium acetate and acetic acid as supporting electrolyte. The electrode reaction is the reduction of elemental sulfur to hydrogen sulfide:

$$S + 2H^+ + 2e \rightleftharpoons H_2S$$

Using tylose as a maximum suppressor Proske found that the diffusion current is directly proportional to the sulfur concentration up to at least 7 millimolar, and he applied the wave to the determination of free sulfur in rubber after pyridine extraction.

Hall[39] independently studied the polarographic reduction of free sulfur in a solvent composed mainly of methanol buffered to an apparent pH of 6 with pyridine and pyridinium hydrochloride (90 ml. of methanol, 9.5 ml. of pyridine, and 0.5 ml. of concentrated hydrochloric acid). In this medium the half-wave potential is −0.50 v. *vs.* the aqueous saturated calomel electrode and the slope of the wave corresponds to a reversible 2-electron reduction to hydrogen sulfide. The diffusion coefficient calculated from the observed diffusion currents and the Ilkovic equation is 2.2×10^{-5} cm.^{2} sec.$^{-1}$ at 25°. When the apparent pH of the solution (as observed with a glass electrode pH meter against an aqueous calomel reference electrode) is greater than about 7 the wave becomes greatly extended

[38] G. Proske, *Naturwissenschaften*, **33**, 220 (1946); *Angew. Chem.*, **A59**, 121 (1947).
[39] M. E. Hall, *Anal. Chem.*, **22**, 1137 (1950).

and unsatisfactory for analytical purposes. Hall developed a simple procedure for the determination of traces of sulfur in gasoline and other petroleum fractions which is superior to ordinary chemical methods. Hall reported that butyl sulfide, propyl disulfide, and thiophene did not interfere with the sulfur wave. Mercaptans interfere by reacting with the free sulfur to produce disulfides.

Sulfide Ion. In common with other substances which form slightly soluble or complex compounds with mercury, sulfide ion in strongly alkaline medium produces a well-developed anodic wave, corresponding to the oxidation of mercury to mercuric sulfide, and the anodic diffusion current is directly proportional to the analytical sulfide ion concentration.[40, 41] From the measurements of Kolthoff and Miller the half-wave potential with 1 millimolar sodium sulfide in 0.1 *M* sodium hydroxide is −0.76 v. *vs.* S.C.E.

Because of the very small second dissociation constant of hydrogen sulfide (1×10^{-15}) the predominant sulfide species in even strongly alkaline medium is hydrogen sulfide ion; in 0.1 *M* hydroxide ion the ratio $(S^{--})/(HS^{-})$ is only 0.01. The predominant electrode reaction in strongly alkaline medium is:

$$Hg + HS^- + OH^- \rightleftharpoons HgS + H_2O + 2e \qquad (16)$$

from which it is evident that the limiting anodic current is governed by the diffusion coefficient of HS^- rather than that of S^{--}. From the standard potential of the half-reaction $Hg + S^{--} \rightleftharpoons HgS + 2e$ (−0.94 v. *vs.* S.C.E.[42]), K_2 for H_2S (1×10^{-15}), and K_w (1×10^{-14}), the standard potential of reaction (16) is computed to be −0.91 v. *vs.* S.C.E. Correspondingly, in 0.1 *M* hydroxide ion the calculated reversible potential is −0.88 v. *vs.* S.C.E. The concentration of HS^- in 0.1 *M* sodium hydroxide containing 1 millimolar sodium sulfide is virtually 1 millimolar, and the HS^- concentration at the midpoint of the anodic wave at the surface of the d.e. is one-half this or 5×10^{-4} *M*. Therefore the theoretical reversible half-wave potential for this particular solution is −0.78 v. *vs.* S.C.E. The value observed by Kolthoff and Miller (−0.76 v.) agrees very well with this theoretical value, proving that the reaction proceeds reversibly. A ten-fold increase in either the total sulfide concentration or the hydroxide

[40] J. Revenda, *Collection Czechoslov. Chem. Communs.*, **6,** 453 (1934).
[41] I. M. Kolthoff and C. S. Miller, *J. Am. Chem. Soc.*, **63,** 1405 (1941).
[42] W. M. Latimer and J. H. Hildebrand, *Reference Book of Inorganic Chemistry*, Macmillan, New York, 1940 (p. 137).

ion concentration shifts the half-wave potential 0.0296 v. in the negative direction.

Furness[37] applied the anodic sulfide wave to the determination of traces of sulfide in commercial hyposulfite preparations. The interference of the anodic hyposulfite wave with the sulfide wave is prevented by converting the former to formaldehyde-sulfoxalate and formaldehyde-bisulfite (*vide supra*), and the anodic sulfide wave is finally recorded in 0.1 *M* disodium hydrogen phosphate as supporting electrolyte.

3. SELENIUM

The polarographic behavior of selenite ion was originally investigated by Schwaer and Suchy.[43] Recently Lingane and Niedrach[44] systematically studied the polarographic chemistry of the various oxidation states of selenium over a wide range of pH and other conditions, and they utilized the coulometric analysis technique based on controlled potential electrolysis with a mercury cathode to identify the electrode reactions.

+6 Selenium. Although selenic acid and selenate ion are less stubborn than the corresponding sulfur compounds toward chemical reductants, the electrolytic reduction rates are still too small to yield reduction waves at the d.e. from either acid or alkaline media.

+4 Selenium. Typical polarograms of +4 selenium at pH values from 0 to 11 are shown in Fig. XXXIII-3. Three main waves are observable at pH 4.85 and 5.90. The first of these divides into several additional parts at lower pH values, at pH 8 to 9 only wave III remains, and at pH values above about 9.5 no reduction is observable.

Lingane and Niedrach found that the changes in the relative heights of waves I, II, and III with pH parallels the changes in the relative proportions of H_2SeO_3, $HSeO_3^-$, and SeO_3^{--}, indicating that the nature of the predominant selenous acid species is partly responsible for the variations with pH. However, Lingane and Niedrach showed that this is not a complete interpretation because the *total* diffusion current decreases with increasing pH up to about 3, remains approximately constant up to about pH = 7, and then increases at higher pH values to reattain approximately the same value as in strongly acid medium. The variation is much larger than can be accounted for by differences in the diffusion coefficients of H_2SeO_3, $HSeO_3^-$, and SeO_3^{--}. Coulometric analysis showed that at both extremes of the pH range the *total* diffusion current results from a 6-electron

[43] L. Schwaer and K. Suchy, *Collection Czechoslov. Chem. Communs.*, **7**, 25 (1935).

[44] J. J. Lingane and L. W. Niedrach, *J. Am. Chem. Soc.*, **70**, 4115 (1948); **71**, 196 1949).

reduction to H_2Se and Se^{--}, according to:

$$H_2SeO_3 + 6e + 6H^+ \rightleftharpoons H_2Se + 3H_2O \tag{17}$$

in strongly acid medium, and:

$$SeO_3^{--} + 6e + 3H_2O \rightleftharpoons Se^{--} + 6OH^- \tag{18}$$

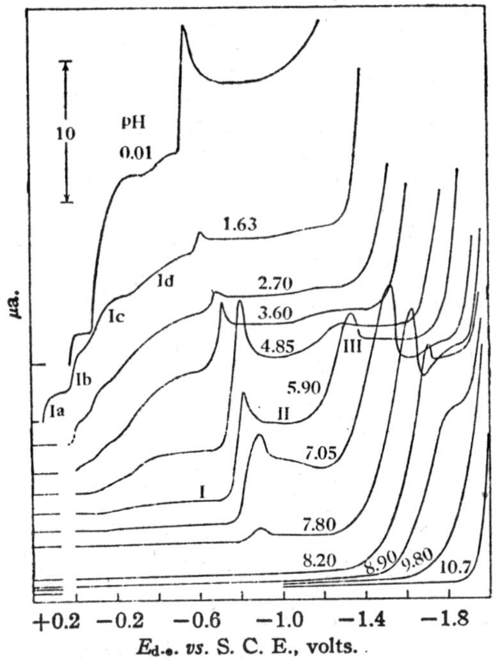

Fig. XXXIII-3. Tracings of polarograms of 1 millimolar +4 selenium at the various pH values indicated for each curve.[44] The galvanometer zero for each curve is indicated by the short lines on the left ordinate. At low pH, hydrochloric acid and citrate buffers were used, at intermediate pH, phosphate buffers, and borate and carbonate buffers were used for the alkaline region. No gelatin or other maximum suppressor was present.

at pH values above about 8. The total diffusion current at intermediate pH values is two-thirds of the extreme values, indicating a net 4-electron reduction at pH values between 3 and 7.

The half-wave potential of wave II was found to be identical at all pH values with that of the anodic wave of Se^{--} (*vide infra*) which results from the reaction:

$$Hg + H_2Se \rightleftharpoons HgSe + 2H^+ + 2e \tag{19}$$

Lingane and Niedrach concluded that wave II is the cathodic reflection of this anodic reaction. The *primary* electrode reaction at potentials more positive than wave II, *i. e.*, on wave I, is the 6-electron reduction to H_2Se according to eq. 17 above, but this is followed by the anodic reaction 19, so that the *net* result is a 4-electron reduction in which mercuric selenide is the final product:

$$H_2SeO_3 + 4e + 4H^+ + Hg \rightleftharpoons HgSe + 3H_2O \quad (20)$$

This interpretation of wave I was substantiated by controlled potential coulometric analysis.

The fact that the anodic reaction 19 is excluded at potentials more negative than wave II—and yet at intermediate pH values the *total* diffusion current corresponds only to a 4-electron reduction—is due to the interaction in the diffusion layer of the Se^{--} produced by reaction 17 with the inwardly diffusing +4 selenium to precipitate elemental selenium:

$$HSeO_3^- + 2H_2Se + H^+ \rightleftharpoons 3Se + 3H_2O \quad (21)$$

One-third of the +4 selenium is thus prevented from reaching the electrode surface, so that the *net* reaction (but not the primary electrode reaction) at pH values between about 3 and 7 is the 4-electron reduction:

$$HSeO_3^- + 4e + 5H^+ \rightleftharpoons Se + 3H_2O \quad (22)$$

At a pH of 8 or above, and at low pH values, reaction 21 (or the analogous reactions involving H_2SeO_3, SeO_3^{--}, HSe^-, and Se^{--}) proceeds so slowly that no appreciable precipitation of elemental selenium occurs during the drop life, and the final product is Se^{--}.

Contrary to the postulate of Schwaer and Suchy, no evidence was obtained that any of the waves at any pH involved reduction to +2 selenium.

At pH values between 3 and 7 the diffusion current quotients i_d/C of the separate waves, as well as that of the total wave, decrease significantly with increasing concentration of +4 selenium, probably due to the fact that the rate of reaction 21, and hence its degree of completeness during the drop life, increases with increasing selenium concentration. Therefore, supporting electrolytes of pH smaller than about 8 are unsuitable for the practical determination of +4 selenium.

For analytical purposes the best supporting electrolyte is 1 *M* ammonium chloride containing sufficient ammonia (0.1 to 1 *M*) to adjust the pH to between 8.0 and 9.5, and 0.001 to 0.003 per cent gelatin. Under these conditions the single wave is excellently developed, reduction proceeds to

Se^{--}, and the diffusion current is directly proportional to the concentration of +4 selenium. The half-wave potential is −1.44 v. *vs.* S.C.E. at pH = 8.0, and −1.54 v. at pH = 9.5. A trace of gelatin is necessary to eliminate a small maximum, but more than 0.003 per cent inhibits the full development of the diffusion current.

In ammoniacal solutions containing copper (or other metals which form insoluble selenides) the apparent height of the +4 selenium wave is decreased. This results from a decreased contribution of copper to the total diffusion current at potentials at which the selenium is reduced due to precipitation of cupric selenide in the diffusion layer (see Chap. VI). Correction for this phenomenon must be made in determinations of selenium in the presence of such metals.

−2 Selenium. Lingane and Niedrach observed that Se^{--}, like S^{--}, depolarizes the dropping electrode to yield an anodic wave:

$$Hg + H_2Se \rightleftharpoons HgSe + 2H^+ + 2e \quad \text{(acid medium)} \tag{23}$$

$$Hg + Se^{--} \rightleftharpoons HgSe + 2e \quad \text{(alkaline medium)} \tag{24}$$

A single well-developed wave is observed at all pH values from 0 to 12, and the reaction proceeds reversibly. In 1 *M* sodium hydroxide a doublet wave is obtained, whose two parts of equal height probably represent the two reactions:

$$2Se^{--} \rightleftharpoons Se_2^{--} + 2e \quad \text{(1st wave)} \tag{25}$$

$$2Hg + Se_2^{--} \rightleftharpoons 2HgSe + 2e \quad \text{(2nd wave)} \tag{26}$$

Because the predominant hydrogen selenide species is a function of pH the anodic half-wave potential shifts from −0.49 v. *vs.* S.C.E. at pH = 0 to −0.94 v. at pH = 12.0.

At pH = 8.0 well-formed waves are obtained only from solutions containing less than about 0.6 millimolar Se^{--}. With larger concentrations the diffusion current is irregular, apparently because the mercuric selenide forms an inhibiting film on the surface of the d.e.

In 0.1 *M* sodium hydroxide the half-wave potential of the anodic selenide wave is *ca.* 0.2 v. more negative than that of sulfide ion, and a simultaneous determination of the two should be possible.

4. TELLURIUM

+6 Tellurium. Neither telluric acid nor tellurate ion is reducible at the d.e.

+4 Tellurium. Tellurous acid and its ions undergo reduction at the d.e., and typical polarograms obtained by Lingane and Niedrach[44] in citrate buffers of pH 0.4 to 6.9 are shown in Fig. XXXIII-4. The limited solubility of tellurium dioxide in neutral and acid solutions limits the choice of supporting electrolyte at pH values less than 7. The solubility in acid medium is greatly increased by citrate or tartrate, apparently because of the formation of telluryl citrate and tartrate complexes, and in citrate medium it is possible to work with tellurium concentrations up to about 5 millimolar. Tellurium dioxide is freely soluble in ammoniacal and alkaline media.

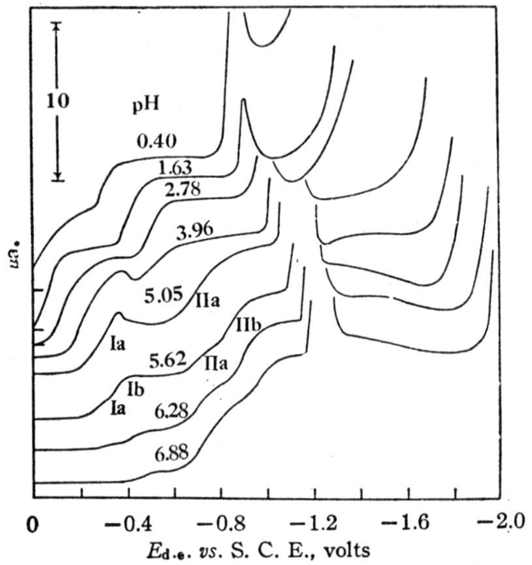

Fig. XXXIII-4. Polarograms of 1 millimolar +4 tellurium in 0.5 *M* citrate buffers at the pH values indicated for each curve.[44] No maximum suppressor was present.

There are two main waves preceding the large maximum, and they are developed best between pH 0.4 and 3. The approximate equality of the heights of these two waves (see Fig. XXXIII-4) is purely fortuitous, and does not correspond to stepwise reduction nor to the production of Te^{++}, as Schwaer and Suchy[43] erroneously concluded. At a pH of 1.6 only a single wave precedes the maximum when the tellurium concentration is below about 0.4 millimolar. With increasing concentration of tellurium the second wave at −0.4 v. develops but the first wave retains a nearly constant height. In spite of this variation in relative wave heights the *total* diffusion current at pH = 1.63 is strictly proportional to the concentration of +4 tellurium between 0.1 and 3 millimolar, and $i_d/Cm^{2/3}t^{1/6}$

is 5.72 at 25°. The virtual constancy of the first wave height suggests the formation of a film of reduction product (elemental tellurium) on the electrode surface, which inhibits further reduction until the potential exceeds −0.4 v.

Coulometric analysis showed that the total wave preceding the maximum results from a 4-electron reduction to elemental tellurium:

$$H_2TeO_3 + 4H^+ + 4e \rightleftharpoons Te + 3H_2O \tag{27}$$

over the pH range from 0.4 to 7. The additional poorly separated waves II*a* and II*b* at pH values from 5.5 to 7.0 may be partly due to reduction to

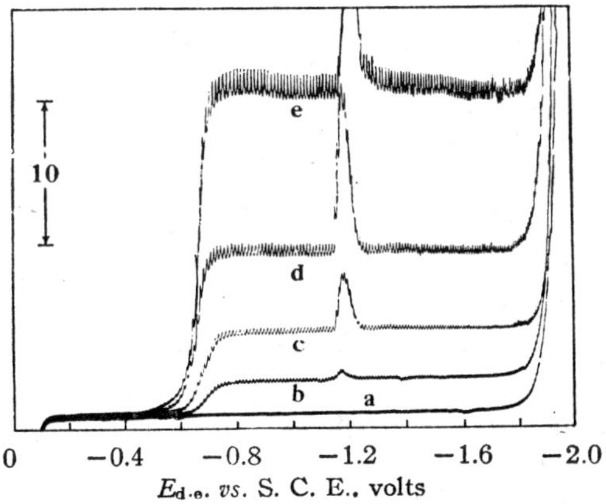

Fig. XXXIII-5. Polarograms of (a) 0, (b) 0.202, (c) 0.500, (d) 1.00, and (e) 2.00 millimolar +4 tellurium in 1 *M* ammonium chloride-ammonia at pH 9.4 in the presence of 0.003 per cent gelatin.[44]

Te^{++}, but film formation and reduction of the different ionization species of tellurous acid are probably more important causes.

In ammoniacal medium only a single, excellently formed wave is observed, as shown in Fig. XXXIII-5. This results from the 4-electron reduction to elemental tellurium. In 1 *M* ammonium chloride containing ammonia and 0.003 per cent gelatin the half-wave potential is −0.63 v. *vs.* S.C.E. at pH = 8.4, and −0.68 v. at pH = 9.4. The ratio i_d/C decreases by about 2 per cent as the concentration of +4 tellurium is increased from 0.2 to 2 millimolar at pH = 9.4. In ammoniacal medium the tellurium wave precedes that of selenium by 0.8 v., which permits the determination of small amounts of the former in the presence of large amounts of the latter.

Single well-formed waves are also obtained in borate and carbonate buffers of pH 7.8 to 11. The half-wave potentials in these media are somewhat more negative than in ammoniacal medium of the same pH; at pH = 9.4 in 0.5 *M* carbonate and borate buffers $E_{1/2}$ is −0.87 v. *vs.* S.C.E. compared to −0.68 v. in ammoniacal medium. The half-wave potentials in 0.5 *M* borate buffers of pH 7.8, 8.8, 10.5, and 11.1 are *seriatim* −0.74, −0.84, −0.96, and −1.18 v. *vs.* S.C.E.

In 0.1 *M* and 1 *M* sodium hydroxide containing 0.003 per cent gelatin a very well-developed single wave is observed, with half-wave potentials of −1.22 v. and −1.19 v., respectively. Lingane and Niedrach[44] proved that this wave results from a 6-electron reduction to telluride ion:

$$TeO_3^{--} + 3H_2O + 6e \rightleftharpoons Te^{--} + 6OH^- \qquad (28)$$

in contrast to the 4-electron reduction observed in ammoniacal media. The ratio i_d/C is constant to ±1 per cent with tellurite ion concentrations from 0.1 to 3 millimolar in 1 *M* sodium hydroxide containing 0.003 per cent gelatin ($i_d/Cm^{2/3}t^{1/6}$ = 9.75 at 25°), but with larger tellurite ion concentrations it decreases significantly. A variation in i_d/C also results when the gelatin concentration is either much smaller or much larger than 0.003 per cent.

Since +4 selenium does not produce a wave in 1 *M* sodium hydroxide this is a most suitable supporting electrolyte for the determination of tellurium in the presence of selenium, especially when selenium predominates.

A striking feature of the polarograms of +4 tellurium is the very large maximum which appears in the diffusion current plateau at all pH values in all supporting electrolytes, except NaOH. This maximum is unusual in that it occurs after the diffusion current plateau has been reached, or in the middle of the plateau in ammoniacal media and in borate and carbonate buffers, rather than as an extension of the rising part of the wave as in the case of ordinary maxima. The maximum is also exceptionally indifferent to the usual maximum suppressors. Significantly the potential at which the maximum develops coincides at all pH values with the half-wave potential of the anodic wave of the reaction $Te^{--} \rightleftharpoons Te + 2e$ discussed below. It is also significant that the only medium in which the maximum is absent is 1 *M* sodium hydroxide, which is probably related to the fact that the cathodic +4 tellurium wave occurs at almost exactly the same potential as the anodic Te^{--} wave in this medium. The occurrence of the maximum is thus intimately related to the Te^{--}–Te couple.

Lingane and Niedrach concluded that the rising part of the maximum reflects the beginning of reduction of +4 tellurium to Te^{--}, but when this starts the normal diffusion layer at the surface of the mercury drops is disrupted, and a vigorous stirring action ensues which causes the current to rise to a very large value. Observations under the microscope revealed formation of a brown turbidity (elemental Te) as a succession of bands which expanded radially from the drop. At potentials before the maximum, as well as after it, the deposited tellurium formed an adherent coating on the drop and no streaming or turbidity near the drop was observed. The band pattern suggests a rhythmically recurrent process during the drop life. A likely mechanism is that the Te^{--} which forms partially dissolves the adherent tellurium deposit, forming polytelluride ion in the diffusion layer and freeing part of the electrode surface. Part of the relatively large concentration of polytelluride ion formed near the electrode surface is rapidly reduced to Te^{--}, causing a large surge of current; part diffuses outward from the electrode and reacts with +4 tellurium to precipitate elemental tellurium and produce the observed bands. The combination of these processes and the resultant density gradients disturbs the diffusion layer and produces stirring which accelerates the transfer of +4 tellurium to the electrode surface. The reduction of the enhanced concentration of +4 tellurium to elemental tellurium proceeds at a faster rate than its reduction to Te^{--}, and than the reaction between the elemental tellurium and Te^{--}, so that the adherent tellurium coating tends to reform, the current surge decreases, and the diffusion layer begins to reestablish itself until the formation of polytelluride again catches up with the decreased rate of reduction of +4 tellurium. The cycle then repeats. The fact that at a slightly more negative potential these processes abruptly cease, so that the current falls to the original value corresponding to only a 4-electron reduction, remains to be explained.

Haus and von Stackelberg[45] corroborated these observations, and presented an interpretation of the maximum similar to that proposed by Lingane and Niedrach. They also reported that in very concentrated potassium chloride medium (4 *M*) the current following the maximum does not decrease to its original value, but is very nearly 1.5 times larger than immediately before the maximum. Haus and von Stackelberg regard this enhanced current as an example of the "water wave" phenomenon reported by Kolthoff and Orlemann[46] in other cases, but they attribute it to streaming at the electrode surface, and claim, contrary to Kolthoff and Orlemann, that hydroxide ion is not produced at the electrode surface in the region of the enhanced current. This latter claim cannot be true in the reduction of tellurite ion from neutral, unbuffered potassium chloride because the hydrogen ion for this reduction has to be produced by ionization of water

[45] W. Haus and M. von Stackelberg, *Z. Elektrochem.*, **54**, 62, 65 (1950).

[46] I. M. Kolthoff and E. F. Orlemann, *J. Am. Chem. Soc.*, **64**, 833 (1942).

with the consequent production of hydroxide ion (in a relatively large concentration) at the electrode surface. The fact that the solution at the electrode surface must become strongly alkaline, coupled with the observation that the current following the maximum is 1.5 times larger than before the maximum, leads us to conclude that the enhanced current after the maximum simply results from the 6-electron reduction to Te^{--}, contrasted to the 4-electron reduction preceding the maximum.

Lingane and Niedrach observed that lead and other metals which produce waves preceding that of +4 tellurium, and which form insoluble tellurides, exert a pronounced suppressive effect on the apparent height of this tellurium wave. This results from a decreased contribution of hydrogen plumbite ion to the total diffusion current at potentials at which TeO_3^{--} is reduced, because of the precipitation of lead telluride in the diffusion layer (see Chap. VI for further details). Correction for this phenomenon must be applied when determining tellurium in the presence of lead and certain other metals.

−2 Tellurium. Lingane and Niedrach[44] observed that telluride ion yields a well-developed anodic wave. In contradistinction to the anodic waves of sulfide and selenide ions, this results from the oxidation of Te^{--} to elemental tellurium rather than "depolarization" of the d.e. to form mercuric telluride.

The anodic wave is obtained at all pH values from 0 to 14, and the oxidation proceeds reversibly. The half-wave potential in 1 *M* hydrochloric acid is −0.72 v. *vs.* S.C.E., and in 1 *M* sodium hydroxide it is −0.08 v., in the presence of 0.003 per cent gelatin. When the telluride ion concentration in 1 *M* sodium hydroxide is smaller than about 0.5 millimolar the anodic diffusion current is perfectly normal, but with larger concentrations the wave shows three distinct parts. The reaction on both the first and second waves is $Te^{--} \rightleftharpoons Te + 2e$, and because the height of the first wave remains nearly constant with increasing telluride ion concentrations above about 0.6 millimolar, this wave probably reflects the formation of a film of tellurium on the surface of the d.e. The third wave at −0.4 v. appears to result from a rate-controlled oxidation to tellurite ion. In spite of the change in wave character, the diffusion current measured between −0.5 v. and −0.9 v. is directly proportional to the concentration of telluride ion up to 1 millimolar.

The half-wave potential of the anodic telluride wave in 1 *M* sodium hydroxide is considerably more negative than the potentials of sulfide and selenide ions, and hence it should be possible to determine small amounts of the former in the presence of much larger amounts of the latter.

CHAPTER XXXIV

Inorganic Halogen Compounds. Anodic Waves of Halide Ions

1. REDUCTION WAVES OF OXYHALOGEN COMPOUNDS

Periodate. The polarography of periodate has been studied independently as a function of pH by Souchay[1] and by Coe and Rogers.[2] In general two waves are observed corresponding to the successive reductions:

$$IO_4^- + 2H^+ + 2e \rightleftharpoons IO_3^- + H_2O$$

$$IO_3^- + 6H^+ + 6e \rightleftharpoons I^- + 3H_2O$$

Coe and Rogers reported that the first diffusion current begins at the anodic dissolution potential of mercury, and hence that the half-wave potential of the periodate to iodate stage could not be measured. Souchay on the other hand claimed that the first wave could be measured by using a mercury-mercurous sulfate electrode, and he reported that its half-wave potential increased from +0.36 v. *vs.* S.C.E. to −0.08 v. as the pH was increased from 0 to 12. The half-wave potential of the second wave corresponds at a given pH to that observed with iodate alone (*vide infra*), and it has a more negative value in the presence of gelatin than in the absence of the latter.

Coe and Rogers found that the height of the first diffusion current decreased somewhat with increasing pH between 0 and 3, but the second wave height nearly doubled over the same pH range, in the presence of 0.01 per cent gelatin. The same authors also reported that i_d/C for the first wave shows a significant variation from constancy over a range of *o*-periodic acid concentration from 0.2 to 1.6 millimolar at a pH of 2 in the presence of 0.1 per cent gelatin. On the other hand, with 0.2 to 1.8 millimolar *m*-periodate at pH = 1.6 they found that i_d/C was constant.

Souchay observed a linear relation between i_d of the first wave and the concentration of periodate from 0.08 to 10 millimolar in a carbonate-bicarbonate buffer containing 0.08 per cent gelatin. This supporting elec-

[1] P. Souchay, *Anal. Chim. Acta.*, **2,** 17 (1948).
[2] R. H. Coe and L. B. Rogers, *J. Am. Chem. Soc.*, **70,** 3276 (1948).

trolyte was recommended for the determination of traces of periodate impurity in iodate salts.

Perchlorate. Perchlorate ion is not reduced at the dropping electrode.

Bromate and Iodate. The reduction of bromate and iodate ions at the dropping electrode was first studied by Rylich,[3] who found only a single wave corresponding to direct reduction to the corresponding halide ions. This was confirmed by the authors.[4, 5]

$$BrO_3^- + 6H^+ + 6e \rightarrow Br^- + 3H_2O$$

The reduction potentials (45° tangent) of iodate and bromate determined by Rylich in various media are summarized in Table XXXIV-1. These

TABLE XXXIV-1

REDUCTION POTENTIALS OF IODATE AND BROMATE IONS AT THE DROPPING ELECTRODE IN VARIOUS MEDIA (RYLICH)

The "reduction potentials" were determined by the 45° tangent method at a concentration of 0.001 *M* at room temperature, and are referred to the normal calomel electrode.

Supporting electrolyte	45° tangent reduction potential, v.	
	IO_3^-	BrO_3^-
0.1 *N* KCl, NaCl, LiCl, or K_2SO_4	−1.05	−1.65
0.1 *N* NH_4Cl	−0.82	−1.52
0.1 *N* $CaCl_2$, $BaCl_2$, or $SrCl_2$	−0.85	−1.32
0.01 *N* $LaCl_3$	−0.40	−0.63
0.1 *N* KOH	−1.08	−1.64
0.1 *N* H_2SO_4	+0.13	+0.13

values may be compared with the reversible standard potentials of the $IO_3^- \rightarrow I^-$ and the $BrO_3^- \rightarrow Br^-$ systems which are +0.80 v. and +1.16 v. in acid medium, and −0.02 v. and +0.33 v. in basic medium, respectively, *vs.* the N.C.E.[6]

It is evident that the reduction of both ions is irreversible and requires a large overvoltage at the dropping electrode; in acid medium the overvoltage amounts to about 0.7 v. for iodate and about 1.3 v. for bromate. In alkaline medium the overvoltages are somewhat larger than in acid medium. Peculiarly enough, the overvoltage is much greater for bromate

[3] A. Rylich, *Collection Czechoslov. Chem. Communs.*, **7,** 288 (1935).
[4] I. M. Kolthoff and J. J. Lingane, *Chem. Revs.*, **24,** 1–94 (1939).
[5] J. J. Lingane and I. M. Kolthoff, *J. Am. Chem. Soc.*, **61,** 825 (1939).
[6] W. M. Latimer, *The Oxidation States of the Elements and Their Potentials in Aqueous Solutions*, New York, 1938.

than for iodate. Consequently, the reduction potential of bromate is more negative than that of iodate, in spite of the fact that the reversible potential of bromate is more positive than that of iodate.

Due to the irreversibility of the electrode reactions, the effect of pH on the reduction potentials does not conform to what one would predict from the net stoichiometric reactions.

Kolthoff and Orlemann[7] found that the half-wave potential of iodate at a constant value of the pH is constant and independent of the iodate ion concentration. The plot of $E_{d.e.}$ *vs.* log $i/(i_d - i)$ produced a straight line at all pH values, but the slope at all pH values is much larger than the theoretical 0.059/6 expected for a reversible reduction. Their results are summarized in Table XXXIV-2. It will be noted that the half-wave potential undergoes a large change between a pH of 5 and 8, and that it is practically constant at pH values larger than about 8.

TABLE XXXIV-2

HALF-WAVE POTENTIAL OF IODATE ION AS A FUNCTION OF THE pH

Supporting electrolyte	pH	$E_{1/2}$ *vs.* S.C.E., v.	$\Delta E_{1/2}/\Delta pH$	$\Delta E_{d.e.}$ / $\Delta \log [i/(i_d - i)]$
$HClO_4$	0.90	−0.042		−0.140
H_2SO_4, 0.2 *M* KNO_3	1.15	− .075	−0.13	.127
0.1 *M* biphthalate, 0.1 *M* KCl	3.2	− .305	.12	.122
0.1 *M* acetate, 0.1 *M* KCl	4.9	− .500	.12	.122
0.1 *M* sodium citrate	5.95	− .650	.14	.220
0.2 *M* sodium phosphate	7.10	−1.050	.35	
0.05 *M* borax, 0.1 *M* KCl	9.2	−1.200	.07	.108
NaOH, 0.1 *M* KCl	13.0	−1.210	.03	.085

The reduction potentials of iodate and bromate are shifted markedly to more positive values in the presence of polyvalent cations, especially lanthanum ion. This effect has been attributed by Heyrovsky to the formation of "ion pairs" between the polyvalent cation and iodate or bromate ion, which causes an electrostatic deformation of the iodate and bromate ions, and renders them more susceptible to reduction.

Coe and Rogers[2] found that addition of gelatin shifts the iodate wave to a more negative potential. For example, the half-wave potentials at a pH of 2.10 with 0, 0.01, and 0.1 per cent gelatin are, respectively, −0.21, −0.30, and −0.35 v. *vs.* S.C.E.

Since the reduction potential of iodate in neutral or alkaline solution is

[7] I. M. Kolthoff and E. E. Orlemann, *J. Am. Chem. Soc.*, **64**, 1044, 1970 (1942).

about 0.6 v. more positive than that of bromate, two well-separated waves are obtained with a solution containing both ions, and their simultaneous determination is a simple matter. Rylich has shown that traces of iodate and bromate in alkali chlorates and perchlorates can be easily detected and determined because the latter two ions are not reducible at the dropping electrode. It is also possible to determine traces of iodate (but not bromate) in the presence of a large excess of nitrate, since the reduction potential of the latter in alkaline medium is more negative than that of iodate. Rylich employed this fact for determining traces of iodate in Chili saltpeter.

Traces of iodide have been determined polarographically by oxidation to iodate with chlorine water, removal of the excess chlorine, and recording of the iodate wave.[3, 8]

Souchay[1] described a procedure for determining small amounts of iodate in sodium or potassium periodate based on precipitation of most of the periodate as sodium periodate from a solution containing acetic acid and a very large concentration of sodium acetate. The iodate wave of the supernatant solution is recorded and correction is made for the small concentration of residual periodate by measuring its first wave (*vide supra*) and subtracting three times the measured value from the iodate wave.

Chlorate. Chlorate ion is not reduced at the dropping electrode.

Hypoiodite. Souchay found that freshly prepared solutions of hypoiodite ion in alkaline medium show a diffusion current that starts from the anodic dissolution potential of mercury corresponding to a 2-electron reduction to iodide ion:

$$IO^- + 2e + H_2O \rightleftharpoons I^- + 2OH^-$$

The rate of disproportionation of hypoiodite ion, according to:

$$3IO^- \rightleftharpoons IO_3^- + 2I^-$$

can be measured by following the decrease in the hypoiodite diffusion current.

Hypochlorite. According to Souchay[1] hypochlorite, like hypoiodite, produces a diffusion current that starts from the anodic dissolution potential of mercury.

2. FREE HALOGENS

Elemental iodine, bromine, and chlorine in acid media are all reduced to the corresponding halide ions at the dropping electrode. Because the

[8] V. Cizek, *Bull. Acad. Sci. Boheme*, (1938).

potentials of the halogen-halide couples are all more positive (more oxidizing) than the oxidation potential of mercury the waves all start from zero applied e.m.f. In alkaline media the free halogens disproportionate to halide and hypohalite (or halate) ions.

3. ANODIC WAVES OF HALIDE IONS

When the dropping electrode is polarized anodically in solutions that do not contain ions that form insoluble salts or complex ions with mercury, *e. g.*, in a potassium nitrate solution, a negative (oxidation) current is observed due to the reaction:

$$2Hg \rightleftharpoons Hg_2^{++} + 2e \quad (1)$$

The negative current increases exponentially on such a "free dissolution wave," and becomes very large in the neighborhood of +0.45 v. *vs.* the S.C.E. The potential of the dropping electrode at any point on such a wave is given by:

$$E = E^0_{Hg} + (0.0591/2) \log C^0_{Hg_2^{++}} \quad (2)$$

where E^0_{Hg} is the standard potential of reaction 1 (+0.553 v. *vs.* the S.C.E.), and $C^0_{Hg_2^{++}}$ is the concentration of mercurous ion produced at the electrode surface. Since $C^0_{Hg_2^{++}} = ki$, the equation of the wave is:

$$E = E^0_{Hg} + 0.0296 \log k + 0.0296 \log i \quad (3)$$

where i is the anodic current.

Since the first two terms in eq. 3 are constants, a plot of E *vs.* log i should produce a straight line with a slope of 0.0296 v. at 25° C. Kolthoff and Miller[9] found that this was true; they found a slope of 0.032 v. in such a log plot, in fairly good agreement with the theoretical value.

When the solution contains ions that form insoluble salts or complex ions with mercurous or mercuric ions the anodic wave of mercury is shifted to a more negative potential, and if the concentration of the depolarizing ion is smaller than about 0.01 M a well-defined anodic diffusion current may result. For example, in a chloride solution the anodic wave starts when the potential has been made sufficiently positive so that the concentration of mercurous ion produced at the electrode surface exceeds the solubility product of mercurous chloride. The solubility product of mercurous chloride is 6×10^{-19}, and hence in a 0.001 M chloride solution the critical concentration of mercurous ion is $6 \times 10^{-19}/(0.001)^2 = 6 \times 10^{-13}$

[9] I. M. Kolthoff and C. S. Miller, *J. Am. Chem. Soc.*, **63**, 1405, 2732 (1941).

M. When this value is substituted into eq. 2 we find that the anodic wave of a 0.001 *M* chloride solution should start at +0.553 + 0.0296 log (6×10^{-13}) = +0.19 v. *vs.* the S.C.E. This is the potential at which the anodic chloride wave is actually observed.

Anodic depolarization waves of the halides, cyanide, thiocyanate, sulfide, hydroxide, and thiosulfate ions were first studied by Revenda,[10] and more recently by Kolthoff and Miller.[9] The characteristics of the anodic halide waves are summarized below, and the anodic waves of the other ions are discussed in the chapters dealing with the elements in question.

Halide ions depolarize the dropping anode with the formation of slightly soluble mercurous halides. In the typical case of chloride the net electrode reaction is:

$$2Hg + 2Cl^- \rightleftharpoons Hg_2Cl_2 + 2e \tag{4}$$

Hence the potential of the dropping anode at any point on the wave should be given by:

$$E = E^0_{Hg_2Cl_2} - 0.0591 \log C^0_{Cl^-} \tag{5}$$

where $C^0_{Cl^-}$ is the concentration of chloride ion at the electrode surface, and $E^0_{Hg_2Cl_2}$ is the standard potential of reaction 4(+0.022 v. *vs.* the S.C.E. at unit activity of chloride ion, or +0.036 v. at unit concentration of chloride ion).

The anodic limiting current is governed by the rate of diffusion of chloride ion up to the electrode surface, and hence is proportional to the concentration of chloride in the body of the solution:

$$i_d = kC_{Cl^-} \tag{6}$$

Therefore at any point on the wave:

$$C^0_{Cl^-} = C_{Cl^-} - (i/k) = (i_d - i)/k \tag{7}$$

and:

$$E = +0.022 + 0.0591 \log k - 0.0591 \log (i_d - i) \tag{8}$$

It is evident that the wave is unsymmetrical about its midpoint, and the half-wave potential should depend on the concentration of chloride in the body of the solution according to:

$$E_{1/2} = +0.022 - 0.0591 \log (C_{Cl^-}/2) \tag{9}$$

[10] J. Revenda, *Collection Czechoslov. Chem. Communs.*, **6**, 453 (1934).

Kolthoff and Miller found that this equation was valid at chloride ion concentrations smaller than about 1.5 × 10^{-3} *M*. With a 0.001 *M* chloride solution they found a half-wave potential of +0.25 v. *vs.* the S.C.E., whereas the value calculated from eq. 9 is +0.22 v. Since the concentration of chloride ion, rather than its activity, is used in eq. 9 this agreement is satisfactory.

Kolthoff and Miller found that the diffusion current of chloride ion is directly proportional to its concentration up to a concentration of about

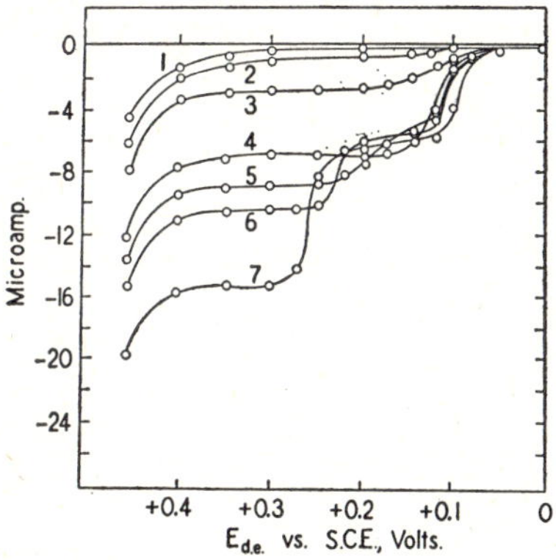

Fig. XXXIV-1. Anodic waves of bromide ion, in 0.1 *N* potassium nitrate. Concentration of bromide ion was: (*1*) none, (*2*) 1 × 10^{-4}, (*3*) 5 × 10^{-4}, (*4*) 1 × 10^{-3}, (*5*) 1.2 × 10^{-3}, (*6*) 1.5 × 10^{-3}, and (*7*) 2 × 10^{-3} *N*.

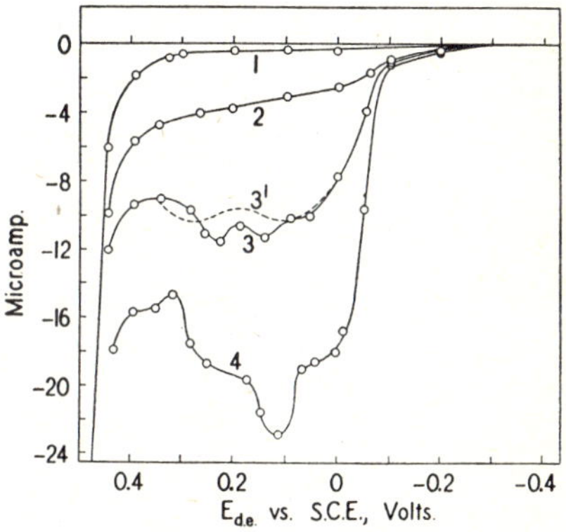

Fig. XXXIV-2. Anodic waves of iodide ion in 0.1 *N* potassium nitrate. Concentration of iodide ion was: (*1*) none, (*2*) 5 × 10^{-4}, (*3*) 1 × 10^{-3}, and (*4*) 2 × 10^{-3} *N*. Curve *3′* was obtained with 1 × 10^{-3} *N* iodide ion in the presence of 10^{-6} *M* in sodium methyl red.

0.002 *M*. With larger chloride concentrations the diffusion current becomes poorly defined.

Some typical anodic waves of bromide ion obtained by Kolthoff and Miller are shown in Fig. XXXIV-1. These waves were obtained with a drop time of 4 sec. When the concentration of bromide ion is smaller than about 0.001 *M* only a single well-defined wave is observed, but with larger concentrations of bromide a double wave is obtained. This double wave was first observed by Revenda. It will be noted that the first limiting current between about 0.1 and 0.2 v. is practically constant with increasing concentration of bromide. However, the total diffusion current between 0.3 and 0.4 v. is directly proportional to the bromide ion concentration up to about 0.005 *M*.

Kolthoff and Miller attributed the first limiting current between 0.1 and 0.2 v. to the formation of a film of mercurous bromide on the surface of the dropping electrode which interferes with the electrode reaction. They were able to observe this film microscopically when the current was of the order of 25 microamp. When a small amount of gelatin is added to the bromide solution the first wave disappears. Apparently the gelatin is so strongly adsorbed on the surface of the mercury drops that it prevents the formation of a coherent film of mercurous bromide. The use of a trace of gelatin is advantageous in the practical determination of bromide.

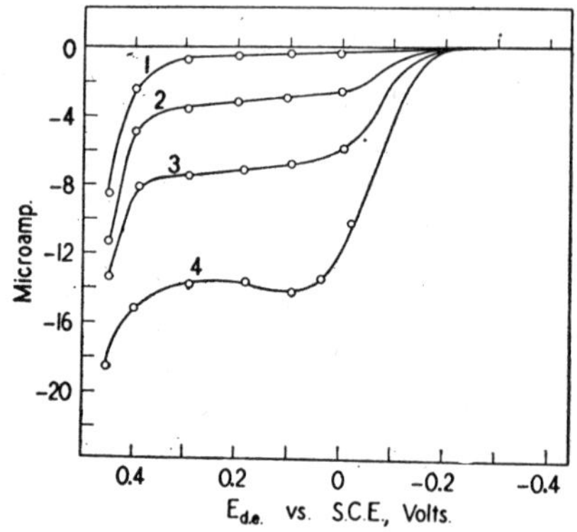

Fig. XXXIV-3. Beneficial effect of gelatin on anodic iodide waves. Concentrations of iodide ion were: (*1*) none, (*2*) 5×10^{-4}, (*3*) 1×10^{-3}, and (*4*) 2×10^{-3} *N* in 0.1 *N* potassium nitrate containing 0.05 per cent gelatin.

The character of the anodic waves of iodide is quite different from that of the bromide waves, as shown in Fig. XXXIV-2. The waves have a normal appearance with very small concentrations of iodide, but when the concentration of iodide is greater than about 5×10^{-4} *M* the diffusion current is very poorly defined and undergoes irregular fluctuations. Kolthoff and Miller noted that the mercury drops grow irregularly, and pulsate during their growth, when the iodide ion concentration is greater than about 0.001 *M*.

The irregularities becomes smaller when a more rapid drop time is used, and with a drop time smaller than 1 sec. the diffusion current is well defined.

The addition of a small amount of gelatin also improves the iodide diffusion current, as shown in Fig. XXXIV-3, and the use of gelatin is recommended for analytical purposes.

Practical analytical applications of the anodic halide waves have been described by Mikhlin[11] and Schönholzer[12] and they have been applied to the determination of chloride and iodide in blood and biological fluids by Šantavy,[13] Vohnout and Pihar,[14] and Zimmerman and Layton.[15]

---

[11] S. G. Mikhlin, *Trans. All-Union Conf. Anal. Chem.*, **2,** 507 (1943).
[12] G. Schönholzer, *Experientia*, **1,** 158 (1945).
[13] F. Santavy, *Casopis Lekaru Ceskych*, **83,** 1312 (1944).
[14] S. Vohnaut and O. Pihar, *Casopis Lekaru Ceskych*, **86,** 1 (1947).
[15] W. J. Zimmerman and W. M. Layton, Jr., *J. Biol. Chem.*, **181,** 141 (1949).

CHAPTER XXXV

Analysis of Alloys and Technical Materials

The determination of alloying constituents in various types of alloys, and the determination of traces of metallic impurities in "pure" metals, is one of the most important practical applications of polarographic analysis. Because time-consuming separations can often be avoided, and since simultaneous determinations are frequently possible, the polarographic methods are especially attractive when the saving of time is an important consideration. The polarographic method generally is not as accurate as classical methods for the determination of large percentages of major alloy constituents, but for the determination of a few per cent or traces of minor constituents it frequently excels orthodox methods.

This chapter presents a selection of procedures that have been developed for the analysis of various important types of alloys and a few other technical materials. In most instances the procedures are given in sufficient detail to permit their application without recourse to the original literature. An attempt has been made to evaluate the methods critically, and to provide information regarding interfering elements and attainable accuracy. The procedures have also been unified, as far as possible, with respect to reagents and other practical details.

1. COPPER-BASE ALLOYS

The following two procedures were recommended by Hohn.[1]

Copper and Zinc. When only copper and zinc are to be determined a 0.1-g. sample of the brass is dissolved in 2 ml. of concentrated nitric acid, the solution is boiled to remove nitric oxide, and then diluted to 50 ml. in a volumetric flask without bothering to remove any precipitate of metastannic acid. One volume of this solution is added to two volumes of a supporting electrolyte solution consisting of 1 *N* ammonia, 2 *N* ammonium chloride, and 0.2 per cent tylose (or 0.05 per cent gelatin) as a maximum suppressor. Air is removed with nitrogen or hydrogen, and the polarogram is obtained in the usual way. The half-wave potentials of the double copper wave are at −0.2 and −0.5 v., and that of the zinc is −1.3 v. *vs.* the S.C.E. It is usually most convenient to measure the second copper wave.

[1] H. Hohn, *Chemische Analysen mit dem Polarographen*, Berlin, 1937; *Z. Elektrochem.*, **43,** 127 (1937).

For increased precision in the zinc determination, Hohn recommends that the diffusion current of the copper be compensated before recording the zinc wave.

Copper, Zinc, Nickel, Lead, and Iron. A 0.5-g. sample of the brass is dissolved in a few milliliters of concentrated nitric acid, and evaporated to dryness to dehydrate metastannic acid. The residue is taken up in 5 ml. of concentrated nitric acid and about 25 ml. of hot water. The metastannic acid is filtered off, washed, ignited, and weighed as SnO_2 in the usual way. The filtrate is treated with 15 ml. of concentrated ammonium hydroxide and 5 ml. of 1 *M* ammonium carbonate, and the small precipitate of hydrous ferric oxide and basic lead carbonate is filtered off, and washed with dilute ammonium hydroxide.

The filtrate and washings are made up to 250 ml. in a volumetric flask. A polarogram of this solution, which contains the copper, zinc, and a small concentration of nickel, is obtained exactly as described in the preceding section. From this polarogram the copper, and the *sum* of the zinc and nickel, are obtained.

In order to determine the nickel, advantage is taken of the fact that nickel gives a good wave, whereas zinc and copper do not produce a wave, in cyanide medium. When excess cyanide is added to a neutral or slightly acid cupric solution the copper is reduced to $Cu(CN)_3^{--}$ and the cyanide is oxidized to cyanogen, C_2N_2. If the solution is ammoniacal the cyanide is oxidized to cyanate ion instead of cyanogen. The formation of cyanate can be avoided by adding an excess of sodium sulfite to the ammoniacal copper-nickel-zinc solution, which reduces the copper to the cuprous state, *before the addition of cyanide*. The addition of sulfite also removes dissolved oxygen and eliminates the necessity of bubbling hydrogen or nitrogen through the solution. 5 ml. of the filtrate from the iron and lead precipitation is treated with 5 ml. each of 2 *M* sodium sulfite, 2 *M* potassium cyanide, and 4 *N* ammonium chloride containing 2 *N* ammonium hydroxide, *in the order named*. A polarogram of this solution shows only the nickel wave, from which the content of nickel in the sample is computed. The zinc content is computed by subtracting the diffusion current corresponding to the known amount of nickel from the combined zinc-nickel wave obtained in the absence of cyanide.

In order to determine the iron and lead, the mixed precipitate of hydrous ferric oxide and basic lead carbonate is dissolved from the filter with a few milliliters of hot 2 *N* nitric acid, and the solution is made up to 25 ml. in a volumetric flask with 2 *N* nitric acid. An aliquot portion of this solution is added to an equal volume of saturated ammonium chloride solution containing tylose or gelatin as a maximum suppressor, and the polarogram is recorded after removing the air. The diffusion current of the ferric ion is obtained right from the start of the polarogram at zero applied e.m.f., and the lead wave is observed at −0.4 v. *vs.* the S.C.E. In order to measure the diffusion current of the ferric ion accurately it is advisable actually to determine the residual current of the supporting electrolyte alone in a separate experiment, rather than merely measure from the galvanometer "zero line."

In the analysis of a typical brass by the foregoing procedures, Hohn obtained the following results: Cu 57.8, Zn 38.5, Ni 1.06, Pb 1.65, Fe 0.42, and Sn (gravimetric) 0.82 per cent. These values give a summation of 100.25 per cent, which is quite satisfactory.

Mnich[2] has criticized Hohn's procedure for the determination of iron and lead in brass. With brasses containing very small amounts of iron (0.05 to 0.1 per cent) Mnich found that the polarographic values were as much as twice as large as the results obtained by the usual volumetric method. He attributed this to the coprecipitation of appreciable amounts of copper with the hydrous ferric oxide, and found that the error could be minimized by prolonged washing of the mixed iron-lead precipitate with ammonia water. With larger amounts of iron the error due to coprecipitated copper is proportionately smaller. Mnich also found that the results for lead by the above procedure tended to be lower than the values obtained by the usual electrodeposition method. He found that the error was minimized when the mixed iron-lead precipitate was allowed to stand for twenty-four hours before filtration.

Lead. The following procedure is recommended by Milner.[3]

Dissolve a 2-g. sample of the copper-base alloy in 25 ml. of nitric acid (sp. gr. 1.20), boil off the oxides of nitrogen, and dilute to 50 ml. Filter through a dry Whatman No. 40 filter paper into a dry beaker, and pipet 10 ml. of the filtrate into second dry beaker. Add 10 ml. of 50 per cent sodium sulfite solution and 16 ml. of 2 *M* potassium cyanide solution, and shake until the precipitated cuprous cyanide dissolves. (These operations should be conducted in a hood, since cyanogen is evolved!) To the resulting solution add 5 ml. of a 25 per cent triethanolamine solution, 10 ml. of 10 *N* sodium hydroxide, and 2 ml. of fresh 0.2 per cent gelatin solution. Mix well and record the polarogram of a portion of the solution. The lead wave appears at −0.7 v. *vs.* S.C.E. When the lead content is less than 0.1 per cent, iron interferes.

Nickel (according to Milner[4]). This method for the determination of nickel in copper-base alloys is based on the elimination of copper interference by use of a very strongly alkaline cyanide supporting electrolyte, in which the copper is converted to the nonreducible cuprous-cyanide complex. Zinc interferes with the nickel wave and is removed by sulfide precipitation. Most of the lead is removed as lead sulfate in the preparation of the sample solution. Amounts of tin up to about 1 per cent do not interfere, but larger amounts should be removed as metastannic acid by preliminary treatment of the sample with nitric acid. Small amounts or iron, manganese, and aluminum do not interfere.

Dissolve a 1-g. sample of the alloy in 20 ml. of 50 per cent sulfuric acid with the aid of the minimal amount of nitric acid. Evaporate to sulfur trioxide fumes, cool, and dilute to 50 ml. in a volumetric flask. Pipet a 25-ml. aliquot of the filtered solution into a dry beaker and add the accurately measured following solutions in the

[2] E. Mnich, *Z. Elektrochem.*, **44,** 132 (1938).
[3] G. W. C. Milner, *Metallurgia*, **36,** 287 (1947); *Analyst*, **73,** 472 (1948).
[4] G. W. C. Milner, *Analyst*, **70,** 468 (1945).

order given: 10 ml. of 50 per cent sodium sulfite, 20 ml. of 10 *N* sodium hydroxide, and 10 ml. of 8 *M* potassium cyanide. Heat the solution just to boiling and add 10 ml. of a sodium polysulfide solution (prepared by dissolving 450 g. of sodium sulfide, 40 g. of sodium hydroxide, and 16 g. of flowers of sulfur in sufficient water to make a total volume of 2 liters). Boil gently for 5 minutes with the beaker covered to minimize evaporation losses. Pour through Whatman No. 42 paper into a dry beaker to remove the zinc sulfide. Cool the filtrate and transfer 25 ml. to a dry beaker. Add accurately 25 ml. of water and 2 ml. of 0.2 per cent gelatin solution, and record the polarogram of nickel.

Milner obtained very satisfactory results by this procedure with copper-base alloys containing from 0.1 to 2 per cent nickel.

Boiling the solutions, even in a covered beaker, is a questionable part of this procedure, and a better technique would be to make up to a definite volume after boiling.

Copper and Zinc in Brass (according to Tyler and Brown[5]). This method was developed specifically for the rapid analysis of electrolytic brass plate containing only copper and zinc. Since no other metals are present the composition of the plate can be determined by measuring only the *ratio* of the diffusion currents of copper and zinc. Since the *ratio* of the diffusion current constants changes relatively little with temperature, no exact temperature control is required. According to Tyler and Brown the temperature may vary over a range of 2° C. Furthermore, the exact weight of sample need not be known. An ammoniacal supporting electrolyte is used and sulfite ion is employed to remove dissolved oxygen.

Strip the brass plate from a desired area of sample by flowing an appropriate quantity, *e. g.*, 10 to 25 ml., of "stripping solution" over the surface until the brass has been entirely removed. The stripping solution contains 80 g. of ammonium persulfate per liter and is 1.5 *M* in ammonium hydroxide. The area stripped should be such that 5 to 25 mg. of brass are present per 5 ml. of the stripping solution. Mix 5 ml. of the brass solution with 15 ml. of a supporting electrolyte comprising 1 *M* ammonium chloride, 1.5 *M* ammonia, 0.54 *M* sodium sulfite, and 0.067 per cent gelatin. Let stand for 10 minutes to insure complete reduction of dissolved oxygen and persulfate ion. Then record the polarogram, and measure the relative heights of the copper and zinc waves.

Tyler and Brown recommend empirical calibration of the procedure with known brass samples under identically the same conditions as in the analysis. The temperature during calibration should not differ by more than about 2° C. from that during an analysis, and should be near room temperature in both cases. With 70-30 brass plate Tyler and Brown obtained results for copper that were accurate to about ±0.5 relative per cent.

Tin, Lead, Nickel, and Zinc (according to Lingane[6]). This method employs electrolysis at a controlled potential with a platinum cathode in a

[5] W. P. Tyler and W. E. Brown, *Ind. Eng. Chem., Anal. Ed.*, **15**, 520 (1943).
[6] J. J. Lingane, *Ind. Eng. Chem., Anal. Ed.*, **18**, 429 (1946).

hydrochloric acid solution to remove copper (and antimony and bismuth) prior to the determination of lead and tin, and to remove copper, lead, and tin prior to the determination of nickel and zinc. Separate aliquot portions of the solution remaining after removal of copper are used for the determination of lead and tin, the former in 1 *N* sodium hydroxide and the latter in a supporting electrolyte composed of 4 *M* ammonium chloride and 1 *M* hydrochloric acid. Nickel and zinc are determined in a second sample from a single polarogram obtained with the solution remaining after electrolytic removal of copper, lead, and tin; an ammoniacal supporting electrolyte is used and small amounts of nickel may be determined accurately in the presence of large amounts of zinc. Minor amounts of antimony, bismuth, iron, and manganese do not interfere. All four determinations can be completed in 2.5 hours, and the method is applicable to all types of copper-base alloys.

The general technique of automatic controlled potential electrolysis has been described in Chapter XVIII. For the present purpose the usual form of cylindrical platinum gauze cathode and a platinum anode are convenient, and the cathode potential is controlled against a saturated calomel reference electrode.

Pertinent half-wave potentials and diffusion current constants for the four metals concerned in the supporting electrolytes used are shown in Table XXXV-1. The concentrations of the components of the supporting electrolytes are not very critical, but should be within about ±10 per cent of the values specified. Since the wave of lead and the second wave of stannic tin coincide in the ammonium chloride-hydrochloric acid supporting electrolyte, a correction must be applied for the lead in computing the concentration of tin. This correction is described below.

The following procedures are applicable to copper-base alloys containing from about 0.05 to 10 per cent or more of the various metals.

For the determination of lead and tin dissolve a 0.5- to 1-g. sample of the alloy in a warm mixture of 6 cc. of 12 *N* hydrochloric acid, 4 cc. of water, and 1 cc. of concentrated nitric acid in a Kjeldahl flask. Dilute to about 50 cc. and boil briefly to remove oxides of nitrogen and chlorine. Transfer the solution to a 250-cc. beaker, add 2 g. of pure hydrazine dihydrochloride, dilute to about 200 cc., and electrolyze with the potential of the platinum cathode at −0.35 v. *vs.* the saturated calomel electrode until the current decreases to a small constant value. Transfer the residual solution to a 250-cc. volumetric flask, cool to room temperature, and dilute to the mark. The final solution will be about 0.4 *N* with respect to hydrochloric acid.

Transfer a 50-cc. aliquot (corresponding to one-fifth of the sample) to a 100-cc. volumetric flask, add 24 cc. of 5 *N* sodium hydroxide (or 4.8 g., 0.12 mole, of the solid) and 2.5 cc. of fresh 0.2 per cent gelatin solution, and dilute to the mark. Transfer a portion of the solution to the polarographic cell in a water thermostat at 25.0° ±

0.2° C., remove dissolved air with nitrogen or hydrogen, record the polarogram, and measure the diffusion current of the lead. Determine the m-value of the dropping electrode, measure the drop time at the potential at which the diffusion current was measured, and compute the concentration of lead from the relation $C = i_d/3.40m^{2/3}t^{1/6}$.

Transfer another 50-cc. aliquot to a 100-cc. volumetric flask, add 21 g. (0.4 mole) of solid ammonium chloride, 6.6 cc. (0.08 mole) of 12 N hydrochloric acid, and 2.5 cc. of 0.2 per cent gelatin solution, and dilute to about 90 cc. Shake until the ammonium chloride is dissolved, warm back to room temperature, and dilute to the mark. Transfer a portion of the solution to the polarographic cell in a water thermostat at 25.0° ± 0.2° C., remove dissolved air with nitrogen or hydrogen, record the polarogram, and measure the diffusion current of the second wave (lead plus tin). Apply the correction for lead, and compute the concentration of tin, in the following manner.

TABLE XXXV-1

HALF-WAVE POTENTIALS AND DIFFUSION CURRENT CONSTANTS PERTINENT TO ANALYSIS OF COPPER-BASE ALLOYS[6]

Temperature = 25° C. The half-wave potentials are referred to the saturated calomel electrode. The diffusion current constants, I, correspond to the observed diffusion current in microamperes, when the concentration is 1 millimolar, and when the value of $m^{2/3}t^{1/6}$ is 1 mg.$^{2/3}$sec.$^{-1/2}$. All solutions contained 0.005 per cent gelatin as a maximum suppressor.

Metal	Supporting electrolyte	$E_{1/2}$, v.	I
Lead	1 M NaOH	−0.76	3.40
	4 M NH_4Cl + 1 M HCl	−0.52	3.52
Tin	4 M NH_4Cl + 1 M HCl	−0.25	2.84[a]
		−0.52	3.49[b]
			6.33[c]
Nickel	0.2 M NH_4Cl + 1 M NH_3	−1.06	3.54
	1 M NH_4Cl + 1 M NH_3	−1.09	3.56
	3 M NH_4Cl + 1 M NH_3	−1.12	3.33
Zinc	0.2 M NH_4Cl + 1 M NH_3	−1.33	3.78
	1 M NH_4Cl + 1 M NH_3	−1.33	3.82
	3 M NH_4Cl + 1 M NH_3	−1.36	3.94

[a] First wave.
[b] Second wave.
[c] Total double wave

The method of applying this correction is best illustrated by a typical example. Aliquot samples (50 cc.) were used for each determination and in each case were finally diluted to 100 cc. in the appropriate supporting electrolyte. The observed diffusion current of lead in 1 N sodium hydroxide was 2.34 microamp. with $m^{2/3}t^{1/6} = 1.57$, and the diffusion current of the second wave in 4 M ammonium chloride–1 M hydrochloric acid was 7.10 microamp. with $m^{2/3}t^{1/6} = 1.60$. From the data in Table XXXV-1 the diffusion current constant of lead in 4 M ammonium chloride–1 M hydrochloric acid is seen to be somewhat larger than in 1 M sodium hydroxide, the ratio being 3.52/3.40 = 1.036. The contribution of lead to the second wave in the former medium was thus $2.34 \times 1.036 \times 1.60/1.57 = 2.47$ microamp., and the diffusion current due to tin was $7.10 - 2.47 = 4.63$ microamp. The concentration of lead was $2.34/(3.40 \times 1.57) = 0.438$ millimolar, and that of tin was $4.63/(3.49 \times 1.60) = 0.829$ millimolar.

For the determination of nickel and zinc dissolve a 0.5- to 1-g. sample, and prepare the solution for electrolysis, exactly as for determination of lead and tin. Electrolyze first at a potential of −0.35 v. *vs.* the saturated calomel electrode, and, after most of the copper has deposited, increase the potential to −0.70 v. to deposit tin and lead. Continue the electrolysis for about 10 minutes after the current decreases to a constant value. Transfer the residual solution to a 250-cc. volumetric flask and dilute to the mark.

Transfer a 50-cc. aliquot to a 100-cc. volumetric flask, add 8 cc. of 15 M ammonia, 1 to 1.5 g. (approximately 0.01 mole) of pure anhydous sodium sulfite to remove oxygen, and 2.5 cc. of 0.2 per cent gelatin, and dilute to the mark. Place a portion of the solution in a polarographic cell in a water thermostat at 25.0° ± 0.2° C., *e. g.*, a 15-cc. Erlenmeyer flask if a lead wire anode wrapped on the dropping electrode is used, and record the polarogram. If the amount of nickel is much smaller than that of zinc, a second polarogram should be recorded at an increased sensitivity to magnify the nickel wave to a value large enough for accurate measurement. Measure the two diffusion currents, the m-value of the capillary, and the drop times at the potentials at which the diffusion currents were measured. Compute the concentration of nickel from the relation $C = i_d/3.54m^{2/3}t^{1/6}$, and the concentration of zinc from the relation $C = i_d/3.78m^{2/3}t^{1/6}$.

TABLE XXXV-2

ANALYSIS OF BUREAU OF STANDARDS COPPER-BASE ALLOYS[6]

Sample and Bureau of Standards analyses	Polarographic analyses, %			
	Pb	Sn	Ni	Zn
Phosphor bronze 63a Cu 78.48, Pb 8.92, Sn 9.76, Ni 0.32, Zn 0.61, Sb 0.49, Fe 0.52, P 0.58, S 0.11	9.04	9.84	0.32	0.60
Cast bronze 52b Cu 88.26, Pb 0.01, Sn 8.00, Ni 0.72, Zn 2.96, Fe 0.034	0.02	7.82	0.70	2.96
Ounce metal 124a Cu 85.05, Pb 4.86, Sn 4.81, Ni 0.001, Zn 5.25, Fe 0.004	4.84	5.00	—	5.29
Brass 37c Cu 70.07, Pb 0.97, Sn 0.96, Ni 0.58, Zn 27.22, Fe 0.17	0.92	0.97	0.59	27.8

Polarograms of some samples show a final current increase at about −1.5 v., which interferes with the full development of the diffusion current of the zinc. This is usually observed with samples that contain relatively large amounts of iron. In the presence of hydrazine the iron in the electrolyzed solution will be present in the ferrous state, and ferrous iron produces a wave in ammoniacal solution whose half-wave potential is −1.49 v. *vs.* the saturated calomel electrode. The interference can easily be eliminated by treating the electrolyzed solution, or an aliquot of it, with 5 cc. of concentrated nitric acid, and evaporating to dryness to oxidize ferrous iron and hydrazine. The residue is taken up in an appropriate amount of dilute hydrochloric acid, and the solution is prepared for analysis as above.

Typical results obtained by the above procedure are shown in Table XXXV-2.

Copper and Lead (according to Reynolds and Rogers[7]). This method utilizes the facts that in a 0.1 *M* sodium pyrophosphate supporting electrolyte of pH between 9 and 11 the cupric-pyrophosphate complex produces a doublet wave with half-wave potentials of −0.40 and −1.33 v. *vs.* S.C.E., and the lead-pyrophosphate complex produces a well-developed wave with a half-wave potential of −0.69 v. *vs.* S.C.E. Hence with mixtures of copper and lead the lead wave lies between the two copper waves and both metals can be determined from a single polarogram. Evidently the simultaneous determination of the lead will not be possible when only very small amounts are present because the large copper diffusion current will mask very small lead waves. Therefore the following procedure is not applicable to copper-base alloys containing less than about 1 per cent lead.

Dissolve a sample of about 0.1 g. in 1 ml. of concentrated nitric acid, and boil to expel oxides of nitrogen completely. Cool, and transfer to a 250-ml. volumetric flask with a 0.1 *M* sodium pyrophosphate solution. Neutralize with a sodium hydroxide solution to a pH of 9 to 11 (indicator test papers) and dilute to the mark with 0.1 *M* sodium pyrophosphate solution. To a 50-ml. aliquot of this solution add 1 ml. of a 2 per cent solution of gum ghatti (maximum suppressor) and 1 g. of solid sodium sulfite to remove dissolved oxygen, and record the polarogram.

The first copper wave shows a maximum that is not completely suppressed by the gum ghatti maximum suppressor, and for this reason Reynolds and Rogers recommend measurement of the second copper wave at −1.33 v. *vs.* S.C.E. Gelatin suppressed the copper maximum, but when sufficient gelatin was used to suppress the maximum completely the lead diffusion current was greatly decreased.

In the 0.1 *M* pyrophosphate supporting electrolyte ferric iron produces a wave with $E_{1/2} = -0.82$ v. *vs.* S.C.E. which coalesceses with the lead wave, and hence ferric iron interferes with the lead determination. Since the amount of iron in copper-base alloys is usually quite small, it should be possible to subtract the appropriate correction for iron from the apparent lead wave when the percentage of iron is known. It may also be possible to eliminate the iron interference by prior reduction to the ferrous state after the alloy is dissolved and before addition of pyrophosphate. Tin, zinc, and nickel do not interfere.

Reynolds and Rogers obtained satisfactory results with alloys containing 66 to 85 per cent copper and 1 to 6 per cent lead.

---

[7] C. A. Reynolds and L. B. Rogers, *Anal. Chem.*, **21**, 176, 758 (1949).

Iron (according to Meites[8]). This procedure utilizes the anodic wave produced by the oxidation of ferrous to ferric iron in an oxalate supporting electrolyte of pH = 5, whose characteristics are described in the section on iron, Chapter XXVII. In 1 *M* potassium oxalate of pH = 4.8, without gelatin or other maximum suppressor, the anodic half-wave potential is −0.25 v. *vs.* S.C.E., and $i_d/Cm^{2/3}t^{1/6}$ is 1.37 at 25°. With the exception of copper, which must be removed, no other metals interfere with the determination of iron under the conditions used. Because the ferrous-oxalate complex is easily air-oxidized care must be exercised to avoid contact of the prepared solution with air. This is accomplished by adding a measured volume of the sample solution (containing sulfurous acid to insure that all the iron is in the ferrous state) to a measured volume of previously de-aerated supporting electrolyte.

A sample of the copper-base alloy containing at least 10 mg. of iron is dissolved by being heated in 10 to 20 ml. of concentrated hydrochloric acid with the aid of the minimal necessary amount of nitric acid. The solution is diluted to 50 to 60 ml., and 20 g. of pure spongy silver is added to precipitate the copper and reduce ferric iron. The mixture is boiled vigorously for about 5 min., then 5 g. more silver is added and the boiling is continued for a minute or two. The excess silver, silver chloride, and precipitated copper is filtered off through paper, and the filtrate and washings are caught in a 100-ml. volumetric flask. 1 g. of anhydrous sodium sulfite is added and the solution is diluted to the mark. An aliquot, *e. g.*, 10 ml., of this sample solution is added to a known volume, *e. g.*, 50 ml., of an air-free 1 *M* potassium oxalate solution, previously adjusted to pH = 5, in a polarographic cell, and the polarogram is recorded or, simply, the diffusion current of the ferrous-oxalate complex is measured at −0.05 v. *vs.* S.C.E.

Meites obtained excellent results by this method with various Bureau of Standards copper-base alloy samples containing 0.2 to 0.8 per cent iron. He also described an alternative procedure for preparing the sample solution, based on preliminary ether extraction of the iron, and extended the method with excellent results to the determination of iron in aluminum alloys, ferrovanadium, and calcium molybdate.

2. ALUMINUM ALLOYS

Iron, Copper, Lead, Nickel, and Zinc. Kolthoff and Matsuyama[9] developed systematic procedures for determining iron, copper, lead, nickel, and zinc in aluminum alloys containing from a few tenths of one per cent up to several per cent of each of these elements.

---

[8] L. Meites, *Anal. Chem.*, **20,** 895 (1948).

[9] I. M. Kolthoff and G. Matsuyama, *Ind. Eng. Chem.*, *Anal. Ed.*, **17,** 615 (1945).

The alloy sample is decomposed by preliminary treatment with sodium hydroxide solution, and the solution is completed with nitric acid. Silicon, tin, and presumably antimony remain as a small residue.

Iron and copper are determined directly in the nitric acid solution of the alloy, the diffusion current of iron preceding the copper wave. Chloride ion must be absent to obtain full development of the ferric diffusion current before the copper wave. When the ratio of iron to copper is large (greater than about 10), the former is reduced to the ferrous state with hydroxylamine before the copper polarogram is obtained.

For the determination of lead preliminary removal of ferric iron and copper is usually necessary because the waves of these precede that of the lead. Iron is rendered harmless by reduction to the ferrous state with hydroxylamine, and copper is removed by precipitation as cuprous thiocyanate. Lead may also be separated from iron and copper by dithizone extraction in carbon tetrachloride.

Nickel and zinc are determined after treating the alloy solution with hydroxylamine (to reduce ferric iron), thiocyanate (to remove copper), citrate (to complex the aluminum and keep it in solution), and pyridine, and adjusting the pH to 4.5. The pyridine is used to complex the zinc and nickel and produce well-separated waves. Since most aluminum alloys contain less nickel than zinc both elements can be determined from a single polarogram, the smaller wave of nickel preceding that of zinc. Nickel may also be determined after removing copper and reducing iron electrolytically. Small amounts of zinc may be determined by preliminary extraction (together with lead) with dithizone in carbon tetrachloride.

The detailed procedures recommended by Kolthoff and Matsuyama follows. For an extensive discussion of the rationale of the procedures and possible alternate schemes the original paper should be consulted.

Preparation of Alloy Solutions. *A. Method.* To 1.000 g. of aluminum alloy drillings in a 150-ml. beaker covered with a watch glass, slowly add 11 ml. of 15 per cent sodium hydroxide. After the reaction subsides, heat the solution to effect completion of the reaction. Add 20 ml. of 1:1 nitric acid, stir, and heat to boil out the lower oxides of nitrogen. The alloy dissolves completely except for a little white residue which is probably metastannic acid and silica. Cool the solution to room temperature and dilute to 50.0 ml. in a volumetric flask. Aliquots of this solution, hereinafter referred to as the "alloy solution," are used for the various determinations.

B. Pure Aluminum Solution. Prepare an aluminum solution following the procedure outlined above, using pure aluminum in place of the alloy. This solution is used to prepare standard samples for determining the residual current and the diffusion current constants in the various procedures.

Determination of Iron and Copper without Separation. Pipet a 10-ml. aliquot sample of the aluminum alloy solution in nitric acid into a 25-ml. volumetric flask, add 0.5 ml. of 0.5 per cent gelatin solution, and make up to volume. Transfer the solution to a polarographic cell, remove the oxygen from the solution by bubbling with nitrogen gas, and run a polarogram or measure the diffusion current at +0.15 and −0.15 v. (*vs.* S.C.E.). Correct the observed diffusion current for the residual current of a pure aluminum solution. Compare the corrected diffusion current at +0.15 v. with that of a standard sample prepared with pure aluminum and standard iron and copper to obtain the amount of iron in the sample. Obtain the diffusion current for copper by subtracting the diffusion current at +0.15 v. from that at −0.15 v. (*vs.* S.C.E.), making corrections for the residual current. Compare this diffusion current for copper with that of a standard sample to obtain the amount of copper in the sample.

If the concentrations of iron and copper are too great and the total diffusion current is greater than *ca.* 20 microamp., use a smaller aliquot of alloy solution. Add a volume of pure aluminum solution such that the total volume of alloy and pure aluminum solutions is 10 ml.

Determination of Copper in Presence of Much Iron. To a 10-ml. aliquot sample of the aluminum alloy solution in a 50-ml. beaker, add 0.3 ml. of 2 *M* hydroxylamine hydrochloride solution and heat to the boiling point. Cool to room temperature and transfer quantitatively to a 25-ml. volumetric flask. Add 0.5 ml. of 0.5 per cent gelatin and dilute to the volume mark with distilled water. Transfer the solution to a polarographic cell and bubble with nitrogen gas for 5 to 10 minutes to remove oxygen. Record the polarogram from +0.2 to −0.2 v. or observe the current at +0.15 and −0.15 v. (*vs.* S.C.E.). Correct the diffusion current for the residual current and calculate the amount of copper present by comparing the corrected diffusion current with that observed with a standard sample. It is not strictly necessary to make the measurements at +0.2 or +0.15 v. The authors recommend that this be done, because in rapid routine work a trace of ferric iron may still be in the solution, giving a diffusion current at +0.2 v.

Determination of Lead in Presence of Much Copper and Iron. *A. Ordinary Method.* Pipet a 10-ml. aliquot of aluminum alloy solution into a 25-ml. volumetric flask. Add one drop of 0.1 per cent thymol blue indicator, then add 1 *M* sodium hydroxide until the indicator changes to an orange color, and 1 ml. in excess. Add 0.5 ml. of 2 *M* hydroxylamine hydrochloride and 0.5 ml. of 2 *M* potassium thiocyanate and shake gently until the red color of the iron thiocyanate disappears. Add a little water to wash down the sides of the flask, 0.5 ml. of 0.5 per cent gelatin solution, and dilute the solution to the volume mark with distilled water. Transfer the solution to a polarographic cell and bubble with nitrogen gas for 5 to 10 minutes to remove oxygen. Record a polarogram from −0.2 to −0.6 v. (*vs.* S.C.E.), or measure the current at −0.3 and −0.5 v. (*vs.* S.C.E.), Correct the observations for the residual current and compare the corrected diffusion current with that of a standard sample.

B. Alternate Method. Extract the lead with dithizone (see below for details).

Rapid Method for Determination of Nickel and Zinc. Pipet a 15-ml. aliquot sample of aluminum alloy solution into a 50-ml. volumetric flask, add 5 ml. of 1 *M* sodium hydroxide, and mix thoroughly to make the precipitated aluminum hydroxide redissolve. Add 0.70 ml. of 2 *M* hydroxylamine hydrochloride and 0.50 ml. of 2 *M* potassium thiocyanate and rinse down the neck of the flask with distilled water. Shake the solution in the flask until the red color of iron thiocyanate disappears; then add 5.0 ml. of 1.25 *M* sodium citrate solution. Mix the solution thoroughly and add 0.50 ml. of pyridine, 2 drops of 0.1 per cent bromocresol green indicator, and 15 per cent sodium hydroxide until the indicator changes to a distinct green color (pH = 4.5). Add 1.0 ml. of 0.5 per cent gelatin solution and dilute the solution to the volume mark with distilled water. Transfer the solution to a polarographic cell, bubble with nitrogen gas for 5 to 10 minutes, and record a polarogram from −0.4 to −1.4 v. (*vs.* S.C.E.). Compare the nickel and zinc diffusion currents with those observed with standard samples in pure aluminum treated in the same way to determine the amount of nickel and zinc present in the unknown.

Separation Method for Determination of Nickel. Electrolyze a suitable aliquot portion of the aluminum solution in nitric acid to remove copper and reduce iron. Transfer the solution to a 50-ml. volumetric flask, add 10 ml. of pyridine, dilute to the volume mark with distilled water, and shake vigorously. Filter the solution through a clean, dry funnel fitted with a coarse-porosity paper, *e. g.*, S. and S. No. 589 "Black Band." Pipet a 20-ml. aliquot of the filtrate and 0.5 ml. of 0.5 per cent gelatin into a 25-ml. volumetric flask and dilute to the volume mark. Transfer the solution to a polarographic cell, bubble with nitrogen for 5 to 10 minutes to remove oxygen, and record a polarogram from −0.6 to −1.0 v. (*vs.* S.C.E.). Compare the diffusion current with that of a standard sample to determine the amount of nickel present.

Extraction Method for Determination of Lead and Zinc. To facilitate the dithizone extraction of zinc, remove copper from the alloy solution before the extraction. Copper can be removed very simply by treating an aliquot portion of the alloy solution with hydroxylamine hydrochloride and potassium thiocyanate or a weighed sample of the alloy may be dissolved in hydrochloric acid. The amount of alloy sample used does not affect the diffusion current constant as in the case of the iron and copper determination, since the zinc is extracted from the solution. For most alloys a 15-ml. aliquot sample of its solution can be treated with 0.5 ml. of 2 M hydroxylamine hydrochloride and 0.5 ml. of 2 M potassium thiocyanate to remove the copper (the copper thiocyanate precipitate need not be removed from the solution). For alloys of low zinc content it is convenient to dissolve a 1-g. sample in 25 ml. of constant-boiling hydrochloric acid and filter the solution to remove the residue of copper, nickel, etc.

To the resulting copper-free solution, add thymol blue indicator, 10 ml. of saturated sodium citrate solution, and ammonia until the indicator changes color to a greenish yellow. Transfer this solution quantitatively to a separatory funnel (the stopcocks of the separatory funnels should not be greased but wetted with water). Add 10 ml. of 0.05 per cent dithizone in carbon tetrachloride and shake for 1 to 2 minutes. (If appreciable amounts of lead and zinc are present, the carbon tetrachloride layer becomes a bright cherry red.) Withdraw the carbon tetrachloride phase into a 125-ml. separatory funnel, add 10 ml. of 0.05 per cent dithizone in carbon tetrachloride to the aqueous solution, and extract as outlined above. Continue the extraction until the carbon tetrachloride phase remains green or becomes a brownish purple; then extract once more to insure the complete extraction of zinc and lead. (If a brownish purple scum forms over the carbon tetrachloride during the extractions, do not transfer it to the 125-ml. separatory funnel. The scum is nickel dithizonate which is slightly soluble in carbon tetrachloride.) Shake the aqueous solution with two small portions of carbon tetrachloride to rinse out droplets containing zinc and lead. Add these to the 125-ml. separatory funnel and discard the aqueous solution. Add 25 ml. of water containing 1 drop of ammonia to the combined extracts and shake to wash the carbon tetrachloride. Withdraw the carbon tetrachloride phase to another 125-ml. separatory funnel, shake the ammonia solution with two small portions of carbon tetrachloride, and add these extracts to the main carbon tetrachloride extract. Discard the ammonia solution and clean the funnel.

To the combined extract add 20 ml. of 0.1 *M* hydrochloric acid and shake for 2 to 3 minutes. (The carbon tetrachloride layer will turn green if no copper is present.) Withdraw the carbon tetrachloride phase to the other 125-ml. separatory funnel. Rinse the hydrochloric acid solution with two small portions of carbon tetrachloride and combine the latter with the carbon tetrachloride extract. Leave the hydrochloric acid solution in the separatory funnel with any scum and droplets of carbon tetrachloride.

Add 10 ml. of fresh 0.1 *M* hydrochloric acid to the carbon tetrachloride extract, shake for 2 to 3 minutes, and discard the carbon tetrachloride phase. Rinse the hydrochloric acid solution with two small portions of carbon tetrachloride and transfer it quantitatively to the separatory funnel containing the 20-ml. portion.

Shake the combined hydrochloric acid solution with several small portions of chloroform until the chloroform remains colorless, and discard the chloroform. (The nickel dithizonate scum is soluble in chloroform and thus only a few droplets of clear chloroform should remain with the acid solution.) Transfer the acid solution quantitatively to a 100-ml. beaker and boil on a hot plate to expel all the chloroform. Cool the solution to room temperature, transfer it to a 50-ml. volumetric flask, add 1 ml. of 0.5 per cent gelatin and 1 ml. of pyridine, and dilute to the volume mark with distilled water. Transfer the solution to a polarographic cell, bubble with nitrogen for 5 to 10 minutes to remove oxygen and record a polarogram. Compare the diffusion currents for lead and zinc with that observed for a standard sample prepared by diluting known amounts of lead and zinc solutions, 30 ml. of 0.1 *M* hydrochloric acid, 1 ml. of pyridine, and 1 ml. of 0.5 per cent gelatin to 50 ml.

Copper, Zinc, and Manganese. Spalenka[10] developed the following rapid procedures for the determination of copper, zinc, and manganese in amounts from a few hundredths of one per cent up to several per cent in aluminum alloys. Since all operations are performed in a single vessel (volumetric flask), and no separations are required, the procedures are particularly suited to rapid routine analysis.

For the determination of copper, a 0.1- to 0.3-g. sample of the aluminum alloy is treated in a 100-ml. volumetric flask with 10 ml. of 6 *N* hydrochloric acid and complete solution is effected by addition of the minimal amount of concentrated nitric acid or 30 per cent hydrogen peroxide, and warming. The solution is diluted to about 50 ml., treated with 25 ml. of 7 *M* (1:1) ammonia, 1 ml. of saturated sodium sulfite solution, and 1 ml. of 0.5 per cent gelatin, cooled to room temperature, and diluted to volume. The aluminum hydroxide precipitate is allowed to settle, and a portion of the supernatant solution is transferred to a polarographic cell and polarographed open to the air (the sulfite present removes oxygen).

[10] M. Spalenka, *Z. anal. Chem.*, **128,** 42 (1947).

To eliminate the error due to coprecipitation of copper with the hydrous aluminum oxide the procedure is calibrated empirically with standard samples containing approximately the same amounts of copper and aluminum as the unknowns.

According to Spalenka the foregoing procedure is also applicable to the simultaneous determination of copper, nickel, and zinc, in those cases in which the amounts of copper and/or nickel do not greatly exceed the amount of zinc.

For the determination of relatively small amounts of zinc in copper-rich or nickel-rich aluminum alloys, Spalenka recommends that a 0.5-g. sample be decomposed in a 50-ml. volumetric flask with 10 ml. of 3.5 *M* sodium hydroxide solution. Decomposition is finally completed by brief boiling, and virtually all the copper, nickel, magnesium, manganese, and iron remain in the residue. The mixture is diluted with 10 ml. of water, cooled to room temperature, and 15 ml. of a solution containing approximately 1 *M* ammonium chloride and 5 *M* ammonia is added. Finally 1 ml. of saturated sodium sulfite solution and 2 ml. of fresh 0.5 per cent gelatin solution are added, and the solution is diluted to the mark. A portion of this solution is then polarographed, and the zinc content computed by comparison with a standard sample carried through the same procedure. The method is applicable to the determination of 0.007 to about 4 per cent zinc. According to Spalenka the zinc wave is better defined in the presence of ammonia than in supporting electrolytes containing only sodium hydroxide, but no data are given on the coprecipitation of zinc with the hydrous aluminum oxide.

For the determination of manganese Spalenka recommends an alkaline cyanide supporting electrolyte, in which copper and zinc do not interfere. Following a suggestion of J. V. A. Novak,[11] Metol (monomethyl paramidophenol sulfate) is add to the solution to prevent the air-oxidation of the manganous-cyanide complex. Aluminum is kept in solution as a tartrate complex, and sulfite ion is used to remove dissolved oxygen.

A 0.5-g. sample is treated in a 50-ml. volumetric flask with 8 ml. of concentrated hydrochloric acid, and complete solution is finally achieved by addition of a few drops of 30 per cent hydrogen peroxide and brief boiling. To the warm solution, 2 ml. of saturated sodium sulfite solution and 3 ml. of a 0.1 per cent Metol solution are added with good mixing. (The Metol solution contains 0.1 g. of Metol, 2 ml. of saturated sodium sulfite, and 5 ml. of 12 *M* hydrochloric acid per 100 ml.) Then 15 ml. of a 50 per cent sodium potassium tartrate solution (containing 10 ml. of saturated sodium sulfite per liter) and 10 ml. of 2.5 *M* potassium cyanide solution (containing 10 ml. of saturated sodium sulfite per liter) are added, and the solution is cooled to room temperature. 1 ml. of fresh 0.5 per cent gelatin solution is added, and the solution is diluted to the mark. A portion of this solution is polarographed open to the air, and the manganese wave appears at −1.3 v. *vs.* S.C.E. The procedure is calibrated with a standard sample of similar composition carried through the same procedure, or by adding a known amount of manganese to the solution after the first polarogram has been recorded and polarographing a second time.

The method yielded good results with aluminum alloys containing from 0.04 to 4 per cent manganese, but is not applicable in the presence of nickel since the wave of the latter coalesces with the manganese wave. This method is not as simple,

[11] J. Heyrovsky, *Polarographie*, Springer, Vienna 1941 (p. 309).

nor probably as accurate, as the method developed by Kolthoff and Watters[12] (described in detail in Chap. XXVI under Manganese), based on the measurement of the diffusion current of the manganic-pyrophosphate complex.

Lead. The following procedure is based on the recommendations of Jablonski and Moritz.[13]

Dissolve a 1-g. sample of the aluminum alloy (0.01 to about 1 per cent lead) in 20 ml. of 6 N (1:1) hydrochloric acid with the aid of a *small* amount of potassium chlorate. Add 10-g. of tartaric acid, and pour the solution into 40 ml. of 8 M sodium hydroxide. Add 0.5 to 1 g. of hydroxylamine hydrochloride and heat to boiling. Cool, add 1 ml. of 1 per cent tylose or gelatin solution, and dilute to exactly 100 ml. Record the polarogram of a portion of this solution. The lead wave appears at −0.75 v. *vs.* S.C.E.

According to Jablonski and Moritz, the hydroxylamine reduces the dissolved oxygen and makes unnecessary the use of an indifferent gas.

Jablonski and Moritz[14] have also described procedures for the determination of nickel and zinc in aluminum alloys.

Zinc. The following procedure for the determination of zinc in aluminum alloys was recommended by Stross.[15]

A weight of sample appropriate to the zinc content is decomposed preliminarily with a small volume of concentrated sodium hydroxide solution and the residue is filtered off. The residue is digested with sulfuric acid and potassium perchlorate, and the solution is made alkaline with sodium hydroxide and combined with the original solution. The combined solution is made up to a known volume with the addition of small amount (0.005 to 0.01 per cent) of gelatin as a maximum suppressor, and the polarogram of the zincate ion is recorded. The zinc content of the unknown is obtained by comparison with standard samples carried through the same procedure.

Relatively large amounts of copper and lead probably will interfere with the determination of zinc by this procedure, so that its usefulness is limited to special alloys.

Coates and Smart[16] have also developed procedures for the polarographic analysis of aluminum and aluminum alloys. The principles utilized in their procedures are outlined below; for complete details the original paper should be consulted.

Lead and Zinc. Lead and zinc are extracted from an ammoniacal tartrate solution of the aluminum sample with dithizone. The dithizone extract is extracted with dilute hydrochloric acid and the extract is divided into two portions. In one

---

[12] I. M. Kolthoff and J. I. Watters, *Ind. Eng. Chem., Anal. Ed.*, **15,** 8 (1943).
[13] F. Jablonski and H. Moritz, *Aluminium*, **26,** 245 (1944).
[14] F. Jablonski and H. Moritz, *Aluminium*, **26,** 38, 97, 321 (1946).
[15] W. Stross, *Metallurgia*, **36,** 163, 223 (1947).
[16] A. C. Coates and R. Smart, *J. Soc. Chem. Ind.*, **60,** 249 (1941).

of these lead is determined, using potassium hydroxide as supporting electrolyte, and zinc is determined in the second portion after neutralizing the free acid with sodium hydroxide (pH = 6.0).

When the zinc content of the alloy exceeds about 0.1 per cent an acid hydrogen sulfide separation is used to remove lead, copper, etc., and the zinc is subsequently separated as the sulfide from a weakly acid formate solution which contains tartrate to keep aluminum in solution. The zinc sulfide is finally dissolved in dilute hydrochloric acid and polarographed in an ammoniacal supporting electrolyte.

Copper. The alloy sample is dissolved in a mixture of hydrochloric and nitric acids, the aluminum is precipitated with ammonia, and the copper is polarographed in the filtrate. Considerable negative error resulting from coprecipitation of the copper with the hydrous aluminum oxide may be expected.

Antimony. The alloy sample is dissolved in a hydrochloric–nitric acid mixture, and fumed with sulfuric acid. The antimony is then distilled as the trichloride in a nitrogen atmosphere, after reduction with pyrogallol. The antimony in the distillate is then polarographed in a dilute hydrochloric acid medium.

3. METALLIC IMPURITIES IN ALUMINUM

Hohn[17] has described a method for the detection and determination of traces of copper, lead, zinc, iron, and indium in technical aluminum. Since aluminum is reduced at a much more negative potential than these metallic impurities, their determination is a relatively simple matter.

According to Hohn, a 5-g. sample of the aluminum is dissolved in concentrated hydrochloric acid, and the solution is evaporated to incipient dryness to remove the excess acid. The residue is taken up with water, transferred to a 50-ml. volumetric flask, and made up to the mark, neglecting any precipitate of aluminum hydroxide that may form by hydrolysis. The polarogram of an aliquot portion of the solution is then recorded in the usual way. The large concentration of aluminum chloride serves as supporting electrolyte. Hohn found that amounts of copper, lead, zinc, iron, and indium as small as 0.001 per cent could be detected in this way. No mention was made of the possible mutual interference of the copper and ferric iron waves. This interference can probably be eliminated by taking precautions to keep the iron in the ferrous state, since ferrous iron is not reduced until about −1.3 v. *vs.* the S.C.E., whereas copper, lead, indium, cadmium, and zinc are reduced at considerably more positive potentials.

Semerano and Capitanio[18, 19] have also developed procedures for the polarographic determination of traces of copper, cadmium, nickel, zinc, iron, and lead in high-purity aluminum. These procedures, which were expecially designed for the determination of 0.00003 to 0.001 per cent of

---

[17] H. Hohn, *Chemische Analysen mit dem Polarographen*, Berlin, 1937 (p. 84).

[18] G. Semerano and V. Capitanio, *Mikrochemie ver. Mikrochim. Acta*, **30**, 71 (1942); *Atti reale 1st. Veneto Sci.*, **Tomo CI, Parte II,** 333 (1942).

[19] G. Semerano, *Die Chemie*, **55**, 351 (1942).

the several metallic impurities, employ a preliminary separation of the trace metals as sulfides.

A 5-g. sample is treated in a platinum dish with a solution of 20 g. of potassium hydroxide in 80 ml. of water, and warmed to affect complete decomposition. The solution is cooled to room temperature and hydrogen sulfide is bubbled into it until hydrous aluminum oxide begins to precipitate. The mixture is then diluted to about 200 ml. and centrifuged. The precipitate is then washed successively by centrifugation with a 1 per cent potassium sulfide solution in 2 per cent potassium hydroxide, then with 1 per cent potassium sulfide, and finally with water. The washed sulfide precipitate is then dissolved in a few drops of hot concentrated nitric acid, and the solution is washed into a small platinum dish and evaporated to dryness. The residue is treated with 1 ml. of an ammoniacal supporting electrolyte containing 1.5 *M* ammonium chloride, 0.1 *M* ammonium carbonate, 0.5 *M* ammonia, 0.1 *M* sodium sulfite, and 0.005 per cent gelatin, which extracts copper, cadmium, nickel, and zinc, and leaves iron and lead insoluble. The extract is centrifuged, and the supernatant solution is transferred to a polarographic cell, and polarographed open to the air. The polarogram shows waves for copper, cadmium, nickel, and zinc, in the order named.

It should be emphasized that the foregoing method will not be applicable if one or more of the more easily reducible metals greatly preponderates.

The iron and lead residue remaining from the extraction is washed with water by centrifugation, and is then dissolved in a few drops of concentrated nitric acid. The solution is transferred to a small platinum dish or platinum crucible and evaporated to dryness. The residue is dissolved in 1 ml. of a supporting electrolyte containing about 4 *M* ammonium chloride, 0.1 *M* hydrochloric acid, and 0.2 per cent tylose, and the polarogram of this solution is obtained after removing dissolved air with nitrogen. Since the diffusion current of the ferric iron begins at the start of the polarogram, the residual current should be determined with the supporting electrolyte alone and subtracted to obtain the diffusion current of the iron. The lead wave is at about −0.45 v. *vs.* S.C.E.

Difficulty may be anticipated in the determination of lead when its amount is much smaller than that of iron. It should be possible to circumvent this difficulty by reducing the iron to the ferrous state with hydroxylamine, as recommended by Kolthoff and Matsuyama,[20] before recording the wave of lead.

Sodium. Urech and Sulzberger[21] developed the following procedure for the determination of 0.001 to 0.01 per cent of sodium in high-purity aluminum.

Dissolve a 2-g. sample in 60 ml. of 6 *N* (1:1) hydrochloric acid in a porcelain casserole or platinum dish, if necessary with the addition of a few drops of 2 per cent ferric chloride solution to accelerate the reaction. Evaporate to crystallization (*ca.* 30 ml.), transfer to a measuring cylinder, and dilute to 35 ml. Cool the solution and saturate it with gaseous hydrogen chloride to precipitate most of the aluminum

[20] I. M. Kolthoff and G. Matsuyama, *Ind. Eng. Chem., Anal. Ed.*, **17**, 615 (1945).

[21] P. Urech and R. Sulzberger, *Helv. Chim. Acta*, **27**, 1074 (1944).

as $AlCl_3 \cdot 6H_2O$. Filter through a medium-porosity, sintered glass filtering crucible, and evaporate 25 ml. of the filtrate to dryness in a platinum dish. Dissolve the residue in 2 drops of 6 *N* hydrochloric acid and water, and dilute to *ca.* 100 ml. Add 1 drop of 30 per cent hydrogen peroxide, heat to boiling, and precipitate the remainder of the aluminum with 3 to 5 drops of concentrated ammonia. Filter after about 30 minutes through paper, and evaporate the filtrate to dryness in a platinum dish. Finally, heat for 3 minutes in a muffle furnace at 350° C. to insure complete expulsion of ammonium salts. Take up the residue in 5 ml. of warm water, filter through paper, and again evaporate to dryness, and finally heat the residue for 3 minutes at 350° C. Dissolve the residue in exactly 2 ml. of water plus 0.2 ml. of 10 per cent tetramethylammonium hydroxide solution plus 1 drop of 2 *N* phosphoric acid. Transfer to a polarographic cell, remove dissolved air with nitrogen, and record the polarogram.

It is necessary to use a polarographic cell with a mercury anode, to avoid introducing alkali metal impurities from a salt bridge.

A blank determination should be run with all the reagents, and the appropriate correction applied. This correction for sodium impurity in the reagents is usually quite appreciable.

4. MAGNESIUM ALLOYS

Gull[22] has described a procedure for the polarographic determination of aluminum, zinc. lead, and manganese in magnesium alloys.

A 1-g. sample of the alloy is treated with 30 ml. of water and a small excess of concentrated hydrochloric acid in a 250-ml. volumetric flask. When the solution of the sample is complete, 100 ml. of water and a few drops of bromphenol blue solution are added, and the solution is carefully neutralized to the transition color of the indicator (pH about 3.8). The solution is made up to 250 ml. and allowed to stand for several hours to insure redissolution of any traces of colloidal aluminum hydroxide that may have been formed. An aliquot portion of the solution is then transferred to a polarographic cell, freed from air with hydrogen or nitrogen, and the polarogram is recorded. Gull does not mention the use of any specific maximum suppressor, and apparently the bromophenol blue suffices for this purpose. The half-wave potential of lead is at −0.4 v., that of zinc at −1.0 v., and that of aluminum at −1.7 v. *vs.* the S.C.E.

The presence of manganese is indicated by an indistinct wave at −1.5 v. just preceding the aluminum wave, but the wave is not well enough defined for quantitative determination. In order to determine the manganese the aluminum is precipitated by adding 0.5 g. of potassium acetate per 15 ml. of solution after the first polarogram is run, and a second polarogram is recorded. The wave of manganese then appears well defined with a half-wave potential of −1.5 v. *vs.* the S.C.E. The presence of relatively large amounts of lead and zinc will interfere with the determination of small amounts of manganese by this procedure. When the proportion of lead and zinc is not too great it may be possible to compensate their diffusion currents before the wave of manganese is recorded; otherwise, a preliminary sepa-

---

[22] H. C. Gull, *J. Soc. Chem. Ind.*, **56**, 177–183 (1937).

ration of the manganese by precipitation as hydrous manganese dioxide will be necessary.

It should also be possible to determine even traces of copper by the above procedure, since the wave of copper is well in advance of those of the other metals. It will probably be necessary to use nitric acid as the solvent for the alloy when copper is to be determined.

Gull stated that the accuracy of the determination of 1 to 10 per cent aluminum, up to 1 per cent zinc, up to 0.5 per cent manganese, and up to 0.1 per cent lead was ± 2 relative per cent by the foregoing procedure.

The polarographic analysis of magnesium alloys has also been described by Semerano,[23] and Heller and Zan'ko.[24]

5. ZINC-BASE ALLOYS

Copper, Lead, and Cadmium. Hawkings and Thode[25] developed the following procedure for the determination of trace amounts of copper, lead, and cadmium in high-purity, zinc-base die casting alloys, which permits the determination of only a few hundredths of one per cent, or less, of these elements with a precision of $\pm 1 \times 10^{-4}$ per cent of the sample weight.

An 8-g. sample in the form of turnings is treated slowly with 25 ml. of concentrated hydrochloric acid in a 125-ml. beaker. After the first violent reaction subsides, add cautiously 2 to 3 ml. of concentrated nitric acid, and warm until solution is complete. Then add 2 to 3 ml. more of nitric acid and evaporate until salts begin to crystallize and the solution becomes syrupy. Cool, wash down with about 10 ml. of water, and add 7 ml. of concentrated hydrochloric acid. Heat until solution is complete (10 to 15 minutes). Transfer to a 50-ml. volumetric flask with a minimal amount of water, and add 2.5 ml. of freshly prepared 0.2 per cent gelatin solution, and 0.1 ml. of 2 *M* hydroxylamine hydrochloride. Shake and warm until the solution becomes colorless, or nearly so, and the ferric iron is reduced. Cool and dilute to volume with freshly boiled, *i. e.*, reasonably air-free, water. Transfer a portion of this solution to a polarographic cell, remove dissolved air with nitrogen, and record the polarogram. The half-wave potential of the copper, lead, and cadmium waves are, respectively, -0.22 v. the, (second copper wave), -0.43 v., and -0.64 v. *vs.* S.C.E.

When the amount of copper in the alloy exceeds about 0.1 per cent, and presumably whenever the amount of copper greatly exceeds the amounts of lead and cadmium, Hawkings and Thode recommend preliminary removal of the copper by electrodeposition on a platinum cathode from a nitric acid solution of the alloy, and weighing the deposit in the ordinary way. Lead and cadmium are then determined in the residual solution by the above procedure after nitric acid has been completely re-

[23] G. Semerano, *Mikrochemie ver. Microchim. Acta*, **25**, 191 (1938).

[24] B. A. Heller and A. M. Zan'ko, *Zvaodskaya Lab.*, **8**, 1030 (1939); **9**, 513 (1940).

[25] R. C. Hawkings and H. G. Thode, *Ind. Eng. Chem., Anal. Ed.*, **16**, 71 (1944).

moved by evaporation to a syrup or incipient crystallization with hydrochloric acid. The evaporation with hydrochloric acid should be repeated to insure complete removal of nitric acid.

Moderate amounts, *e. g.*, 0.05 per cent, of nickel, cobalt, manganese, silver, arsenic, mercury, indium, and relatively large amounts of aluminum, iron, and magnesium do not interfere with the determination of trace amounts (0.005 to 0.05 per cent) of copper, lead, and cadmium. Bismuth, antimony, thallium, and tin all produce waves in the neighborhood of, or preceding, the copper, lead, and cadmium waves, and hence these elements interfere to a greater or lesser degree, depending on their relative amounts in the sample. Germanium in small amounts would not be expected to interfere because of the volatility of germanium tetrachloride and hence its removal during the preparation of the sample solution. Gallium has a more negative reduction potential than cadmium, and hence should not interfere.

Lead and Cadmium. The following procedure is recommended by Kilian[26] for the determination of traces of lead and cadmium in high-purity electrolytic zinc in the absence of tin and thallium. It appears to be applicable to samples containing from about 0.001 to a few hundredths of one per cent of lead and cadmium.

Dissolve a 25-g. sample in about 125 ml. of concentrated hydrochloric acid, and, if necessary, add a small amount of 30 per cent hydrogen peroxide dropwise to attain complete solution. Cool and dilute to 250 ml. in a volumetric flask. Evaporate a 25-ml. portion of this solution to dryness. Dissolve the residue in about 10 ml. of water and 1 to 2 drops (no more!) of 12 *M* hydrochloric acid, transfer to a 25-ml. volumetric flask, add 2 ml. of 1 per cent gum arabic solution (as a maximum suppressor), and dilute to volume. Transfer a portion of this solution to a polarographic cell, remove dissolved air with nitrogen or hydrogen, and record the polarogram. The lead wave is observed at −0.4 v., and the cadmium wave occurs at −0.6 v., *vs.* S.C.E.

The lead and cadmium contents of the unknown can be determined by adding known amounts of lead and cadmium (about equivalent to the lead and cadmium in the unknown) to another portion of the sample solution, and recording a second polarogram.

This procedure is not applicable to samples containing tin and/or thallium, whose waves coincide with those of lead and cadmium. Copper and iron will also interfere when present in larger amounts than the lead and cadmium. It should be possible to eliminate the interference of iron by reducing it to the ferrous state with hydroxylamine.

A procedure very similar to those outlined above for the determination of lead and cadmium in zinc-base alloys has been recommended by the American Society for Testing Materials.[27]

---

[26] W. Kilian, in *Analyse der Metalle, Bd. I, Schiedsverfahren*, Springer, Berlin, 1942 (p. 442).

[27] *A. S. T. M. Methods of Chemical Analysis of Metals*, 1946, (p. 270).

Copeland and Griffith[28] employed essentially this same method and they obtained excellent results with samples of zinc containing about 0.001 per cent each of lead and cadmium. These authors also demonstrated that interference by tin and indium could be eliminated by precipitating them as the hydrous oxides by adjusting the pH of the sample solution to 5.6 with ammonia before measuring the lead and cadmium waves. The precipitated tin and indium hydroxides need not be filtered off.

Copper and Manganese. Spalenka[29] described the following rapid procedures for the determination of copper and manganese in zinc-base alloys containing from 0.03 to 10 per cent of each of these elements.

The procedure for the copper determination is the same as that for aluminum alloys described in the section on Aluminum Alloys. According to Spalenka this procedure is applicable to zinc-base alloys containing 0.03 to about 10 per cent copper.

For the determination of manganese in zinc-base alloys, a 0.5-g. sample is dissolved in a 50-ml. volumetric flask with 8 ml. of 6 *N* (1:1) hydrochloric acid and 5 to 20 drops of 30 per cent hydrogen peroxide. Complete solution is obtained by brief boiling. While the solution is still warm, 2 ml. of saturated sodium sulfite, and 3 ml. of Metol solution are added. (The Metol solution contains 0.1 g. of Metol, 2 ml. of saturated sodium sulfite, and 5 ml. of 12 *N* hydrochloric acid per 100 ml.) The solution is then treated with 10 ml. of 7 *M* (1:1) ammonia solution (containing 20 ml. of saturated sodium sulfite per liter), and cooled to room temperature. In the order named, 15 ml. of 5 *M* potassium cyanide solution (0.5 *M* in sodium hydroxide), and 5 ml. of 10 *M* sodium hydroxide solution (containing 5 ml. of saturated sodium sulfite per 100 ml.) are added, with thorough mixing after each addition. 1 ml. of fresh 0.5 per cent gelatin is added, and the solution is diluted to the mark. The polarogram of a portion of this solution is recorded, and the manganese wave appears at −1.3 v. *vs.* S.C.E. The method is calibrated by carrying a standard sample through the same procedure. The procedure is not applicable to alloys containing nickel since the nickel wave coalesces with the manganese wave.

The procedure developed by Kolthoff and Watters,[12] which is described in detail in Chapter XXVI in the section on manganese, can doubtless be applied advantageously to zinc alloys. It is less subject to interference by other metals than the foregoing method.

Zinc-Thorium Alloys. A method for the determination of 20 p.p.m. to 1 per cent zinc in zinc-thorium alloys has been developed by Patterson and Banks.[30] The supporting electrolyte contains sulfosalicylate to complex the thorium and keep it in solution and the pH is adjusted to 8.5.

A sample of the alloy weighing 10 g. or less (depending on the zinc content) is placed in a 600-ml. beaker, covered with water, and treated with 50 ml. of concentrated hydrochloric acid. After the vigorous reaction subsides, about 1 mg. of

[28] L. C. Copeland and F. S. Griffith, *Anal. Chem.*, **22,** 1269 (1950).

[29] M. Spalenka, *Z. anal. Chem.*, **128,** 42 (1947).

[30] J. H. Patterson and C. V. Banks, *Anal. Chem.*, **20,** 897 (1948).

sodium fluosilicate is added, and the solution is boiled until the black residue dissolves. It is then evaporated to 20 ml. and treated with 25 ml. of 2 *M* sulfosalicylic acid, 4 ml. of 0.5 per cent gelatin, and 20 ml. of concentrated (15 *M*) ammonia. The solution is stirred until all the precipitate has dissolved, the pH is then adjusted to 8.5 ± 0.2 by addition of either ammonia or hydrochloric acid, and the solution is transferred to a 100-ml. volumetric flask and diluted to the mark. The polarogram of a portion of this solution is then recorded after removing dissolved air with nitrogen. The zinc wave occurs at −1.3 v. *vs.* S.C.E.

Patterson and Banks obtained very good results with synthetic samples containing from 19 to 141 p.p.m. of zinc. Nitrate interferes, apparently because of its catalyzed reduction in the presence of the thorium.

6. ZINC ORES

Kraus and Novak[31] recommended the following procedures for the determination of copper, lead, cadmium, and zinc in crude zinc ores.

Copper, Cadmium, and Zinc. A 1-g. sample of the finely powdered ore, *e. g.*, zinc blende, is dissolved in a hot mixture of 10 ml. of concentrated hydrochloric acid and 5 ml. of concentrated nitric acid; after solution is complete, 1 ml. of concentrated sulfuric acid is added and the solution is evaporated nearly to dryness. The residue is transferred to a 50-ml. volumetric flask with 20 to 25 ml. of water and 20 ml. of concentrated ammonia is added to remove iron, etc. After adding 2.5 ml. of 0.2 per cent gelatin as a maximum suppressor, the solution is made up to the mark and mixed thoroughly. The polarogram of an aliquot portion of the supernatant solution is then recorded. (A small quantity of solid sodium sulfite may conveniently be used to remove dissolved oxygen.) The double copper wave is at −0.2 v. and −0.5 v., that of cadmium is at about −0.7 v., and the zinc wave is at about −1.4 v. *vs.* the S.C.E. Since the amounts of copper and cadmium are relatively small their waves are first recorded with a relatively great sensitivity of the galvanometer, and the sensitivity is then reduced to record the zinc wave.

Lead and Cadmium. A 1-g. sample of the zinc ore is dissolved in 10 ml. of hot concentrated hydrochloric acid in a 50-ml. volumetric flask, with the dropwise addition of the minimal amount of concentrated nitric acid that is necessary for complete solution. After cooling, and diluting with an equal volume of water, 1 g. of very pure aluminum foil is added to precipitate copper and reduce iron to the ferrous state. When the solution of the aluminum is complete, 2.5 ml. of 0.2 per cent gelatin is added and the solution is made up to the mark with boiled, air-free water. The polarogram of an aliquot portion of the solution is then obtained with care to avoid air-oxidation of the ferrous iron. The wave of lead is at −0.45 v. and that of cadmium is at −0.65 v. *vs.* the S.C.E.

The polarographic determination of copper, lead, cadmium, iron, aluminum, and zinc in zinc ores and other zinc metallurgical products has also been described by Kozlova and Portnov.[32]

---

[31] R. Kraus and J. Novak, *Collection Czechoslov. Chem. Communs.*, **10,** 534 (1938).

[32] A. A. Kozlova and M. A. Portnov, *Zavodskaya Lab.*, **9,** 287 (1940); *Chem. Abstracts*, **34,** 5783 (1940).

7. LEAD AND ZINC IN PAINTS

The following procedure was developed by Abraham and Huffman.[33]

A 0.3-g. sample of the extracted paint pigment is digested for 30 minutes with 2 to 3 ml. of concentrated hydrochloric acid, evaporated to dryness, and the residue is taken up in 5 ml. of hot 0.5 *N* hydrochloric acid. The solution is diluted with 50 ml. of hot water, filtered, and the residue is washed with boiling water. The combined filtrate and washings are cooled, transferred to a 250-ml. volumetric flask, treated with 25 ml. of 1 *N* lithium chloride (potassium chloride should serve as well), and diluted to the mark. The polarogram of this solution is then recorded, and the lead and zinc contents are determined from their well-separated waves. With this procedure Abraham and Huffman obtained results that agreed very well with those obtained by ordinary chemical methods.

8. LEAD- AND TIN-BASE ALLOYS

Zinc. The following procedure for the determination of traces of zinc (0.001 to 0.05 per cent) in lead-tin solder and white metal bearing alloys is based on the recommendations of Committee E-3 of the American Society for Testing Materials.[34] Tin (and antimony) are removed by volatilization—by dissolving the sample in a hydrobromic acid–perchloric acid–bromine mixture and evaporating to dryness. The bulk of the lead is removed by precipitation as lead chloride, and the remainder by hydrogen sulfide precipitation, which also removes other metals whose sulfides are insoluble in dilute hydrochloric acid. The zinc is finally polarographed in an ammoniacal supporting electrolyte.

Treat a 2-g. sample of the solder in a 200-ml. flask with 20 ml. of a solution prepared by mixing 10 ml. of pure bromine with 90 ml. of 40 per cent hydrobromic acid. Heat gently until dissolution is complete, avoiding excessive loss of bromine. Add 10 ml. of 70 per cent perchloric acid and heat cautiously by rotating over an open flame to expel the hydrobromic acid (and stannic and antimonic bromides) and convert all salts to perchlorates. (This evaporation should be performed slowly to avoid excessive loss of perchloric acid, and the usual safety precautions in evaporating perchloric acid should be observed.) If the solution is turbid (incomplete expulsion of tin and antimony), add 10 ml. of the hydrobromic acid–bromine mixture and repeat the evaporation. Finally evaporate to 1 to 2 ml. and cool.

Add 100 ml. of hot water and, when the soluble salts are dissolved, add 10 ml. of 6 *N* (1:1) hydrochloric acid and cool in an ice bath. Filter off the lead chloride and wash with cold 1 *M* (1:10) hydrochloric acid. Pass hydrogen sulfide into the filtrate and washings for 15 minutes and, without delay, filter through close-texture paper, and wash with 1 *M* hydrochloric acid saturated with hydrogen sulfide. Evaporate the filtrate to about 20 ml., add 1 ml. of 1:1 sulfuric acid and evaporate on a hot plate

[33] B. M. Abraham and R. S. Huffman, *Ind. Eng. Chem., Anal. Ed.*, **12,** 656 (1940).
[34] *A. S. T. M. Methods of Chemical Analysis of Metals*, 1946, p. 239.

having a surface temperature of about 300° C. until the sulfuric acid has been evaporated and white fumes are no longer evolved.

Cool, and dissolve the residue in exactly 5 ml. of a supporting electrolyte containing 1 *M* ammonium chloride, 1 *M* ammonia, and 0.005 per cent gelatin. Transfer the solution to a polarographic cell, remove dissolved air with nitrogen, and record the polarogram. The zinc wave is observed at −1.3 v. *vs.* S.C.E. The procedure may be calibrated with synthetic samples prepared from known amounts of zinc and very pure lead and tin (and copper and antimony if these are present in the unknown) to approximate the composition of the unknown.

Nickel, if present, will produce a wave at −1.06 v. *vs.* S.C.E., which will not interfere unless the amount of nickel exceeds the amount of zinc present. When excessive amounts of nickel are present the final residue obtained after the evaporation with sulfuric acid is dissolved in a mixture of 1 ml. of 6 *N* (1:1) hydrochloric acid and 10 ml. of water. The solution is heated to 50° to 60°, 1 ml. of a 1 per cent alcoholic solution of dimethylglyoxime is added, and the solution is made slightly alkaline with ammonia to precipitate nickel dimethylglyoximate. After 1 hour the precipitate is filtered off on paper and washed with cold water. The filtrate is treated with 1 ml. of 1:1 sulfuric acid, evaporated to dryness, and the residue is finally heated in a muffle furnace at 450° to 500° C. to eliminate ammonium salts and and organic material completely. The residue is then dissolved in 5 ml. of the ammoniacal solution, and the solution is polarographed as above.

Blank determinations should be run with all the reagents.

Shaikind[35] recommended the following procedures for the determination of small amounts of copper, zinc, cadmium, nickel, and manganese in lead alloys.

Copper and Zinc. A 10-g. sample is dissolved in nitric acid and, after dilution, the lead is precipitated by addition of dilute sulfuric acid. The lead sulfate is filtered off, washed, and the filtrate and washings are evaporated to dryness and ignited gently. The residue is moistened with 3 to 4 drops of dilute sulfuric acid, and taken up in 15 to 20 ml. of hot water to precipitate basic bismuth salts. The solution is filtered into a 100-ml. volumetric flask, 10 ml. of 4 *N* ammonium chloride and one drop of methyl orange indicator are added, and the solution is neutralized with a few drops of concentrated ammonia. 10 ml. of concentrated ammonia is added in excess, followed by 2 ml. of 0.5 per cent gelatin, and the solution is made up to the mark. The polarogram of an aliquot portion is recorded in the usual way.

According to Shaikind, results by this method are accurate to about ±3 per cent with amounts of copper and zinc from 0.05 to 0.4 per cent. Apparently, antimony and bismuth are removed completely in the above procedure and do not interfere, but more information on this point would be desirable. The possible interference of arsenic also should be investigated.

Cadmium, Nickel, and Manganese can also be determined in lead alloys by the foregoing procedure, according to Shaikind.

Procedures for the polarographic determination of copper, bismuth,

[35] S. P. Shaikind, *Zavodskaya Lab.*, **8,** 567 (1939); *Chem. Abstracts*, **34,** 5372 (1939).

lead, cadmium, and zinc in lead-tin alloys have also been described by Pletenev, Aref'eva, Tal, and Dubovitskaya.[36]

9. IMPURITIES IN REFINED LEAD

The determination of a few thousandths up to a few hundredths of one per cent of copper, cadmium, nickel, zinc, iron, bismuth, and antimony in highly refined lead (purity above 99.9 per cent) has been discussed by Zotta,[37] and the following procedures are based on her recommendations.

A 10-g. sample of the lead is dissolved in 50 ml. of 1:4 nitric acid in a 100-ml. volumetric flask. The solution is treated with 5 ml. of concentrated sulfuric acid to precipitate the lead, and is diluted to volume, and allowed to stand for about one-half hour until the lead sulfate settles out.

For the determination of antimony a 5-ml. aliquot of the supernatant sample solution is polarographed without any further treatment. According to Zotta the antimony wave follows the combined waves of copper, bismuth, and iron. (It seems doubtful that this method is reliable because the antimony will have been more or less oxidized to the +5 state by the solution of the sample in nitric acid, and only +3 antimony and not +5 antimony produces a reduction wave from a sulfuric-nitric acid medium. Furthermore, antimony will tend to remain insoluble as Sb_2O_5 when the sample is dissolved, especially if tin is also present.)

For the simultaneous determination of copper, cadmium, nickel, and zinc a 20-ml. aliquot of the supernatant sample solution is evaporated completely to dryness. The residue is taken up in exactly 3 ml. of a supporting electrolyte comprising 1.5 *N* ammonium sulfate, 0.5 *N* ammonia, 0.1 *N* ammonium carbonate, and 0.005 per cent gelatin, and the solution is polarographed. In addition to the waves of copper, cadmium, nickel, and zinc, which appear in that order, a small wave of lead is observed between the copper and cadmium waves.

For the simultaneous determination of iron and bismuth, a 40-ml. aliquot of the sample solution is evaporated to sulfur trioxide fumes. The residue is transferred to a small centrifuge tube with 2 ml. of water and concentrated ammonia is added in slight excess. The mixture is heated on a steam bath for 20 minutes and then centrifuged. The precipitate (containing the iron and bismuth) is washed by centrifugation with dilute ammonia. The washed precipitate is dissolved in 1.5 ml. of 4 *N* sulfuric acid, 3 ml. of 1 *M* sodium citrate and one drop of methyl red indicator are added, the solution is neutralized to a pH of 5 to 5.5 and polarographed. The wave of ferric iron occurs just prior to that of bismuth.

According to Zotta the minimum detectable amounts of the various metals by these procedures are 0.0005 per cent of cadmium, bismuth, or iron, 0.0004 per cent of copper, 0.0002 per cent of zinc, and 0.001 per cent of antimony.

[36] S. A. Pletenev, T. V. Aref'eva, E. M. Tal, and E. I. Dubovitskaya, *Zavodskaya Lab.*, **12,** 38 (1946).

[37] M. Zotta, *Gazz. chim. ital.*, **78,** 143 (1948). See also S. P. Scharkind and A. Ya. Kil'ter, *J. Applied Chem., U. S. S. R.*, **13,** 455 (1940).

10. STEEL AND FERRO-ALLOYS

Procedures for the polarographic determination of copper, nickel, cobalt, manganese, chromium, molybdenum, vanadium, and tungsten in steel and other iron products have been described by Thanheiser and Maassen,[38] Thanheiser and Willems,[39] von Stackelberg, Klinger, Koch, and Krath,[40] Lingane and Kerlinger,[41] and Kolthoff and Watters.[42] The determination of nickel and cobalt in steel by amperometric titration has been described by Kolthoff and Langer[43] (see Chap. XLVII).

Since ferric iron is reduced at a more positive potential than the alloying elements in steel it must be removed before the latter can be determined. Thanheiser and Maassen employed ammonia precipitation of the iron (and chromium) when only copper and nickel were to be determined. Coprecipitation of copper and nickel with the hydrous ferric oxide is minimized by having a large concentration of ammonium chloride present. The copper and nickel can be determined simultaneously in the filtrate, since the former is reduced at a much more positive potential than the latter.

The ammonia separation of iron cannot be used when cobalt and manganese are to be determined because these metals partially coprecipitate with the hydrous ferric oxide.[44] Thanheiser and Maassen found that the barium carbonate method for the removal of iron is satisfactory when nickel and manganese, or nickel and cobalt, are to be determined in a single sample. With steels containing relatively large proportions of cobalt and nickel, the manganese must be separated as manganese dioxide prior to its polarographic determination.

Nickel and cobalt can be determined simultaneously in the filtrate from the barium carbonate separation of iron, provided that they are present in

---

[38] G. Thanheiser and G. Maasen, *Mitt. Kaiser-Wilhelm Inst. Eisenforsch., Düsseldorf*, **10**, 27 (1937), *Arch. Eisenhüttenw.*, **10**, 441 (1937). G. Maassen, *Angew. Chem.*, **50**, 375 (1937).

[39] G. Thanheiser and J. Willems, *Mitt. Kaiser-Wilhelm Inst. Eisenforsch., Düsselforf*, **21**, 65 (1939).

[40] M. von Stackelberg, P. Klinger, W. Koch, and E. Krath, *Tech. Mitt. Krupp Forschungsber.*, **2**, 59 (1939).

[41] J. J. Lingane and H. Kerlinger, *Ind. Eng. Chem., Anal. Ed.*, **13**, 77 (1941).

[42] I. M. Kolthoff and J. I. Watters, *Ind. Eng. Chem., Anal. Ed.*, **15**, 8 (1943); **16**, 187 (1944).

[43] I. M. Kolthoff and A. Langer, *J. Am. Chem. Soc.*, **62**, 211 (1940).

[44] For a study of the coprecipitation of divalent metal ions with hydrous ferric oxide see I. M. Kolthoff and L. G. Overholser, *J. Phys. Chem.*, **43**, 767, 909 (1939).

about equal amounts. When the amounts of nickel and cobalt are quite disproportionate their simultaneous determination in a supporting electrolyte containing barium chloride, or in ammoniacal medium, is no longer feasible because the two waves tend to coalesce.

Lingane and Kerlinger have shown that an excellent separation of the nickel and cobalt waves is obtained in a supporting electrolyte containing pyridine or thocyanate. They also showed that pyridine can be used advantageously for separating large amounts of iron and chromium from copper, nickel, and cobalt. Pyridine has the double advantage of permitting the precipitation of iron and chromium at a relatively low pH (about 5.5) and of forming stable complexes with copper, nickel, and cobalt; both of these factors prevent the coprecipitation of copper, nickel, and cobalt. Lingane and Kerlinger found that there is no appreciable coprecipitation of nickel and cobalt, even with a 100-fold excess of iron, when the hydrous ferric oxide is precipitated at a pH of about 5 to 5.5 from a solution containing about equal concentrations of pyridine and pyridine hydrochloride. In the absence of iron, chromic ion is not precipitated from a pyridine solution of pH = 5.4, but in the presence of excess iron the chromium is completely coprecipitated with the hydrous ferric oxide. Hence the presence of chromium in a steel does not interfere with the determination of nickel and cobalt. Manganous ion in pyridine solution is reduced at a considerably more negative potential than nickel and cobalt, and hence moderate amounts of manganese do not interfere with the nickel-cobalt determination.

Watters and Kolthoff[45] demonstrated that pyridine precipitation serves well to separate chromium, vanadium, cerium, and large amounts of iron from manganese, prior to the determination of the latter in steel by the manganic pyrophosphate method described in Chapter XXVI.

Von Stackelberg *et al.* and Thanheiser and Willems developed methods for the determination of chromium in steel, based on its conversion to chromate. Chromate produces a well-defined wave in strongly alkaline medium, and none of the other metals present in steel interferes with the chromium determination by this method. The basic metals are removed as hydroxides, and the more acidic elements (vanadium, tungsten, molybdenum, and silicon) remain in solution in the form of anions which are more difficultly reducible than chromate.

For the determination of molybdenum in steel, von Stackelberg *et al.* first separated it from the other metals by precipitation as the sulfide, MoS_2. The sulfide is then roasted to the trioxide, MoO_3, the latter is

[45] J. I. Watters and I. M. Kolthoff, *Ind. Eng. Chem., Anal. Ed.*, **16**, 187 (1944).

dissolved in strong sulfuric acid, and the polarogram of this solution is obtained. Molybdenum is present in this final solution in the +6 state, and gives a well-defined wave at −0.3 v *vs.* the S.C.E. When W is present it tends to precipitate along with the MoS_2. This is prevented by having phosphoric acid present, which forms soluble phosphotungstic acid from which the tungsten is not precipitated by hydrogen sulfide. None of the other metals present in steel interferes with the molybdenum determination by this procedure. The amperometric titration of molybdenum in steel is discussed in Chapter XLVII.

The procedure developed by von Stackelberg *et al.* for the determination of vanadium in steel is based on its conversion to vanadate, in which form it gives a fairly good wave in strongly ammoniacal medium. The conversion of the vanadium to vanadate, and its separation from the other elements in steel, is effected by treating a hydrochloric acid solution of the steel with excess sodium hydroxide, which precipitates the other elements and leaves vanadate in solution. According to von Stackelberg *et al.* none of the other elements in steel interferes with the determination of vanadium by this procedure.

For the determination of tungsten in steel von Stackelberg *et al.* recommend its preliminary separation as tungstic acid from the iron and other elements. The tungstic acid is ignited, dissolved in concentrated potassium hydroxide, and the solution is finally made strongly acid with hydrochloric acid. From this solution the +6 tungsten produces a well-defined wave at −0.4 v. *vs.* the S.C.E. The only other element in steel that interferes is molybdenum, and then only when the amount of molybdenum is greater than that of the tungsten. When the amount of molybdenum is about the same as that of the tungsten it does not interfere seriously, since the molybdate wave in hydrochloric acid is well in advance of that of tungsten.

Copper and Nickel (according to Thanheiser and Maassen). A 0.1-g. sample of the steel is dissolved in 3 to 5 ml. of 6 *N* hydrochloric acid, with the addition of a few drops of concentrated nitric acid to oxidize the iron, and the solution is evaporated to dryness. The residue is taken up with 1 ml. of 6 *N* hydrochloric acid, transferred to a 50-ml. volumetric flask with 10 to 15 ml. of water, and then 20 to 25 ml. of a 2 *N* ammonium chloride-2 *N* ammonia solution is added with thorough shaking to precipitate the iron and chromium. After addition of 1 ml. of 0.5 per cent gelatin or methyl cellulose solution as a maximum suppressor, the solution is diluted to the mark with the 2 *N* ammonium chloride-2 *N* ammonia solution. The polarogram of an aliquot portion of the supernatant solution, or filtrate, is recorded in the usual manner. The double copper wave is at −0.2 v. and −0.5 v., and the nickel wave is at −1.0 v. *vs.* the S.C.E.

Cobalt is partially coprecipitated with the iron and chromium in this procedure, and hence *small amounts* of it do not interfere seriously with the determination of the nickel. The cobalt wave is at a potential about 0.2 v. more negative than that of the nickel in ammoniacal solution. Considerable interference from cobalt may be expected, however, with high cobalt steels. With such steels, the pyridine method for the removal of the iron and chromium is to be preferred, especially since this method also permits the simultaneous determination of both the nickel and cobalt.

Nickel and Cobalt (according to Thanheiser and Maassen). The following procedure is only suitable when the amounts of nickel and cobalt are of about the same order of magnitude. A 0.1-g. sample of the steel is dissolved in 3 to 5 ml. of 6 *N* hydrochloric acid, oxidized with a few drops of concentrated nitric acid, evaporated to dryness, and the residue is taken up with 1 ml. of 6 *N* hydrochloric acid and transferred to a 50-ml. volumetric flask with 10 ml. of 4 *N* ammonium chloride solution. Then 10 ml. of a suspension of barium carbonate (250 g. of $BaCO_3$; per liter) is added in small portions with thorough shaking to precipitate the iron and chromium. After adding 1 ml. of 0.5 per cent gelatin solution, the solution is made up to the mark and mixed thoroughly. The polarogram of an aliquot portion of the filtrate or supernatant solution is then obtained. The nickel wave is at −1.0 v. and that of the cobalt is at −1.2 v. *vs.* the S.C.E.

Copper is partially precipitated with the iron and chromium. Manganese remains in solution with the nickel and cobalt, but its half-wave potential is about 0.3 v. more negative than that of the cobalt, and it does not interfere unless it is present in very much larger amounts than the nickel and cobalt.

With this procedure Thanheiser and Maassen obtained results accurate to about ±2 per cent with steels whose nickel content ranged from 0.1 to 25 per cent, and cobalt content from 1 to 37 per cent. The method is not suitable when the cobalt content is much greater than fifteen times that of the nickel, or the nickel content greater than about five times that of the cobalt, due to the coalescence of the two waves.

Von Stackelberg *et al.* employed a similar procedure.

Copper, Nickel, and Cobalt (according to Lingane and Kerlinger). These three metals can be determined simultaneously by using pyridine to precipitate the iron and chromium. The nickel and cobalt waves are separated by about 0.3 v. in pyridine solution, and hence even traces of nickel can be determined in the presence of large amounts of cobalt. Neither manganese nor the other alloying elements in steel interfere.

A 0.1- to 0.5-g. sample of the steel is dissolved in 3 to 5 ml. of concentrated hydrochloric acid, oxidized with a few drops of concentrated nitric acid, and evaporated to dryness. The residue is taken up with exactly 2 ml. of concentrated hydrochloric acid, and transferred to a 100-ml. volumetric flask with 50 to 75 ml. of water. To precipitate the iron and chromium, 5 ml. of pure pyridine (about 13 *M*) is added; the resulting solution contains about equal concentrations of pyridine and pyridinium chloride, and has a pH of 5 to 5.5. Sufficient gelatin solution is added to give a final concentration of 0.01 per cent, and the mixture is made up to the mark and mixed thoroughly. After allowing a few minutes for the hydrous ferric oxide to settle, an aliquot portion of the supernatant solution is taken with a pipet, transferred to the polarographic cell, and the polarogram is recorded in the usual way. The double copper wave is at +0.05 v. and −0.25 v., that of nickel is at −0.78 v., and that of cobalt is at −1.07 v. *vs.* the S.C.E.

Relatively large amounts of copper would interfere with the determination of nickel and cobalt by this procedure, but this would be a rare occurrence in steel analysis. Normally, the amount of copper in steel is considerably smaller than that of cobalt and cobalt.

Nickel (according to West and Dean[46]). This method utilizes the nickel waves in fluoride medium of pH 4 to 6. The bulk of the iron is precipitated by fluoride and the remainder is complexed so strongly that it does not yield a reduction wave.

A sample equivalent to 10 to 50 mg. of nickel is decomposed by heating with 50 ml. of 6 *M* (1:1) hydrochloric acid, and 5 ml. of concentrated nitric acid is added to oxidize the iron. The solution is evaporated until most of the free acid is removed, and it is then transferred to a 50-ml. volumetric flask and diluted to the mark with water. A 5-ml. aliquot of this sample solution is transferred to a 50-ml. volumetric flask, 25 ml. of 1 *M* sodium fluoride and 1 ml. of fresh 0.2 per cent gelatin are added, and the solution is diluted to the mark. The pH at this point should be about 4.5 or slightly higher. The solution is filtered (centrifugation would be simpler), and a portion of the filtrate is polarographed. The nickel wave is at −1.12 v. *vs.* S.C.E.

West and Dean reported very good results with simple nickel steels containing only small amounts of other alloying elements. The influence of large amounts of other elements was not investigated. The method cannot be applied in the presence of much cobalt because the reduction wave of the latter in fluoride medium follows the nickel wave too closely.

Manganese (according to von Stackelberg, Klinger, Koch, and Krath). The following procedure is applicable to manganese steels that contain only relatively small amounts of nickel and cobalt.

A 0.2-g. sample is dissolved in 10 ml. of 6 *N* hydrochloric acid, and oxidized with a few drops of concentrated nitric acid. An excess of nitric acid must be avoided because nitrate acts deleteriously on the manganous wave. The solution of the sample is evaporated to a volume of 1 ml. and transferred to a 100-ml. volumetric flask with about 70 ml. of water. The solution is then cooled to 20° C. or lower (to prevent partial coprecipitation of the manganese with the hydrous ferric oxide) and a small excess of barium carbonate suspension (250 g. of $BaCO_3$ per liter) is added in small portions with thorough shaking to precipitate the iron and chromium. After addition of 2 ml. of 0.5 per cent gelatin, the solution is made up to the mark, and the polarogram of an aliquot portion of the supernatant solution or filtrate is recorded. The manganese wave is at −1.55 v. *vs.* the S.C.E.

When the amounts of either nickel or cobalt in the steel are greater than about five times that of the manganese, their waves precede, and interfere with, the manganous wave. In such cases von Stackelberg *et al.* recommend that the manganese first be separated as hydrous manganese dioxide prior to its polarographic determination, but when this is done there is little advantage in concluding the determination polarographically.

Manganese (according to Kolthoff and Watters). This procedure, which is described in detail in the section on Manganese in Chapter XXVI, utilizes the diffusion current produced by the manganic pyrophosphate complex, and it has a distinct advantage over the foregoing method of von Stackelberg *et al.* in that nickel and cobalt do not interfere. Amounts

[46] P. W. West and J. F. Dean, *Ind. Eng. Chem., Anal. Ed.*, **17**, 686 (1945).

of iron up to 0.2 g. do not interfere, and the interference of large amounts of iron, and of chromium, vanadium, and cerium is circumvented by precipitating these elements with pyridine as described by Watters and Kolthoff.[45]

The solution of the sample (volume about 25 ml.) in a 100-ml. volumetric flask is acidified with 5 ml. of 1:3 nitric acid. 1 ml. of 20 per cent sodium bisulfite solution is added to reduce any dichromate, and the solution is boiled for several minutes to expel the excess sulfur dioxide and reoxidize ferrous iron. The excess acid is neutralized with ammonia, and the solution is diluted to about 80 ml. and acidified with exactly 1 ml. of 1:1 (9 *M*) sulfuric acid. 15 ml. of 1:2 pyridine is then added slowly with thorough swirling, after which the solution is diluted to the mark and mixed. The manganese in a filtered aliquot portion of this solution is then determined by the pyrophosphate procedure described in Chapter XXVI in the section on Manganese.

Chromium (according to von Stackelberg, Klinger, Koch, and Krath). This method is based on the conversion of the chromium to chromate by fusion with sodium peroxide, and the measurement of the well-defined chromate wave in strongly alkaline medium.

A 0.2-g. sample of the steel is dissolved in a few milliliters of 6 *N* hydrochloric acid, in a porcelain crucible, oxidized with nitric acid, and evaporated to dryness. The residue is fused with 4 g. of sodium peroxide. After cooling, the melt is disintegrated in 25 to 50 ml. of water in a small beaker. The solution, which contains the chromium in the form of chromate, is transferred to a 100-ml. volumetric flask without bothering to filter off the precipitated ferric hydroxide, and is made up to the mark. The polarogram of an aliquot portion of the solution is then recorded. The half-wave potential of the chromate is at −0.85 v. *vs.* the S.C.E. None of the other metals in steel interfere.

Chromium (according to Thanheiser and Willems). In this method the chromium is converted to chromate by dissolving the steel sample in hot, concentrated perchloric acid. A 0.1- to 0.2-g. sample is dissolved in 2 to 4 ml. of 60 per cent perchloric acid, and heated until dense fumes are evolved. The solution is cooled, diluted with 10 ml. of water, and boiled for a short time to remove free chlorine. The solution is cooled, neutralized with 2 *N* potassium hydroxide,[47] and then poured into 20 ml. of 4 *N* sodium hydroxide that contains 25 drops of 30 per cent hydrogen peroxide. After adding a small amount of graphite to prevent bumping, the solution is heated at the boiling point for 10 minutes, then cooled, transferred to a 50-ml. volumetric flask, and made up to the mark after addition of 1 ml. of 0.5 per cent gelatin. The polarogram of an aliquot portion of the supernatant solution or filtrate is then recorded.

From the data given by von Stackelberg *et al.* and Thanheiser and Willems the

[47] Potassium perchlorate precipitates during this neutralization, and thus the salt concentration in the final solution is maintained approximately constant in spite of small variations in the concentration of free perchloric acid remaining after the solution of the sample.

accuracy of these two different methods appears to be about the same. Results obtained by both procedures on a variety of steels agreed very well with the chromium contents determined by the usual volumetric methods.

Molybdenum (according to von Stackelberg, Klinger, Koch, and Krath). The following procedure permits the determination of molybdenum in the presence of excess tungsten. A 1-g. sample of the steel is dissolved in an acid mixture consisting of 5 ml. of concentrated nitric acid, 10 ml. of 6 *N* hydrochloric acid, and 10 ml. of 85 per cent phosphoric acid. The phosphoric acid forms complex phosphotungstic acid with the tungsten, and thus prevents its precipitation with the molybdenum sulfide later on. The solution is evaporated to a syrupy consistency, and then treated with 5 ml. of concentrated hydrochloric acid and re-evaporated to insure complete removal of the nitric acid. The syrupy residue is diluted to about 500 ml. with water and heated to boiling, disregarding any precipitate of ferric phosphate. A small amount of solid ferrous sulfate is added, and the solution is boiled gently for five minutes to reduce the molybdenum to the +4 state. The molybdenum is then precipitated as MoS_2 by the careful dropwise addition of a saturated sodium sulfide solution, until the iron is reduced and a permanent dark brown precipitate forms. A large excess of sulfide is avoided in order to prevent excessive coprecipitation of iron. Dilute hydrochloric acid is then added until any precipitated iron sulfide and iron phosphate are redissolved, and the solution is boiled gently for a few minutes to coagulate the MoS_2. The MoS_2 is filtered off, washed with 0.5 *N* hydrochloric acid, and is then transferred to a procelain crucible and roasted at 450° to convert it to MoO_3. The residue, which may contain a little ferric oxide and silica, is dissolved in a few milliliters of warm 18 *N* sulfuric acid, and the solution is transferred to a 100-ml. volumetric flask and made up to the mark with 18 *N* sulfuric acid. The polarogram of an aliquot portion of this solution is then recorded. The molybdate wave is at −0.3 v. *vs.* the S.C.E.

By this procedure von Stackelberg *et al.* obtained results that checked the potentiometric and gravimetric determination of molybdenum to about ±3 relative per cent with steels that contained from 0.2 to 2.5 per cent molybdenum. A large excess of tungsten did not interfere. Iron itself is the only metal in steel that interferes, because the diffusion current of ferric iron is obtained before the molybdate wave. For this reason care must be taken to avoid excessive coprecipitation of iron sulfide with the MoS_2. However, a small amount of iron coprecipitated with the molybdenum sulfide does not interfere.

Tungsten (according to von Stackelberg, Klinger, Koch, and Krath). A 0.5- to 3-g. sample of the steel, depending on the tungsten content, is digested near the boiling point with about 150 ml. of 6 *N* hydrochloric acid until action ceases, and the solution is then evaporated to a syrupy consistency. The syrupy residue is taken up in 80 ml. of 6 *N* hydrochloric acid, and a small excess of concentrated nitric acid is added to oxidize the iron and precipitate most of the tungsten as tungstic acid. In order to insure complete precipitation of the H_2WO_4 the solution is evaporated to a volume of about 25 ml., and then diluted to about 100 ml. with hot water and allowed to stand for about one-half hour. The tungstic acid is filtered off on hard paper, washed with 0.1 *N* hydrochloric acid, and converted to WO_3 by ignition at 800° in a platinum crucible. The ignited tungstic oxide is then dissolved in the crucible in 5 ml. of 4 *N* potassium hydroxide, with gentle warming, and the solution is transferred to a 100-ml. volumetric flask with no more than 20 ml. of water. 10 to

20 ml. of concentrated hydrochloric acid is then added quickly. The precipitate of tungstic acid that first forms redissolves when the hydrochloric acid is in excess. The solution is finally cooled to room temperature and made up to the mark with concentrated hydrochloric acid. The polarogram of an aliquot portion of this solution shows a well-defined wave of the tungstic acid at −0.6 v. *vs.* the S.C.E.

Large amounts of molybdenum interfere with the tungsten determination by this procedure, for the reason already explained above. None of the other metals in steel interfere, according to von Stackelberg *et al.* These authors obtained results in close agreement with those obtained by the gravimetric procedure (weighing of WO_3) with steels whose tungsten content ranged from 0.6 to 20 per cent, and whose molybdenum content did not exceed 0.35 per cent.

Vanadium (according to von Stackelberg, Klinger, Koch, and Krath). A 0.4- to 0.8-g. sample of the steel is dissolved in 10 ml. of 6 *N* hydrochloric acid, oxidized with a few drops of nitric acid, and evaporated to a volume of about 2 ml. The solution is then diluted to about 30 ml., heated to boiling, and poured into 30 ml. of a hot 2.5 *N* sodium hydroxide solution contained in a 100-ml. volumetric flask. The vanadium remains in solution as vanadate. After cooling to room temperature, the solution is made up to the mark, and filtered through a dry filter. 50 ml. of the filtrate (corresponding to one-half the original sample) is pipeted into a 100-ml. volumetric flask and carefully neutralized with concentrated hydrochloric acid. After addition of 1 g. of ammonium chloride, the solution is cooled and made up to the mark with concentrated ammonia (15 *N*). The solution is allowed to stand until it becomes colorless, and the polarogram of an aliquot portion is then recorded. The vanadate wave is at −1.3 v. *vs.* the S.C.E.

According to von Stackelberg *et al.* none of the other elements in steel interfere with the vanadium determination. They obtained results in close agreement with the potentiometrically determined vanadium content with steels containing from 0.1 to 2.5 per cent vanadium.

Vanadium (according to Lingane and Meites[48]). This method is based on the measurement of the anodic wave produced by the oxidation of +4 to +5 vanadium in strongly alkaline solution. The anodic wave is very well developed in a supporting electrolyte comprising 1 *M* sodium hydroxide and 0.1 *M* sodium sulfite to remove oxygen, and its half-wave potential is −0.42 v. *vs.* S.C.E.

Iron and other interfering elements are removed by electrolyzing a phosphoric–sulfuric acid solution of the sample with a mercury cathode in a Melaven type cell.[49] Of the elements likely to be present in ferro-alloys this procedure leaves in solution, in addition to +3 vanadium, only molybdenum, tungsten, titanium, uranium, columbium (niobium), tantalum, aluminum, and a trace of manganese. None of these interfere with the

[48] J. J. Lingane and L. Meites, *Anal Chem.*, **19,** 159 (1947).

[49] A. D. Melaven, *Ind. Eng. Chem., Anal. Ed.*, **2,** 180 (1930). See also N. J. Chlopin, *Z. anal. Chem.*, **107,** 104 (1936); and G. E. F. Lundell, J. I. Hoffman, and H. A. Bright, *Chemical Analysis of Iron and Steel*, Wiley, New York, 1929.

vanadium determination. A complete determination requires about 90 minutes, and very accurate results are obtainable with all types of steels and ferro-alloys containing from 0.01 to 2 per cent vanadium.

Transfer a 0.5- to 2.5-g. sample to a small Kjeldahl flask, and add 20 ml. of 6 *N* (1:1) hydrochloric acid, 5 g. of disodium phosphate dodecahydrate, and 3 ml. of concentrated nitric acid in small portions. When the action moderates, add 3 ml. of concentrated sulfuric acid and evaporate to fumes of sulfur trioxide to remove oxides of nitrogen and chloride. Disregarding any graphite and/or siliceous residue, transfer the solution to a Melaven cell with water, and dilute to about 100 ml. This solution will be about 0.3 *M* with respect to sulfuric acid and about 0.14 *M* in phosphoric acid.

Electrolyze with a current of 3 to 5 amp. for a length of time corresponding to at least 3 ampere-hours per gram of sample, *i. e.*, for 45 minutes if a 1-g. sample was used and the current is 4 amp. If the amount of molybdenum in the sample is more than about 100 times greater than the amount of vanadium (a most unusual circumstance), the electrolysis should be prolonged for a time corresponding to 30 ampere-hours per gram of molybdenum present. (Molybdenum is only partially deposited in this procedure. As long as the amount remaining in solution is not more than about 100 times the amount of vanadium present it does not interfere, but larger amounts produce erratic results in the polarographic determination of the vanadium.)

Transfer the residual solution to a Kjeldahl flask, add 2 to 5 ml. of 30 per cent H_2O_2 to oxidize any +3 vanadium, and boil gently for a minute or two. Add 2 g. sodium sulfite to reduce the vanadium to the +4 state, evaporate the solution to about 75 ml., cool, transfer to a 100-ml. volumetric flask, add 1 g. sodium sulfite, and dilute to volume. Add a suitable aliquot portion of this solution, *e. g.*, 10 ml., to a suitable volume, *e. g.*, 50 ml., of air-free 1 *N* sodium hydroxide-0.1 *M* sodium sulfite supporting electrolyte in a polarographic cell, and measure the anodic diffusion current at 25° C. at −0.25 v. *vs.* the saturated calomel electrode, *i. e.*, at the middle of the plateau. Subtract the previously measured residual current from the observed diffusion current. Measure the drop time at −0.25 v., and the *m*-value of the capillary, and compute the concentration of vanadium from the relation $C = i_d/1.466m^{2/3}t^{1/6}$.

The constant 1.466 is the experimentally determined diffusion current constant. Instead of employing this constant directly the dropping electrode may be calibrated with standard steel samples. The diffusion current constant is not affected by changing the sodium hydroxide concentration from 0.5 to 3 *M*. Since +4 vanadium is so easily air-oxidized in alkaline media the combined use of sulfite ion and an inert gas (nitrogen or hydrogen) is advisable to insure complete exclusion of oxygen.

Titanium. The following procedure developed by Zan'ko, Geller, and Nikitin[50] utilizes the cathodic wave of +4 titanium in tartrate medium. Interference of iron is eliminated by reduction to the ferrous state with aluminum powder. Chromium does not interfere. The method is stated

[50] A. M. Zan'ko, B. A. Geller, and A. D. Nikitin, *Zavodskaya Lab.*, **13**, 299 (1947).

to be suitable for steels and cast irons containing 0.1 per cent or more titanium.

A 0.2 g.-sample is dissolved by heating with 5 ml. of 9 N (1:4) sulfuric acid, followed by addition of 1 ml. of 7.5 N (1:1) nitric acid and evaporation to sulfur trioxide fumes. The residue is dissolved in about 15 ml. of water, 3.5 g. of sodium potassium tartrate is added, and the solution is diluted to exactly 25 ml. A portion of this solution is warmed with aluminum powder until the ferric iron is reduced (which may be indicated by adding a small amount of potassium thiocyanate and noting the disappearance of the red color), after which it is cooled, swept with nitrogen to remove dissolved air, and the polarogram is recorded. The titanium wave is at about −0.4 v. *vs.* S.C.E.

Tin. The following procedure for the determination of a few thousandths up to several hundredths of one per cent of tin in tool steels was developed by Allsopp and Damerall.[51] The tin is first separated by hydrogen sulfide precipitation (along with copper, molybdenum, etc.), ignited to the oxide, fused with potassium pyrosulfate, taken into solution in hydrochloric acid, and precipitated with ammonia with hydrous ferric oxide as collector. The mixture of hydrous stannic and ferric oxides is finally taken into solution in hydrochloric acid, ferric iron is reduced to the noninterfering ferrous state with hydroxylamine, the solution is made up to 1 M in hydrochloric acid and 4 M in ammonium chloride, and the polarogram of the chlorostannate complex is recorded.

Dissolve a 1.5-g. sample in 100 ml. of 1 to 4 sulfuric acid in an 800-ml. beaker with the aid of heat. A larger sample may be used if the molybdenum and copper concentration will permit the sulfide precipitate to be handled easily. (If the steel contains less than 0.5 per cent molybdenum, add a solution of molybdic oxide in 1:4 sulfuric acid to bring the molybdenum concentration to that of a 0.5 per cent molybdenum steel. The molybdenum sulfide acts as a collector during the hydrogen sulfide precipitation.) After the sample has dissolved, add 15 ml. of 1:1 nitric acid; after the reaction has subsided add 10 ml. of a 1.5 per cent solution of potassium permanganate or add until manganese dioxide appears, boil for 10 minutes, and add 8 per cent potassium nitrite dropwise until the precipitate disappears. Boil the solution until salts appear and then carefully fume with the aid of gentle heat and an infrared lamp.

Cool and dilute to approximately 100 ml. with distilled water and add 10 g. of tartaric acid. Make the resulting solution ammoniacal and warm until all the tungstic acid is in solution.

Dilute to 400 ml., add 24 ml. of 1:1 sulfuric acid, and heat to boiling. Any insoluble chromium salts will go into solution at this point. Pass hydrogen sulfide through the solution for 45 minutes. Let the precipitate digest in a warm place for another 45 minutes, then filter and wash it with a 1 per cent sulfuric acid solution saturated with hydrogen sulfide.

---

[51] W. E. Allsopp and V. R. Damerall, *Anal. Chem.*, **21,** 677 (1949).

Ignite the precipitate in a 35-ml. porcelain crucible and fuse with 3 g. of potassium pyrosulfate. Cool and leach the fused mass with 2 *N* hydrochloric acid, keeping the volume less than 100 ml.

Transfer to a 250-ml. beaker and add a slight excess of ammonium hydroxide. Boil for a few minutes to coagulate the precipitate and filter. If less than 30 mg. of ferric hydroxide is present, add ferric ion as a collector. Wash the precipitate well with 2 per cent ammonium hydroxide and twice with water.

Dissolve the precipitate from the paper with 50 ml. of 2 *N* hydrochloric acid and catch the filtrate in the original 250-ml. beaker used in the ammonia precipitation. Wash the paper twice with water. Add 0.75 g. of hydroxylamine hydrochloride and boil gently for 5 minutes.

Transfer the warm solution to a 100-ml. volumetric flask that contains 21 g. of ammonium chloride. Add 2 ml. of a 0.5 per cent gelatin solution and dilute to the mark with water.

Remove the dissolved oxygen from the solution with pure nitrogen and obtain a polarogram over the range 0 to −1.4 v. *vs.* the saturated calomel electrode.

According to Allsopp and Damerall this method is quicker and more convenient than the standard method of Scherrer,[52, 53] and they obtained excellent results with a variety of Bureau of Standards steel samples containing from 0.007 to 0.036 per cent tin.

Lead. Haim and Barnes[54] described the following very simple method for determining 0.1 to 0.4 per cent lead in steel.

A 1.0-g. sample is dissolved in 10 ml. of 12 *M* hydrochloric acid in a 50-ml. volumetric flask in an atmosphere of nitrogen to keep the iron in the noninterfering ferrous state. When complete solution is obtained (gentle heating if necessary), 2 ml. of 2 per cent gelatin is added, the solution is diluted to the mark with previously boiled (air-free) water, and the polarogram of lead is recorded.

Haim and Barnes obtained very satisfactory results by this method with steels containing from 0.14 to 0.28 per cent lead. However, they did not identify the types of steels analyzed, nor make any mention of possible interference by other elements except the statement that the method was not applicable to 18-8 chromium-nickel stainless steels.

Aluminum. The following procedure recommended by Willard and Dean[55] utilizes the fact that aluminum ion complexes the reducible azo dye Pontochrome Violet SW, and the reduction potential of the complexed dye is about 0.2 v. more negative than that of the free dye. In buffered solutions of pH 4.6 containing aluminum and excess dye two waves result: the first at −0.3 v. *vs.* S.C.E. corresponds to the reduction of the free excess

---

[52] J. A. Scherrer, *J. Research Natl. Bur. Standards*, **8,** 309 (1931).

[53] G. E. F. Lundell, J. I. Hoffman, and H. A. Bright, *Chemical Analysis of Iron and Steel*, Wiley, New York, 1931 (p. 383).

[54] G. Haim and W. C. E. Barnes, *Ind. Eng. Chem., Anal. Ed.*, **14,** 867 (1942).

[55] H. H. Willard and J. A. Dean, *Anal. Chem.*, **22,** 1264 (1950).

dye, and the second at −0.5 v. results from the reduction of the dye that is complexed with aluminum. The second wave height is directly proportional to the aluminum concentration. The principle of this method is described in detail in Chapter XXX in the section on Aluminum. Since iron and most other metals interfere these are first removed by electrolysis with a mercury cathode.

For steels containing 0.1 to 1 per cent aluminum, a 1-g. sample is dissolved in 20 ml. of hot 5 *M* perchloric acid, and any insoluble residue is removed by filtration through paper. The filtrate is diluted to about 50 ml. and electrolyzed with a mercury cathode. The residual solution is diluted to exactly 100 ml. in a volumetric flask. An aliquot of this solution corresponding to 0.01 to 0.3 mg. of aluminum is transferred to a 50-ml. volumetric flask and neutralized to methyl red (pH *ca.* 6) with 2.5 *M* sodium hydroxide solution. The solution is then acidified with exactly 1 ml. of 5 *M* perchloric acid, and 5 ml. of 2 *M* sodium acetate followed by 20 ml. of a 0.05 per cent (1.4 millimolar) aqueous solution of Pontochrome Violet SW (Color Index 169) are added; the solution is then diluted to the mark. The flask is immersed in water at 55° to 70° C. for 5 minutes and then cooled to room temperature. A portion of the solution is transferred to a polarographic cell, dissolved air is removed with nitrogen, and the polarogram is recorded. The aluminum content is computed from the height i_2 of the second wave at −0.5 v. *vs.* S.C.E., and the measured m and t values of the dropping electrode, by the relation $C = i_2/Im^{2/3}t^{1/6}$. The diffusion current constant I is evaluated with known amounts of aluminum carried through the same procedure.

Willard and Dean obtained excellent results with Bureau of Standards Cr-Mo-Al Steel 106 containing 1.06 per cent aluminum, and with High Silicon Steel 125 containing 0.261 per cent aluminum. Because titanium and vanadium interfere and are not removed by electrolysis with the mercury cathode the method is not applicable to steels containing these elements.

An exactly similar procedure may be used for determining aluminum in copper-base alloys.

11. GLASS

The following procedures for the polarographic determination of calcium, barium, zinc, lead, aluminum, and the sum of sodium and potassium in glass have been described by Vandenbosch.[56] Silicon and boron are eliminated by decomposing the sample with hydrofluoric and perchloric acids.

Preparation of Sample Solution. A 0.25-g. sample of the powdered glass is treated in a platinum crucible with 3 ml. of 40 per cent hydrofluoric acid and 1 ml. of 70 per cent perchloric acid. The solution is evaporated to dryness in air bath, with care not to heat above the fusion point of the residual metallic perchlorates, as otherwise a considerable loss of aluminum by volatilization occurs. The evaporation with

[56] V. Vandenbosh, *International Congress on Analytical Chemistry*, June 1948; *Anal. Chim. Acta*, **2**, 566 (1948).

perchloric acid is repeated to insure complete expulsion of hydrofluoric acid. The residue is taken up in about 5 ml. of water and 2 drops of perchloric acid, transferred to a 50-ml. volumetric flask, and diluted to volume.

If the glass contains sulfate, a corresponding amount of barium will remain insoluble. This does not interfere with the determination of the other metals and the precipitated barium sulfate need not be removed. In such a case the special procedure described below must be used for the determination of barium.

Barium, Sodium, and Potassium. An exactly measured portion, *e. g.*, 4 ml. of the sample solution is mixed with one-fourth its volume, *e. g.*, 1 ml., of pure 1 *M* tetramethylammonium hydroxide solution, and the polarogram of this solution is obtained. The barium wave appears at −1.9 v. *vs.* S.C.E., and sodium, potassium, and calcium produce a combined wave at −2.1 v. A second polarogram is recorded after adding 0.2 ml. of 85 per cent phosphoric acid to precipitate the calcium, and the sum of sodium and potassium is computed from the remaining wave at −2.1 v. The difference between the heights of the total-alkali-plus-calcium wave before and after addition of the phosphoric acid provides an approximate determination of calcium.

The tetramethylammonium hydroxide used must, of course, be free of alkali metal impurities, and a blank polarogram should be run to test its purity.

Calcium. An approximate determination of calcium is obtained by the preceding procedure, and a more exact result is obtained as follows. A 20-ml. portion of the sample solution is heated to boiling and 5 ml. of 4 per cent ammonium oxalate solution is added dropwise to precipitate calcium oxalate. The precipitate is collected on ashless paper, washed with 0.1 per cent ammonium oxalate solution, and ignited to the oxide or carbonate. The residue is dissolved in a minimal amount of 1:1 hydrochloric acid, and the solution is diluted to a known volume in a volumetric flask, the volume being chosen so that the final solution corresponds to less than 80 mg. of calcium oxide per liter. A 5-ml. portion of this solution is mixed with 1 ml. of 1 *M* tetramethylammonium hydroxide, and the polarogram is obtained. This polarogram also shows the barium wave preceding the calcium wave.

Vandenbosch recommends that a second polarogram be obtained after addition of a small excess of phosphoric acid to correct for any alkali metals that might have been coprecipitated with the calcium oxalate.

Zinc and Lead. A 5-ml. portion of the sample solution is mixed with 1 ml. of 1 *M* ammonium chloride, and the polarogram is recorded. The lead wave is at −0.4 v. *vs.* S.C.E., and the zinc wave is at −1.0 v.

Vandenbosch recommends obtaining two polarograms: the first for zinc without any maximum suppressor present, and the second for lead after adding a small amount of methyl red as maximum suppressor. There is no evident reason why a maximum suppressor (methyl red or gelatin) could not be added and both metals determined from a single polarogram.

Since the sample solution is slightly acid it would also seem to be advisable to check the pH after addition of the ammonium chloride, and if necessary adjust it to a value of about 5, to avoid interference of hydrogen ion reduction with the full development of the zinc wave.

Aluminum. Dimethyl yellow indicator is added to a 10-ml. portion of the sample solution, and the solution is carefully neutralized to a pH of 4 with dilute sodium hydroxide solution. The solution is then treated with 0.4 ml. of 0.05 *N* hydrochloric acid, 5 ml. of 1 *M* lithium chloride, and diluted to exactly 25 ml. The polarogram of this solution is then obtained. Since the solution is slightly acidic, the aluminum wave at −1.7 v. *vs.* S.C.E. is preceded by a wave due to reduction of hydrogen ion.

The proper control of pH has long been recognized as one of the most critical factors in the polarographic determination of aluminum (see Chap. XXX). According to Vandenbosch the height of the aluminum wave increases markedly with increase in the hydrogen ion concentration over the range from 10^{-4} to about 5×10^{-3} *M*, and therefore the concentration of free hydrogen ion must be controlled carefully. It is impossible, of course, to employ a buffered solution because the acid constituent of the buffer, even though not dissociated, would produce a large hydrogen reduction wave.

Small amounts of iron do not interfere with the aluminum determination, but amounts equal to the amount of aluminum do.

Barium (in Glasses Containing Sulfate). A 0.1- to 0.15-g. sample of the powdered glass is decomposed in a platinum crucible with 1.5 ml. of 40 per cent hydrofluoric acid and 1 ml. of 70 per cent perchloric acid, and the solution is evaporated to dryness. The residue is fused with 1 g. of sodium carbonate, the melt is leached with water, and the insoluble residue is collected on paper and washed with 1 per cent sodium carbonate solution. The washed residue is then dissolved in the minimal amount of dilute hydrochloric acid, and the solution is diluted to 50 ml. A portion of this solution is mixed with one-fourth its volume of 1 *M* tetramethylammonium hydroxide solution, and the polarogram is obtained.

Results obtained by Vandenbosch by these procedures with several types of glasses were in very good agreement with the results obtained by the classical gravimetric methods.

Part Four

ORGANIC POLAROGRAPHY

With the Collaboration of
Dr. Stanley Wawzonek
of the *University of Iowa*

CHAPTER XXXVI

Oxidation and Reduction of Organic Compounds

1. GENERAL CHARACTERISTICS OF C.V. CURVES OF ORGANIC COMPOUNDS

Most polarographic reductions of organic compounds are irreversible and the general characteristics of these waves—also, the relationship to structure—are considered in this chapter. The characteristics of reversible waves of organic compounds (quinones and anodic hydroquinone waves) have been discussed at length in Chapters XIV and XV.

The field of organic polarography has seen considerable advancement since the polarographic study of the first organic compound, nitrobenzene.[1] The "reduction potentials" used in the older work have been abandoned and replaced by the half-wave potential. A considerable amount of the older work based upon the reduction potentials has been reinvestigated. For the old work which has not yet been repeated and should be, the reduction potentials referred to were derived by the following empirical methods.

(1) A 45° tangent is drawn to the curve and the applied e.m.f. corresponding to the point of contact is taken as the reduction potential. (This method was used by Heyrovsky and his school before the development of the concept of the half-wave potential.)

(2) The applied e.m.f. at the point of maximum curvature of the c.v. curve has been used (tangent at 35° 16′) (Semerano). From the change of this reduction potential with the concentration of the substance the "molar reduction potential," corresponding to the potential at which reduction starts at a concentration of 1 molar, was often calculated and reported.

(3) That value of the applied e.m.f. is taken at which an increase of 10 millivolts causes an increase in current of 0.019 microamp. (1.9×10^{-8} amp.) (Shikata). It has been pointed out[2] that the most outstanding shortcomings of these methods are that these potentials change with the concentration of the reducible substance, with the sensitivity of the galvanometer, and with the drop time and the rate of flow of the mercury.

In future work it is desirable that the half-wave potential be reported, preferably in terms of a standard reference electrode, together with the

[1] M. Shikata, *Trans. Faraday Soc.*, **21,** 42, 53 (1925); *J. Agr. Chem. Soc. Japan*, **1,** 533 (1925).

[2] O. H. Müller, *Chem. Revs.*, **24,** 95 (1939).

diffusion current constant, $I_d = i_d/Cm^{2/3}t^{1/6}$, for the wave. If possible, the equation of the polarographic wave should be included. If the half-wave potentials change with the concentration, the variation should be reported. In such cases, and also when the waves are drawn out and the equations of the waves depend upon the concentration, it may be of practical importance to report the reduction potentials obtained by one of the above methods. Such measurements are the basis for the potentials reported for the catalytic waves of the alkaloids and are reproducible only under specified conditions. They are useful only for comparative purposes.[3]

The well-known LeBlanc method[4] of extrapolation to zero current is of little use in polarographic work because no portion of the wave is actually a straight line and the extrapolation is uncertain.

In studies of the reduction of organic compounds it is desirable to work with buffer solutions. Frequently the pH of the medium has a great effect upon the half-wave potential since hydrogen ions often participate in the electrode reaction. Consequently, in an unbuffered solution the pH at the electrode surface changes considerably and this causes the waves to be drawn out or sometimes even the appearance of two waves. When the half-wave potential is dependent upon the pH, the use of buffer solutions becomes imperative if results are to be duplicated. This has been often overlooked and the interpretation of many reported c.v. curves is greatly obscured by the fact that the solutions used were unbuffered. For a preliminary survey of the reducible groups present in a compound unbuffered solutions can be used, *e. g.*, with tetraalkylammonium salts as supporting electrolytes. These salts are not reduced until potentials considerably more negative than 2 v. (*vs.* S.C.E.) are attained.

The type and concentration of the buffer system may affect the shape of c.v. curves of irreversible waves and the half-wave potential in the reduction of organic substances. Both cations and anions may exert an effect. For example, anthraquinone 2-sodium sulfonate in an acetate buffer of pH 8 gives only one reduction wave, while in a phosphate buffer of pH 8 two distinct waves are observed, the first one corresponding to a reduction to the semiquinone and the second wave to a reduction of the semiquinone.[5]

For the effect of the concentration of the buffer in the reduction of maleic acid see Elving and Teitelbaum.[6]

---

[3] H. F. W. Kirkpatrick, *Quart. J. Pharm. Pharmacol.*, **20,** 87 (1947).
[4] Applied by A. Winkel and G. Proske, *Ber.*, **69,** 693, 1917 (1936).
[5] N. H. Furman and K. G. Stone, *J. Am. Chem. Soc.*, **70,** 3055 (1948).
[6] P. J. Elving and C. Teitelbaum, *J. Am. Chem. Soc.*, **71,** 3916 (1949).

The effect of the medium upon the half-wave potential is a very important one. The insolubility of organic compounds in water makes it necessary that their reduction and oxidation be studied in mixtures of water with organic solvents, such as alcohols, glycols, dioxane, Cellosolve, and glacial acetic acid, or in nonaqueous media. If a layer of mercury in the cell is used as an anode, its potential is also affected by a change in the medium, and it should be measured separately. In this measurement and also with an outside depolarized anode an uncertainty is introduced by a liquid junction potential.

In completely anhydrous media, such as acetic acid, formamide, ethyl alcohol, methyl alcohol, glycerol, and ethylene glycol, waves are obtained which are lower in height and sometimes more drawn out than those obtained for the same concentration in aqueous media.

Glacial acetic acid has several disadvantages and is limited to the study of substances which have half-wave potentials between -0.3 and -1.4 v. Substances which are reduced at more positive values than -0.3 v. give a discontinuous current whereas above -1.4 v. the hydrogen wave interferes. The half-wave potentials observed are more negative than those in water.[7] The Ilkovic equation, however, applies to this medium with satisfactory accuracy.[8]

Anhydrous formamide, ethyl alcohol, methyl alcohol, glycerin,[9] and ethylene glycol[10] have very little effect upon the half-wave potentials of various inorganic ions as compared to water (see Chapter V).

Benzene, toluene, and aniline are unsatisfactory solvents for polarographic studies since most inorganic electrolytes are insoluble in these media while most organic electrolytes have a relatively large resistance. A mixture of equal volumes of methanol and benzene is fairly suitable since this combination not only dissolves many water-insoluble organic substances and some inorganic salts ($LiCl$ and $LiOCH_3$) but also has a reasonable resistance.[11]

The solubility of oxygen in mixtures of water and organic solvents or in anhydrous organic solvents is much greater than in water and a longer time is required to remove the oxygen from the solution than from an aqueous medium.

It is of special importance to emphasize that, upon addition of an aque-

---

7 G. B. Bachman and M. J. Astle, *J. Am. Chem. Soc.*, **64,** 1303, 2177 (1942).

8 E. Hala, *Chem. Obzor.*, **23,** 145 (1948); *Chem. Abstracts*, **43,** 2874 (1949).

9 A. M. Sanko and F. A. Manussova, *J. Gen. Chem. U. S. S. R.*, **10,** 1171 (1940).

10 C. H. R. Gentry, *Nature*, **157,** 479 (1946).

11 W. R. Lewis, F. W. Quackenbush, and T. DeVries, *Anal. Chem.*, **21,** 762 (1949).

ous buffer solution to an organic solvent or to a mixture of water and the organic solvent, the pH of the buffer may change considerably. This has been often overlooked in the literature, and causes confusion.

For reproducible polarographic results, the mixtures of water and organic solvents should be prepared with care so that they always have the same composition. The required accuracy in the preparation of such mixtures depends upon the slope of the viscosity–composition curve.

There is still another effect which has to be considered when dealing with irreversible reductions. Frequently, the half-wave potentials are shifted to more positive potentials and the wave height is decreased by the addition of multivalent cations. The effect increases with increasing cation concentration and increasing atomic weight. Thus, cations shift the half-wave potential of the second wave of benzophenone in the following order: $Cs^+ > Rb^+, Ca^{++}, Ba^{++} > K^+ > Na^+ > Li^+ > NR_4^+$. They decrease the height of the second wave in the following order: $NR_4^+ \gg Ca^{++}$, $Ba^{++}, Cs^+ > Rb^+ > K^+, Na^+, Li^+$. The effect depends upon the compound being reduced. For example, the shift of the half-wave potential of fluorenone is in the following order: $(C_2H_5)_4N^+ > (CH_3)_4N^+$; $Mg^{++} > Ba^{++} > Ca^{++} > Cs^{++} > Rb^+ > K^+ > Na^+ > Li^+$, while the decrease in wave height is in the order: $Mg^{++} > R_4N^+ > Ca^{++} > Ba^{++} > Cs^+$.[12] This electrolyte effect does not occur when dealing with reversible reductions.

It need scarcely be mentioned that maxima should be eliminated in the analysis of the waves.

In irreversible reductions the nature of the reduction product is often unknown but is important in formulating the electrode reaction. The amount of the reduction product formed is so small that it escapes analytical identification. Amounts that are sufficient for identification may be made by prolonged electrolysis at a large mercury cathode under conditions similar to those used for the polarographic analysis. Such electrolytic preparations are essential when complex organic molecules are used and when reactions are postulated that have no chemical or polarographic precedent.[12a] Controlled potentials are necessary only when several waves appear and the nature of an earlier reduction is desired. Electrolytic preparations at controlled potentials are not as yet suitable on a large scale since long periods of time are required in such work. The electrolysis product formed under these conditions may not be identical with that formed at the dropping mercury electrode but may be a disproportionation

[12] M. Ashworth, *Collection Czechslov. Chem. Communs.*, **13**, 229 (1948).
[12a] J. J. Lingane, G. Swain, and M. Fields, *J. Am. Chem. Soc.*, **65**, 1348 (1943).

product or a dimer from the intermediate formed. In any case the nature of the compound isolated is valuable in interpreting the electrode reaction.

When the isolation of electrolytic products is difficult, the factors important in deducing a reaction mechanism at the dropping mercury electrode in their order of importance are: (1) the determination of the number of electrons, *n*, by coulometric measurements[13] and from the Ilkovic equation or from a comparison with the understood polarographic behavior of a compound of similar size; (2) the chemical properties of the compounds concerned; (3) the variation of half-wave potentials with pH; and (4) the slope of the wave.

The determination of *n* by coulometric measurements is the most accurate method and is especially useful when the number of electrons involved is large, as in the case of picric acid. In reductions involving only a few electrons per molecule, *n* can be determined by substituting estimated values of the diffusion coefficient in the Ilkovic equation. The value of *n* may give an indication of the final product formed at the electrode. Thus, if *n* is 4 in the reduction of nitrobenzene, the product formed probably is phenylhydroxylamine.

The chemical properties of the compounds concerned are important. The products postulated as being formed at the dropping mercury electrode should not be reducible further at the electrode or should not react with the electrolyte used. An example in which this point may be important is the reduction of picric acid in 0.1 *N* hydrochloric acid. Coulometric measurements[13] indicate a 17-electron change; reduction to triaminophenol requires 18 electrons. To explain these facts, it has been suggested that the reduction proceeds to the substituted hydrazobenzene. Since the hydrazobenzene represents a compound which can react with the electrolyte and undergo the benzidine rearrangement, its chemical behavior should be studied in support of such a mechanism.

The shift of half-wave potentials to more negative values with an increase in pH in acid solution sometimes can be interpreted as a direct reaction of the compound with a proton to form an intermediate ion which is more easily reduced (see the effect of constitution on half-wave potentials).

$$R_2C{=}O + H^+ \rightleftarrows R_2\overset{+}{C}{-}OH$$

$$R_2\overset{+}{C}{-}OH + e \rightleftarrows R_2\dot{C}{-}OH$$

and at higher pH values by a reduction of the molecule directly followed

[13] J. J. Lingane, *J. Am. Chem. Soc.*, **67**, 1916 (1945).

by a reaction of the anion with water or hydrogen ions in the following manner:[13a]

$$R_2C{=}O + e \rightarrow R_2\dot{C}{-}O^-$$
$$R_2\underset{\cdot}{C}{-}O^- + H_2O \rightarrow R_2\underset{\cdot}{C}{-}OH + OH^-$$

Kinetic waves and adsorption waves observed with organic compounds are discussed in Chapters XIV and XV.

Summarizing, it is recommended that in irreversible reductions the following data be reported:

(1) Half-wave potentials (referred to a standard electrode) and their dependence upon concentration, pH, and the medium.
(2) Equation of the wave(s).
(3) The diffusion current constant.
(4) Number of electrons involved in the reduction.
(5) The effect of pH (buffer solutions are required).
(6) The effect of the kind and concentration of the buffer and of indifferent ions on the half-wave potential.
(7) The composition of the solvent.
(8) Concentration of maximum eliminator at given concentration of substance being studied.

In the following chapters a condensed review of the polarography of organic compounds is given. Half-wave potentials are usually referred to the saturated calomel electrode. If no reference electrode is mentioned, it is not stated in the original publication. "Reduction potentials" are reported only if the half-wave potential is not available. The reported data are of practical importance in organic chemistry and of special interest in connection with:

(a) Qualitative and quantitative analysis of organic compounds.
(b) Structure determination.
(c) The relation between structure and half-wave potentials.
(d) The mechanisms of electroreductions.
(e) The effects of organic substances upon the overvoltage of hydrogen.

2. APPARENT REDUCTION POTENTIALS AND HALF-WAVE POTENTIALS

A general discussion of the analysis of c.v. curves of irreversible reductions has been given in Chapters XIV and XV. In many instances the relation between the potential on the one hand and the current i, and $(i_d - i)$ on

[13a] P. J. Elving, O. H. Müller, S. Wawzonek, M. J. Astle, and L. Meites, *Anal. Chem.*, **22**, 482 (1950).

the other, is given by an equation which has the same form as that for reversible reductions. In irreversible reductions the value of n calculated from the slope of log $i/(i_d - i)$ *vs.* E line has no exact significance. The constant in the equation of the polarographic wave also includes one or more rate constants.

Several attempts have been made to relate the half-wave potentials to potentials determined by other methods for irreversible systems.

"Apparent reduction potentials" for a number of irreversible reductions were assigned by Conant[14] by determining the rate at which the oxidized form reacts with a reversible oxidation-reduction system of known oxidation potential. In this reaction it was assumed that the first step in the reduction is reversible and instantaneous while the next irreversible step is rate determining. The irreversible system was characterized by a potential called the "apparent reduction potential" (A.R.P.). This is the potential corresponding to a reduction of 20 to 30 per cent of the oxidized form in 5 minutes. The A.R.P. of some compounds are given in Table XXXVI-1.

Müller's comparison[15] of polarograms of nitrobenzene, dinitrobenzene azobenzene, and maleic acid with Conant's A.R.P. values indicated that the potentials at which the currents started to rise coincided approximately with the A.R.P. potentials, whereas the half-wave potentials were found at more negative values. The data of Pearson[16] on various nitro compounds at pH 0.5 are in Table XXXVI-1. As long as a thermodynamically satisfactory interpretation of apparent reduction potentials cannot be given[17] it remains questionable whether any close relation exists between A.R.P. and the half-wave potential.

The true oxidation potentials have been calculated by Adkins[18] for ninety carbonyl compounds from experimentally determined equilibrium constants of reactions of the following type:

$$R_2C{=}O + R'_2CHOH \xrightleftharpoons{Al(OC(CH_3)_3)_3} R_2CHOH + R'_2C{=}O$$

The establishment of equilibrium is very slow and, therefore, it is to be expected that no simple correlation could be found between the oxidation potentials of these compounds and their half-wave potentials. The oxi-

[14] J. B. Conant, *Chem. Revs.*, **3,** 1 (1926).

[15] O. H. Müller, *Chem. Revs.*, **24,** 95 (1939). O. H. Müller and J. P. Baumberger, *J. Am. Chem. Soc.*, **61,** 590 (1939).

[16] J. Pearson, *Trans. Faraday Soc.*, **44,** 683 (1948).

[17] L. Michaelis, *Oxydation-Reductions Potentiale*, 2nd ed., Berlin, 1933 (p. 133).

[18] H. Adkins *et al.*, *J. Am. Chem. Soc.*, **71,** 3622 (1949).

TABLE XXXVI-1
APPARENT REDUCTION POTENTIALS

Substance	Solvent	A.R.P. (N.H.E.)		$E_{1/2}$(N.H.E.) at pH 0.5
Unsaturated Compounds				
Dibenzoylethylene, BzCH=CHBz	(*a*)	+0.27	(±0.02)	
Benzoylacrylic acid, BzCH=CHCOOH	(*a*)	+0.06	(±0.02)	
Maleic acid, HOOCCH=CHCOOH	(*a*)	−0.25	(±0.06)	
	(*b*)	−0.25	(±0.06)	
Nitro Compounds				
1,3,5-Trinitrobenzene	(*a*)	+0.26	(±0.02)	+0.206
1,3-Dinitrobenzene	(*a*)	+0.16	(±0.01)	+0.216
Nitrobenzene	(*a*)	−0.06	(±0.04)	+0.086
2,4-Dinitrobenzoic Acid	(*b*)	+0.23	(±0.02)	
3-Nitrobenzoic Acid	(*b*)	+0.06	(±0.04)	
Phenylnitromethane	(*a*)	−0.08	(±0.06)	
Azo Dyes (c)				
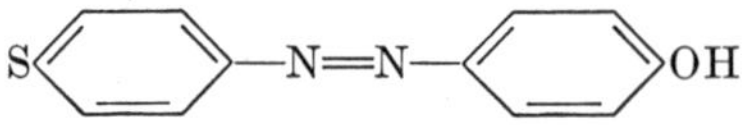	(*b*)	+0.42		
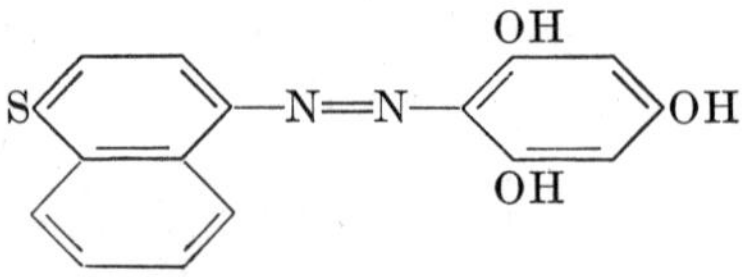	(*b*)	+0.36		
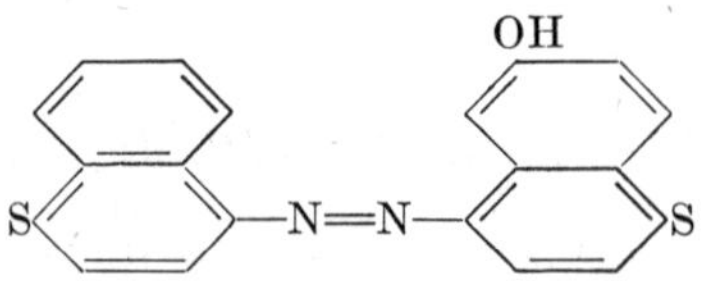	(*b*)	+0.32		
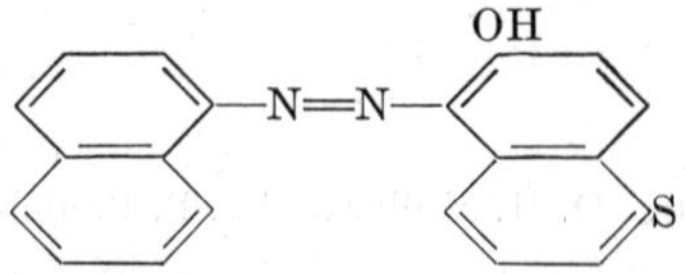	(*b*)	+0.29		

(*a*) Mixture of 75 per cent acetone and 25 per cent dilute HCl, total acidity 0.2 *N*.
(*b*) 0.2 *N* hydrochloric acid in water (25° C.).
(*c*) S = SO_3H.

dation potentials were calculated for toluene solutions while the polarographic determinations were always carried out in a more polar solvent, usually a mixture of alcohol and water. Polarographic data correspond to the thermodynamic data only when the process which determines the potential is thermodynamically reversible.

3. HALF-WAVE POTENTIALS AND CONSTITUTION

Though the half-wave potential of an irreversible reaction always involves some rate constants, there seems to be a fairly close relation between the polarographic half-wave potential and chemical constitution. This relationship is similar to that observed between "reduction potentials" (by the tangent method) and structure, which Shikata and Tachi[19] formulated into the "electronegativity rule of reduction potentials"—that organic compounds are more easily reduced as more electronegative groups are substituted in the same compound. There seems to be a parallelism between the effect of various substituents on the reduction potential and on the ionization constant,[20] the Raman spectrum, and adsorption spectrum.[21]

The relative effects produced by various groups are dependent not only upon their tautomeric (T) and inductive (I) effects* but also upon their

[19] M. Shikata and I. Tachi, *J. Chem. Soc. Japan*, **53,** 834 (1932); *Collection Czechoslov. Chem. Communs.*, **10,** 368 (1938). I. Tachi, *Mem. Coll. Agr. Kyoto Imp. Univ.*, **42,** 1 (1938).

[20] Compare, *e. g.*, C. G. Derick, *J. Am. Chem. Soc.*, **33,** 1125, 1162 (1911). H. Lucas, *ibid.*, **48,** 1827 (1926).

[21] A. Winkel and G. Proske, *Ber.*, **69,** 1917 (1936).

* The two ways in which electron displacements or polarizations in an organic molecule can be expressed are the inductive (I) effect and the tautomeric (T) effect. The inductive effect is dependent upon electrostatic forces and arises from an unequal sharing of the electron pair in a covalent bond. Groups which are electron attracting are considered to be —I groups (*e.g.*, $—NH_3^+$, $—NO_2$, —Cl) whereas groups which are electron-repelling or releasing are +I groups (*e.g.*, $—COO^-$, $—CH_3$).

The tautomeric (T) effect is the electron displacement which is associated with systems containing multiple links. The displacement occurs whenever electrically dissimilar atoms are joined by a double bond ($>$C=O, $—NO_2$) or whenever a carbon-carbon double bond is joined to such a group, another double bond, or to an atom that has at least one unshared pair of electrons ($—NH_2$, —OH, —Cl). This effect is often known as the resonance effect. Groups which attract electrons are considered to have a —T effect (*e.g.*, $—NO_2$, —COOH); groups which donate or release electrons are +T groups (*e.g.*, $—\ddot{\underset{..}{Cl}}:$, $—\ddot{N}H_2$, $—\ddot{\underset{..}{O}}H$).

The terminology used in the above is that of Ingold and the English school (see C. K. Ingold, *Chem. Revs.*, **15,** 225 (1934).

position in the molecule and the buffer system used. Thus in the reduction of substituted nitrobenzenes the reduction becomes more difficult (half-wave potential becomes more negative) with the introduction of substituents in the order: $NO_2 > COOH > Cl > CH_3$. The nitro and carboxyl groups show both an inductive effect (−I) and a tautomeric effect (−T) which decreases the electron density around the nitrogen of the nitro group and facilitates its reduction. Since the tautomeric effect is greatest in the ortho and para positions, the meta isomer is the most difficult to reduce. Similarly, chlorine shows a (−I) and (+T) effect, while the methyl group only exhibits a (+I) effect which again is most pronounced if present in the ortho and para positions to the nitro group. In agreement with this the meta isomer is the easiest to reduce.[22]

Ortho-substituted derivatives usually show abnormal behavior. The hydroxyl group, which is an electron-donating group (+T), decreases the half-wave potential, *i. e.*, to more positive potentials, of the nitro group if it is present in the ortho position. This opposite effect is due to hydrogen bonding which interferes with the normal resonance of the nitro group and makes the reduction easier.[23] A similar observation has been made in the reduction of *o*-hydroxybenzaldehyde.[24]

The pH of the medium influences the intensity with which the (I) effect and (T) effect will operate in a molecule by changing the nature of the molecule which is being reduced. Thus many organic compounds which are very weak bases, in acid buffers will combine with hydrogen ions to form the corresponding Brönsted acid.

$$RC_6H_4CHO + H^+ \rightleftharpoons RC_6H_4\overset{+}{C}HOH$$

This proton addition is considered to be very rapid and reversible and results in a more easily reducible complex in which the (T) effect would be more important than the (I) effect. This postulation has been verified by a study of substituted benzaldehydes; the depolarization potentials were found to increase (become more positive) in the order: *p*-Cl $<$ H $<$ Me $\approx$ Me_3C $\approx$ Et $<$ Me_2CH $<$ OCH_3 in agreement with the behavior predicted from the inductive (I) and (T) effects of the groups.

In a medium approaching neutrality, pH 4, the concentration of the protons is small and the depolarization potential is mainly determined by the polarizability of the aldehyde molecule. The (I) effect remains the same but the (T) effect becomes less. The order of increasing depolariza-

---

[22] S. F. Dennis, A. S. Powell, and M. J. Astle, *J. Am. Chem. Soc.*, **71,** 1484 (1949).
[23] J. Pearson, *Trans. Faraday Soc.*, **44,** 692 (1948).
[24] G. Semerano and V. Capitanio, *Gazz. chim. ital.*, **70,** 490 (1940).

tion potentials observed is now: p-Cl < Me_3C < Me_2CH < H ≈ Et < Me < OCH_3 and is in agreement with the order postulated. In this region of pH if the acid form is produced and involved in the reduction it would be expected that the wave would be a kinetic one. No study of this type has been made however.

In slightly alkaline or neutral solutions the reduction becomes independent of pH and the tautomeric effect is further reduced. The order of increasing depolarization potentials is: p-Cl < H < Me_3C ≈ Me_2CH ≈ C_2H_5 < CH_3 < OCH_3.

The dimethylamino group behaves in an unusual way, since it changes its nature depending upon the pH. In acid medium it is electron attracting, $N^+(CH_3)_2H$, and at a pH of 1.4 its effect is between that of the p-chloro derivative and benzaldehyde. In alkaline medium (pH 8) it shows a (+T) effect and has the most negative depolarization potential: $Me_3C \ll OCH_3 < N(CH_3)_2$.[25]

4. POLAROGRAPHICALLY REDUCIBLE AND OXIDIZABLE GROUPS

Groups reducible at the dropping mercury electrode under appropriate conditions are: (1) the carbon-carbon double bond when conjugated with another double bond, an aromatic ring, or another unsaturated group; (2) the carbonyl group in aldehydes, ketones, and quinones: (3) nitro, nitroso, amine oxide,[26] and azo groups; (4) phenylhydroxylammonium, diazonium, and quaternary ammonium groups; and (5) most halogens, disulfides, peroxides, and epoxides.

Groups or compounds which are oxidizable at the dropping mercury electrode are hydroquinones and related compounds, ene-diols, and substances which depolarize mercury by formation of slightly dissociated or insoluble compounds (—SH in thiol compounds).

Amines and acids which lack the above groupings are not reducible but may give catalytic hydrogen waves.

[25] J. W. Baker, W. C. Davies, and M. L. Hemming, *J. Chem. Soc.*, **1940,** 692.

[26] E. Ochiai, *Proc. Imp. Acad. Tokyo*, **19,** 309 (1943); *Chem. Abstracts*, **41,** 5880 (1947).

CHAPTER XXXVII

Unsaturated Hydrocarbons

Compounds containing isolated ethylenic and acetylenic linkages, such as *o*-allylanisole (I) and heptyne-1 (II),

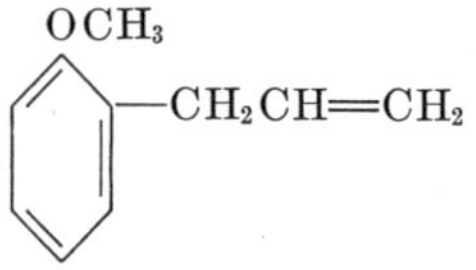

(I)

$CH_3CH_2CH_2CH_2CH_2C{\equiv}CH$

(II)

are not reducible at the dropping mercury electrode in a 0.175 *M* tetrabutylammonium iodide–75 per cent dioxane solution.[1] The reported reduction of the double bond in allyl alcohol,[2] geraniol, and phytol probably must be attributed to the reduction of peroxide present as an impurity. A purified sample of allyl alcohol does not give a wave in 0.175 *M* tetrabutylammonium iodide–75 per cent dioxane solution.[3]

Hydrocarbons in which the double or triple bond is conjugated with an aromatic ring, polynuclear aromatic hydrocarbons, and aliphatic hydrocarbons such as allene and those containing conjugated double bonds are reducible at the dropping mercury electrode.

Phenyl-Substituted Olefins and Acetylenes. Phenyl-substituted olefins and acetylenes give well-defined waves in a medium of 50 to 75 per cent dioxane with tetrabutylammonium iodide as the supporting electrolyte.[1]

The half-wave potentials which are independent of the concentration are given in Table XXXVII-1. The diffusion coefficients (D) were calculated with the aid of the Ilkovic equation. The substances listed can be determined polarographically.

The equation of the waves is given by:

$$E = E_{1/2} + \frac{RT}{F} \ln \frac{i_d - i}{i}$$

[1] H. A. Laitinen and S. Wawzonek, *J. Am. Chem. Soc.*, **64,** 1765 (1942).

[2] P. van Rysselberghe, J. M. McGee, A. H. Gropp, and R. W. Lane. *J. Am. Chem. Soc.*, **69,** 809 (1948).

[3] S. Wawzonek and J. H. Fossum, Private communication. See also I. Tachi, *Mem. Coll. Agr. Kyoto Imp. Univ.*, **17,** 35 (1931).

Since the half-wave potentials are independent of the pH and the logarithmic analysis of the reduction wave of stilbene gives a slope of 0.062, the following reaction mechanism is suggested:

$$R + e \leftrightarrows R^- \text{ (reversible, potential determining)}$$

$$R^- + e \rightarrow R^{--} \text{ (irreversible, fast)}$$

$$R^{--} + 2H_2O \rightarrow RH_2 + 2OH^- \text{ (irreversible, rapid)}$$

Substitution of electron-donating groups in one or both nuclei of stilbene results in a slightly more negative half-wave potential. The substituted stilbene analogs containing a naphthalene nucleus give waves at potentials equal to that of stilbene or more positive than this compound. The

TABLE XXXVII-1

HALF-WAVE POTENTIALS (S.C.E.) AND DIFFUSION COEFFICIENTS (D) OF PHENYL-SUBSTITUTED OLEFINS AND ACETYLENES AT 25° C. IN 0.175 M TETRABUTYLAMMONIUM IODIDE–75 PER CENT DIOXANE

Compound	Formula	$-E_{1/2}$	$D \times 10^6$
Styrene	$C_6H_5CH{=}CH_2$	2.35	5.6
β-Methylstyrene	$C_6H_5CH{=}CHCH_3$	2.54	4.7
1,1-Diphenylethylene	$(C_6H_5)_2C{=}CH_2$	2.14	3.3
Stilbene	$C_6H_5CH{=}CHC_6H_5$	2.26	3.3
1,4-Diphenylbutadiene-1,3	$C_6H_5CH{=}CH{-}CH{=}CHC_6H_5$	1.98	3.0
Triphenylethylene	$(C_6H_5)_2C{=}CHC_6H_5$	2.11	2.6
Tetraphenylethylene	$(C_6H_5)_2C{=}C(C_6H_5)_2$	2.05	2.5
Phenylacetylene	$C_6H_5C{\equiv}CH$	2.37	4.4
Diphenylacetylene	$C_6H_5C{\equiv}CC_6H_5$	2.20	2.9

second wave shown by these compounds at −2.50 v. is due to the reduction of the naphthalene nucleus (see p. 637). The half-wave potentials[4] are given in Table XXXVII-2.

Stilbenes with monosubstituents on the α carbon atom give half-wave potentials approximately the same or slightly higher than that of stilbene itself. Stilbene derivatives with alkyl groups on both α and β carbon atoms are not reducible with the exception of α,β-diethylstilbene. The wave obtained for 4,4′-diacetyl-α,β-diethylstilbene is for the reduction of the ketone groups. A summary of the results of Goulden and Warren[4] is given in Table XXXVII-3.

Values enclosed in parentheses are for ill-defined waves. The higher melting isomer of α,β-diethylstilbene, however (see last value in table)

[4] F. Goulden and F. L. Warren, *Biochem. J.*, **42**, 420 (1948).

TABLE XXXVII-2

HALF-WAVE POTENTIALS (S.C.E.) AND i_d/C OF SUBSTITUTED STILBENES AT 25° IN 0.1 *M* TETRAETHYLAMMONIUM IODIDE–75 PER CENT DIOXANE

Compound	$-E_{1/2}$	i_d/C, microamp. millimole^{-1} liter^{-1}
Stilbene	2.17	4.12
4-Aminostilbene	2.31	3.43
4-Methoxystilbene	2.24	3.77
4-Hydroxystilbene	2.34	3.14
4-Dimethylaminostilbene	2.29	3.70
3-Dimethylaminostilbene	2.19	3.58
4-Amino-4-hydroxystilbene	2.45	2.27
4,4′-Dimethoxystilbene	2.32	3.69
4-Dimethylamino-2′-methylstilbene	2.31	3.47
4-Dimethylamino-3′-methylstilbene	2.33	3.33
4-Dimethylamino-4′-methylstilbene	2.34	3.30
4-Dimethylamino-2′,5′-diethylstilbene	2.35	3.01
4-Amino-4′-hydroxy-2′-methylstilbene	2.43	2.35
2,4,6,2′,4′,6′-Hexamethylstilbene	2.47	1.38
α-(4-Dimethylaminophenyl)-β-(1-naphthyl)ethylene	2.12; 2.50	4.12
α-(4-Dimethylaminophenyl)-β-(2-naphthyl)ethylene	1.81; 2.49	4.06
α-(4-Dimethylaminophenyl)-β-1-(2-methylnaphthyl)ethylene	2.17; 2.54	3.14
α-Phenyl-β-1-(dimethylaminonaphthyl)ethylene	2.08; 2.53	

TABLE XXXVII-3

HALF-WAVE POTENTIALS (S.C.E.) AND i_d/C OF STILBENE WITH SUBSTITUENTS ON THE ETHYLENIC CARBON ATOMS AT 25° IN 0.1 *M* TETRAETHYLAMMONIUM IODIDE–75 PER CENT DIOXANE

Compound	$-E_{1/2}$	i_d/C
α-Ethylstilbene	2.31	3.84
4,4′-Diacetyl-α,β-diethylstilbene	1.78	7.85
4,4′-Dihydroxy-α-*n*-propylstilbene	2.52	1.47
4,4′-Dihydroxy-α-*n*-amylstilbene	2.53	1.15
4,4′-Dihydroxy-α-isopropylstilbene	(2.61)	
Triphenylethylene	2.10	3.86
Triphenylmethylethylene	2.24	3.64
α-Hydroxyphenylstilbene	2.17	3.29
α-Phenyl-β,β-di-4-hydroxyphenylethylene	2.28	3.56
α-Ethyl-β-4-hydroxyphenylstilbene	2.36	4.72
α,β-Diethylstilbene (m.p. 74–75°)	(2.59)	

gives a well-defined wave at −2.55 v. (S.C.E.)[5] in 0.175 *M* tetrabutylammonium iodide–75 per cent dioxane.

[5] S. Wawzonek and J. W. Fan, Private communication.

According to Goulden and Warren, the lower melting isomer of α,β-diethylstilbene, stilbestrol, dienestrol, 4,4′-dihydroxy-α-methyl-β-ethylstilbene, 4,4′-dihydroxy-α-methyl-β-ethylstilbene, 4,4′-dihydroxy-α-methyl-β-cetylstilbene, 4,4′-dihydroxy-α-phenyl-β-ethylstilbene, 2,2′-dihydroxy-α,β-diethylstilbene, and 4-methoxy-α,β-diethylstilbene give no waves in 0.1 *M* tetraethylammonium iodide–75 per cent dioxane. The geometrical isomers of α,β-dimethylstilbene and 4,4′-dimethoxy-α,β-diethylstilbene give no waves in 0.1 *N* tetraethylammonium chloride–70 per cent dioxane solution. In this medium isostilbene, the *cis* isomer, is reported to reduce at the same potential, −2.34 v., as the *trans* isomer, stilbene. The substitution of 80 per cent alcohol for the dioxane causes isostilbene to be reduced at a slightly more negative value (−2.18 v.) than stilbene (−2.13 v.).[6]

TABLE XXXVII-4
HALF-WAVE POTENTIAL (S.C.E.) AND i_d/C OF AROMATIC POLYNUCLEAR HYDROCARBONS AT 25° IN 0.175 *M* TETRABUTYLAMMONIUM IODIDE–75 PER CENT DIOXANE

Compound	$-E_{1/2}$	i_d/C
Naphthalene	2.50	5.15
1,2-Dihydronaphthalene	2.57	4.44
Acenaphthene	2.58	4.90
Indene	Wave starts at 2.54	
3-Phenylindene	2.33	4.65
Fluorene	2.65	5.65
Biphenyl	2.70	6.96
Phenanthrene	2.46; 2.71	8.84
9,10-Dihydrophenanthrene	2.62	7.18
Chrysene		12.88
Pyrene	2.10; 2.46; 2.68	3.22; 5.5
Anthracene	1.94	4.47
1,2-Benzanthracene	2.03; 2.54	4.10; 4.54
1,2,5,6-Dibenzanthracene	2.07; 2.53	4.12; 7.55
9,10-Dimethyl-1,2-benzanthracene	2.05; 2.53	4.14; 4.32
3-Methylcholanthrene	2.11; 2.51	3.70; 4.15
3,4-Benzpyrene	1.88	2.72; 4.54

Polynuclear Hydrocarbons. Polynuclear aromatic hydrocarbons are electroreducible in 75 per cent dioxane solution which contains 0.175 *M* tetrabutylammoniun iodide.[7] A summary of the results is given in Table XXXVII-4.

[6] F. Wessely and J. Wratil, *Mikrochemie ver. Mikrochim. Acta*, **33**, 248 (1947).
[7] S. Wawzonek and H. A. Laitinen, *J. Am. Chem. Soc.*, **64**, 2365 (1942).

In general all the hydrocarbons give well-defined reduction waves with the exceptions of phenanthrene (I), pyrene (II), chrysene (III), and 3,4-benzpyrene (IV):

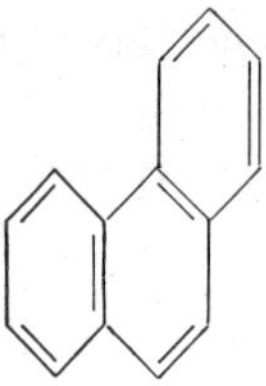

(I)

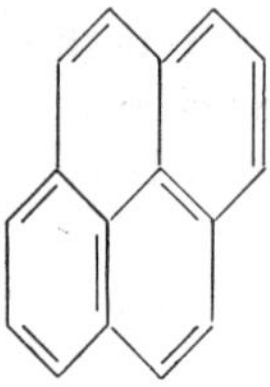

(II)

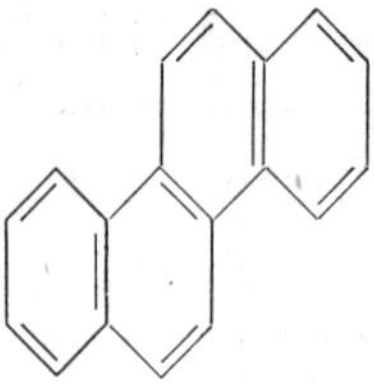

(III)

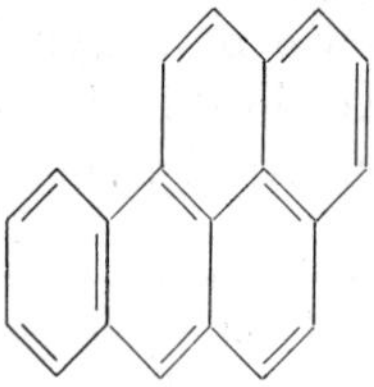

(IV)

Phenanthrene and pyrene give two waves close together and of equal heights. In Table XXXVII-4 only the total diffusion current of phenanthrene is given even though two waves were obtained. Chrysene gives ill defined waves with a total diffusion current corresponding to a 6-electron reduction. 3,4-Benzpyrene gives a composite wave, but the first wave is fairly well defined and the half wave potential can be measured.

The half-wave potentials were found to be independent of the concentrations of the hydrocarbons. Since unbuffered solutions were used, $E_{1/2}$ apparently is not dependent upon pH.

The diffusion current is proportional to the concentration and can be used for the determination of the various hydrocarbons.

The reduction of the polynuclear hydrocarbons at the dropping mercury electrode resembles the phenomenon of the addition of alkali metals to these compounds or of their reduction by alkali metals and alcohols or alkali metal amalgams and alcohols. This resemblance is not surprising since the reduction at the dropping electrode occurs at such negative potentials. The reduction of naphthalene involves only two electrons and proceeds by a 1,4-mechanism since the resulting 1,4-dihydronaphthalene is not further reducible. The product of a 1,2-reduction would be 1,2-dihydronaphthalene which is further reducible at −2.57 v. This latter value is very close to that of β-methylstyrene (−2.54 v.) and indicates that the electroreduction of a double bond conjugated with an aromatic ring is apparently not affected by its presence in a ring.

The wave for anthracene (−1.94 v.) represents a 1,4-reduction at the 9,10-positions. The course of this reduction is best demonstrated in the reduction of substituted anthracenes. 1,2-Benzanthracene, for example, shows one wave at approximately the same point (−2.03 v.) as anthracene and a second wave at −2.54 v. The latter wave approximates that of naphthalene (−2.50 v.) and points to the following reduction for 1,2-benzanthracene:

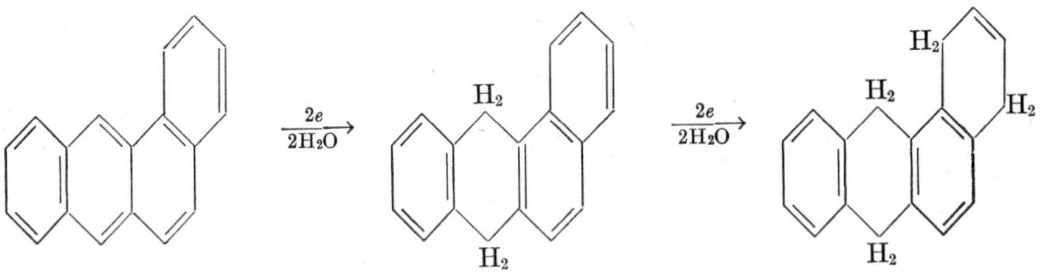

Similar behavior is shown by 3,4-benzpyrene and is in agreement with the anthracene structure proposed for it from absorption spectra work.[8] The first wave at −1.88 v. and involving two electrons is similar to the wave observed with anthracene; the second wave of two electrons represents the reduction of the resulting naphthalene:

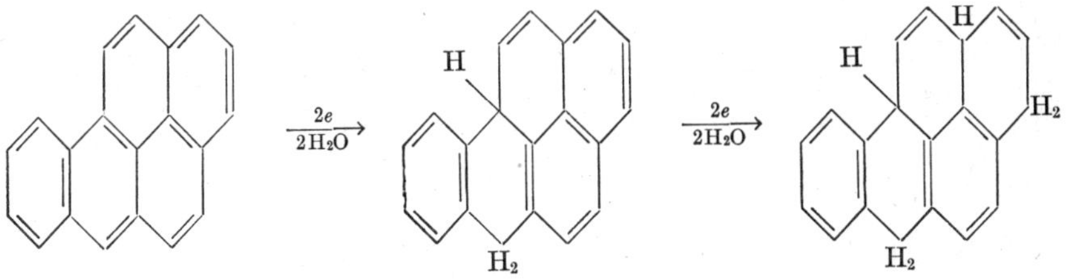

Aliphatic Hydrocarbons. According to von Stackelberg[9] aliphatic hydrocarbons such as allene and those containing conjugated double and triple bonds are reducible at the dropping mercury electrode. Half-wave potentials (S.C.E.) and diffusion currents for 0.001 *M* solutions in 0.05 *M* tetraethylammonium bromide—75 per cent dioxane solution are given in Table XXXVII-5.

It is worth noticing that the diffusion currents obtained for these compounds are very small, indicating incomplete reaction at the electrode. It would be of interest to investigate whether these currents are kinetically controlled.

Cyclic conjugated olefins[10] that are electroreducible are dimethylfulvene (I) which gives a wave at −1.89 v. (S.C.E.) in 0.175 *M* tetrabutylam-

---

[8] J. W. Cook and C. L. Hewett, *J. Chem. Soc.*, **1933,** 400.
[9] M. von Stackelberg and W. Strache, *Z. Electrochem.*, **53,** 118 (1949).
[10] S. Wawzonek and J. W. Fan, *J. Am. Chem. Soc.*, **68,** 2541 (1946).

monium iodide–75 per cent dioxane and cycloöctatetraene (II) which gives a wave at −1.51 v. (S.C.E.)[11]

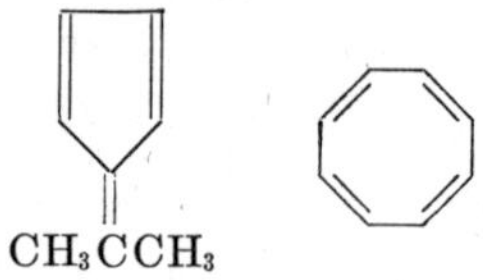

(I) (II)

in 0.1 *M* tetramethylammonium hydroxide–50 per cent ethanol. The number of electrons involved in the reduction is two in each case. The half-wave potentials are independent of pH and for cycloöctatetraene the

TABLE XXXVII-5

POLAROGRAPHIC BEHAVIOR OF ALIPHATIC HYDROCARBONS

Compound	$-E_{1/2}$	i_d/C
Ethylene	No wave	
Butadiene	2.59	0.6
Allene	2.29	0.5
Acetylene, methylacetylene	No wave	
Diacetylene	2.27	0.9
Vinylacetylene	2.40	1.5
	2.58	0.8
Divinylacetylene	2.07	1.1
	2.55	0.6–0.3

diffusion current is proportional to the concentration in the range 4.5×10^{-6} to 2.07×10^{-3} *M*.

The reduction of cycloöctatetraene is unusual in that tetramethylammonium ions are necessary for the formation of a well-defined wave. It has been suggested that this ion stabilizes the anion formed in the reduction in the following manner:

$$C_8H_8 + 2e + 2(CH_3)_4N^+ \rightarrow ((CH_3)_4N)_2^{++}C_8H_8^{--}$$

Uses. The polarographic method has been used to determine naphthalenes and substituted naphthalenes in petroleum fractions.[12] Monocyclic aromatics, olefins, naphthenes, and monoolefins do not interfere. Higher polynuclear aromatics can usually be removed by distillation. The half-wave potentials of α-methylnaphthalene and β-methylnaph-

[11] R. M. Elofson, *Anal. Chem.*, **21,** 916 (1949).

[12] R. E. Burdett and B. E. Gordon, *Anal. Chem.*, **19,** 843 (1947).

thalene occur at −2.51 v., while that of acenaphthene is at −2.60 v. The diffusion current is measured at −2.74 v. Biphenyl (−2.70 v.) interferes if large amounts are present.

The method has been used[13] to establish the location of the double bond in 2,6-diphenylbicyclo[3,0,3]octene(1-5) (I).

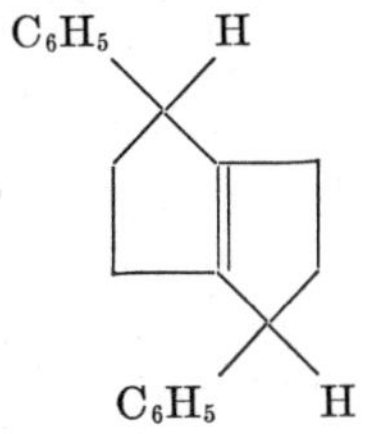
(I)

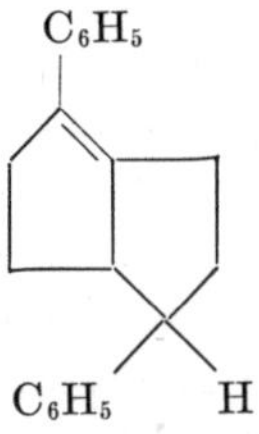
(II)

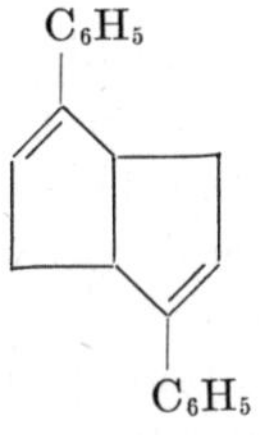
(III)

This hydrocarbon (I) contains an isolated double bond and gives no wave in a 0.175 *M* tetrabutylammonium iodide–75 per cent dioxane solution. Other members of the series in which the bond is conjugated with the aromatic ring are reducible. Values obtained are listed here.

Compound	$-E_{1/2}$ (S.C.E.), v.	i_d/C, microamp.
II	2.57	3.63
III	2.53	7.09

The similarity of the behavior of unsaturated hydrocarbons at the dropping mercury electrode with their behavior with alkali metals in ether solution was used as a basis for investigating the mechanism involved in the dimerization of 1,1-diphenylethylene and related compounds by alkali metals in absolute ether.[10]

$$2(C_6H_5)_2C{=}CH_2 + 2Na \rightarrow (C_6H_5)_2\underset{\displaystyle Na}{\underset{|}{C}}CH_2CH_2\underset{\displaystyle Na}{\underset{|}{C}}(C_6H_5)_2$$

(IV)

Ziegler and co-workers[14] have suggested that the reduction follows 1,2-addition with the resulting organo-alkali compound adding to another molecule of olefin:

$$(C_6H_5)_2C{=}CH_2 \xrightarrow{2Na} (C_6H_5)_2\underset{\displaystyle Na}{\underset{|}{C}}{-}\underset{\displaystyle Na}{\underset{|}{CH_2}} \xrightarrow{(C_6H_5)_2C{=}CH_2} \text{(IV)}$$

[13] S. Wawzonek, *J. Am. Chem. Soc.*, **65**, 839 (1943).
[14] K. Ziegler, H. Colonius, and O. Schafer, *Ann.*, **473**, 36 (1929).

Schlenk and Bergmann[15] have proposed free radical formation as the initial step followed by polymerization:

$$2(C_6H_5)_2C{=}CH_2 \xrightarrow{2Na} 2(C_6H_5)_2\underset{\displaystyle Na}{\underset{|}{C}}{-}\dot{C}H_2 \rightarrow \text{(IV)}$$

Work with 1,1-diphenylethylene indicates that the mechanism of electroreduction (see p. 635) is in essential agreement with that proposed by Ziegler (1929). The second electron (or sodium) adds to the intermediate free radical before it can dimerize. The resulting products at the dropping mercury electrode differ from those obtained with sodium under anhydrous conditions in that the bivalent anion formed reacts preferentially with the water present rather than with another molecule of olefin. At the dropping mercury electrode a wave is observed involving two electrons.

To determine whether this mechanism is a general one, other unsaturated hydrocarbons which undergo dimerization with alkali metals[16] were studied at the dropping mercury electrode. The results obtained are summarized in Table XXXVII-6.

A comparison of the diffusion current constants, $i_d/Cm^{2/3}t^{1/6}$, obtained indicates that two electrons are involved in the first reduction wave of all compounds with the exception of cinnamylidenefluorene, fluoranthene, and dibiphenyleneethylene. Accordingly, 1,1-diphenylethylene (4.74), dimethylfulvene (4.81), and dimethylbenzofulvene (5.20) have similar diffusion current constants to those of stilbene (5.20), diphenylfulvene (4.65), and diphenylbenzofulvene (4.38), compounds which undergo normal 1,2-addition. Benzalfluorene (4.52) and furfuralfluorene (4.11) in turn have diffusion current constants approximately equal to that of biphenylenediphenylethylene (4.46) which undergoes 1,2-addition. The half-wave potentials obtained for these waves indicate that the same double bonds are involved in these reductions as in the addition of sodium. Thus compounds with similar structures have approximately the same half-wave potentials. In the fluorene series the appearance of a wave at approximately −1.63 v. for all the compounds points to the reduction of the 9,10-double bond. A second wave at approximately −2.64 v. which is characteristic of the fluorene ring system is obtained with every compound.

Cinnamylidenefluorene, fluoranthene, and dibiphenyleneethylene give diffusion current constants which point to a different type of reduction.

Cinnamylidenefluorene gives four waves with diffusion current constants

[15] W. Schlenk and E. Bergmann, *Ann.*, **463,** 1 (1928).
[16] C. B. Wooster, *Chem. Revs.*, **11,** 1 (1932).

TABLE XXXVII-6

HALF-WAVE POTENTIALS AND DIFFUSION CURRENT CONSTANTS OF VARIOUS COMPOUNDS IN 0.175 *M* TETRABUTYLAMMONIUM IODIDE–75 PER CENT DIOXANE

Compound	$-E_{1/2}$ (S.C.E.)	$\frac{i_d}{Cm^{2/3}t^{1/6}}$
Stilbene	2.14	5.20
1,1-Diphenylethylene	2.22	4.74
	2.17	4.54[a]
Dimethylfulvene	1.89	4.81
	1.85[a]	4.92[a]
Diphenylfulvene	1.57	4.65
Dimethylbenzofulvene	1.98	5.20
	1.94[a]	4.92[a]
Diphenylbenzofulvene	1.61	4.38[b]
	2.74	4.23[b]
Benzalfluorene	1.67	4.23[b]
	2.65	8.69[b]
	1.67[a]	4.52[a]
	2.69[a]	4.69[a]
Furfuralfluorene	1.62	4.11[b]
	2.62	4.52[b]
	1.66[a]	4.74[a]
	2.63[a]	10.50[a]
Biphenylenediphenylethylene	1.03	4.40[b]
	2.64	15.63[b]
9-Benzhydrylfluorene	2.64	7.77
Cinnamylidenefluorene	1.46	2.19[b]
	1.62	2.62[b]
	1.91	2.48[b]
	2.66	8.03[b]
	1.43[a]	2.29[a]
	1.61[a]	2.18[a]
	1.30[a]	2.05[a]
	2.66[a]	1.91[a]

TABLE XXXVII-6—*Continued*

Compound	$-E_{1/2}$ (S.C.E.)	$\frac{i_d}{Cm^{2/3}t^{1/6}}$
γ-Phenylpropylidenefluorene	1.87	3.67
	2.67	4.65
1-Fluorenyl-2-benzylethylene	2.63	5.63
Fluoranthene	1.79	2.40[b]
	2.11	3.49[b]
	2.62	9.31[b]
	1.72[a]	2.72[a]
	2.08[a]	3.87[a]
	2.63[a]	3.66[a]
Dibiphenyleneethylene	1.04	2.35[b]
	1.39	2.13[b]
	2.42	3.02[b]
	2.67	16.09[b]
	0.99[a]	2.35[a]
	1.37[a]	1.93[a]
	2.41[a]	3.30[a]
	2.67[a]	9.09[a]
9,9-Difluorenyl	2.40	3.94
	2.69	9.60
Triphenylmethyl	1.05	
	1.72	
	2.00	

[a] In 0.175 *M* tetrabutylammonium iodide–0.052 *M* tetrabutylammonium hydroxide–75 per cent dioxane.

[b] Capillary number one. All other values were obtained with capillary number two.

of 2.19, 2.62, 2.48, and 8.03, respectively. The first three values point to 1-electron reduction steps since γ-phenylpropylidenefluorene, which has a similar molecular weight and adds 2 electrons, gives a wave with a diffusion current constant of 3.67. The only apparent explanation at the present time for this behavior is that 1 electron adds to cinnamylidenefluorene and the radical formed is stabilized toward further reduction by resonance and dimerizes.

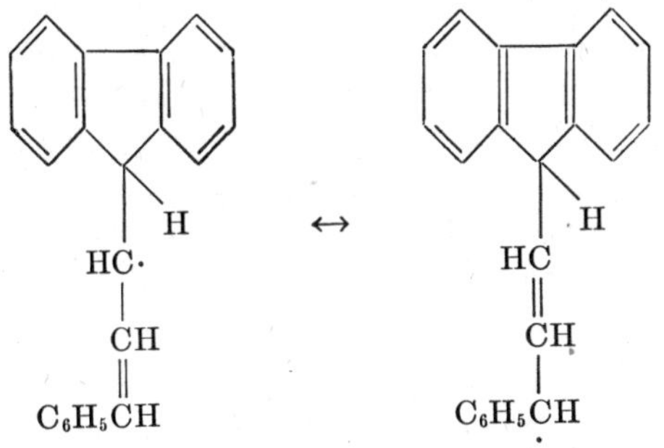

Fluoranthene gives three waves with diffusion current constants of 2.40, 3.49, and 9.31, respectively. The first wave, in analogy to cinnamylidenefluorene, points to a 1-electron reduction. The mechanism of reduction in this step must be similar to that in the first step of cinnamylidenefluorene. This similarity is not surprising, since both compounds involve a conjugated system:

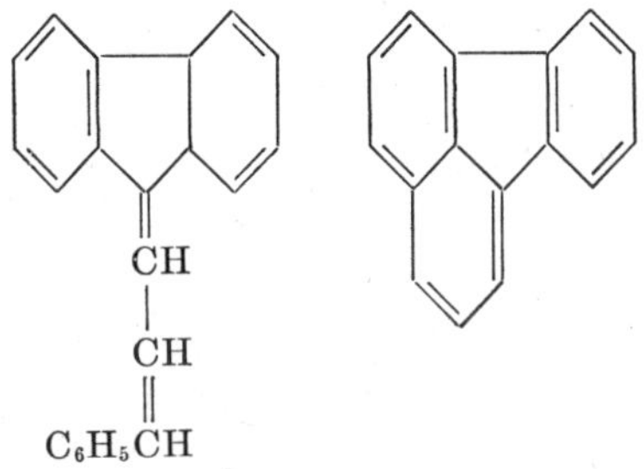

Dibiphenyleneethylene gives four waves with diffusion current constants of 2.35, 2.13, 3.02, and 16.09, respectively. The first two waves point to two 1-electron reductions and the formation of 9,9′-difluorenyl since the last two waves at −2.42 and −2.67 v. are identical with those obtained for this compound. The first reduction wave at −1.04 v. suggests that dibiphenyleneethylene may be a diradical since this value is similar to that observed with the first wave of triphenylmethyl at −1.05 v.

Logarithmic analyses of the first reduction wave of 1,1-diphenylethylene, cinnamylidenefluorene, fluoranthene, and dibiphenyleneethylene give straight lines in each case, with slopes of 0.079, 0.0481, 0.055, and 0.0604 v., respectively. These results are in essential agreement with the theoretical slope of 0.059 v. for a reversible, potential-determining reaction involving 1 electron.

Attempts to correlate the polarographic data obtained with various stilbenes and their ability to produce tumors have proved unsuccessful. The determination of these compounds in blood and urine of rats treated with these compounds failed because of the difficulty encountered in removing the interfering Na^+, K^+, and NH_4^+ ions.[17]

Use of the suppression of the oxygen maximum is made in a highly unspecific assay (see section on adsorption analysis, p. 184) of diethylstilbestrol.[18]

[17] F. Goulden and F. L. Warren, *Biochem. J.*, **42,** 420 (1948).

[18] L. E. Biegenheimer, Jr., and J. E. Christian, *J. Am. Pharm. Assoc.*, **38,** 117 (1949).

CHAPTER XXXVIII

Organic Halogen Compounds

Various types of organic halogen compounds are reducible at the dropping mercury electrode. The word reducible is used here in the sense of a substitution of halogen by hydrogen according to the equations:

$$RCH_2Cl + H^+ + 2e \rightarrow RCH_3 + Cl^-$$

From a study of a large number of compounds the following general rules can be made.

(1) Iodides reduce more easily than bromides and these in turn more easily than chlorides.

(2) The more halogens there are on a carbon atom the more positive is the half-wave potential and the greater the number of waves.

(3) Allylic halides are more easily reduced than alkyl halides and these in turn more easily than vinyl halides. Controlled potential electrolysis indicates that the unsaturation is not affected in these compounds.

(4) The longer the aliphatic chain the more difficult the reduction.

(5) The ease of reduction of a halogen attached to an aromatic ring is similar to that of a halogen attached to a large alkyl group.

(6) The half-wave potential is independent of pH.

A list of the half-wave potentials and diffusion currents observed for various organic halogen compounds in 0.05 *M* tetraethylammonium bromide–75 per cent dioxane solution is given in Table XXXVIII-1.[1]

Usually reduction is that of the halide to the hydrocarbon:

$$RX + 2e + H^+ \rightarrow RH + X^-$$

This mechanism has been confirmed by von Stackelberg[1] by controlled potential electrolysis on a large scale. The products obtained are listed below.

Methyl iodide	Methane
Ethyl bromide	Ethane
α-Bromonaphthalene	Naphthalene
Allyl bromide	Propylene
1,2-Diiodoethylene	Ethylene
Diiodoacetylene	Acetylene
1,2-Dibromoethane	80% ethylene and 20% ethane
2,3-Dibromobutane	Butylene

[1] M. von Stackelberg and W. Stracke, *Z. Elektrochem.*, **53**, 118 (1949).

Table XXXVIII-1

Half-Wave Potentials (S.C.E.) and Diffusion Currents of Various Organic Halogen Compounds

Compound	$-E_{1/2}$	i_d/C
Methyl chloride	2.23	0.9
Methylene chloride	2.33	1.4
Chloroform	1.67	1.4
	Start of 2nd wave	
Carbon tetrachloride	0.78	1.6
	1.71	1.4
Methyl bromide	1.63	1.0
Methylene bromide	1.48	2.6
Bromoform	0.64	1.7
	1.51	2.7
Carbon tetrabromide	0.3	2.4
	0.75	0.8
	1.49	2.4
Methyl iodide	1.63	1.7
Methylene iodide	1.12	1.5
	1.53	1.6
Iodoform	0.49	1.9
	1.09	1.0
	1.50	1.1
Ethyl chloride	No wave	
Diiodoacetylene	0.25	0.8
	0.61	2.0
Chlorobenzene	No wave	
o-Dichlorobenzene	2.51	1.9
m-Dichlorobenzene	2.48	2.0
p-Dichlorobenzene	2.49	1.7
Hexachlorobenzene	1.44	2.0
	1.69	1.9
	1.90	1.5
	2.10	1.5
Bromobenzene	2.32	1.4
p-Dibromobenzene	2.10	2.3
1,2,4,5-Tetrabromobenzene	1.45	3.3
	2.05	1.9
Hexabromobenzene	0.75 } satd. soln.	
	1.45 } satd. soln.	
	Start of 3rd wave	
Iodobenzene	1.62	2.0
α-Chloronaphthalene	2.10	2.0
	2.38	1.3
α-Bromonaphthalene	1.96	1.9
	2.38	1.1
Benzyl chloride	1.94	1.6
Benzal chloride	1.81	3.2

TABLE XXXVIII-1—*Continued*

Compound	$-E_{1/2}$		i_d/C
Benzotrichloride	0.68		1.1
	1.65		2.3
	2.00		1.1
Hexachloroethane	0.62		2.4
	1.73		2.4
	1.96		2.3
Ethyl bromide	2.08		1.0
1,2-Dibromoethane	1.52		2.3
1,1-Dibromoethane	1.62		2.8
Tribromoethane	0.70		2.3
Pentabromoethane	Max.		1.7
	0.68		3.7
	1.18		2.0
Ethyl iodide	1.67		1.0
n-Butyl bromide	2.27		0.9
n-Octyl bromide	2.38		0.6
Vinyl chloride	No wave		
1,2-Dichloroethene	No wave		
1,1-Dichloroethene	*ca.* 2.4		0.9 max.
Trichloroethylene	2.14		2.0
Tetrachloroethylene	1.88		3.6
Vinyl bromide	2.47		1.2
Allyl chloride	1.91		1.2
Allyl bromide	1.29		1.9
Allyl iodide	0.23	(Concn. dependent)	0.8–0.5
	1.16		0.9

The reduction of α-bromonaphthalene was carried out at −2.05 v. to prevent reduction of the naphthalene nucleus.

1,2-Dihalogen compounds are apparent exceptions and may form intermediate diradicals which give olefins. The diffusion currents are sometimes abnormally small, but are reproducible and proportional to concentration. The temperature coefficient is about 1 per cent per degree. No reason is offered for the small diffusion currents; kinetic effects may be involved.

The independence of pH was established with dibromoethane and allyl bromide in buffers of pH 8, 10, and 13.

Values obtained for other organic halogen compounds in 0.01 *N* tetramethylammonium bromide–80 per cent alcohol are listed in Table XXXVIII-2.[2]

Some of the results are questionable. Triphenylmethyl chloride hydrolyzes in 80 per cent ethanol and no doubt gives a wave which is for the

[2] H. Keller, M. Hochweber, and H. von Halban, *Helv. Chim. Acta*, **29**, 761 (1946).

hydrogen chloride formed. The close similarity in half-wave potentials of 1,1-di-*p*-chlorophenyl-2,2-dichloroethene and 1,1-di-*p*-chlorophenyl-1,2,2,2-tetrachloroethane suggests that the mercury may remove two

TABLE XXXVIII-2

HALF-WAVE POTENTIALS (S.C.E.) AND WAVE HEIGHTS FOR ORGANIC HALOGEN COMPOUNDS

Compound	$-E_{1/2}$	Electrons used (from wave height)
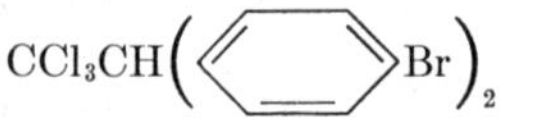	0.86	1
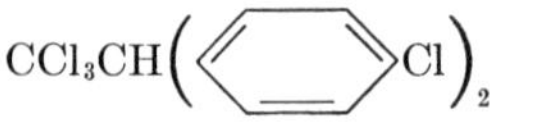	0.88	1
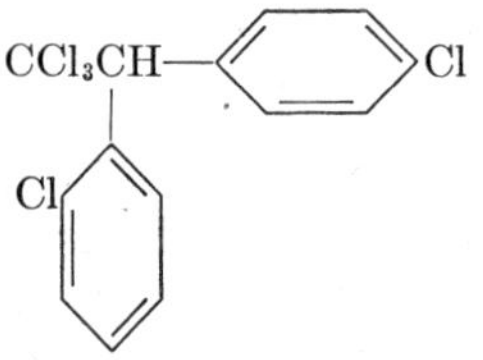	0.94	1
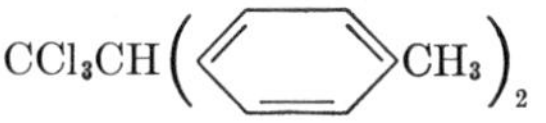	0.95	1.25
$CCl_3CH(C_6H_5)_2$	0.97	1.5
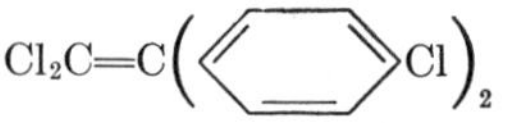	2.02	2.0
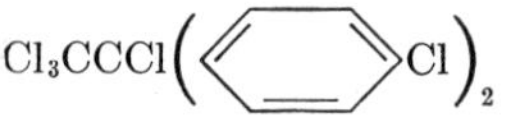	2.06	2.0
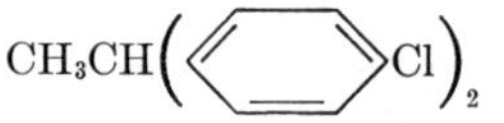	Not reducible	
$(C_6H_5)_3CCl$	1.29	1
$C_6H_6Cl_6$	1.89	2

chlorine atoms from the latter and form the former. The reduction of 1,1-di-*p*-chlorophenyl-2,2-dichloroethene may be of the double bond and not of the halogen.

The polarographic behavior of 1,2,3,4,5,6-hexachlorocyclohexane has been studied by several workers because of its use as an insecticide. In 80 per cent ethanol with 0.1 *N* tetraethylammonium iodide as the electrolyte half wave potentials (S.C.E.) of −2.06 v. (α), −2.19 v. (β), −1.65 v. and −2.62 v. (γ) (Gammexane) are reported. Impure (80 per cent) δ-isomer gives a wave at −2.08 v.[3]

In a 50 per cent ethanol solution containing 1 per cent potassium iodide only the γ-isomer gives a wave at −1.35 v.[4] The other forms are reduced at potentials more negative than that of the potassium ion reduction. The diffusion current in this medium is proportional to the concentration in the range of 10^{-3} to 10^{-5} M. The range found most satisfactory for quantitative determination was between 0.005 and 0.04 per cent.[5] It is claimed that the other isomers in the crude material and talcum and mercury compounds that are added to some commercial products have no effect on the wave height for the γ-isomer.

Heptachlorocyclohexanes, if present, increase the height of the residual current in a potassium chloride–sodium acetate solution in an alcohol-water mixture. This impurity can be corrected for empirically.[6]

The polarographic method for Gammexane is reported to give better accuracy than the infrared spectroscopic method. The electrolyte used was 0.1 N $(CH_3)_4NCl$ or 0.1 N $(CH_3)_4NBr$ in Kolthoff's buffer solution (pH 7.07) and 40 per cent dioxane.[7]

Chloramine T gives a wave in 0.5 N K_2SO_4 solution at −0.13 v. (S.C.E.) while hypochlorite ion in the same medium is reduced at +0.08 v.[8]

Uses. The rate of hydrolysis of benzyl bromide has been followed polarographically.[9]

The polarographic method can be used to determine: (1) carbon tetrachloride in a mixture with tetrachloroethylene, and (2) carbon tetrachloride and chloroform in a mixture with methylene chloride and methyl chloride.[10]

Alkyl halides, haloforms, carbon tetrachloride, allyl chloride, and 1-chloro-2-bromocyclopropane are reducible polarographically using a fritted glass disc electrode. Half-wave potentials are reported, without indicating the reference electrode.[11]

---

3 K. Schwabe, *Z. Naturforsch.*, **3,** 217 (1948).

4 G. B. Ingram and H. K. Southern, *Nature*, **161,** 437 (1948).

5 H. Grass and E. Y. Spencer, *Can. J. Research*, **27,** 368 (1949).

6 G. Dragt, *Anal. Chem.*, **20,** 737 (1948).

7 M. Suzuki and M. Nakajima, *Biochem. Kagaku* (*Sci. Insect Control*), No. 10, 31 (1948); *Chem. Abstracts*, **43,** 5896 (1949).

8 K. Heller and E. N. Jenkins, *Nature*, **158,** 706 (1946).

9 R. Pasternak and H. von Halban, *Helv. Chim. Acta*, **29,** 191 (1946).

10 M. von Stackelberg and W. Stracke, *Z. Elektrochem.*, **53,** 118 (1949).

11 M. B. Neiman, A. V. Ryabov, and E. M. Sheyanova, *Doklady Akad. Nauk S. S. S. R.*, **68,** 1065 (1949); *Chem. Abstracts*, **44,** 1360 (1950).

CHAPTER XXXIX

Carbonyl Compounds

1. ALIPHATIC COMPOUNDS

(a) Aliphatic Aldehydes

Formaldehyde. In Chapter XV on Kinetic Currents a discussion is given of the work of Vesely and Brdicka[1] who presented evidence that the aldehyde form of formaldehyde is reducible and that the polarographic limiting currents of formaldehyde are governed by the rate of dehydration of the hydrated form of formaldehyde:

$$H_2C(OH)_2 \rightleftarrows H_2C{=}O + H_2O$$

This reaction is acid-base catalyzed. Factors such as temperature, pH, solvents, and buffers greatly influence the equilibrium and, more important, the rate and consequently the limiting current. Hydroxyl ion, for example, greatly catalyzes the dehydration of the methylene glycol to the aldehyde. At room temperature Vesely and Brdicka found a maximum value of the limiting current in solutions of about 0.1 N hydroxyl ion. With further increase in the alkalinity the limiting current drops because of the transformation of the methylene glycol into the corresponding anion:

$$H_2C(OH)_2 + OH^- \rightleftarrows H_2C(O^-)(OH) + H_2O$$

$$H_2C(O^-)(OH) \rightleftarrows H_2C{=}O + OH^-$$

[1] K. Vesely and R. Brdicka, *Collection Czechoslov. Chem. Communs.*, **12**, 313 (1947).

This interpretation accounts for many phenomena dealing with the polarography of formaldehyde that were previously described in the literature. For example, the height of the formaldehyde wave was found to be much smaller with a neutral salt as supporting electrolyte than in a solution of sodium hydroxide.[2] The difference in the wave heights in neutral and alkaline media becomes smaller with increasing temperature and disappears near 80° C. The diffusion current at this temperature corresponds to the reaction:

$$HCHO + 2H_2O + 2e \rightarrow CH_3OH + 2OH^-$$

Empirically the use[3–5] of solutions about 0.1 N in hydroxyl ion was recommended for the quantitative polarographic determination of formaldehyde. An increase in the limiting current with temperature of about 6.5 per cent per 1° C. was observed in these solutions, with values as high as 11 per cent per 1° C. reported[6] in a buffer of pH 4. From the temperature *vs.* limiting current curve, the heat of hydration of formaldehyde has been calculated[7] to be 13,000 cal. mole^{-1}.

The half-wave potential is constant with temperature[8] but varies with pH in the following manner:[9]

pH	8	9.45	10.7	12.7
$E_{1/2}$(N.H.E.)	−1.222	−1.28	−1.35	−1.465

The last value is comparable to the value −1.67 v. (S.C.E.) found in 0.05 M KOH and 0.1 N KCl solution.[5]

For analytical purposes the best supporting medium for the determination of formaldehyde is 0.1 N lithium hydroxide plus 0.01 N lithium chloride.[10] The supporting electrolyte contains 2×10^{-4} per cent methyl red and 3×10^{-5} per cent bromocresol green as maximum suppressors. The formaldehyde wave appears in the voltage range between −1.50 and −1.65 and can be used for determination of formaldehyde in the presence of reasonable amounts of acetaldehyde and acrolein, since formaldehyde is re-

[2] F. G. Jahoda, *Collection Czechoslov. Chem. Communs.*, **7**, 415 (1935).
[3] G. C. Whitnack and K. W. Moshier, *Ind. Eng. Chem., Anal. Ed.*, **16**, 496 (1944).
[4] E. C. Barnes and H. W. Speicher, *J. Ind. Hyg. Toxicol.*, **24**, 10 (1942).
[5] M. J. Boyd and K. Bambach, *Ind. Eng. Chem., Anal. Ed.*, **15**, 314 (1943).
[6] R. Bieber and G. Trümpler, *Helv. Chim. Acta*, **30**, 1286 (1947).
[7] M. B. Neiman and M. I. Gerber, *Zhur. Anal. Khim.*, **2**, 135 (1947); *Chem. Abstracts*, **43**, 5702 (1949).
[8] R. Bieber and G. Trümpler, *Helv. Chim. Acta*, **30**, 706 (1947).
[9] R. Bieber and G. Trümpler, *Helv. Chim. Acta*, **30**, 1109 (1947).
[10] G. C. Whitnack and K. W. Moshier, *Ind. Eng. Chem., Anal. Ed.*, **16**, 496 (1944).

duced first. With 6 mg. of formaldehyde, no more than 0.75 mg. of acrolein may be present, while with 1.2 mg. of formaldehyde, 2.2 mg. of acrolein and 2.74 mg. of acetaldehyde do not interfere. Propionaldehyde interferes slightly. The formaldehyde wave is eliminated by either sodium bisulfite or dimedon since the addition products are not reducible.

Proportionality between the diffusion current and concentration is obtained only if the solution is well buffered and if enough maximum suppressor, such as lanthanum ions, are present.[11]

The polarographic method has been used[12] to determine formaldehyde in air by passing the air through a 1.25 per cent potassium hydroxide solution. The diffusion current is measured in this medium and is proportional to the formaldehyde concentration. At this alkalinity the Cannizarro reaction:

$$2HCHO + KOH \rightarrow HCOOK + CH_3OH$$

was found to be so slow that the diffusion current was unchanged after the solution stood for one week.

The polarographic formaldehyde determination has been used in a procedure for estimating serine.[13] The method consists essentially of quantitatively separating the formaldehyde by distillation formed from the action of periodic acid on protein hydrolyzates containing serine.[14] To 3 ml. of the distillate 1 ml. of supporting electrolyte (0.4 *N* in KCl and 0.2 *N* in KOH) is added. Small amounts of acetaldehyde do not interfere in the above procedure.

A similar procedure was used to determine ethylene glycol and 1,2-propylene glycol in the presence of one another,[15] and glycerol in fermentation residues.[16] Upon oxidation with periodic acid ethylene glycol and glycerol yield 2 moles of formaldehyde, while 1,2-propylene glycol gives 1 mole of formaldehyde and 1 mole of acetaldehyde. The aldehydes are distilled from the reaction mixture and determined simultaneously in a supporting electrolyte which is 0.1 *N* in lithium hydroxide and 0.01 *N* in lithium chloride. Since the aldol condensation of aldehydes is rapid in a strongly alkaline medium, the polarographic measurements must be made soon after the addition of the supporting electrolyte.

Diethylene glycol or dipropylene glycol if present as contaminants in

---

[11] R. Bieber and G. Trümpler, *Helv. Chim. Acta*, **30,** 1286 (1947).

[12] E. C. Barnes and H. W. Speicher, *J. Ind. Hyg. Toxicol.*, **24,** 10 (1942).

[13] M. J. Boyd and K. Bambach, *Ind. Eng. Chem., Anal. Ed.*, **15,** 314 (1943).

[14] M. J. Boyd and M. A. Logan, *J. Biol. Chem.*, **146,** 279 (1942).

[15] B. Warshowsky and P. J. Elving, *Ind. Eng. Chem., Anal. Ed.*, **18,** 253 (1946).

[16] P. J. Elving, B. Warshowsky, E. Shoemaker, and J. Margolit, *Anal. Chem.*, **20,** 25 (1948).

mixtures of ethylene glycol and propylene glycol may be determined by reducing the excess periodic acid to iodic acid and oxidizing the diglycols with potassium dichromate. The chromate is measured polarographically before and after the oxidation in 0.1 *N* sodium hydroxide.[17]

The polarographic method has been used to demonstrate that an equilibrium exists between monomethylolurea and urea and formaldehyde in 0.05 *N* lithium hydroxide,[18] to study the effects of pH and temperature on this reaction,[19] and to determine formaldehyde in lysoform, formamint,[20] water, milk, sake, various commercial products,[21, 22] and various pharmaceutical preparations.[23]

Acetaldehyde and Propionaldehyde. Acetaldehyde[24] and propionaldehyde[25] give waves that are less dependent on external factors since these aldehydes are less hydrated than formaldehyde. The temperature coefficient, for example, for acetaldehyde is only 1.8 per cent per degree.[26] Part of the current is rate controlled, which makes the interpretation of the reduction mechanism difficult.

Half-wave potentials (*vs.* S.C.E.) are reported by Elving and Rutner for acetaldehyde of −1.89 v. in lithium acetate-acetic acid buffer of pH 6.8 (this is scarcely buffered), −2.04 v. in unbuffered lithium acetate solutions (pH 8.8), and −1.89 v. in a 0.06 *M* LiOH–0.05 *M* LiCl solution (pH 12.7).[26]

The following half-wave potentials (*vs.* N.H.E.) are reported for acetaldehyde and propionaldehyde.[24, 25]

Solution	0.00131 *M* LiCl 0.00019 *M* HCl	pH 9.1	pH 10	0.1 *M* LiOH	0.1 *M* LiCl
$E_{1/2}$ CH_3CHO		−1.51	−1.80[26a]	−1.65	−1.55
$E_{1/2}$ CH_3CH_2CHO	−1.59			−1.68	−1.57

A 0.1 *M* lithium chloride solution is recommended by Elving and Rutner as a supporting electrolyte for the range of 0.50×10^{-4} to 5.0×10^{-4} *M*

17 C. V. Francis, *Anal. Chem.*, **21,** 1238 (1949).

18 G. A. Crowe, Jr., and C. C. Lynch, *J. Am. Chem. Soc.*, **70,** 3795 (1948).

19 G. A. Crowe, Jr., and C. C. Lynch, *J. Am. Chem. Soc.*, **71,** 3731 (1949).

20 I. Smoler, *Collection Czechoslov. Chem. Communs.*, **2,** 699 (1930).

21 R. Akano and M. Watanabe, *Mitt. med. Akad. Kioto*, **30,** 629 (1940); *Chem. Abstracts*, **35,** 4705 (1941).

22 R. Portillo and G. Varela, *Anales fís. y quím. Madrid*, **40,** 794 (1944); *Chem. Abstracts*, **42,** 6705 (1948).

23 R. Portillo and V. Mosquera, *Farm. nueva Madrid*, **10,** 659 (1945); *Chem. Abstracts*, **41,** 5259 (1947).

24 R. Bieber and G. Trümpler, *Helv. Chim. Acta*, **30,** 2000 (1947)

25 R. Bieber and G. Trümpler, *Helv. Chim. Acta*, **31,** 5 (1948).

26 P. J. Elving and V. Rutner, *Ind. Eng. Chem., Anal. Ed.*, **18,** 176 (1946).

26a This value is against N.C.E.; see footnote 28.

acetaldehyde.[26] For less than 0.0014 *M* solutions, a 0.1 *M* lithium hydroxide solution is suitable. Sodium hydroxide may be substituted for the lithium hydroxide as the supporting electrolyte.[27]

The polarographic method has been used by Elving and Rutner[26] to determine amount of acetaldehyde produced in the conversion of ethanol to butadiene, and by Dirscherl and Bergmeyer[28] to determine lactic acid. In the latter method lactic acid is oxidized with potassium permanganate

TABLE XXXIX-1

POLAROGRAPHIC REDUCTION POTENTIALS OF ALIPHATIC ALDEHYDES

Name	Formula	Reduction potential (N.C.E.)	Half-wave potential[a]
Formaldehyde	HCHO	−1.38[b]	[e]
Acetaldehyde	CH_3CHO	−1.60[c]	−1.87[e]
Propionaldehyde	C_2H_5CHO	−1.60[c]	−1.92[e]
n-Butyraldehyde	*n*-C_3H_7CHO		−1.90
Isobutyraldehyde	*i*-C_3H_7CHO		−1.91
n-Valeraldehyde	*n*-C_4H_9CHO	−1.36 (molar)[d]	
n-Capronaldehyde	*n*-$C_5H_{11}CHO$		−1.98
n-Heptaldehyde	*n*-$C_6H_{13}CHO$		−1.90

[a] H. Adkins and F. W. Cox, *J. Am. Chem. Soc.*, **60**, 1151 (1938). They used 0.001 *M* solutions of aldehyde in 0.2 *N* tetramethylammonium hydroxide; these authors did not measure the anode potential, but it was probably close to −0.2 v (S.C.E.) and hence the true half-wave potentials are more negative by this value.

[b] A. Winkel and C. Proske, *Ber.*, **69**, 693 (1936). They used 0.1 *M* formaldehyde (partly polymerized) in 0.1 *M* ammonium chloride.

[c] Same as [b], except with 0.001 *M* aldehyde solutions.

[d] M. Shikata and I. Tachi, *Mem. Coll. Agr. Kyoto Imp. Univ.*, **4**, 9 (1927); 0.0187 *M* aldehyde solution in 0.1 *N* ammonium chloride. In 0.1 *N* hydrochloric acid they found −1.20 v., in 0.1 *N* potassium chloride −1.35 v., in 0.1 *N* sodium hydroxide −1.82 v.

[e] For more complete values see the discussion on these compounds.

in the presence of manganese sulfate and the acetaldehyde formed is distilled into a bisulfite solution. The acetaldehyde is liberated for analysis by the addition of lithium hydroxide.

Paraldehyde does not yield a wave.[29]

Values for several other aliphatic aldehydes are listed in Table XXXIX-1.

(b) Unsaturated Aliphatic Aldehydes

Unsaturated aldehydes in which the aldehyde group is conjugated with one or more double bonds give two reduction waves.

[27] I. Smoler, *Collection Czechoslov. Chem. Communs.*, **2**, 699 (1930).

[28] W. Dirscherl and H. U. Bergmeyer, *Naturwissenschaften*, **35**, 220 (1948).

[29] A. Winkel and G. Proske, *Ber.*, **69**, 693 (1936).

Acrolein. Acrolein has been studied in various buffers and found to give two waves.[30] The following half-wave potentials (against the mercury pool) are reported for the first wave:

pH	4.8	5.8	7.0–11.0
$E_{1/2}$(volt)	−0.83	−0.98	−1.04

A value of −1.44 v. is obtained for the second wave in buffers with pH 8.7–11.0. In more acidic buffers the second wave is obscured by the residual current. The diffusion current of the first wave decreases with an increase in pH whereas that of the second wave increases with an increase in pH. A change of one degree in temperature causes a change of 0.8 per cent. This type of behavior indicates that the first wave is due to the reduction of the double bond while the second wave is due to the reduction of the aldehyde group.

For analytical purposes a lithium phosphate buffered solution at approximately pH 7 containing 0.01 *M* lithium chloride is best for determining acrolein, using its first wave. Methyl red or bromocresol green is used as a maximum suppressor. Since acrolein is volatile at room temperature, the oxygen is not removed from the solution. Acrolein may be determined, using the first wave, in the presence of formaldehyde and acetaldehyde. If formaldehyde is the only aldehyde present, either wave may be used since the formaldehyde wave comes in between the two acrolein waves.

Gerber *et al.*[31] suggest that acrolein be determined in acid (pH 3–4) or weakly alkaline solution (pH 8). At pH 4–6 the diffusion current of acrolein changes rapidly with the change in pH. In the presence of saturated aldehydes, acid solutions are recommended, for the saturated aldehydes are not reduced in this range. If peroxides are present, the determination is best carried out at pH 5–6 to avoid interference with the peroxide wave.[32]

Crotonaldehyde. Crotonaldehyde behaves in a fashion similar to that of acrolein and gives two waves[32a] with half-wave potentials of −1.37 v. (1st wave) and −1.80 v. (2nd wave) in 0.2 *N* tetramethylammonium hydroxide, and −1.25 v. (1st wave) and −1.55 v. (2nd wave) in 0.1 *N* ammonium chloride solution.*

---

[30] R. W. Moshier, *Ind. Eng. Chem., Anal. Ed.*, **15**, 107 (1943).

[31] M. L. Gerber, A. A. Dobrinskaya, and M. B. Neuman, *Trudy Vsesoyuz. Konferents. Anal. Khim.*, **2**, 585 (1943); *Chem. Abstracts*, **39**, 3760 (1945).

[32] M. B. Neiman and M. I. Gerber, *Zhur. Anal. Khim.*, **2**, 135 (1947).

[32a] H. Adkins and F. W. Cox, *J. Am. Chem. Soc.*, **60**, 1151 (1938).

* The last two potentials have been referred by us to the S.C.E.

The work of Fields and Blout[33] on crotonaldehyde is discussed in the next section.

Polyene Aldehydes. Aliphatic polyene aldehydes of the type $CH_3(CH{=}CH)_xCHO$ where x equals 1, 2, 3, 4, and 5 have been studied in the presence of buffers in a mixture containing 50 per cent dioxane and water.

Each of the compounds studied gave at least one well-defined wave in the pH range of 1 to 11. In addition, a second wave of approximately the same height as the first one is observed over a more limited pH range. The initial appearance of the second wave occurs at progressively lower pH values as the number of double bonds is increased. From the height of the first waves and the characteristics of the capillary the following values of the diffusion coefficients at 25° C. were calculated: crotonaldehyde 6.1×10^{-6} (cm.^{2} sec.$^{-1}$); 2,4-hexadienal 5.3×10^{-6}; 2,4,6-octatrienal 5.1×10^{-6}; 2,4,6,8-decatetraenal 4.3×10^{-6}; and 2,4,6,8,10-dodecapentaenal 4.2×10^{-6}.

In Figs. XXXIX-1 and XXXIX-2 the half-wave potentials of the first wave $(E_{1/2})_1$ and of the second wave $(E_{1/2})_2$ are plotted as a function of the pH. For the five polyene aldehydes studied $(E_{1/2})_1$ is an approximately linear function of pH; in each case a unit increase in pH shifts the half-wave potential to more negative values by 0.06 v. In the measured pH range 6–8 two points on the crotonaldehyde curve and one on the hexadienal curve deviate markedly from the straight lines which fit all the remaining data.

The second wave is well defined in most cases only in solutions more alkaline than pH 6; generally, in acid solutions this wave merges with or is obscured by the wave due to electrolysis of the supporting electrolyte. For 2,4,6,8,10-dodecapentaenal a distinct second wave is obtained over the pH range of 4 to 10 and in acid solutions $(E_{1/2})_2$ seems to be independent of pH. For all the compounds studied the plot of $(E_{1/2})_2$ *vs.* pH shows a minimum in the region of pH 8.

From Figure XXXIX-1 on page 659 it can be seen that the addition of another carbon-carbon double bond to an aliphatic polyene aldehyde causes a shift in $(E_{1/2})_1$ toward less negative potentials; that is, the ease of reduction increases with the number of double bonds conjugated with the carbonyl group. The increment per double bond decreases as the chain is lengthened. Thus, while the $(E_{1/2})_1$ value for 2,4-hexadienal is, on the average, 0.23 v. more positive than that for crotonaldehyde, the $(E_{1/2})_1$ value for 2,4,6,8,10-dodecapentaenal is on the average only 0.07 v. more positive than that for 2,4,6,8-decatetraenal.

---

[33] M. Fields and E. R. Blout, *J. Am. Chem. Soc.*, **70**, 930 (1948).

Blout and Fields offer an explanation for the direction of the shift of $(E_{1/2})_1$ with the number (x) of carbon-carbon double bonds based upon the following reaction at the electrode:

$$CH_3(CH{=}CH)_xCHO + H^+ + e \rightarrow CH_3(CH{=}CH)_x\dot{C}HOH$$

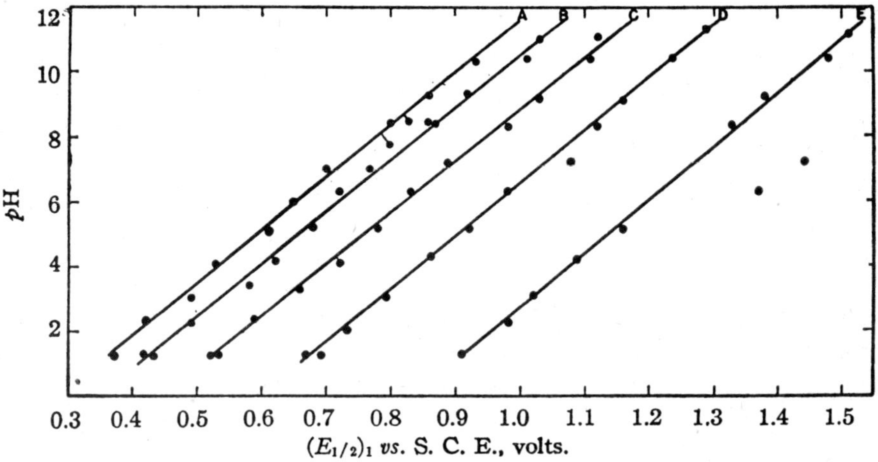

Fig. XXXIX-1. Half-wave potential of first wave *vs*. pH for: (A) 2,4,6,8,10-dodecapentaenal; (B) 2,4,6,8-decatetraenal; (C) 2,4,6-octatrienal; (D) 2,4-hexadienal; and (E) crotonaldehyde.[33]

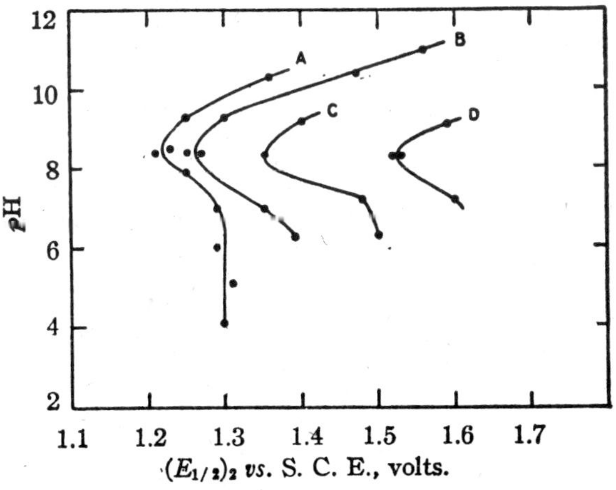

Fig. XXXIX-2. Half-wave potential of second wave *vs*. pH for: (A) 2,4,6,8,10-dodecapentaenal; (B) 2,4,6,8-decatetraenal; (C) 2,4,6-octatrienal; and (D) 2,4-hexadienal.[33]

For $x = 2$ as an example:

$$\mathrm{CH_3(CH{=}CH)_2\dot{C}HOH} \leftrightarrow \mathrm{CH_3CH{=}CH{-}\underset{\cdot}{CH}{-}CH{=}CHOH}$$
$$\updownarrow$$
$$\mathrm{CH_3\underset{\cdot}{CH}{-}CH{=}CH{-}CH{=}CHOH}$$

$$\mathrm{2CH_3\underset{\cdot}{CH}CH{=}CH{-}CH{=}CHOH} \rightarrow \begin{array}{l}\mathrm{CH_3CH{-}CH{=}CH{-}CH_2CHO}\\ \quad\ \ \vert \\ \mathrm{CH_3CH{-}CH{=}CH{-}CH_2CHO}\end{array}$$

or:

$$\mathrm{2CH_3CH{=}CH{-}\underset{\cdot}{CH}{-}CH{=}CHOH} \rightarrow \begin{array}{l}\mathrm{CH_3CH{=}CH{-}CH{-}CH_2CHO}\\ \qquad\qquad\quad \vert \\ \mathrm{CH_3CH{=}CH{-}CH{-}CH_2CHO}\end{array}$$

Dimerization must take place to form an aldehyde since a second reduction wave is obtained. The actual position of dimerization for the polyene aldehydes is not known.

Terpene Aldehydes. Citral, jasmine aldehyde, and citronellal have been studied polarographically.[34] Half-wave potentials obtained in various solutions are listed in Table XXXIX-2.

TABLE XXXIX-2

HALF-WAVE POTENTIALS OF TERPENE ALDEHYDES

	Supporting electrolyte						
	HCl		LiCl		KOH		1 *M* NH_4Cl
Compound	$-E_{1/2}$	K*	$-E_{1/2}$	K*	$-E_{1/2}$	K*	$-E_{1/2}$
Citral	0.90	1.45	1.55	1.25	1.40	1.30	1.93
Jasmine aldehyde	0.75	1.55	1.55	1.44	1.40	1.55	
Citronellal	No wave		1.9	1.15	1.9	1.15	1.72

* K is equal to $i_d/Cm^{2/3}t^{1/6}$.

The first three determinations were carried out in 66 per cent alcohol while the last was carried out in 75 per cent ethanol.[35] Citral and citronellal can be determined quantitatively in essential oils using 1 *M* ammonium chloride.

[34] M. I. Gerber, Z. B. Kuznetsova, and M. B. Neiman, *Zhur. Anal. Khim.*, **4,** 103 (1949); *Chem. Abstracts*, **44,** 2416 (1950).

[35] B. Y. Ogrin and A. A. Braun, *Zavodskaya Lab.*, **10,** 467 (1941); *Chem. Abstracts*, **35,** 8206 (1941).

(c) Aliphatic Ketones

Acetone and methyl ethyl ketone are reduced at −2.20 v. and −2.25 v., respectively, in 0.025 *N* tetramethylammonium iodide solution.[36] In 0.05 *N* tetraethylammonium iodide–75 per cent dioxane the following values are reported for acetone and cyclohexanone.[37] The determination can be used quantitatively with an accuracy of 5 per cent. No waves are obtained in solutions containing hydrochloric acid, lithium chloride, or sodium hydroxide as the inert electrolyte. Similar negative results[38] are reported for acetone, methyl ethyl ketone, ethyl isobutyl ketone, methyl hexyl ketone, methyl nonyl ketone, cyclopentanone, and cyclohexanone in ammonium chloride and lithium chloride solutions. The presence of acetone moves the polarograms of lithium chloride solutions to lower *E* values with a moderate change in the shape of the curve.[39]

Compound	$-E_{1/2}$(S.C.E.)	i_d/C
Acetone	2.46	2.2–1.5*
Cyclohexanone	2.45	1.9

* Concentration dependent.

Aliphatic ketones can also be determined indirectly. One method is dependent upon the diminution produced in the sulfur dioxide curve of a sulfuric acid solution of sodium sulfite and has been used for acetone, methyl ethyl ketone, and acetylacetone. The drop in current is a function of concentration but is not a linear one.[39]

The other method of determination is dependent upon the use of carbonyl derivatives. These derivatives are not isolated but are studied in solution. Thus, isobutyraldehyde, acetone, methyl isobutyl ketone, and methyl *n*-amyl ketone,[40] are not reduced in ammonium chloride solution but give double waves in acid hydrazine solution in which they are present as hydrazones. The half-wave potentials vary somewhat with concentration but for the ketones the first wave occurs near −1.1 v. (S.C.E.), and for isobutyraldehyde near −0.9 v. For both types the half-wave potential of the second wave is in the range −1.3 to −1.4 v. Acetone and isobutyraldehyde give reduction waves in acid phenylhydrazine solution

[36] M. B. Neiman and Z. V. Markina, *Zavodskaya Lab.*, **13,** 1177 (1947); *Chem. Abstracts*, **43,** 4183 (1949).

[37] M. von Stackelberg and W. Stracke, *Z. Electrochem.*, **53,** 118 (1949).

[38] A. Winkel and G. Proske, *Ber.*, **69,** 693 (1936).

[39] R. Portillo and G. Varela, *Anales fís. y quím. Madrid*, **41,** 1429 (1945); *Chem. Abstracts*, **42,** 7199 (1948).

[40] J. M. Lupton and C. C. Lynch, *J. Am. Chem. Soc.*, **66,** 697 (1944).

which are similar to those obtained in hydrazine solution. The hydrazonium and phenylhydrazonium ions are not responsible for the second wave in hydrochloric acid and sulfuric acid solutions (0.1 *N*).

To obtain a constant diffusion current thirty minutes must be allowed for reaction of the carbonyl compounds with the reagent before carrying out the analysis. The diffusion current thus obtained is a linear function of the concentration over the range 0.0001 to 0.010 *M*. Below 0.0001 *M* the configuration of the waves is such that a highly empirical method for determination of the diffusion current must be used. The second wave cannot be used for quantitative measurements because it gives a maximum. Precision of the method varies between 2 and 5 per cent.

The nature of the group being reduced is not known but, since similar waves are obtained in both phenylhydrazine and hydrazine solutions, it has been suggested that more likely the hydrazone (I) is reduced rather than the azine (II).

$$R_2C{=}NHHX \quad X{=}H \text{ or } C_6H_5 \text{ (I)} \qquad RC{=}N{-}N{=}CR_2 \text{ (II)} \qquad R_2C{=}N\overset{+}{N}H_2R \text{ (III)}$$

Analysis of the first wave gives no information because a slope equivalent to a 0.62-electron transfer is obtained. Since acid solutions are used, salts of the hydrazone (III) are more likely responsible for the reduction waves.

Girard derivatives of ketones (IV)

$$R_2C{=}NNHCOCH_2\overset{+}{N}(CH_3)_3\ Cl^- \text{ (IV)} \qquad H_2NNHCOCH_2\overset{+}{N}(CH_3)_3\ Cl^- \text{ (V)}$$

are reducible at the dropping mercury electrode. The Girard reagent (V),[41] or more specifically betaine hydrazide hydrochloride itself, is discharged at such a negative potential that it does not interfere with the polarographic characterization of the condensation product.

A half-wave potential of about −1.56 v. (N.C.E.) is obtained for the Girard derivatives from acetone, methyl ethyl ketone, diethyl ketone, methyl *n*-undecyl ketone, and di-*n*-octyl ketone in a buffer of pH 8.2. Half-wave potentials obtained for the Girard derivatives of cyclic ketones at pH 8.2 are shown in Fig. XXXIX-3.[42] Gelatin solution (0.5 per cent) is used to suppress maxima.

[41] *Organic Syntheses, Collective Volume II*, Wiley, New York, 1943, p. 85.
[42] V. Prelog and O. Hafliger, *Helv. Chim. Acta*, **32**, 2088 (1949).

The stability of the Girard derivative is greatly influenced by the pH of the solution used. The reaction involved is one of hydrolysis since addition of excess Girard reagent increases the height of the diffusion current. The use of an excessively large amount of reagent is to be avoided since it lowers the hydrogen overvoltage and causes interference with the reduction wave. The limiting current of the Girard derivative of cyclohexanone is diffusion controlled in a buffer of pH 8.2 and rate controlled in a buffer of pH 12.2.

The Girard derivative of cyclohexanone is the least stable derivative in the cyclic ketone series. This compound gives one small wave at pH 3.4, two waves at pH 4.3–6, and one wave in more alkaline buffers which

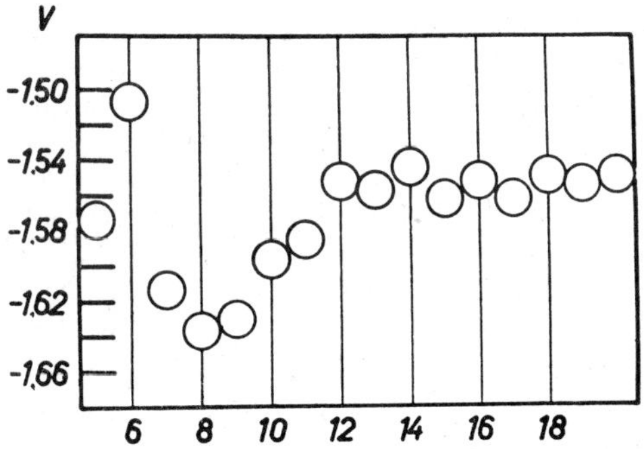

Fig. XXXIX-3. Half-wave potentials (N.C.E.) of the Girard derivatives of various cyclanones.[42] Abscissa is number of carbon atoms. Ordinate, $E_{1/2}$.

starts to decrease at pH 10.7. The effects of time and pH upon the wave height are shown in Figs. XXXIX-4 and XXXIX-5. This behavior explains the reported non reduction of cyclohexanone derivatives.[43]

The wave height of the Girard derivatives corresponds to a 2-electron reduction while the analysis of the wave points to a potential-determining step of 1 electron. These values indicate a reduction of the Girard derivative to a hydrazine:

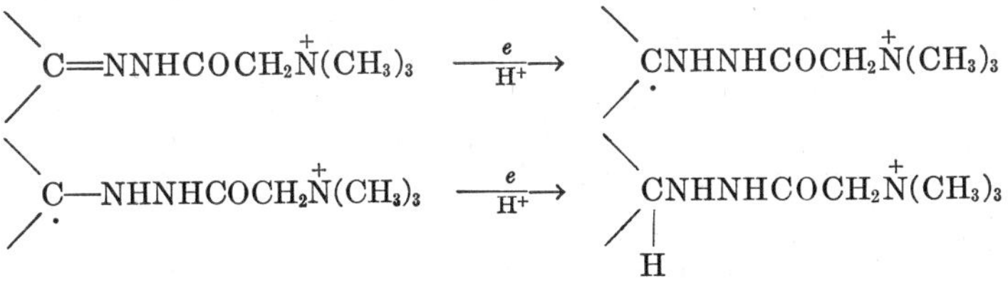

[43] J. K. Wolfe, F. B. Hershberg, and L. F. Fieser, *J. Biol. Chem.*, **136**, 653 (1940).

Unsaturated Aliphatic Ketones. Unsaturated aliphatic ketones in which the double bond is conjugated with the carbonyl group are more easily reduced at the dropping mercury electrode than saturated aliphatic ketones.

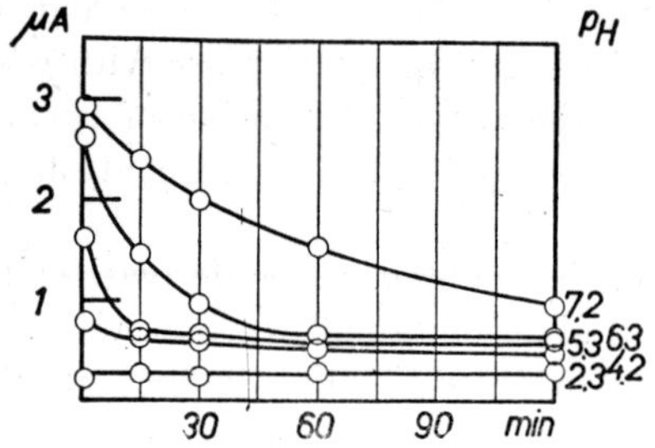

Fig. XXXIX-4. The effect of time upon the wave height of the Girard derivative of cyclohexanone in acid media.[42]

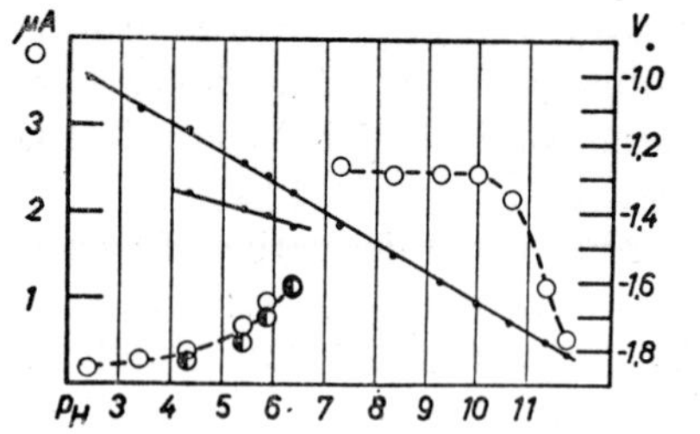

Fig. XXXIX-5. The effect of pH upon wave height and half-wave potential of the Girard derivative of cyclohexanone.[42]

The reduction proceeds by a 1,4-mechanism and gives a bimolecular ketone in acid solutions and the saturated ketone in alkaline solutions:

$$\mathrm{RCH_2CH_2COR} \xleftarrow[\mathrm{2H_2O}]{2e} \mathrm{RCH{=}CHCOR} \xrightarrow[\mathrm{2H^+}]{2e} \begin{array}{l}\mathrm{RCHCH_2COR}\\ \quad|\\ \mathrm{RCHCH_2COR}\end{array}$$

Half-wave potentials reported for certain unsaturated aliphatic ketones are listed in Table XXXIX-3.

The course of the reduction has been demonstrated at pH 1.3 for mesityl oxide by controlled potential electrolysis; the bimolecular ketone was isolated as one of the products:

$$\mathrm{2(CH_3)_2C{=}CHCOCH_3} + 2e \xrightarrow{\mathrm{2H^+}} \begin{array}{l}\mathrm{(CH_3)_2CCH_2COCH_3}\\ \qquad|\\ \mathrm{(CH_3)_2CCH_2COCH_3}\end{array}$$

At pH 8.6 the product obtained was not identified.[48]

[44] E. I. Fulmer, J. J. Kolfenbach, and L. A. Underkofler, *Ind. Eng. Chem., Anal. Ed.*, **16,** 469 (1944).

[45] M. I. Gerber, Z. B. Kuznetsova, and M. B. Neiman, *Zhur. Anal. Khim.*, **1949,** 103; *Chem. Abstracts*, **44,** 2416 (1950).

[46] H. Adkins, R. M. Elofson, A. C. Rossow, and C. C. Robinson, *J. Am. Chem. Soc.*, **71,** 3622 (1949).

[47] H. Adkins and F. W. Cox, *J. Am. Chem. Soc.*, **60,** 1151 (1938).

[48] R. Pasternak, *Helv. Chim. Acta*, **31,** 753 (1948).

[49] G. Scaramelli, *Boll. sci. facoltà chim. ind., Bologna*, **1941,** 129; *Chem. Zentr. II*, **1942,** 269.

The method has been used to determine methyl vinyl ketone in the presence of a mixture of methyl vinyl carbinol and methyl ethyl ketone obtained from the catalytic vapor-phase oxidation of methyl vinyl carbinol.

TABLE XXXIX-3

HALF-WAVE POTENTIALS OF UNSATURATED ALIPHATIC KETONES

Compound	$-E_{1/2}$	Electrolyte	Ref.
Methyl vinyl ketone....	1.42 (S.C.E.)	0.1 *N* KCl	44
Methylionone..........	0.95 (K = 1.80)[a]	HCl	45
	1.75 (K = 1.75)	LiCl	
	1.70 (K = 1.80)	KOH	
Pseudomethylionone....	0.80 (K = 1.67)	HCl	45
	1.40 (K = 1.67)	LiCl	
	1.40 (K = 1.67)	KOH	
2-Cyclohexenone........	1.55	0.1 *N* NH_4Cl	46
Isopropylidenacetone....	1.61	0.1 *N* NH_4Cl	47
	1.64	0.2 *N* $(CH_3)_4NOH$	
Heptadiene-3,5-one-2....	1.25	0.1 *N* NH_4Cl	47
Mesityl oxide..........	1.05 (N.C.E.) (2.4)[b]	pH 1.3	48
	1.28 (1.4)[b]	pH 4.9	
	1.60 (1.9)[b]	pH 7.2	
	1.62 (2.0)[b]	pH 8.6	
	1.64 (1.8)[b]	pH 11.3	
Carvone...............	1.37[c]		49

[a] $K = i_d/(Cm^{2/3}t^{1/6})$.
[b] Values are for i_d/C.
[c] Reduction potential.

The diffusion current is reported to be a linear function of the molar concentration.[44]

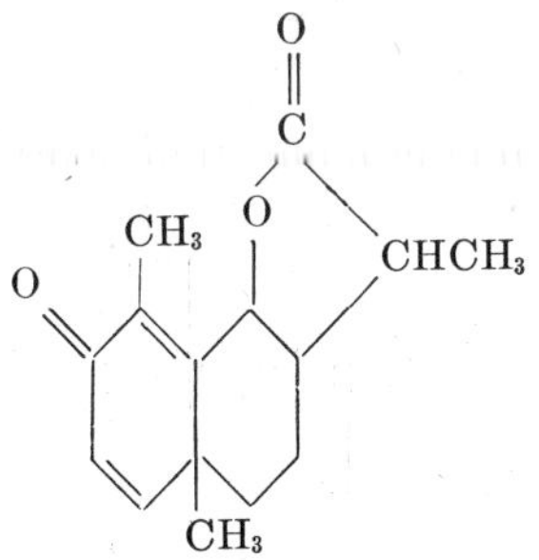

(I)

Santonin (I) can be determined polarographically since it gives well-defined waves in a water–alcohol medium.[50] The experimental results are

[50] F. Santavy, *Collection Czechoslov. Chem. Communs.*, **12**, 422 (1947).

not tabulated, but from the figures given by Santavy[50] it appears that in a pH range between 4 and 7 two waves occur. At a pH smaller than 4 only one wave is observed, as the second wave merges with the hydrogen discharge.

The half-wave potential of the first wave is estimated to be −1.1 v. (S.C.E.) at pH 2.5 and −1.45 v. at pH 5. In the pH range between 4 and 7 the height of the first wave decreases with increasing pH, but the height of the total wave remains constant. At a pH of 8 or greater only one wave is observed with a half-wave potential of −1.6 v. (S.C.E.) which is independent of the pH. The diffusion current is also independent of the pH in the range between 8 and 12.

In the determination of santonin when mixed with phenolphthalein, sucrose, lactose, or coca, the solid is extracted with alcohol. The alcoholic extract which contains santonin and phenolphthalein is allowed to stand for 4 to 6 hours, then 2 ml. of 0.2 *N* lithium sulfate solution is added to one volume of the extract. The mixture is made air-free and the diffusion current is measured at −2.0 v. No mention is made of a reduction wave of phenolphthalein.[51]

For determination of santonin in *Flores cinae* the drug is extracted in a Soxhlet extractor with chloroform, which is then distilled off; the residue is treated with warm ethanol and transferred with ethanol into a volumetric flask and filled to the mark. To 1 ml. of the solution is added 2 ml. of 0.2 *N* lithium sulfate solution and the polarogram is determined after removal of oxygen. If suspended material interferes it is recommended to filter the mixture. The amount of santonin is found from a calibration curve.

Steroids. The polarographic method offers a simple method for the determination of 17-ketosteroids, 3-ketosteroids, 3-keto-Δ^4-unsaturated steroids and the corresponding hydroxysteroids. 17-Ketosteroids (II) are determined polarographically in the form of their Girard derivatives. The ketone itself is not reducible in ammonium chloride solutions.[52]

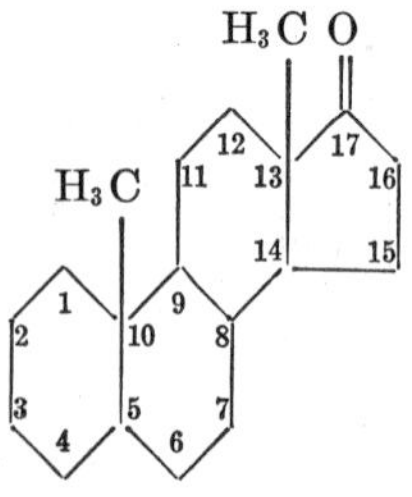

(II)

[51] See I. M. Kolthoff and D. Lehmicke, *J. Am. Chem. Soc.*, **70,** 1879 (1948).

[52] J. K. Wolfe, E. B. Hershberg, and L. F. Fieser, *J. Biol. Chem.*, **136,** 653 (1940).

(d) Procedure for 17-Ketosteroids[52]

A suitable volume of a known solution of a pure hormone (usually 1 ml.) in isopropyl alcohol is placed in a 12 × 75 mm. Pyrex test tube and evaporated to dryness under diminished pressure (water pump vacuum). The oily residue is treated with 0.05 ml. of a fresh solution of 100 ml. of Girard reagent in 1 ml. of glacial acetic acid, measured with a 0.10-ml. pipet, and the test tube is rotated to insure mixing. The tube is then closed with a cork stopper immersed in a water or steam bath at 100° and heated for 2 minutes. The sample is cooled, diluted with 0.95 ml. of pure water, and thoroughly mixed. A 0.25-ml. portion of the solution is measured promptly with a 1-ml. pipet into the semimicro cell, followed by 0.50 ml. of 0.5 *N* ammonium chloride solution and 0.625 ml. of 0.20 *N* sodium hydroxide solution. The solution is made up to a volume of 2.50 ml. by the addition of 1.125 ml. of water. Oxygen is removed in the usual way and the c.v. curve is determined, using a pool of mercury as an anode.

A wave starting at about −1.4 v. and yielding a diffusion current at about −1.5 v. characterizes the 17-ketosteroid. The reagent itself does not yield a wave until a potential of about −1.65 v. is reached. Working with solutions of *androsterone* (III), *isoandrosterone* (IV), and *dehydroisoandrosterone* (V), the diffusion currents were found to be almost proportional to the concentration in the lower concentrations, but marked deviations occurred when the amount of hormone was increased. A satisfactory linear relation between i_d and the concentration of hormone (0.05

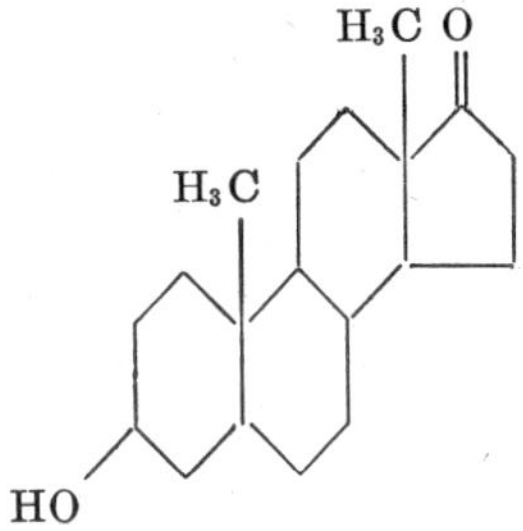

(III)

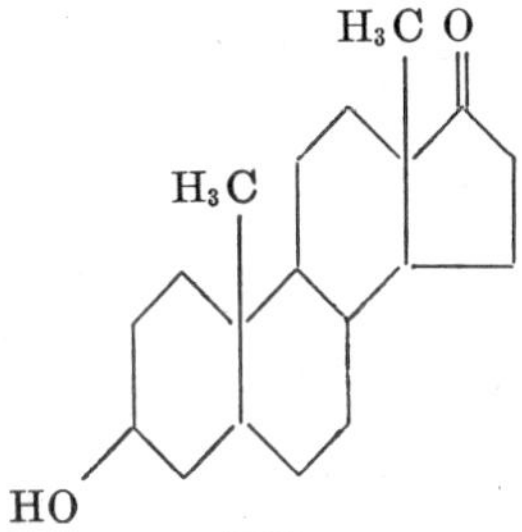

(IV)

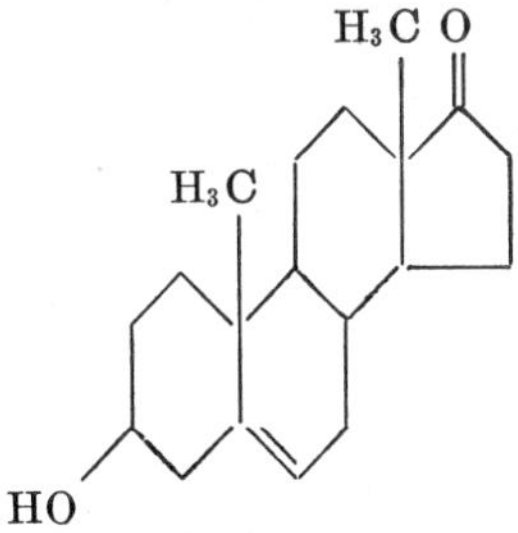

(V)

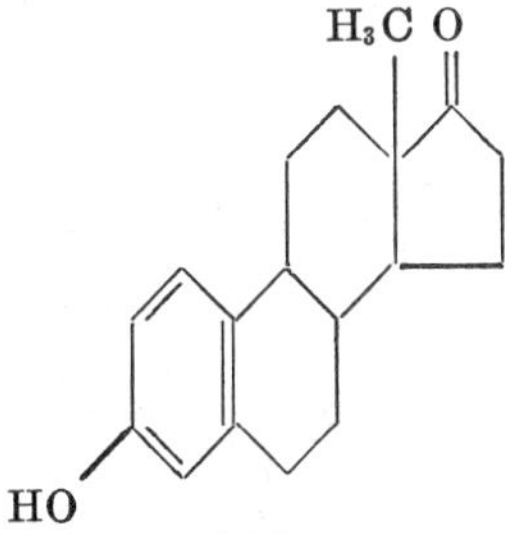

(VI)

to 1.0 mg.) was found upon the addition of traces of urine extracts. Apparently, the urine extract stabilizes the hormones. The behavior of *estrone* (VI) is almost identical with that of the above androgens. A 0.05 per cent glue solution was used to suppress maxima.

(e) Procedure for Determination of 17-Ketones in Androgen Fraction of Urine[53]

A 1-liter sample of urine bearing a known ratio to the daily output is extracted with solvent. The estrogen fraction is removed from a solution of the extract by four washings with 2.5 *N* sodium hydroxide and the neutral fraction is washed twice with 2.5 *N* sodium hydroxide containing 10 per cent of sodium hydrosulfite, each time with vigorous shaking for 5 minutes. The solution is then washed once with water, evaporated to dryness, and the residue made up to a volume of 4 ml. with 95 per cent ethanol. When a 0.2-ml. sample of the solution is converted to the Girard derivative and a portion polarographed by the standard procedure, determinations can be made covering the range from 1 to 20 mg. of hormone per liter of urine.

Certain urinary extracts of the neutral 17-ketosteroids may contain material which distorts the wave formed in two respects. The diffusion current is ill defined and a small wave extending from about −1.0 to −1.3 v. makes it difficult to find the shape and value of the "residual" current. The interfering material is ketonic and can be destroyed by oxidation of the extract, with aqueous permanganate in aqueous dioxane, under conditions which do not affect 17-ketosteroids.[53]

Another improvement can be obtained by decreasing the rate of rise of applied voltage.[54]

No improvement was obtained by substituting other reagents for Girard reagent T. A list of the compounds tried with isodehydroandrosterone is given in Table XXXIX-4. Only maleic acid monohydrazide and aminoguanidine are possible alternatives to Girard reagent but offer no advantages over this reagent.[55]

The polarographic method gives lower results and shows a smaller variation due to errors of technique than the colorimetric method.[56]

The polarographic method has been used to determine estrone in the urine of pregnant mares[57] and, in combination with digiton precipitation, has been used to estimate amounts of α- and β-17-ketosteroids.[58]

(f) 3-Ketosteroids

3-Ketosteroids contain a cyclohexanone ring and, similar to cyclohexanone, form Girard derivatives which are not stable in acid media. At

[53] J. Barnett, A. A. Henly, and C. J. O. R. Morris, *Biochem. J.*, **40,** 445 (1946).
[54] N. T. Werthessen and C. F. Baker, *Endocrinology*, **36,** 351 (1945).
[55] J. Barnett and C. J. O. R. Morris, *Biochem. J.*, **40,** 450 (1946).
[56] J. Barnett, A. A. Henly, C. J. O. R. Morris, and E. L. Warren, *Biochem. J.*, **40,** 778 (1946).
[57] O. Bjornson and N. Ottesen, *Quart. J. Pharm. Pharmacol.*, **19,** 519 (1946).
[58] W. R. Butt, A. A. Henly, and C. J. O. R. Morris, *Biochem. J.*, **42,** 447 (1948).

TABLE XXXIX-4

HALF-WAVE POTENTIALS OF VARIOUS DERIVATIVES OF ISODEHYDROANDROSTERONE[55]

Reagent	$-E_{1/2}$ (S.C.E.)	Wave height for 100 mg. of isodehydroandrosterone, mm.
Glycine hydrazide	1.32	20
Monomethylaminoacethydrazide	1.36	10.5
Dimethylaminoacethydrazide	1.39	9.5
Girard reagent T	1.35	20
Maleic acid monohydrazide	1.32	20
Aminoguanidine	1.41	14
p-Aminobenzoic acid hydrazide	1.25	7.5
Thiosemicarbazide	0.9	3.1
p-Dimethylaminobenzenesulfonhydrazide	1.09[a]	6.2

[a] *vs.* Hg pool.

higher pH values c.v. curves are obtained which can be used for the quantitative determination of these compounds. This method has been used for the determination of cholestanone (I).[59]

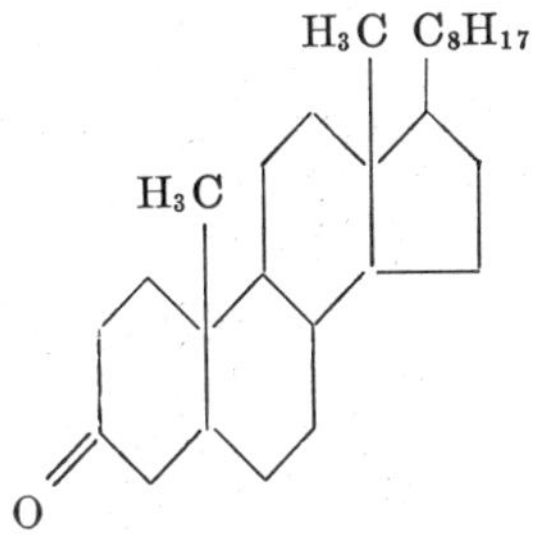

(I)

3-Keto-4-Unsaturated Steroids. Steroids containing an α,β-unsaturated ketone grouping are reducible at the dropping mercury electrode.

Testosterone (II), progesterone (III), corticosterone (IV), and desoxycorticosterone (V) give a wave between E_a = 1.6 and 1.8 v. in a solution which is 90 per cent ethanol and 0.1 *N* in lithium chloride. The height of the wave was found to be proportional to the concentration of the hormone.[60]

Using a solution of the sample in 1 milliliter of isopropyl alcohol into the cell and adding 0.5 milliliter of aqueous 0.75 normal tetraethylammonium hydroxide and 1 milliliter of water, half-wave potentials

[59] V. Prelog and O. Häfliger, *Helv. Chim. Acta*, **32,** 2088 (1949).
[60] J. H. Eisenbrand and H. Picher, *Z. physiol. Chem.*, **260,** 83 (1939).

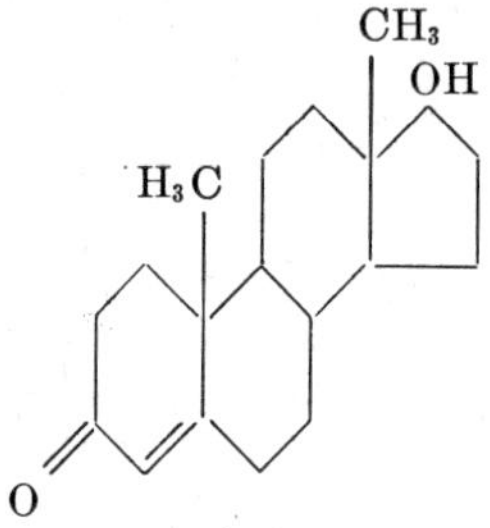

(II)

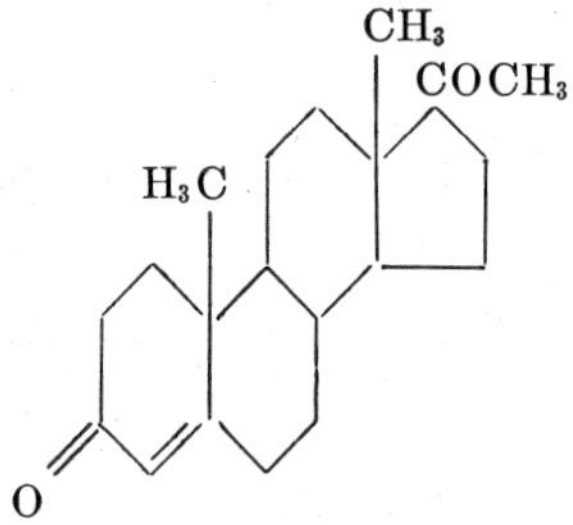

(III)

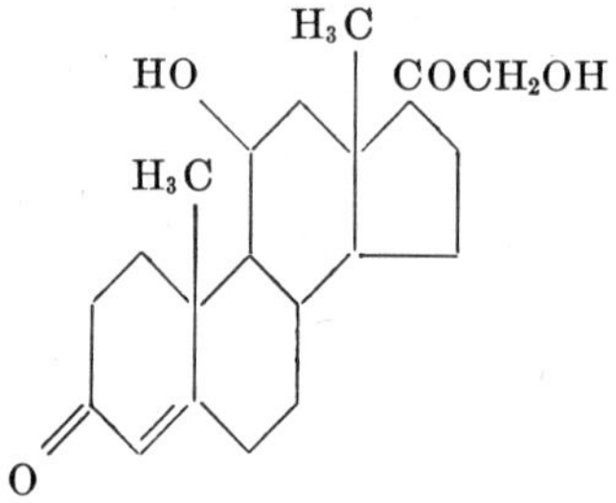

(IV)

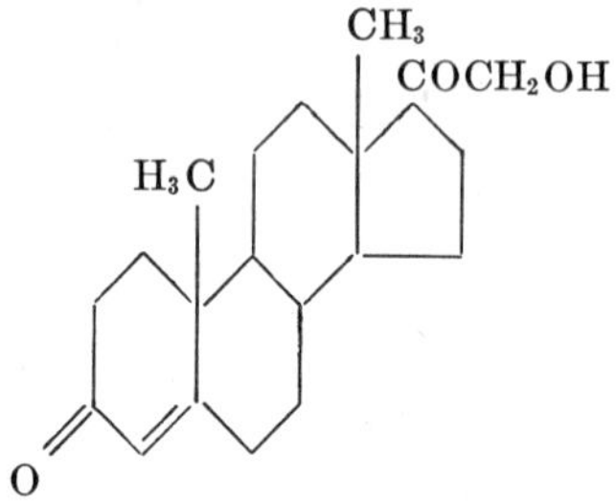

(V)

(against a mercury pool) between −1.53 v. and −1.63 v. are obtained for testosterone (II) and progesterone (III), corticosterone (IV), desoxycorticosterone (V), Δ^4-cholestenonc (VI) and cortisone (VII):

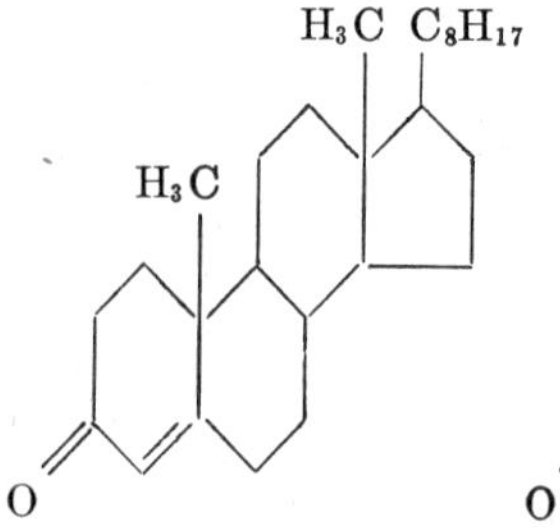

(VI)

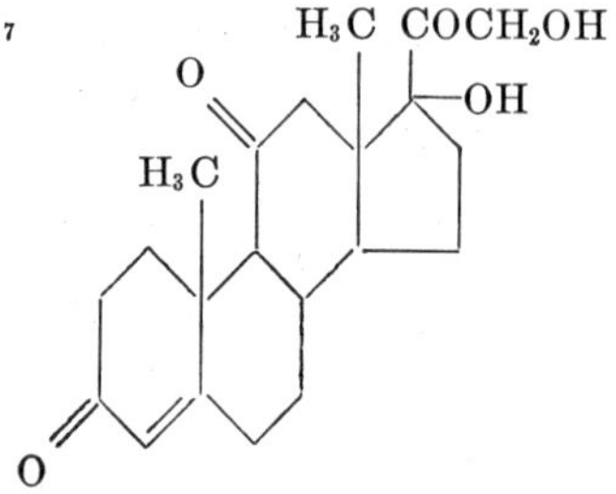

(VII)

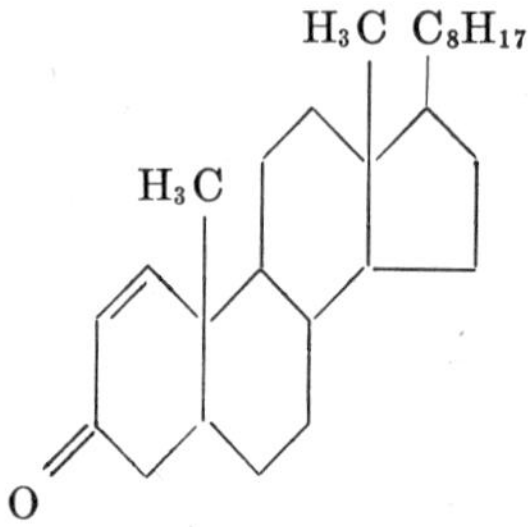

(VIII)

Δ^1-Cholestenone (VIII) in which the double bond is less hindered than in Δ^4-cholestenone (VI) gives a half-wave potential of −1.43 v. The half-wave potentials of the above compounds become more negative with increasing concentration. The diffusion currents are inversely proportional to the molecular weights of the compounds.[59]

In 0.1 *N* ammonium chloride the half-wave potential of Δ^4-cholestenone (VI) is −1.30 v.[61]

[61] H. Adkins, R. M. Elofson, A. G. Rossow, and C. C. Robinson, *J. Am. Chem. Soc.*, **71**, 3622 (1949).

The Girard derivatives of 3-keto-Δ^4-unsaturated ketones give a wave with a half-wave potential of -1.23 v., whereas the wave for the 17-ketosteroid derivative is at -1.44 v. Thus, it is possible to determine the unsaturated ketone in the presence of the other type. The wave height of the Girard derivative of testosterone (II) is 6.3 times as great as that of the free hormone.

Unsaturated ketosteroids having a second keto group in the side chain such as desoxycorticosterone (V), corticosterone (IV), and cortisone (VII) give Girard derivatives which have waves at -1.23 v. and -1.45 v., respectively. The value at -1.23 v. is for the Δ^4-3-ketonic structure, while the second value at about -1.45 v. is for the carbonyl group in the 20-position. The latter reduction is demonstrated more clearly by the reduction of the Girard derivative of Δ^5-pregnenol-3-one-20 (IX) at -1.45 v.

Mixtures of androsterone (X) with testosterone (II) and with desoxycorticosterone (V) were successfully analyzed polarographically.

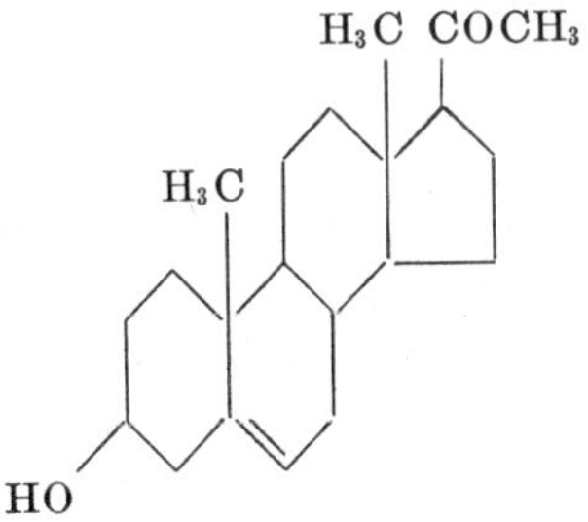

(IX)

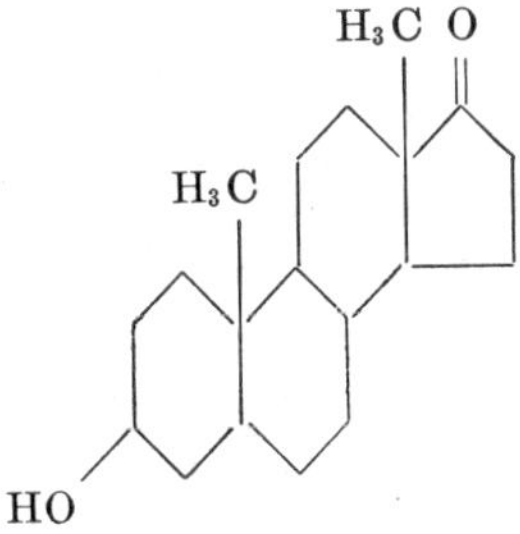

(X)

The method has also been applied to the determination of methyltestosterone (XI) and pregnenin-17-ol-3-one (XII) in pharmaceutical preparations.[62]

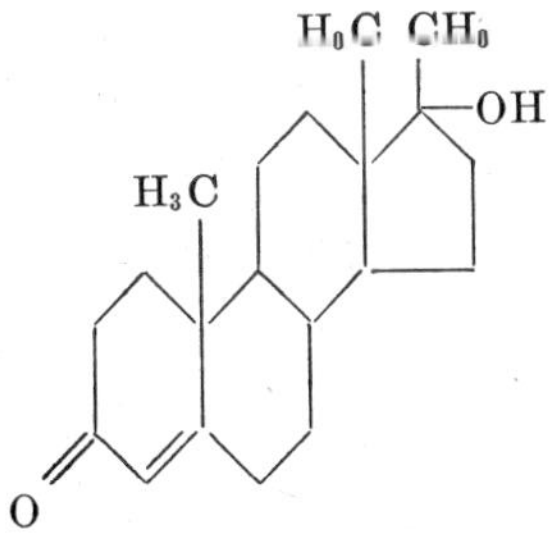

(XI)

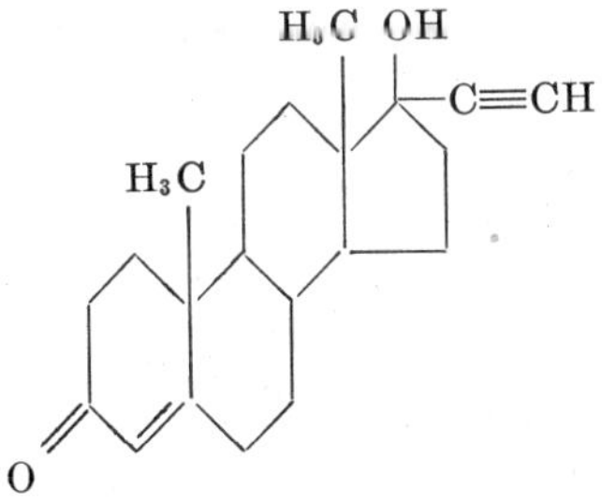

(XII)

[62] G. Sartori and E. Bianchi, *Gazz. chim. ital.*, **74,** 8 (1940); *Chem. Abstracts*, **41,** 1266 (1947).

(g) Hydroxysteroids

The polarographic method is suitable for determining 3-hydroxy-Δ^5-steroids,[63] such as dehydroisoandrosterone (XIII), after first oxidizing and rearranging these to the corresponding 3-keto-Δ^4-unsaturated steroid (XIV) with aluminum *tert*-butoxide according to the method of Oppenauer.[64]

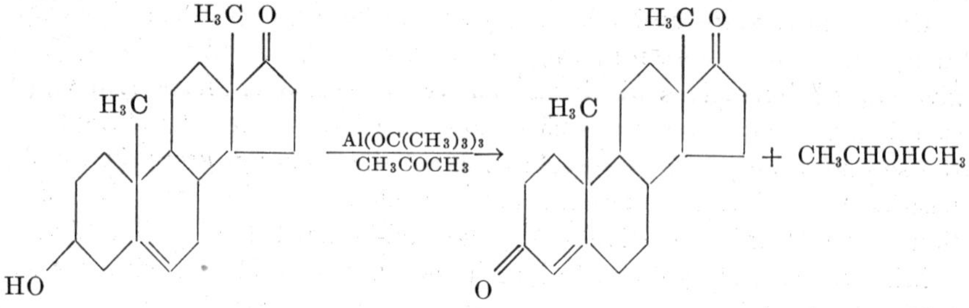

(XIII) (XIV)

After conversion of the oxidized material to the Girard derivative one wave is found at −1.2 to −1.3 v. corresponding to the 3-keto-Δ^4-unsaturated grouping and another at about −1.5 v. characteristic of the 17-keto grouping.[65]

The oxidation is found to proceed best with a minimum of side reaction if the reaction is carried out in benzene with only a moderate excess of acetone at 100° for 1¼ hours.

Procedure (by Hershberg, Wolfe, and Fieser—for the purification of the reagents and the apparatus used see the original literature): A volume of an alcoholic solution of the sample equivalent to 0.5 mg. of 17-ketosteroid is measured with a pipet into a pressure vessel. In the case of an androgen fraction from a urine extract the total 17-ketosteroid content is determined polarographically by the method previously outlined. The solution is evaporated by first applying suction gradually until the tube is cold to the touch and then warming the tube gently. For the removal of the last traces the tube is heated for 10 to 15 minutes on the steam bath at a pressure of 10 to 15 mm. To the residue are then added 14 to 55 mg. of aluminum *tert*-butoxide, 0.40 ml. of benzene, and 0.10 ml. of a mixture of equal volumes of benzene and acetone. The closed vessel is heated in an oven at 100° for 1¼ hours. After the tube has been allowed to cool the cap is removed and the contents transferred quantitatively with a pipet to a separatory funnel of special design; the tube is then rinsed by alternate washings with 1 *N* hydrochloric acid (total 1 ml.) and benzene (total 1 ml.). The funnel is stoppered, shaken thoroughly, and centrifuged, after which

---

[63] E. B. Hershberg, J. K. Wolfe, and L. F. Fieser, *J. Am. Chem. Soc.*, **62**, 3516 (1940).
[64] R. V. Oppenauer, *Rec. trav. chim.*, **56**, 137 (1937).
[65] E. B. Hershberg, J. K. Wolfe, and L. F. Fieser, *J. Biol. Chem.*, **140**, 215 (1941).

the aqueous layer is drawn off and discarded. The process of washing is repeated once with the 1 N hydrochloric acid and twice with distilled water. Finally, the lower level of the benzene layer is adjusted to the mark and fresh benzene is added to bring the volume to 3 ml. 1 ml. of the resulting solution is pipetted into a 12 × 75 mm. test tube equipped with a 14/20 standard taper joint with which connection is made to an all-glass adapter.

The benzene is largely removed by heating the vessel on the steam bath while a stream of air is drawn through the adapter; finally, the residue is fully evaporated at the vacuum of the water pump.

The sample is prepared for polarographic analysis by adding to the residue 0.02 ml. of a fresh solution of 100 mg. of Girard's reagent T in 1 ml. of acetic acid, warming the mixture for 2 minutes on the steam bath, cooling and adding 0.48 ml. of water, 0.50 ml. of 0.5 M ammonium chloride solution, and 1.00 ml. of 0.2 N sodium hydroxide. The flask is stoppered, the contents shaken, and the solution poured into the polarographic cell. The diffusion current is measured at about −1.3 v., and the amount of dehydroisoandrosterone is found by reference to a calibration curve applicable to the sensitivity of the galvanometer in question.

The procedure can be combined with that described for determining 17-ketosteroids in the androgen fraction from urine extracts. Thus, an initial polarographic analysis of the Girard derivative of the androgen fraction permits determination of the total 17-ketosteroid content by measurement of the single wave at −1.45 v. Analysis after the Oppenauer oxidation gives the dehydroisoandrosterone content as indicated by the wave at −1.25 v.

Cholesterol (XV), stigmasterol (XVI), and phytosterol mixtures, react in a similar manner to dehydroisoandrosterone and can be determined polarographically.

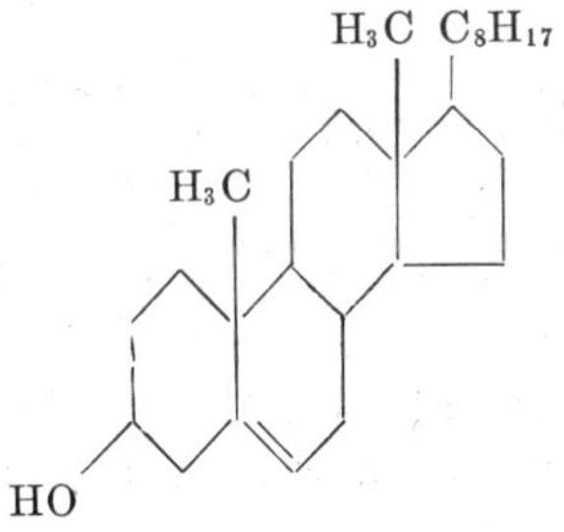

(XV)

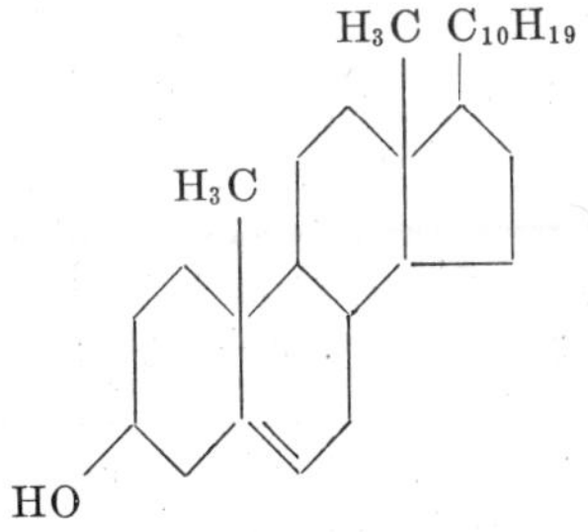

(XVI)

Cholesterol can be freed from other interfering ketonic substances by the Girard separation.

The polarographic variations have been used as an aid in identifying 17-

ketosteroids from the acid-hydrolyzed urine of a girl with a corticoadrenal tumor.[66]

(h) Hydroxy Carbonyl Compounds

Glycolaldehyde[67] gives a wave similar to that for formaldehyde in that its height is strongly dependent on pH and temperature. In a buffered

TABLE XXXIX-5

HALF-WAVE POTENTIALS AND CURRENTS FOR GLYCOLALDEHYDE IN VARIOUS BUFFERS

pH	i_d, mm.	$-E_{1/2}$ (N.H.E.)
7.0	6.0	1.26
7.5	6.5	1.29
8.0	7	1.30
8.9	7	1.31
9.9	7	1.33
10.7	12	1.34
(0.1 *N* NaOH) 12.8	36	1.45
	36	1.45 (after 15 min.)
	25	1.45 (after 50 min.)

TABLE XXXIX-6

REDUCTION POTENTIALS OF SOME HYDROXY CARBONYL COMPOUNDS

Formula	Substance	W. and P. (a)	W. and P. (b)	T.
$CH_2OHCHOHCHO$	Glyceraldehyde	−1.30		
$CH_2OHCOCH_3$	Hydroxyacetone	−1.75	−1.85	
$CH_3COCHOHCH_3$	Acetoin	−1.7		−1.14

W. and P.—A. Winkel and G. Proske, *Ber.*, **69,** 1917 (1936); 0.001 *M* solution (a) in 0.1 *N* NH_4Cl, (b) in 0.1 *N* LiCl. T.—I. Tachi, *Mem. Coll. Agr. Kyoto Imp. Univ.*, **42,** 1 (1938); 0.0001 *M* solution in mixture which is 0.01 *N* in HCl and 0.1 *N* in KCl.

solution of pH 6.6 the wave height has a temperature coefficient of 6 per cent per degree between 20 and 30°C. In Table XXXIX-5[67] half-wave potentials and currents are listed for a 0.00059 *M* solution in various buffers. A saturated calcium chloride solution gives a wave at a more positive potential than that in a lithium chloride solution.

The reduction potentials of some other hydroxy carbonyl compounds are given in Table XXXIX-6.

[66] J. K. Wolfe, L. F. Fieser, and H. B. Friedgood, *J. Am. Chem. Soc.*, **63,** 592 (1941).
[67] R. Bieber and G. Trümpler, *Helv. Chim. Acta*, **31,** 5 (1948).

Dihydroxyacetone gives a half-wave potential at −1.60 (S.C.E.) in a buffer of pH 4.5.[68]

(i) Sugars (Aldoses and Ketoses)

Aldoses[69] and ketoses give waves which are dependent on the rate of transformation of the nonreducible cyclic hemiacetal form to the aldehydo or keto form at the electrode. The concentration of the reducible form actually present is negligibly small. Therefore the height of the small waves (glucose, xylose and galactose) are not diffusion controlled and independent of the height of the mercury in the reservoir. They increase with increasing pH (between pH 6 and 8).[70] In Table XXXIX-7 some of

TABLE XXXIX-7
"APPARENT PERCENTAGE" REDUCIBLE FORM IN HEXOSE AND PENTOSE SUGARS IN 0.25 *M* SOLUTIONS AND CORRESPONDING $E_{1/2}$ (*vs.* N.C.E.)

	Apparent percentage of reducible form			Half-wave potential		
Sugar	pH 6.5	7.0	7.5	pH 6.5	7.0	7.5
D-Glucose	0.016	0.024	0.004	−1.58	−1.59	−1.60
D-Galactose	0.062	0.082	0.18	−1.57	−1.60	−1.65
D-Mannose	0.039	0.064	0.14		−1.57	−1.63
L-Allose	(2.6)	(1.4)	(0.7)	−(1.71)	−1.79	
D-Lyxose	0.31	0.40	0.72	−1.57	−1.60	−1.63
D-Xylose	0.10	0.17	0.37	−1.50	−1.54	−1.59
L-Arabinose	0.16	0.28	0.46	−1.58	−1.61	−1.64
D-Ribose[a]	(10)	8.5	30	−1.79	−1.81	−1.87

[a] In 0.1 *M* solution.

the results are summarized. The "apparent percentage" reducible form gives the value of the kinetic current (apparent diffusion current) expressed in percentage of the total current which would have been observed if all the sugar were reduced. The wave height obtained with 5 hydroxymethyl furfural, an aldehyde of about the same molecular weight as the sugars used, was taken as equivalent to that which was expected of the aldehyde form and calculations were based upon this. The data refer to 0.25 *M* sugar solutions in buffers with pH values of 6.5, 7.0, and 7.5. The half-wave potentials (*vs.* N.C.E.) are also listed. Similar data were reported by Cantor and Peniston for 0.1 and 0.5 *M* sugar solutions.

Ketoses (fructose and sorbose) differ from aldoses and give well-de-

[68] G. MacKinney and O. Temmer, *J. Am. Chem. Soc.*, **70**, 3586 (1948).
[69] K. Wiesner, *Collection Czechoslov. Chem. Communs.*, **12**, 64 (1947).
[70] S. M. Cantor and D. P. Peniston, *J. Am. Chem. Soc.*, **62**, 2113 (1940).

fined waves in 0.02 *N* lithium chloride or hydroxide starting at −1.80 v. (N.C.E.).[71] In an acid medium the hydrogen wave starts before the ketone wave.

The diffusion current was found to increase three-fold between 4° and 22° C. and twice between 20° and 40° C. This behavior together with the fact that the decrease in wave height produced by lowering the mercury reservoir was smaller than expected indicates that the current is diffusion and kinetically controlled. At 40° C. Heyrovsky and Smoler[71] found the height of the diffusion current the same as for an equimolar barium solution and suggested a 2-electron reduction.

The height of the fructose wave is not a linear function of the fructose concentration but is affected by the hydroxide ion concentration and reaches a maximum at 0.002 *N*.

A half-wave potential of −1.8 v. is obtained for fructose in solutions of barium, ammonium, calcium, rubidium, potassium, sodium, strontium, cesium, lithium, and quaternary amine salts. The most distinct wave occurs with calcium or lithium salts at pH 7.0.[72]

Disaccharides (sucrose, maltose, and lactose) are reported not to be reduced at the dropping mercury electrode. Maltose and lactose, however, should show behavior similar to that of the aldoses.

Analytical use of the reduction of fructose can be made in the determination of this substance in the presence of sucrose and glucose in honey, urine, and candy.[73]

In determinations of this type if small amounts of ketones are present with large amounts of aldoses, the presence of the reducible aldehyde form and the pH must be considered. In a 0.25 *M* glucose solution the apparent concentration of the reducible form is about 10^{-4} *M* at a pH of 7.5; in a galactose solution of the same concentration it is 4×10^{-4} *M*.

Coloring matter in sucrose is surface active and can be determined polarographically by its effect on the oxygen maximum in a 0.002 *M* K_2SO_4 solution[74] (see p. 185).

(j) Aliphatic Dicarbonyl Compounds

Glyoxal $(CHO)_2$. Glyoxal gives a half-wave potential at −1.41 v. (S.C.E.) in a buffer of pH 3.4.

---

[71] J. Heyrovsky and I. Smoler, *Collection Czechoslov. Chem. Communs.*, **4,** 521 (1932).

[72] I. Vavruch, *Listy Cukrovar.*, **65,** 171 (1949); *Chem. Abstracts*, **43,** 7869 (1949).

[73] I. Vavruch and E. Rubes, *Listy Cukrovar.*, **64,** 185 (1948); *Chem. Abstracts*, **42,** 4690 (1948).

[74] C. A. Browne and F. W. Zerban, *Physical and Chemical Methods of Sugar Analysis*, Wiley, New York, 1941.

Methylglyoxal (CH_3COCHO). Methylglyoxal gives waves at -0.83 v. and -1.04 v. (S.C.E.) in buffers of pH 4.5 and 7.1, respectively.[75]

Biacetyl ($CH_3COCOCH_3$). Biacetyl gives two waves of about equal height in 0.1 *N* ammonium chloride at -0.84 and -1.59 v. (S.C.E.).[76] In a 0.1 *N* potassium chloride solution the half-wave potential for the first wave is -0.83 v. (S.C.E.). The limiting current of the first wave is not a linear function of the concentration but is given by the hyperbolic relation:[77]

$$\log i_d = 0.8766 \log (M \times 10^4) - 0.3967$$

In 0.1 *N* lithium chloride solution reduction potentials of -0.81 and -1.93 v. were obtained which in neutral and alkaline media shifted to less negative potentials in the presence of divalent cations, such as calcium and barium.[78]

The effect of the concentration of biacetyl upon reduction potentials and the heights of the waves has been studied systematically in various solutions.[79]

Acetylacetone ($CH_3COCH_2COCH_3$). Acetylacetone gives two waves in 0.1 *N* lithium chloride with molar reduction potentials of -1.07 and -1.37 v. Since acetylacetone is present as 76 per cent enol, the first wave was attributed to the reduction of the enol form and the second to the keto form according to the following equations:[80]

$$CH_3C(OH){=}CHCOCH_3 + 2H^+ + 2e \rightarrow CH_3CHOHCH_2COCH_3$$

and:

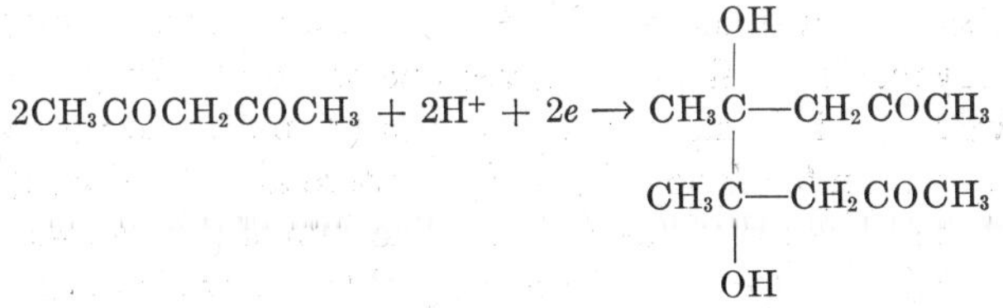

In a mixture of 0.01 *N* hydrochloric acid and 0.1 *N* potassium chloride a

[75] G. Mackinney and O. Temmer, *J. Am. Chem. Soc.*, **70**, 3586 (1948).
[76] H. Adkins and F. W. Cox, *J. Am. Chem. Soc.*, **60**, 1151 (1938).
[77] E. I. Fulmer, J. J. Kolfenbach, and L. A. Underkofler, *Ind. Eng. Chem., Anal. Ed.*, **16**, 469 (1944).
[78] A. Winkel and G. Proske, *Ber.*, **71**, 1785 (1938).
[79] G. Semerano and B. Polaczek, *Gazz. chim. ital.*, **70**, 448 (1940).
[80] G. Semerano and A. Chisini, *Gazz. chim. ital.*, **66**, 504 (1936).

reduction potential of −1.20 v. is reported,[81] while in a medium of 0.1 *N* ammonium chloride a value of −1.69 v. is reported.[82]

The polarographic method has been used to investigate whether glyoxal or methylglyoxal are products of fruit deterioration,[75] to determine the amount of biacetyl produced in the vapor phase oxidation of 2,3-butylene glycol,[77] and to determine (indirectly) the amount of acetoin in blood and urine.[83] The acetoin can be oxidized by ferric chloride to biacetyl and by the following procedure, 0.5 milligram per cent of acetoin can be determined in 2 ml. of blood within one hour.

A 1:10 tungstic acid, protein filtrate is prepared from 2 ml. of blood by the Folin-Wu method. 10 ml. of the filtrate is placed in a 125-ml. distillation flask and 10 ml. of 50 per cent ferric chloride solution is added and an ebullition tube inserted. The flask is connected to a small vertical condenser, the lower end of which extends into a receiving tube marked at 10 ml. The distillation flask is immersed in a hot water bath for 30 minutes; the bath is then removed and the flask is heated directly with the flame from a small microburner. Distillation is continued until exactly 10 ml. is collected.

A few milliliters of the distillate is placed in the polarographic cell, some sodium sulfite is added, and the diffusion current is determined (air need not be removed when sulfite is added). The half-wave potential is found at an applied e.m.f. of −0.86 v. A calibration curve with a known amount of biacetyl has to be made, since the diffusion current is not proportional to the concentration.

2. AROMATIC COMPOUNDS

(a) Aromatic Aldehydes

The behavior of aromatic aldehyde at the dropping mercury electrode is strongly dependent upon the pH of the medium and the electrolyte used.

Benzaldehyde. Benzaldehyde gives one wave at low pH values (1.2–2.2), two waves at intermediate pH values (3 to 5), and one wave at pH values greater than 6. The height of the wave at pH values greater than 6 is twice that at low pH values (1–2). The sum of the two waves at intermediate pH values between 3 and 5 is equal to the wave obtained at a pH greater than 6.[84]

The half-wave potentials (S.C.E.) of benzaldehyde in phosphate and citrate buffers of pH 2.33 to 7.8 in 30 per cent ethanol are given in Figure XXXIX-6.[85] See also Table XXXIX-8.[86] Increasing the alcohol content

[81] I. Tachi, *Mem. Coll. Agr. Kyoto Imp. Univ.*, **42,** 1 (1938).
[82] A. Winkel and G. Proske, *Ber.*, **69,** 1917 (1936).
[83] L. A. Greenberg, *J. Biol. Chem.*, **147,** 11 (1943).
[84] M. Tokuoka, *Collection Czechoslov. Chem. Communs.*, **7,** 392 (1935).
[85] I. A. Korshunov and L. N. Sazanova, *Zhur. Fiz. Khim.*, **23,** 202 (1949).
[86] R. Pasternak, *Helv. Chim. Acta.* **31,** 753 (1948).

of the electrolyte to 50 per cent by volume apparently changes $E_{1/2}$ only slightly.

It is surprising, and not as yet explained, that in solutions of salts of calcium and strontium, and of tetramethylammonium hydroxide, two waves of equal height are obtained. The behavior in neutral alkali salts is normal but, upon the addition of dipositive ions such as calcium and

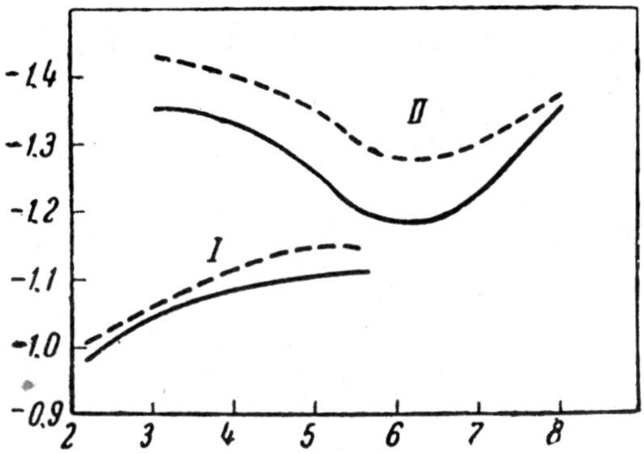

Fig. XXXIX-6. Half-wave potentials (S.C.E.) of benzaldehyde at various pH values: I = first wave; II = second wave. The solid lines represent values for 2.8×10^{-3} *M*; the broken lines represent values for 10.86×10^{-3} *M*.[85] Ordinate is $E_{1/2}$. Abscissa is pH.

TABLE XXXIX-8

HALF-WAVE POTENTIALS OF 0.001 *M* BENZALDEHYDE[86]

$-E_{1/2}$ (N.C.E.)	i_d/C	pH
0.98	2.3	1.3
1.20	2.4	4.9
1.36	4.2	7.2
1.45	4.0	8.6
1.48	3.8	11.0

strontium, the potential of the wave is shifted to less negative values and two waves appear. The mechanism of this reduction is not completely understood.

The relative heights of the two waves at low pH values depend upon the concentration used. The relationship is 1:1 only if the concentration of benzaldehyde is of the order of 10^{-2} mole per liter. At lower concentrations the first wave is larger than the second. Concentration also affects the half-wave potentials; the higher the concentration, the more negative is the half-wave potential.[85]

Analysis of the slope of the waves gives a value of 0.055 and points to a potential-determining step of 1 electron. The combined height of the two waves from the Ilkovic equation is equivalent to a 2-electron change. These data indicate the following mechanism for the reduction of benzaldehyde.

First Wave:

$$2C_6H_5CHO + 2H^+ + 2e \rightleftharpoons 2C_6H_5\dot{C}(H)OH \quad \text{(free radical)}$$

$$2C_6H_5\dot{C}(H)OH \rightarrow C_6H_5C(H)(OH)\text{—}C(H)(OH)C_6H_5 \quad \text{(hydrobenzoin)}$$

Second Wave:

$$C_6H_5CHO + 2H^+ + 2e \rightarrow C_6H_5CH_2OH \quad \text{(benzyl alcohol)}$$

This mechanism has been verified by controlled potential electrolysis; at pH 1.3 only hydrobenzoin is isolated as a product. At pH 8.6 a mixture of hydrobenzoin and benzyl alcohol is obtained, according to Pasternak.

From the above data it is apparent that the reduction potentials of benzaldehyde measured in unbuffered solutions[87] have no exact significance.

p-Dimethylaminobenzaldehyde. *p*-Dimethylaminobenzaldehyde differs in its behavior from benzaldehyde in that below pH 5 and above pH 6.7 only one wave is obtained. Between these two values two waves are observed. The change in half-wave potential (S.C.E.) with pH is given in Figure XXXIX-7.[85]

Salicylaldehyde. Salicylaldehyde gives two waves in the region of pH 2.56 to 3.29. Its methyl ether gives two waves between a pH of 3.29 and 4.10. The results obtained[88] for these compounds and related compounds are listed in Table XXXIX-9.

The behavior of other aromatic aldehydes reported in the literature is given in Table XXXIX-10.

The "reduction potentials" (depolarization potentials) of various para-substituted benzaldehydes were determined at various pH values and are

[87] M. Shikata and K. Shoji, *Mem. Coll. Agr. Kyoto Imp. Univ.*, **4**, 75 (1927). K. Shoji, *Bull. Ins. Phys. Chem. Research Tokyo*, **9**, 69, 621 (1930). A. Winkel and G. Proske, *Ber.*, **69**, 1917 (1936). H. Adkins and F. W. Cox, *J. Am. Chem. Soc.*, **60**, 1151 (1938). M. Shikata and I. Tachi, *Microchemie*, **12**, 62 (1933).

[88] F. Santavy, *Collection Czechoslov. Chem. Communs.*, **14**, 145 (1949).

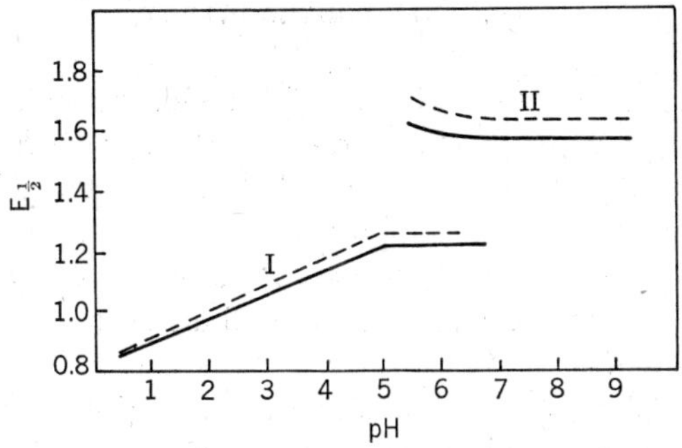

Fig. XXXIX-7. Half-wave potentials of *p*-dimethylaminobenzaldehyde at various pH values: I = first wave; II = second wave. The solid lines represent values for 2×10^{-3} *M*; the broken lines represent values for 2.6×10^{-3} *M*.[85]

TABLE XXXIX-9

HALF-WAVE POTENTIALS OF SALICYLALDEHYDE AND RELATED COMPOUNDS

		$-E_{1/2}$ (S.C.E.)			
Substance	Solvent	pH 1.81	6.80	11.98	0.1 *N* LiOH
Salicylaldehyde................	H_2O	1.02	1.35	1.63	1.64
o-Methoxybenzaldehyde.........	H_2O	1.03	1.49	1.53	
p-Hydroxybenzaldehyde.........	50% alc.	1.16	1.45	1.85	
p-Methoxybenzaldehyde.........	H_2O	1.07	1.47	1.60	
4-Hydroxynaphthaldehyde......	50% alc.	0.98	1.35	1.82	
4-Methoxynaphthaldehyde......	50% alc.	1.02	1.40	1.51	

TABLE XXXIX-10

HALF-WAVE POTENTIALS AND REDUCTION POTENTIALS OF OTHER AROMATIC ALDEHYDES

		Half-wave potential	
Substance	Reduction potential (N.C.E.)[a] in 0.1 *N* NH_4Cl	In 0.1 *N* NH_4Cl (N.C.E.)[b]	In tetramethylammonium hydroxide[c]
m-Hydroxybenzaldehyde.....	1.21		
o-Chlorobenzaldehyde........	−1.07		
m-Chlorobenzaldehyde.......	−1.12		
p-Chlorobenzaldehyde........	−1.18		
m-Bromobenzaldehyde.......	−1.19		
p-Bromobenzaldehyde........	−1.21		
Vanillin....................	−1.33	−1.41	−1.73
Piperonal....................	−1.30	−1.38	−1.39
Phenylpropionaldehyde......	−1.38		
Cinnamaldehyde.............	−0.78		

[a] A. Winkel and G. Proske, *Ber.*, **69,** 1917 (1936). Compounds which have been studied more exactly in buffered solutions have been omitted from their list.

[b] H. Adkins and F. W. Cox, *J. Am. Chem. Soc.*, **60,** 1151 (1938). We have assumed that the anode potential in 0.1 *N* ammonium chloride was equal to that of the 0.1 *N* calomel electrode and recalculated the values with reference to the N.C.E.

[c] G. Semerano and A. Chisini, *Gazz. chim. ital.*, **66,** 510 (1936); the molar reduction potential is given. These two aldehydes take up one electron per molecule in the reduction.

summarized in Table XXXIX-11. The depolarization potentials were found by the tangent method and refer to the *normal hydrogen electrode.* The experiments were carried out at 21° C. with 0.0067 *M* solutions in 33 per cent ethanol to which the supporting electrolyte was added.[89]

The behavior of these compounds parallels very closely the behavior of benzaldehyde. In general maxima were eliminated by the addition of a trace of methyl red.

The behavior of protocatechuic aldehyde and *p*-nitrobenzaldehyde has been studied at the dropping mercury electrode. The method has been

TABLE XXXIX-11

DEPOLARIZATION POTENTIALS (N.H.E.) OF PARA-SUBSTITUTED BENZALDEHYDES, RC_6H_4CHO IN 33 PER CENT ETHANOL

Supporting electrolyte	Substituent							
	Cl	H	Me	Et	Pr	Bu	OMe	NMe_2
Buffer pH 1.4	−0.59	−0.64	−0.66	−0.66	−0.70	−0.67	−0.72	−0.63
Buffer pH 4 (a)	−0.95	−1.02	−1.04	−1.02	−1.08	1.05	−1.09	−1.07
(b)	−1.24	−1.27	−1.31	−1.32	−1.40	1.48	−1.39	
Buffer pH 7	−1.05	−1.14	−1.16	−1.14	−1.12	−1.11	−1.22	−1.18
Buffer pH 8	−1.11	−1.19	−1.24	−1.22	−1.22	−1.22		
0.1 *N* Me_4NBr								
(a)	−1.12	−1.19	−1.25	−1.24	−1.24	−1.25	−1.31	−1.38
(b)	−1.66	−1.74	−1.96	−1.99	−2.10	−2.11	−2.16	−2.18
0.1 *N* KCl	−1.14	−1.20	−1.25	−1.26	−1.25	−1.25	−1.30	−1.40

(a) is first wave. (b) is second wave.

used to determine benzaldehyde and cinnamaldehyde in pharmaceutical preparations.[90]

o-, *m*-, and *p*-Deuteriohydroxybenzaldehyde have the same reduction potential as ordinary benzaldehyde.[91]

(b) Aromatic Ketones

Aromatic ketones resemble quite closely the aromatic aldehydes in their behavior in that the waves are strongly dependent upon the pH and upon the electrolyte used.

---

[89] J. W. Baker, W. C. Davies, and M. L. Hemming, *J. Chem. Soc.*, **1940,** 692.

[90] R. Portillo and G. Varela, *Anales fís. y quím. Madrid*, **40,** 839 (1944): *Chem. Abstracts*, **42,** 6704 (1948).

[91] R. M. Ansidei and M. Carissimi, *Boll sci. facoltà chim. ind. Bologna*, **1940,** 167, **36,** 5469 (1942).

Benzophenone. Half-wave potentials and diffusion currents observed with benzophenone in 50 per cent ethanol are listed below.[92] Two waves

$-E_{1/2}$ (N.C.E.)	i_d/C	pH
0.94	2.2	1.3
1.16	4.3	4.9
1.29	4.0	7.2
1.36	4.1	8.6
1.42	4.1	11.3

have been observed in some alkaline buffers by Ashworth.[93] The height of the first increases, and that of the second decreases, with rise in temperature, decrease in pH, and addition of various cations. The effect of the cations in decreasing the height of the second wave is in the following order: $NR_4^+ \gg Ca^{++} > Ba^{++}, Cs^+ > Rb^+ > K^+, Na^+, Li^+$. The half-wave potential of the second wave is determined mainly by the nature and concentrations of the cations present in the solution. It shifts to less negative values with increasing cation concentration, the alkali ions of greater atomic weight having a greater effect than those of smaller atomic weight. Fluorenone behaves in a similar fashion.[93]

The wave in acid solution represents a reduction of benzophenone to benzpinacol:

$$2(C_6H_5)_2C{=}O \xrightarrow[2H^+]{2e} 2(C_6H_5)_2\dot{C}{-}OH \qquad \text{(free radical)}$$

$$2(C_6H_5)_2\dot{C}{-}OH \rightarrow (C_6H_5)_2C(OH){-}C(OH)(C_6H_5)_2$$

The wave in more alkaline solutions represents a reduction of benzophenone to benzohydrol:

$$(C_6H_5)_2C{=}O \xrightarrow[2H]{2e} (C_6H_5)_2CHOH$$

This mechanism has been confirmed by Pasternak by controlled potential electrolysis; at pH 1.3 only benzpinacol is isolated, at pH 4.3 a mixture of benzpinacol (25 per cent) and benzohydrol, and at pH 8.6 mainly benzohydrol with only a small amount of benzpinacol.

It has been suggested that metallic cations influence the reaction by

[92] R. Pasternak, *Helv. Chim. Acta,* **31,** 753 (1948).

[93] M. Ashworth, *Collection Czechoslov. Chem. Communs.,* **13,** 229 (1948).

forming metallic ketyls which would influence the ease of the dimerization of the free radical:

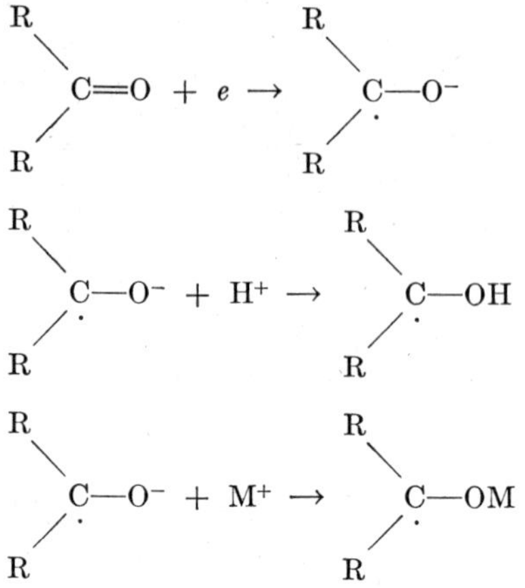

This behavior is in agreement with the dimerization tendency of metal ketyls which decreases from lithium to caesium.

Substitution of hydroxyl groups into the ring shifts the half-wave potentials of benzophenone to more negative values. Half-wave potentials for a number of substituted benzophenones in an aqueous buffer of pH 7 at 60° are given below.[94]

Compound	$-E_{1/2}$ (N.C.E.)
Benzophenone	1.24
4,4′-Dihydroxybenzophenone	1.47
3,3′-Dimethyl-4,4′-dihydroxybenzophenone	1.48
3,3′-Dimethoxy-4,4′-dihydroxybenzophenone	1.46

Acetophenone. Half-wave potentials obtained for acetophenone are listed below.[92]

$-E_{1/2}$ (N.C.E.)	i_d/C	pH
1.12	2.2	1.3
1.33	1.7	4.9
1.58	3.7	7.2
1.62	3.6	8.6
1.64	3.1	11.3

[94] A. G. Stromberg and L. M. Reinus, *J. Gen. Chem. U. S. S. R.*, **16**, 1431 (1946).

The wave in acid medium represents the reduction of acetophenone to the pinacol while the wave in alkaline medium is for the reduction to phenylmethylcarbinol.

The effect of substituents upon the half-wave potential varies and depends upon the group and the position involved. Substitution in the side chain has very little effect. Benzylacetophenone has a half-wave potential quite similar to that of acetophenone.[92]

$-E_{1/2}$ (N.C.E.)	i_d/C	pH
1.09	2.0	1.3
1.52	3.4	7.2
1.64	3.5	11.3

The "reduction" potentials (N.C.E.) for phenyl alkyl ketones vary only slightly in 0.1 *N* lithium hydroxide–50 per cent ethanol solutions[95] (Table XXXIX-12).

TABLE XXXIX-12

REDUCTION POTENTIALS (N.C.E.) OF PHENYL ALKYL KETONES

Compound	Reduction potential
$C_6H_5COCH_3$	−1.63
$C_6H_5COCH_2CH_3$	−1.65
$C_6H_5COCH_2CH_2CH_3$	−1.64
$C_6H_5COCH(CH_3)_2$	−1.70
$C_6H_5COCH_2CH_2CH_2CH_3$	−1.66

Substituents in the ring have a greater effect on the half-wave potential. Results obtained for various methoxy and hydroxy acetophenones in 0.1 *M* ammonium chloride–50 per cent ethanol are listed in Table XXXIX-13.[96] A hydroxyl group in the para position has a greater effect than one in the ortho position in shifting the half-wave potential to more negative values. Other ketones studied and the values obtained are listed in Table XXXIX-14.

Reduction potentials of −1.43, −1.60 are reported for 2,4-diacetoresorcinol and −1.41, −1.63 for 4,6-diacetoresorcinol.[97]

Benzanthrone gives a wave in 70 per cent methanol containing 0.1 *N*

[95] W. P. Davies and D. P. Evans, *J. Chem. Soc.*, **1939,** 546.

[96] N. A. Valyashko and Y. S. Rozun, *Zhur. Obschcheĭ Khim.*, **18,** 710 (1948).

[97] G. Scaramelli, *Boll. sci. facoltà chim. ind. Bologna*, **2,** 122 (1941); *Chem. Zentr.*, *II*, **1942,** 269.

H_2SO_4 with a half-wave potential of −0.96 v. and can be analyzed for the presence of anthraquinone since the latter is reduced at −0.36 v.[98]

The polarographic method has been used in studies of the rate of tautomerization of an optically active azomethine by determining the reduction waves of the ketones formed upon hydrolysis of mixtures of isomeric azomethines:[99]

$$RC_6H_4C(C_6H_5){=}N{-}CH(C_6H_5)CH_3 \rightleftharpoons RC_6H_4CH(C_6H_5){-}N{=}C(C_6H_5){-}CH_3$$

TABLE XXXIX-13

HALF-WAVE POTENTIALS OF SUBSTITUTED ACETOPHENONES

Compound	$-E_{1/2}$ (S.C.E.)
$C_6H_5COCH_3$	1.34
2-$HOC_6H_4COCH_3$	1.36
4-$HOC_6H_4COCH_3$	1.45
2-$CH_3OC_6H_4COCH_3$	1.37
4-$CH_3OC_6H_4COCH_3$	1.43
2,4-$(HO)_2C_6H_3COCH_3$	1.49
2,4-$(CH_3O)_2C_6H_3COCH_3$	1.48
2,4-$HO(CH_3O)C_6H_3COCH_3$	1.44
2,4-$CH_3O(HO)C_6H_3COCH_3$	1.51

(c) Unsaturated Aromatic Ketones

Unsaturated ketones in the aromatic series show behavior at the dropping mercury electrode which is similar to that observed with aliphatic unsaturated ketones. The wave heights and half-wave potentials for a number of these are given in Table XXXIX-15.

In an acid medium reduction proceeds by a 1,4-bimolecular mechanism to give the dimeric ketone formed:

$$2RCH{=}CHCOR \xrightarrow[2H^+]{2e} \begin{array}{c} RCHCH_2COR \\ | \\ RCHCH_2COR \end{array}$$

In neutral or alkaline media bimolecular reduction is accompanied by the formation of the saturated ketone in varying amounts:

$$RCH{=}CHCOR + 2e + 2H^+ \rightarrow RCH_2CH_2COR$$

The greater the alkalinity the greater the formation of the saturated ketone.

[98] Y. I. Vainshstein, *Zadvodskaya Lab.*, **15**, 411 (1949); *Chem. Abstracts*, **43**, 6948 (1949).

[99] G. T. Borcherdt and H. Adkins, *J. Am. Chem. Soc.*, **60**, 3 (1938).

Table XXXIX-14

Polarographic Data for Aromatic Ketones

Compound	Reduction potential (*1*)	Half-wave potential (*2–4*)
$C_6H_5COCH_3$	−1.44[a] (N.C.E.)	−1.51[a] (N.C.E.)
	−1.55[b] (N.C.E.)	−1.52[c]
p-$CH_3C_6H_4COCH_3$	−1.48[a] (N.C.E.)	−1.66[d]
	−1.63[b] (N.C.E.)	
p-$ClC_6H_4COCH_3$	−1.39[a] (N.C.E.)	
	−1.52[b] (N.C.E.)	
α-$C_{10}H_7COCH_3$	−1.35[a] (N.C.E.)	
	−1.51[b] (N.C.E.)	
β-$C_{10}H_7COCH_3$		−1.65[d]
m-$CH_3C_6H_4COCH_3$		−1.57[a]
o-$HOC_6H_4COCH_3$	−1.46 (*4*)	−1.56[a] (N.C.E.)
		−1.64[c]
p-$BrC_6H_4COCH_3$		−1.43[a] (N.C.E.)
		−1.52[c]
m-$HOC_6H_4COCH_3$	−1.42 (*4*)	−1.57[d]
p-$HOC_6H_4COCH_3$	−1.54 (*4*)	
o-$CH_3OC_6H_4COCH_3$		−1.63
m-$CH_3OC_6H_4COCH_3$		−1.65[d]
p-$CH_3OC_6H_4COCH_3$		−1.70[d]
n-$C_3H_7COC_6H_5$		−1.50[a] (N.C.E.)
		−1.60[c]
i-$C_3H_7COC_6H_5$		−1.53[a] (N.C.E.)
		−1.61[c]
$C_6H_4COCH_2CH_2Br$		−1.50[a] (N.C.E.)
		−1.60
3,5-$(CH_3O)_2C_6H_3COC_4H_9$		−1.56[a]
$C_6H_5CH_2CH_2COC_6H_5$		−1.84[c]
$C_6H_5CH_2COC_6H_5$		−1.42[a] (N.C.E.)
		−1.48[c]
$C_6H_5COCH_2OCH_3$		−1.43[d]
p-$ClC_6H_4COC_6H_5$		−1.16[a] (N.C.E.)
		−1.35[c]
1-$C_{10}H_7COC_6H_5$		−1.46[a]
2-$C_{10}H_7COC_6H_5$	−1.12 (*4*)	
α-Hydrindone		−1.55[d]
α-Tetralone		−1.67[d]
Fluorenone		−1.07[a]
2-Fluorenyl methyl ketone		−1.51[a]
2-Chlorofluorenone		−1.06[a]
2,7-Dichlorofluorenone		−0.84[a]
		−1.07[d]

[a] 0.1 N NH_4Cl. [b] 0.1 N LiCl. [c] 0.2 N $(CH_3)_4NOH$. [d] 0.05 N $(CH_3)_4NOH$.

[1] A. Winkel and G. Proske, *Ber.*, **69**, 693 (1936).

[2] H. Adkins and F. W. Cox, *J. Am. Chem. Soc.*, **60**, 1151 (1938).

[3] R. B. Baker and J. G. Schafer, *J. Am. Chem. Soc.*, **65**, 1675 (1943).

[4] G. Scaramelli, *Atti accad. nazl. Lincei Classe sci. fis. mat. e nat.*, [7], **1**, 768 (1940); *Chem. Abstracts*, **37**, 1408 (1943). H. Adkins, R. M. Elofson, A. G. Rossow, and C. C. Robinson, *J. Am. Chem. Soc.*, **71**, 3622 (1949). R. Manzoni-Ansidei and G. Scaramelli, *Boll. sci. facoltà chim. ind. Bologna*, **5**, 1 (1944–47); *Chem. Abstracts*, **44**, 1941 (1950).

This behavior has been verified by controlled potential electrolysis of benzalacetone and benzalacetophenone at pH 1.3 and 8.6. At pH 1.3 only the bimolecular ketone is obtained while at pH 8.6 a mixture of the bimolecular compound and the saturated ketone is formed. If the resulting saturated ketone has the ketone group conjugated with an aromatic ring, a third reduction wave appears for the reduction of this grouping. Thus benzalacetophenone at pH 11.3 gives waves at −1.19 v. (N.C.E.) and −1.68 v. (N.C.E.). The product of the first wave contains some benzylacetophenone which in the same buffer gives a wave at −1.64 v.

TABLE XXXIX-15
HALF-WAVE POTENTIALS (N.C.E.) AND WAVE HEIGHTS OF AROMATIC UNSATURATED KETONES

Compound	pH 1.3		pH 4.9		pH 7.2		pH 8.6		pH 11.3	
	$-E_{1/2}$ (N.C.E.)	i_d/C	$-E_{1/2}$	i_d/C	$-E_{1/2}$	i_d/C	$-E_{1/2}$	i_d/C	$-E_{1/2}$	i_d/C
Benzalacetone......	0.76	2.4	1.00	2.3	1.15	4.2	1.31	4.0	1.36	3.9
	—	—	1.32	2.2	1.33	—	—		—	—
Dibenzalacetone....	0.4(2)	1.6	0.6(5)	1.7	0.77	1.7	0.92	1.9	0.97	1.8
	0.9(7)	2.3	1.1(1)	1.9	1.13	2.1	1.22	3.3	1.31	4.5
Benzalacetophenone.	0.57	1.9	0.78	2.1	0.93	2.0	1.12	3.9	1.19	3.9
	1.00	2.6	1.11	2.3	1.13	1.9	1.64	2.6	1.68	3.1
Ethylideneacetophenone ($C_6H_5COCH{=}CHCH_3$).........	0.71	2.0	0.97	1.9	1.18	3.4	1.25	3.9	1.28	3.5
	—	—	—	—	1.38		1.40		1.39	
	—	—	—	—	1.62	0.5	1.61	1.2	1.62	0.8

Dibenzalacetone reduces only partially at pH 1.3, as indicated below,

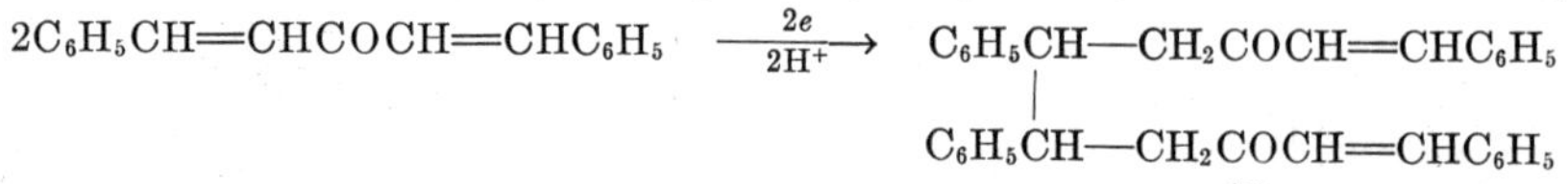

(I)

and is reported to give a mixture of this compound (I) and dibenzylacetone at pH 4.9.[100]

Substitution of hydroxy groups in benzalacetophenone (or chalcone) affects the amounts of bimolecular reduction and reduction to the saturated ketone by conferring varying degrees of stability upon the intermediate free radical:

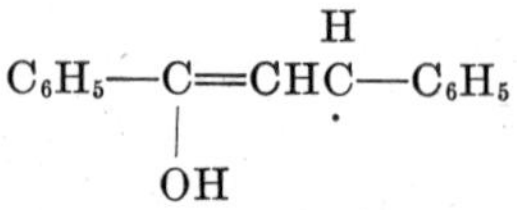

The half-wave potentials (S.C.E.) and wave heights obtained for a number

[100] R. Pasternak, *Helv. Chim. Acta*, **31**, 753 (1948).

TABLE XXXIX-16
HALF-WAVE POTENTIALS (S.C.E.) AND WAVE HEIGHTS OF CHALCONES (BENZALACETOPHENONES)

Chalcone	pH 6.1		pH 7.5		pH 8.6		pH 9.6	
	$-E_{1/2}$	i_d/C	$-E_{1/2}$	i_d/C	$-E_{1/2}$	i_d/C	$-E_{1/2}$	i_d/C
Unsubstituted	0.89	1.71	0.93	1.38	1.08	3.58	1.10	3.55
	1.12	1.82	1.12	1.60	1.62	2.71	1.63	2.62
	1.44	0.68	1.60^b	0.53	—	—	—	—
2′-Hydroxy	0.93	2.10	0.98	1.87	1.01	1.70	1.02	1.40
	1.12	1.77	1.20	1.72	1.25	1.49	1.25	1.34
	1.47	2.62	1.58	1.98	1.64	0.73	1.63	0.80
2′,4-Dihydroxy	1.04	1.43	1.11	1.40	1.16	1.27	1.16	1.22
	1.19	1.31	1.24	1.43	1.35	1.34	1.35	1.35
	1.44	2.07	1.55	2.25	1.64	0.76	1.64	0.85
2′,3,4-Trihydroxy	—	—	1.18^c	2.56	—	—	—	—
	—	—	1.55	1.75	—	—	—	—
2′,4′,4-Trihydroxy	1.10	0.75	1.13	1.40	1.24	1.02	1.27	0.83
	1.25	1.59	1.28	1.42	1.43	1.15	1.44	1.32
	1.55	1.67	1.66	1.80	1.90	0.96	1.92	0.92
2′,4′,3-Trihydroxy	—	—	1.16^a	2.15	—	—	—	—
	—	—	1.69	1.89	—	—	—	—
2′,4′,3,4-Tetrahydroxy	1.15^a	2.54	1.25^a	2.37	1.25	1.25	—	—
	1.57	1.94	1.69	1.82	1.40	1.40	—	—
	—	—	—	—	1.85	1.87	—	—
Hydroxy	—	—	1.13^{a,c}	2.70	—	—	—	—
	—	—	1.79	—	—	—	—	—
4′,4-Dihydroxy	—	—	1.10	1.35	—	—	—	—
	—	—	1.24	1.32	—	—	—	—
	—	—	1.72	0.65	—	—	—	—
4′-Methoxy	—	—	1.01	1.35	—	—	—	—
	—	—	1.21	1.32	—	—	—	—
	—	—	1.68	0.65	—	—	—	—
4′,4-Dimethoxy	0.94	—	1.04	—	1.13	—	1.10	—
	1.18	1.22	1.22	—	1.26	—	1.23	—
	1.45	—	1.62	—	1.73	—	1.73	—
2′-Hydroxy-4,4′,6′-trimethoxy	1.13^a	—	1.19^a	—	1.19	—	1.20	—
	—	—	—	—	1.34	—	1.34	—
	1.58	—	1.72	—	1.85	—	1.85	—
2′,4′,6′,4-Tetramethoxy	1.03^a	0.85	1.09^a	0.88	1.28	1.71	1.28	1.32
	1.24	0.88	1.24	1.17	—	—	—	—
2′,3′,4′-Trihydroxy	—	—	1.15^{a,c}	3.55	—	—	—	—
	—	—	1.77	1.15	—	—	—	—
2′-Hydroxy-6′-methoxy 3,4-Methylene dioxy	0.93	—	—	—	—	—	—	—
	1.05	—	—	—	—	—	—	—
	1.47	—	—	—	—	—	—	—
2′-Hydroxy-4′,6′,3,4-tetramethoxy	1.03^a	—	—	—	—	—	—	—
	1.13	—	—	—	—	—	—	—

^{a} Single (fused) wave. ^{b} Small symmetrical wave. ^{c} Slight maximum in wave.

of these in 50 per cent isopropyl alcohol are listed in Table XXXIX-16.[101]

Cinnamalacetophenone, $C_6H_5COCH{=}CHCH{=}CHC_6H_5$ gives waves at −0.93 and −1.11 v. in 0.2 *N* tetramethylammonium hydroxide.[102]

[101] T. A. Geissman and S. L. Friess, *J. Am. Chem. Soc.*, **71**, 3893 (1949).
[102] H. Adkins and F. W. Cox, *J. Am. Chem. Soc.*, **60**, 1151 (1938).

Fuchsones behave like unsaturated ketones and give two waves. Values obtained for a number of these in a buffer of pH 7 containing 30 per cent ethanol at 60° are listed below.[103]

Compound and formula		$-E_{1/2}$ (N.C.E.)
Aurin	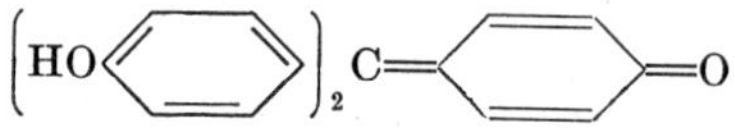	0.80, 1.24
o-Cresaurin	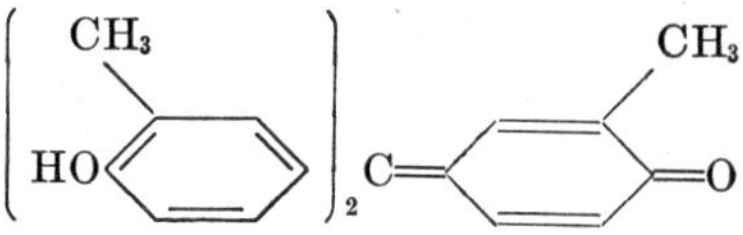	0.82, 1.24
Rubrophen	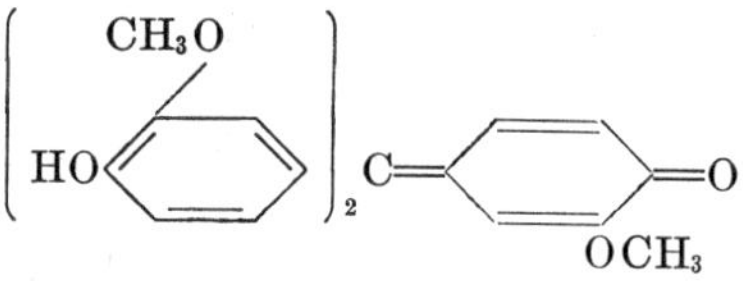	0.82, 1.24
Hexamethoxyaurin	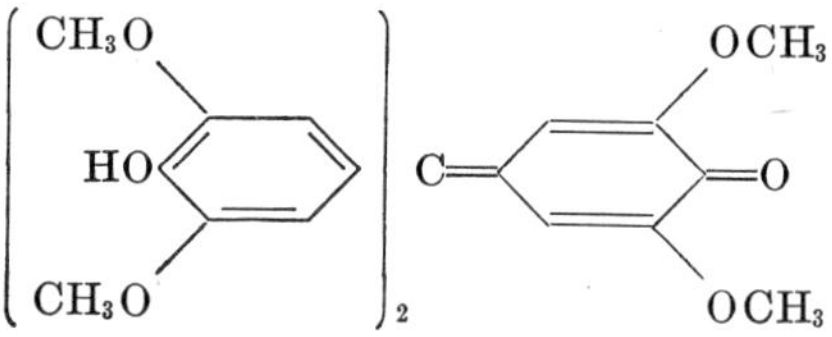	0.82, 1.26
Aurintricarboxylic acid	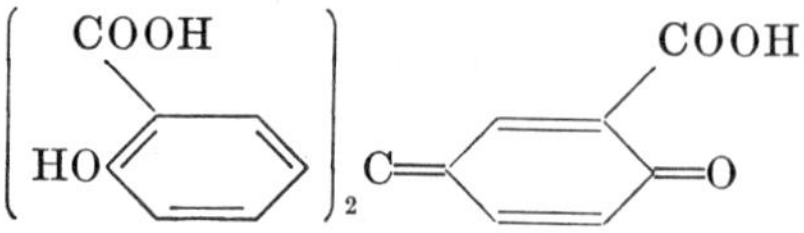	0.81, 1.27
Triiodoaurin	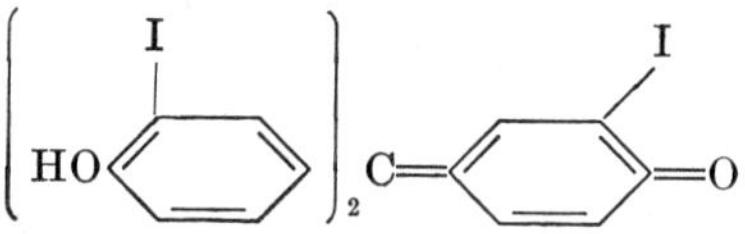	0.81, 1.49

Benzalacetone has an unusual effect on the height of the waves of acetophenone in a 50 per cent isopropyl alcohol solution of 0.2 *N* tetramethylammonium hydroxide when a mixture of the two is studied. The first acetophenone wave is decreased while the second wave is increased; the sum of the two waves, however, remains constant. Normal behavior occurs in 0.1 *N* ammonium chloride, 0.1 *N* lithium chloride, and 0.01 *N* tetramethylammonium hydroxide–0.1 *N* tetramethylammonium bromide solutions. The reason for the effect is not clear. Chemical interaction

[103] A. G. Stromberg and L. M. Reinus, *J. Gen. Chem. U. S. S. R.*, **16**, 1431 (1946).

was eliminated as a possible explanation. No other analogs of benzalacetone were studied to see whether the effect was general or not.[104]

The polarographic method has been used to follow the formation of 2,2′-dihydroxy-5-bromochalcone from *o*-hydroxyacetophenone and *m*-bromobenzaldehyde.[105] Values reported for various chalcones by this investigator are given in Table XXXIX-17 (p. 692).

The polarographic method has been used to demonstrate the presence of a double bond in the reduction product from the cyclic tetrabenzoylethylene isomer.[106]

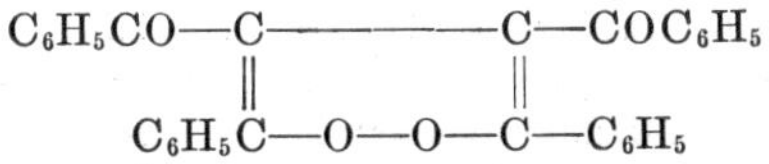

(d) Aromatic Diketones

The behavior of aromatic diketones depends to a great extent upon the relationship of the ketone groups to each other.

Benzil. Benzil, an α-diketone, gives a wave involving two electrons in various buffers.[107] The results are given below.

$-E_{1/2}$ (N.C.E.)	i_d/C	pH
0.31	4.0	1.3
0.54	3.9	4.9
0.68	3.9	7.2
0.76	3.9	8.6
0.79	3.9	11.3

The wave represents a 1,4-reduction to benzoin:

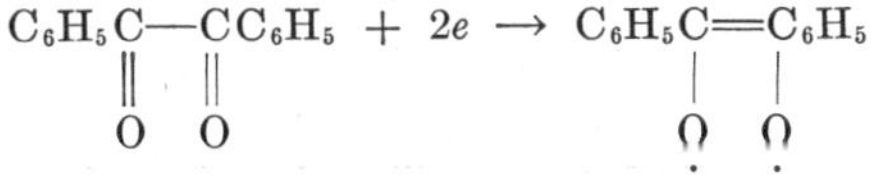

This mechanism has been confirmed by controlled potential electrolysis at pH 1.3 and 8.6; a quantitative yield of benzoin is obtained. The rearrangement to benzoin is apparently very slow at the dropping mercury electrode since no wave for benzoin is observed.[107]

In alkaline borate solution benzil forms a complex which is not reducible

[104] S. Wawzonek and H. A. Laitinen, *J. Am. Chem. Soc.*, **63**, 2342 (1941). F. Takazawa and L. Smith, *Proc. Iowa Acad. Sci.*, **51**, 301 (1944).

[105] E. Schraufstatter, *Experientia*, **4**, 192 (1948).

[106] H. Keller, E. Brandenberger, and H. von Halben, *Helv. Chim. Acta*, **29**, 1466 (1948).

[107] R. Pasternak, *Helv. Chim. Acta*, **31**, 753 (1948).

TABLE XXXIX-17

HALF-WAVE POTENTIALS OF SUBSTITUTED CHALCONES

Chalcone	$-E_{1/2}$ 0.4 N $(NH_4)_2SO_4$ 40% isopropyl alcohol		$-E_{1/2}$ 0.1 N NaOH 50% isopropyl alcohol	
Chalcone	1.18	1.48	1.18	1.50
2′-Hydroxychalcone	1.28	1.50	—	—
2-Hydroxychalcone	1.27	1.46	—	—
2,2′-Dihydroxychalcone	1.36	1.50	—	—
2,2′-Dihydroxy-5-bromochalcone	1.27	1.46	1.26	1.50
Furfurylidene-5-hydroxyacetophenone	1.32	1.60	—	—
Cinnamylidene-2-hydroxyacetophenone	—	—	1.07	1.45
2′-Hydroxy-5′-bromobenzylidene-2-hydroxy-1-acetonaphthone	1.30	1.90	—	—

at the dropping mercury electrode and which consists of one mole of benzil and two of borate ion. The equilibrium constant of the complex formation reaction was determined using polarographic measurements.[108]

Dibenzoylmethane. Dibenzoylmethane, a typical β-diketone, shows the following behavior in buffered solutions.[107]

pH	$-E_{1/2}$ (N.C.E.)	i_d/C
1.3	0.63	1.6
4.9	0.85	1.7
	1.25	2.3
7.2	1.10	3.5
	1.29	
	1.61	1.2
8.6	1.28	3.7
	1.62	2.5
11.3	1.34	3.9
	1.66	2.6

The actual course of the reduction is not known but the relatively positive half-wave potential suggests reduction of the enol form to a pinacol (I) in acid solution and to the keto alcohol (II) in more alkaline solution:

(I) 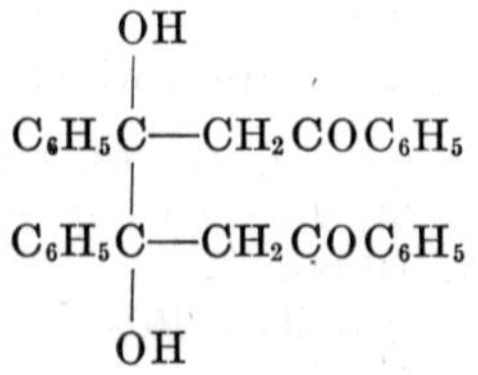

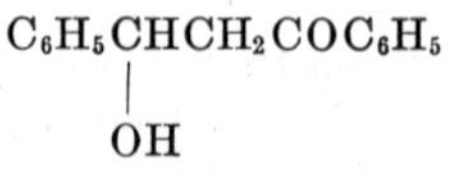 (II)

[108] R. Pasternak, *Helv. Chim. Acta*, **30**, 1984 (1947).

The third wave at −1.61 is similar to that for acetophenone and must be for the reduction of the above compounds. The product of controlled potential electrolysis at pH 4.9 was not identified.[107]

The polarographic method has been used to follow the tautomeric shift of the keto form of dibenzoylmethane to the enol form.[109]

TABLE XXXIX-18

HALF-WAVE POTENTIALS (N.C.E.) OF BENZOYLACETONE AS FUNCTION OF pH

pH	1	2	3	4	5	6	7	8	9	10	11	12
$-E_{1/2}$	1.0	1.08	1.16	1.21	1.25	1.28	1.34	1.40	1.47	1.54	1.62	1.63

TABLE XXXIX-19

HALF-WAVE POTENTIALS OF AROMATIC DIKETONES

Substance	Formula	$-E_{1/2}$ 0.1 *N* NH_4Cl	$-E_{1/2}$ 0.2 *N* $(CH_3)_4NOH$
Acetylbenzoyl-ethyl-methane	$CH_3COCHCOC_6H_5$ | C_2H_5		1.52 1.79
Acetylbenzoyl-*n*-butylmethane	$CH_3COCHCOC_6H_5$ | C_4H_9		1.55 1.81
o-Hydroxybenzoyl-benzoyl-methane	*o*-$HOC_6H_4COCH_2COC_6H_5$	1.37 (N.C.E.)	1.51 1.79
Benzoylfuroyl-methane	$C_4H_3OCOCH_2COC_6H_5$		1.53 1.88
Neovalerylbenzoyl-methane	$C_6H_5COCH_2COC(CH_3)_3$		1.89
Acetylfuroyl-methane	$CH_3COCH_2COC_4H_3O$		1.79

The half-wave potentials (N.C.E.) of benzoylacetone at various pH values can be estimated from a graph given by Tachi and are listed in Table XXXIX-18.[110] A summary of the data on other aromatic diketones is given in Table XXXIX-19.[111]

The triketone, diphenyltriketone $C_6H_5COCOCOC_6H_5$ (0.001 *M*), gives a reduction potential of −0.25 v. (N.C.E.) at a pH of 2.4.[110]

Dibenzoylethylene. Dibenzoylethylene, an unsaturated diketone, differs in its behavior from α,β-unsaturated ketones in that reduction occurs

[109] R. Pasternak and H. von Halban, *Helv. Chim. Acta,* **29,** 190 (1946).
[110] I. Tachi, *Mem. Coll. Agr. Kyoto Imp. Univ.,* **42,** 27 (1938).
[110a] H. Keller, R. Pasternak, and H. von Halban, *Helv. Chim. Acta,* **29,** 512 (1946).
[111] H. Adkins and F. W. Cox, *J. Am. Chem. Soc.,* **60,** 1151 (1938).

to the saturated ketone. Half-wave potentials (N.C.E.) and wave heights for both isomers are given below.

Substance	pH	$-E_{1/2}$	i_d/C	pH	$-E_{1/2}$	i_d/C	pH	$-E_{1/2}$	i_d/C
Dibenzoylethylene									
(*trans*)................	1.3	0.16	3.8	4.9	0.36	3.9	7.2	0.48	3.9
								1.5	1.2
(*cis*)..................	1.3	0.34	3.9	4.9	0.45	3.9	7.2	0.51	3.7
								1.58	0.5

Substance	pH	$-E_{1/2}$	i_d/C	pH	$-E_{1/2}$	i_d/C
Dibenzoylethylene						
(*trans*)................	8.6	0.57	3.8	11.3	0.61	1.4
		1.57	1.9		1.57	(3.5)
(*cis*)..................	8.6	0.57	3.9	11.3	0.66	3.6
		1.60	0.8		1.69	1.9

The first wave represents, no doubt, reduction of both isomers to *sym.*-dibenzoylethane. This has been demonstrated by controlled potential electrolysis at pH 4.9 and formation of dibenzoylethane in a quantitative yield. The second wave observed is approximately at the same point as the wave for dibenzoylethane but the diffusion currents are much lower.

Thirty-seven copper derivatives of salicylaldehydes and their nitrogen derivatives, and of several β-diketones have been studied in 50 per cent pyridine solution containing 0.1 *M* potassium nitrate, to determine the effect of the structure of the chelating group upon the relative stability of the copper chelates.[112]

(e) Hydroxy Aromatic Ketones

Benzoin. Benzoin shows the following behavior in buffered solutions:[107]

$-E_{1/2}$ (N.C.E.)	i_d/C	pH
−0.94	4.0	1.3
−1.25	4.0	4.9
−1.40	3.9	7.2
−1.49	3.9	8.6
−1.55	3.9	11.3

The wave corresponds to two electrons and the formation of hydrobenzoin. This mechanism has been verified at pH 8.6 by production of hydrobenzoin under controlled electrolysis. At pH 1.3, due to effect of acid, hydro-

[112] M. Calvin and R. H. Bailes, *J. Am. Chem. Soc.*, **68**, 949 (1946).

benzoin is dehydrated to desoxybenzoin which, on further reduction, produces the following pinacol:

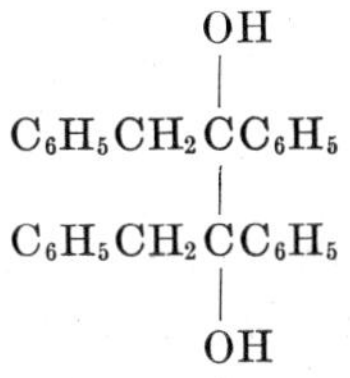

Similar values for benzoin are reported by Santavy in buffered solutions and are listed below for 50 per cent ethanol:[113]

pH	1.81	6.80	11.98
$-E_{1/2}$ (S.C.E.)	0.96	1.42	1.55

Colchicine and Related Compounds. Colchiceine (I) and related 7-membered-ring hydroxy ketones (II–IV) show behavior which is different from that of benzoin at the dropping mercury electrode:

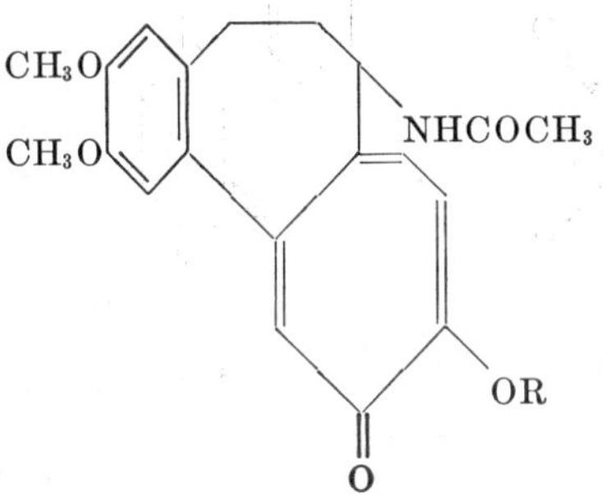

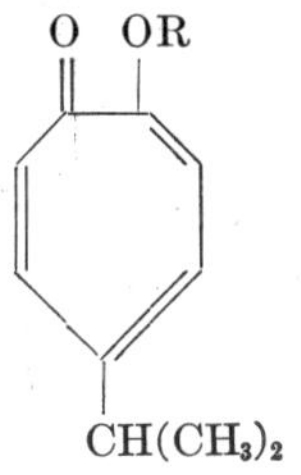

(I) R = H
(II) R = CH_3

(III) R = H
(IV) R = CH_3

Values obtained for these compounds are listed below:

Compound	Solvent	$-E_{1/2}$ (S.C.E.) at pH 1.81	6.80	11.98	0.1 *N* LiOH
Colchiceine (I)	50% alcohol	0.96	1.42	1.78	1.80
Colchicine (II)	H_2O	0.96	1.40	1.42	1.43
γ-Thujaplicine (III)	50% alcohol	0.93	1.22	—	2.10
Methoxy-γ-thujaplicine (IV)	50% alcohol	0.96	1.35	1.55	1.55

[113] F. Santavy, *Collection Czechoslov. Chem. Communs.*, **14**, 145 (1949).

The behavior of colchiceine (I) is very similar to that of salicylaldehyde, *p*-hydroxybenzaldehyde, and γ-thujaplicine, while that of colchicine (II) is similar to that of the corresponding methylated derivatives. This similarity has been offered as evidence for the presence of a 7-membered ring in colchiceine. The other possibility, which is a 6-membered ring containing a hydroxymethylene keto grouping, is discussed later.

Evidence against a hydroxy aldehyde structure for colchiceine (I) is the formation of a complex with boric acid in neutral solutions which changes the polarographic behavior. A similar complex is formed with γ-thujaplicine (III) but not with salicylaldehyde.[114]

Hydroxymethylene Ketones. Hydroxymethylene ketone groups attached to an aromatic ring as in 2-hydroxymethylene-1-ketotetrahydrophenanthrene (V) show a different behavior from that of colchicine, thujaplicine, or benzoin. This difference is offered as evidence against a hydroxymethylene ketone structure of colchiceine. Values obtained in buffers are listed below:[113]

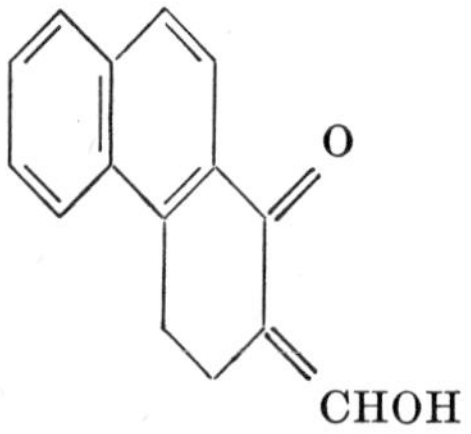

(V)

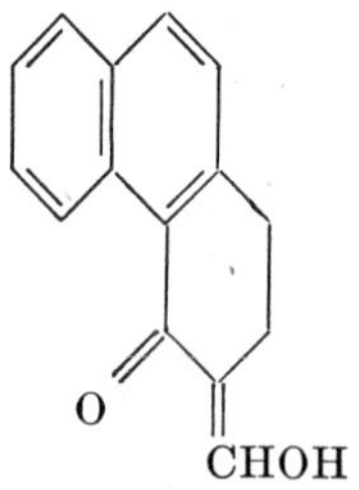

(VI)

Compound	Solvent	$-E_{1/2}$ (S.C.E.) at pH		
		1.81	6.80	11.98
3-Hydroxymethylene-4-ketotetrahydrophenanthrene (VI)	50% alcohol	0.96	1.35	1.75
2-Hydroxymethylene-1-ketotetrahydrophenanthrene (V)	50% alcohol	0.96	1.32	1.75

The reduction potentials for several hydroxymethylene derivatives of several ketones in the terpene series are listed on facing page:[115]

[114] R. Brdicka, *Arkiv. Kemi, Mineral. Geol.*, **B26,** No. 19, 7 pp. (1948); *Chem. Abstracts*, **43,** 3728 (1949).

[115] G. Scaramelli, *Boll. sci. facoltà chim. ind. Bologna*, **1941,** 129; *Chem. Zentr., II*, **1942,** 269.

Compound	Reduction potentials, v.	
3-Hydroxymethylenecamphor	−1.75	
2-Hydroxymethylenementhone	−1.57,	−1.73
	−1.28,	−1.59
2-Hydroxymethylenethujone	−1.59	

The polarographic method has been used to follow the action of sodium methoxide upon colchicine and to prove that colchiceine and colchicic acid are formed.[116]

(f) Halogenated Carbonyl Compounds

Halogenated ketones are reducible in both the aliphatic and aromatic series. Results obtained with derivatives of acetone are given in Table XXXIX-20.[117] It is possible to determine the three monoderivatives, iodo-, bromo-, and chloroacetone, simultaneously.[117] From the analytical viewpoint it should be remembered that the dissolved iodo- and bromoacetones hydrolyze on standing and that the solutions should be electrolyzed soon after their preparation.

TABLE XXXIX-20

REDUCTION POTENTIALS (N.C.E.) OF HALOGEN DERIVATIVES OF ACETONE

Substance	(a)	(b)
Monochloroacetone	−1.13	−1.18
Monobromoacetone	−0.30	−0.29
Monoiodoacetone	−0.14	−0.14
Dichloroacetone (sym)	−0.83	0.00

(a) In 0.1 *N* ammonium chloride. (b) In 0.1 *N* lithium chloride.

From the polarogram[117] given for dichloroacetone a second wave can be seen to start at −1.16 v. This second wave is much smaller than the first and may be due to the reduction of the monochloroacetone produced. It is also remarkable that the first wave of dichloroacetone is about five times smaller than that of the monohalogen derivatives.

Chloral hydrate is reported to give a wave at −0.8 v. in 0.1 *N* potassium chloride solution.[118]

---

[116] F. Santavy, *Helv. Chim. Acta*, **31**, 821 (1948).

[117] A. Winkel and G. Proske, *Ber.*, **69**, 693 (1936).

[118] M. B. Neiman, A. V. Ryabov, and E. M. Sheyanova, *Doklady Akad. Nauk. S. S. S. R.*, **68**, 1065 (1949); *Chem. Abstracts*, **44**, 1360 (1950).

From experiments carried out by Pasternak[119] it is quite evident that the waves are not due to reduction of the carbonyl group, but to reduction of the halogen to halide ion:

$$CH_3COCH_2X + 2H^+ + 2e \rightarrow CH_3COCH_3 + X^- + H^+$$

The mechanism of reduction can be demonstrated more clearly with ω-bromoacetophenone. This compound gives two waves, the second of which is at the same point as acetophenone.[120]

The iodine in iodoaurin is removed at $E_{1/2} = -1.40$ (N.C.E.).[121]

[119] R. Pasternak, *Spektrographische und polarographische Untersuchungen von Ketonen und anderen ungestattigten Verbindungen*, Thesis, University of Zurich, 1946.

[120] R. Pasternak and H. von Halban, *Helv. Chim. Acta*, **29**, 190 (1946).

[121] A. G. Stromberg and L. M. Reinus, *J. Gen. Chem. U. S. S. R.*, **16**, 1431 (1946).

CHAPTER XL

Quinones

The reduction of quinones and the oxidation of hydroquinones are strictly reversible at the dropping mercury electrode. The half-wave potentials, after correction for the iR drop, correspond to the quinhydrone potentials and are independent of the concentration of hydroquinone or quinone. The diffusion currents are proportional to the concentrations of the hydroquinone or quinone,[1] and are slightly larger for the quinone than for the corresponding hydroquinone at the same concentration. The relation between $\log[i/(i_d - i)]$ and the potential corresponds to the theoretical one.[2]

Benzoquinones and Naphthoquinones. A great number of substituted benzoquinones has been studied polarographically. The half-wave potentials obtained are in agreement with standard oxidation-reduction potentials. The values obtained for a number of these in 50 per cent methanol (at pH = 0) are compared with the standard oxidation potentials in 50 per cent ethanol[3] in Table XL-1.

TABLE XL-1

COMPARISON OF NORMAL OXIDATION POTENTIALS (N.H.E.) OF QUINONES CALCULATED FROM HALF-WAVE POTENTIALS WITH VALUES OF CONANT AND FIESER

Compound	$E_{1/2}$ (polarographic)	E_0 (Conant and Fieser)
Quinone	0.711	0.711
Toluquinone	0.654	0.656
p-Xyloquinone	0.599	0.597
Duroquinone	0.471	0.466
2,3-Dimethylnaphthoquinone	0.349	0.340

A similar agreement is found for other compounds in 50 per cent ethanol[4] as shown on the following page.

[1] O. H. Müller and J. P. Baumberger, *Trans. Electrochem. Soc.*, **71,** 181 (1937).

[2] L. I. Smith, I. M. Kolthoff, S. Wawzonek, and P. M. Ruoff, *J. Am. Chem. Soc.*, **63,** 1018 (1941).

[3] J. B. Conant and L. F. Fieser, *J. Am. Chem. Soc.*, **45,** 2194 (1923).

[4] R. T. Arnold and H. E. Zaugg, *J. Am. Chem. Soc.*, **63,** 1317 (1941).

Compound	$E_{1/2}$ (N.H.E.)	E_0 (potentiometric)
o-Xyloquinone	0.589	0.588
4,7-Hydrindenequinone	0.638	0.641
1,2,3,4-Tetrahydronaphthoquinone	0.585	0.585

Half-wave potentials of some compounds in 75 per cent ethanol containing an acetate buffer of pH 6.24 are listed below.[5]

Compound	$E_{1/2}$ (S.C.E.)
Toluquinone	+0.05
Pseudo-cumoquinone	−0.07
Duroquinone	−0.14
6-Bromo-pseudocumoquinone	−0.03
5-Methoxytoluquinone	−0.06
3,5-Dimethoxy-1,4-toluquinone	−0.16
3,5-Dimethoxytoluhydroquinone	−0.16
2-Methyl-1,4-naphthoquinone	−0.17
3,6-Dihydroxy-5-methoxy-1,4-toluquinone (spinulosin)	−0.30
5-Methoxy-6-hydroxy-1,4-toluquinone (fumigatin)	−0.16

The mold metabolites citrinin and penicillic acid have half-wave potentials of −1.14 v. (S.C.E.) (citrinin) and −0.69 v. (penicillic acid) in an aqueous buffer of pH = 4.63. The value obtained for 2-methylnaphthoquinone is +0.43 v. (N.H.E.), as compared to the value of +0.408 obtained by the classical method.[6]

The following half-wave potentials (N.C.E.) are reported for substituted naphthoquinones at pH 4.17: 2-methyl- −0.184 v., 2,3-dimethyl- −0.232 v., and 2-ethyl-1,4-naphthoquinone −0.197 v. 2-Hydroxy-1,4-naphthoquinone gives a value of −0.182 v. at pH 3.30. At higher pH values these compounds give two waves and apparently form semiquinones as intermediates which dimerize (see p. 254).[7]

A list of other quinones studied and the half-wave potentials obtained are listed on the following page.

The polarographic method has been used to study photochemical changes in solutions of benzoquinone, naphthoquinone, and 2-methylnaphthoquinone. The half-wave potentials of the products were found to agree

[5] J. E. Page and F. A. Robinson, *J. Chem. Soc.*, **1943,** 133.

[6] M. Fieser and L. Fieser, *J. Am. Chem. Soc.*, **57,** 491 (1935).

[7] G. Sartori and C. Cattaneo, *Gazz. chim. ital.*, **71,** 713 (1941).

Quinone	$E_{1/2}$	Solvent	pH	Reference
Benzoquinone.............	+0.015 (N.C.E.)	H_2O	7	*1*
2-Methoxy-..............	−0.085 (N.C.E.)	H_2O	7	*1*
2,6-Dimethoxy-...........	−0.17 (N.C.E.)	H_2O	7	*1*
2,3,5-Trimethyl-6-(3′-hydroxy-3′-methylbutyl)-..	−0.90 (S.C.E.)	50% CH_3OH	5.4	*2*
2,3,5-Trimethyl-6-(3′-hydroxybutyl)-...........	−0.092 (S.C.E.)	50% CH_3OH	5.4	*2*
2,3,5-Trimethyl-6-(2′-hydroxypropyl)-..........	−0.080 (S.C.E.)	50% CH_3OH	5.4	*2*
2,3,5-Trimethyl-6-(2′-hydroxybutyl)-...........	−0.083 (S.C.E.)	50% CH_3OH	5.4	*2*
α-Tocopherolquinone......	−0.158 (S.C.E.)	75% C_2H_5OH	6.24	*3*
Vitamin K (synthetic)....	+0.328 (N.H.E.)		0	*4*
Phthiocol................	+0.256 (N.H.E.)		0	*4*

(1) A. G. Stromberg and I. M. Reinus, *J. Gen. Chem. U. S. S. R.*, **16,** 1431 (1946).
(2) L. I. Smith, I. M. Kolthoff, S. Wawzonek, and P. M. Ruoff, *J. Am. Chem. Soc.*, **63,** 1018 (1941).
(3) L. I. Smith, L. J. Spillane, and I. M. Kolthoff, *J. Am. Chem. Soc.*, **64,** 644 (1942).
(4) E. L. McCawley and C. Gurchot, *Univ. Calif. Pub. Pharmacol.*, **1,** 325 (1940); *Chem. Abstracts*, **35,** 1466 (1941).

with the values for hydroxybenzoquinone, 2-hydroxynaphthoquinone, and 2-methyl-3-hydroxynaphthoquinone (phthiocol).[8]

The influence of the size of the polymethylene bridge on the redox potential of the following system has been studied polarographically:

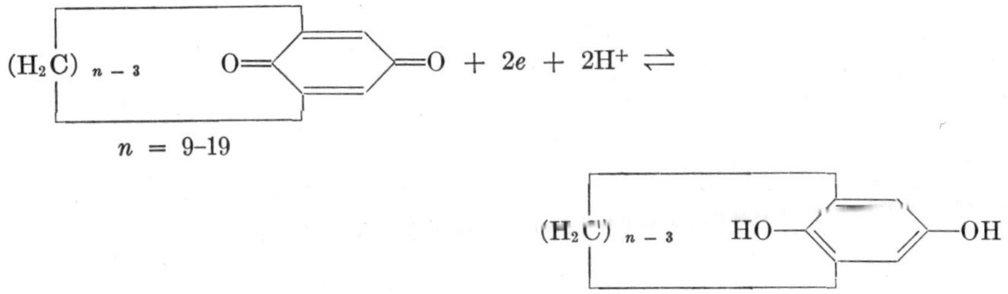

The compounds were studied in a solution containing 4 volumes of alcohol, one volume of water and one of acetic acid (pH = 2.56). The reversibility of the system was established by a study of the oxidation of the corresponding hydroquinones. Rings with n equal to and greater than 13 were found to be strain-free since half-wave potentials similar to that of 2,6-

[8] F. Poupe, *Collection Czechoslov. Chem. Communs.*, **12,** 225, (1947).

diethylbenzoquinone were found. Values for the oxidation potential of the corresponding 2,6-polymethylene-4-aminophenols were determined and found to be about 0.060 v. more positive. The half-wave potentials reported are listed below:[9]

n	Quinone $E_{1/2}$ (N.C.E.)	Hydroquinone $E_{1/2}$ (N.C.E.)	Aminophenol $E_{1/2}$ (N.C.E.)
9	−0.029	....	....
10	+0.026	....	+0.081
11	+0.086	....	....
12	+0.117	+0.111	+0.178
13	+0.139	+0.139	+0.200
14	+0.143	+0.136	+0.192
15	+0.160	+0.151	+0.211
16	+0.151	+0.147	+0.205
17	+0.158	....	+0.202
18	+0.144	....	+0.190
19	+0.144	....	+0.200
Benzoquinone	+0.265	....	....
m-Xyloquinone	+0.140	....	....
2,6-Diethylquinone	+0.140	....	....

Vitamin K, with a half-wave potential of −0.58 v., can be determined in 50 per cent isopropyl alcohol containing 0.05 *N* potassium chloride.[10]

Anthraquinones. A detailed study of a number of anthraquinones has been carried out by Furman and Stone[11] in various buffers. Many instances of specific buffer effects were found; *e.g.*, the slopes of waves of the same compound were often different at the same pH in buffers of different composition. In a number of cases the slopes of the waves indicated semiquinone formation and in some the semiquinone waves were separated. In other cases the slopes of the single waves were close to the theoretical for a process in which $n = 2$. Complex formation was observed in phosphate and borate buffers with various hydroxyanthraquinones. Sulfonic acid groups alpha to one or both of the oxygen atoms apparently interact (pseudo acid formation?) since the waves in acidic buffers are of peculiar long drawn-out form.

The quantitative estimation of the majority of the anthraquinones is possible over part of the pH range. In aqueous media, buffers of high pH (9–12) must be used to dissolve many of the compounds.

[9] V. Prelog, O. Hafliger, and K. Wiesner, *Helv. Chim. Acta*, **31,** 877 (1948).
[10] E. B. Hershberg, J. K. Wolfe, and L. F. Fieser, *J. Am. Chem. Soc.*, **62,** 3516 (1940).
[11] N. H. Furman and K. G. Stone, *J. Am. Chem. Soc.*, **70,** 3055 (1949).

Half-wave potentials (S.C.E.) for substituted anthraquinones in 1 per cent ethanol are listed below in a buffer of pH 7:[11]

Substituent	2-$NaSO_3$	2,6-$(NaSO_3)_2$	2,7-$(NaSO_3)_2$	1-NH_2-2-$NaSO_3$	3-NH_2-2-HSO_3	1-NH_2-4-Br-2-$NaSO_3$
$-E_{1/2}$ (S.C.E.)	0.45	0.45	0.44	0.59	0.52	0.56

Substituent	1-$NaSO_3$	1,5-$(NaSO_3)_2$	1-$NaSO_3$-5-NO_2	1-$NaSO_3$-8-NO_2	1,2-$(HO)_2$(alizarin)
$-E_{1/2}$	0.56	0.63	0.43, 0.70, 1.02	0.45, 0.64	0.61

Substituent	1,4-$(HO)_2$ (quinizarin)	1,5-$(HO)_2$ (anthrarufin)	1,2,4-$(HO)_3$ (purpurin)	1,2-$(HO)_2$-3-$NaSO_3$ (Alizarin Red S)
$-E_{1/2}$	0.78*	0.77*	0.42, 0.66	0.56

Substituent	1,8-$(HO)_2$-3-CH_2OH (aloe emodin)	1,5-$(HO)_2$-4,8-$(NH_2)_2$-2-$(NaSO_3)$	1,3,4,6-$(HO)_4$-2-$C_6H_{11}O_5$-8-CH_3
$-E_{1/2}$	0.77*	0.65, 1.16	0.68

* In ammonia-ammonium chloride buffer, pH 10.

Values reported for some typical anthraquinones in a 40 per cent dioxane solution of an acetate buffer of pH 7.4 are given below:[11]

Compound	$-E_{1/2}$ (S.C.E.)
Anthraquinone	0.54
1-NH_2-2-$NaSO_3$-	0.63
1,2,4-$(HO)_3$-	0.70

In connection with studies of the aloins, barbaloin, and isobarbaloin, a number of compounds related to anthraquinone have been studied.

Anthraquinone monoxime shows waves in the anthraquinone region. Anthrone gives poorly defined waves while 10-hydroxy-10-methylanthrone gives well-defined waves. 10-Vinylanthranyl acetate shows a reduction wave which is probably for the anthracene nucleus. The aloins show waves which resemble those of 10-hydroxy-10-methylanthrone. The close similarity among members in this series points to position isomers. Values obtained for these compounds in an acetate buffer of pH 7.4 in 40 per cent dioxane are listed below:[12]

Compound	$-E_{1/2}$ (S.C.E.)
Anthraquinone monoxime	0.64
Anthrone	0.85
10-Methyl-10-hydroxyanthrone	1.10

[12] K. G. Stone and N. H. Furman, *J. Am. Chem. Soc.*, **70,** 3062 (1948).

Values for barbaloin and isobarbaloin in a buffer of pH 7 in 1 per cent ethanol are −1.20 and −1.23 v. (S.C.E.), respectively.

The polarographic method has been used to estimate aloins and aloe emodin in Curacao aloes.[13] The medium consists of 8 per cent ethanol which is 0.1 *M* in an acetate buffer of pH 4. The half-wave potential is −1.34 v. If aloe emodin, which is a hydroxyanthraquinone, is present with the aloes it is more advantageously determined spectrophotometrically since less steps are involved in preparing the sample.

The reducibility of anthraquinones is made use of in the amperometric titration of diamidines.[14]

***o*-Quinones.** *o*-Quinones have not been studied polarographically as intensely as the *p*-quinones because of the difficulty involved in preparing them. Only the four compounds shown here have been studied thus far:

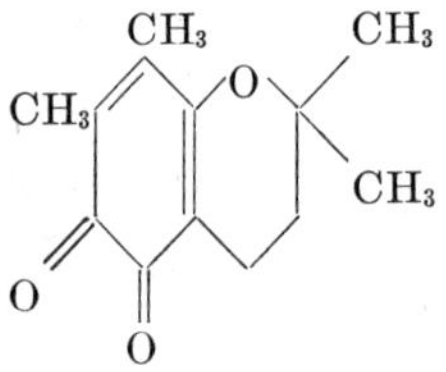

(I)

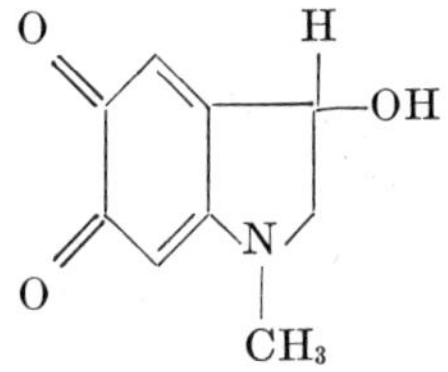

(II)

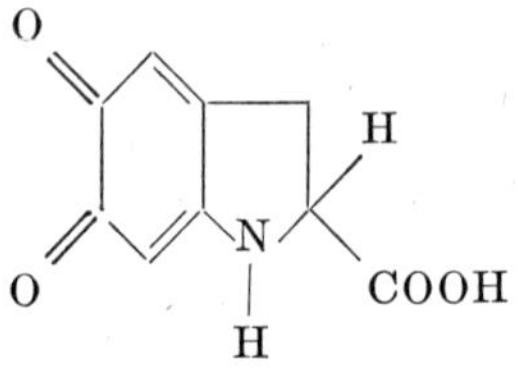

(III)

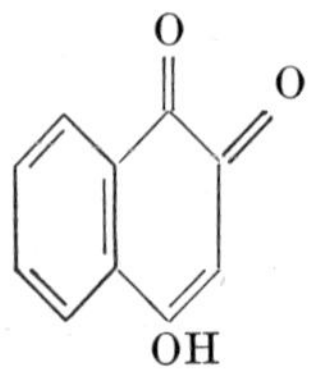

(IV)

2,2,7,8-Tetramethylchroman-5,6-quinone (I) gives a well-defined wave in 75 per cent ethanol containing an acetate buffer of pH 6.24 with $E_{1/2}$ at −0.078 v. (S.C.E.). $E_{1/2}$ shifts 0.059 v. with pH as is expected for a reversible reduction. The analogous compound from α-tocopherol could not be purified sufficiently to give conclusive results. An approximate value of −0.073 (S.C.E.) was reported.[15]

[13] K. G. Stone and N. H. Furman, *Anal. Chem.*, **19,** 105 (1947). K. G. Stone, *J. Am. Pharm. Assoc.*, **36,** 391 (1947).

[14] J. B. Conn, *Anal. Chem.*, **20,** 585 (1948).

[15] L. I. Smith, L. J. Spillane, and I. M. Kolthoff, *J. Am. Chem. Soc.*, **64,** 644 (1942).

Adrenochrome (II) and the *o*-quinone (III) obtained by enzymic oxidation of adrenaline and tyrosine, respectively, give well-defined reduction waves in acid and alkaline media. At a pH of 7, the half-wave potential for adrenochrome is −0.204 v. (S.C.E.) and becomes 58 millivolts more negative with each unit change of pH, as shown below.

pH	$-E_{1/2}$ (S.C.E.)	pH	$-E_{1/2}$ (S.C.E.)
4.53	0.082	6.47	0.181
5.30	0.119	6.64	0.191
5.91	0.149	6.81	0.200

The reduction of adrenochrome is polarographically reversible and involves two electrons. The reversibility was demonstrated by the observation of an anodic wave with the same half-wave potential as the cathodic wave on a sample of adrenochrome which had been reduced with hydrosulfite or hydrazine. The leuco form of adrenochrome is very unstable and disappears slowly. It apparently reacts with the reducing agents since it gives two new cathodic waves. The polarographic method is very suitable for the determination of the reduction potential of such unstable systems.[16]

The behavior of the *o*-quinone (III) from tyrosine approximates that observed for adrenochrome. A half-wave potential of −0.348 v. (S.C.E.) is obtained at a pH of 6.48.[17] 4-Hydroxy-1,2-naphthoquinone (IV) gives two waves below a pH of 7.00 and one in solutions above this value.[18]

o-Dihydroxybenzenes, the hydroquinones in this series, give anodic waves at the dropping mercury electrode. Catechol, pyrogallic acid, and adrenaline give anodic waves with half-wave potentials which are a linear function of pH. For catechol in the pH range from 3.7 to 9.74 the potential[19] is given by $E_{1/2} = -0.058 \text{ pH} + 57$; for pyrogallic acid the half-wave potential[20] follows the equation $E_{1/2} = -0.062 \text{ pH} + 0.50$. The half wave potential of the anodic wave of adrenaline[21] follows the equation $E_{1/2} = -0.058 \text{ pH} + 0.527$.

Resorcinol and phloroglucinol give no waves.

16 K. Wiesner, *Biochem. Z.*, **313,** 48 (1942).

17 K. Wiesner, *Biochem. Z.*, **314,** 214 (1943).

18 G. Sartori and C. Cattaneo, *Gazz. chim. ital.*, **71,** 713 (1941).

19 A. K. Vlcek, V. Mansfield, and D. Krkoskova, *Collegium*, **1943,** 231; *Chem. Abstracts*, **40,** 6279 (1946).

20 A. K. Vlcek and E. Spalek, *Tech. Hlidka Kozeluzska*, **19,** 165 (1943); *Chem. Abstracts*, **42,** 7073 (1948).

21 G. Sartori and C. Cattaneo, *Gazz. chim. ital.*, **72,** 525 (1942).

The polarographic method is suitable for the quantitative estimation of catechol and pyrogallic acid.

(a) Nitrogen Derivatives Related to Quinones

***p*-Aminophenols.** Aminophenols produce anodic waves at the dropping mercury electrode at more positive potentials than the corresponding hydroquinones. The oxidation yields the quinoneimine and represents a reversible reaction. Polarographically, for example, a value of +0.739 v. is found for E_0 of *p*-aminophenol, whereas potentiometrically a value of +0.733 v. is reported for *p*-quinoneimine.

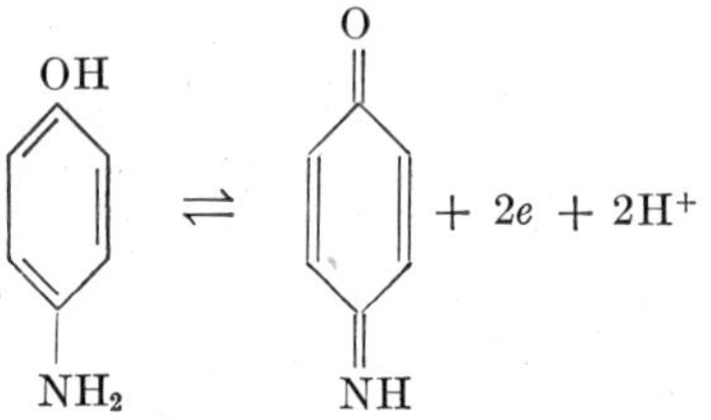

The study of this system is not possible in all buffers since the quinoneimine is hydrolyzed very easily to the quinone. The stability varies and depends upon the substituents present. 4-Amino-2-methyl-1-naphthol (vitamin K_5) yields an oxidation product at a potential of +0.510 v. which decomposes slowly enough so that the reaction can be followed polarographically.

m-Aminophenol gives no reproducible wave whereas *o*-aminophenol gives a drawn-out anodic wave in a borate buffer (pH = 9).

The polarographic method has been used to identify the products of oxidation of vitamin K_5 (4-amino-2-methyl-1-naphthol) by oxygen and by human red blood cells. In this connection the half-wave potentials (S.C.E.) of 2-methyl-1,4-naphthoquinone (−0.272 v.) and of 2-methyl-1,4-naphthoquinone oxide (−0.544 v., −0.990 v.) were determined at pH 7 and found to agree with the values obtained for the oxidized products.[22]

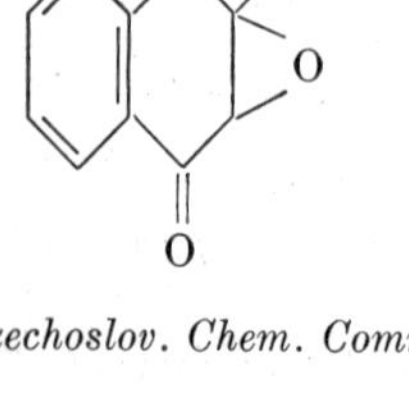

[22] E. Knobloch, *Collection Czechoslov. Chem. Communs.*, **14**, 508 (1949).

(b) Indophenols

Indophenols, 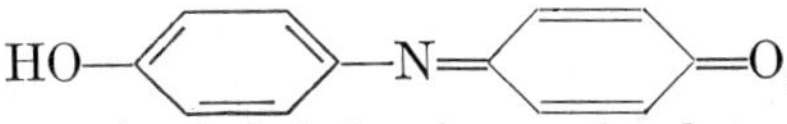, are dyestuffs which parallel the quinones in their behavior at the dropping mercury electrode. Half-wave potentials obtained for 2,6(or 3′,5′)-dichloroindophenol and 2,6(or 3′,5′)-dibromoindophenol in buffered solutions agree with the values obtained by the classical method.[23]

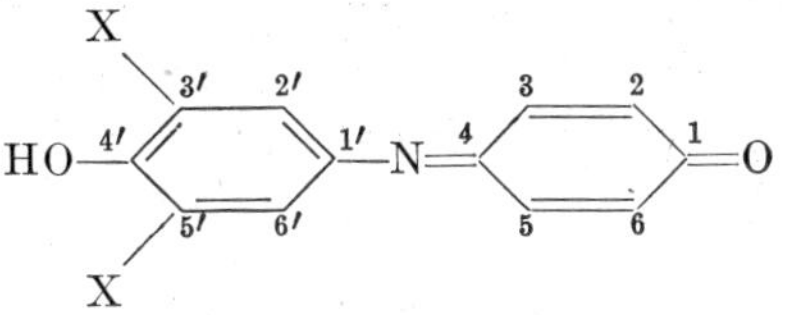

The method has been used to demonstrate that the violet dye formed from 4-amino-2-methyl-1-naphthol (vitamin K_5) on contact with air, especially in concentrated solutions, is an indophenol formed from the hydrolysis product, 2-methyl-1,4-naphthoquinone, and vitamin K_5.[22]

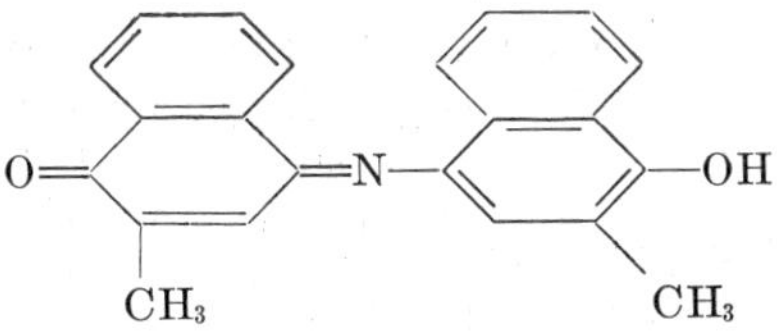

This substance exhibits a wave with a half-wave potential of −0.2 v. (S.C.E.) in a 0.1 *N* ammonium chloride in 50 per cent ethanol solution.

The polarographic method has been used to study qualitatively the reactions of leucylglycine, carbobenzoxyalanine, casein, and gelatin with quinone. The half-wave potentials of the addition products of quinone and glycine, alanine, valine, leucine, tyrosine, serine, tryptophan, proline, and glutamic acid have been measured and their dependence on pH determined.[24]

---

[23] O. H. Müller and J. P. Baumberger, *Trans. Electrochem. Soc.*, **71**, 181 (1937).

[24] K. Wiesner, *Chem. Listy*, **36**, 313 (1942); *Chem. Abstracts*, **44**, 1364 (1950).

CHAPTER XLI

Organic Acids and Acid Derivatives

Acids which contain a reducible group can give rise to two different reducible species at the dropping mercury electrode because of their ability to ionize. The anion form is reduced at more negative potentials than the undissociated molecule.

1. ALIPHATIC ACIDS

Simple aliphatic acids give a wave at the dropping mercury electrode that is due to the discharge of the hydrogen ions.[1] No wave is obtained with the esters (ethyl acetate).[2]

TABLE XLI-1

POLAROGRAPHIC BEHAVIOR OF SIMPLE ALIPHATIC ACIDS

Acid	K	Concn., millimoles per liter	i_d/C	$-E_{1/2}$ (S.C.E.)
Formic	1.77×10^{-4}	1–10	2.42	1.74–1.85
Acetic	1.75×10^{-5}	2–15	1.95	1.76–1.86
Isobutyric	1.44×10^{-5}	3–25	1.64	1.81–1.87
Isovaleric	1.7×10^{-5}	4–27	1.35	1.75–1.82

The half-wave potential becomes more negative when the concentration of the acid increases. Values obtained for a number of acids in 0.1 N LiCl, 0.1 N Li_2SO_4 and 0.05 N tetramethylammonium iodide solutions in concentrations of 1–10 millimoles per liter are given in Table XLI-1.[1]

Simple polybasic acids also exhibit only one wave in a neutral supporting electrolyte. A list of these is given in Table XLI-2.[1]

The diffusion current with a given capillary is represented by Korshunov *et al.*[1] in terms of the dissociation constant K by the following equation:

$$i_d/C = a + b \log K$$

where i_d = diffusion current, C = concentration of the organic acid,

[1] I. A. Korshunov, E. B. Kuznetsova, and M. K. Shchennikova, *Zhur. Fiz. Khim.*, **23,** 1292 (1949).

[2] M. von Stackelberg and W. Stracke, *Z. Electrochem.*, **53,** 118 (1949).

K = dissociation constant of the acid or the first dissociation constant of the polybasic acid, and a and b are constants depending on the nature and concentration of the supporting electrolyte.

In view of the work of Korshunov and co-workers on oxalic acid and their inability to duplicate the results of Winkel and Proske[3] in acid media, the data presented on oxalic acid [3,4] are to be questioned.

Reduction potentials are reported in 0.1 N ammonium chloride for ethyl oxalate (−1.36 v.) and oxamide (−0.55 v.) (N.C.E.).[3] A value of −1.50 v. is given for oxamic acid in sodium acetate solution.[5] The possibility of hydrolysis was not eliminated for any of these compounds.

TABLE XLI-2

POLAROGRAPHIC BEHAVIOR OF SIMPLE POLYBASIC ACIDS

Acid	K	Concn., millimoles per liter	i_d/C	$-E_{1/2}$ (S.C.E.)
Oxalic	5.90×10^{-2}	1–8	4.45	1.66–1.80
Malonic	1.49×10^{-3}	2–13	2.70	1.69–1.74
Tartaric	1.04×10^{-3}	1–8	3.12	1.64–1.77
Citric	8.40×10^{-4}	1–6	3.30	1.64–1.77
Malic	3.88×10^{-4}	1.5–10	2.85	1.66–1.74
Succinic	6.89×10^{-5}	1–9	2.63	1.80
Adipic	3.7×10^{-5}	2–11	2.07	1.76–1.80

Saturated fatty acids may be determined indirectly by precipitation with an excess of cadmium sulfate and determination of the excess cadmium ions polarographically.[6]

2. AROMATIC ACIDS

Aromatic acids show a similar behavior to that observed for aliphatic acids in 0.05 N tetramethylammonium iodide solution, 0.1 N lithium chloride, or 0.1 N lithium sulfate.[1] Values obtained are listed for a number of these acids and their derivatives in Table XLI-3.

The values are reported in Table XLI-4 for benzoic acid and its derivatives in dioxane–water mixtures.[2]

Phthalic Acid. The polarographic behavior of phthalic acid is greatly dependent upon the pH of the medium. Results obtained by Furman

[3] A. Winkel and G. Proske, *Ber.*, **69,** 1917 (1936).
[4] P. Herasymenko, *Z. Electrochem.*, **34,** 129 (1928).
[5] J. Brezina and J. Heyrovsky, *Collection Czechoslov. Chem. Communs.*, **8,** 114 (1936).
[6] S. Fiala and V. Janick, *Collection Czechoslov. Chem. Communs.*, **13,** 30 (1948).

and Bricker[7] in unbuffered solutions are therefore very complex since the pH increases during the polarographic reduction.

Unbuffered potassium biphthalate solutions give two waves. Upon

TABLE XLI-3

POLAROGRAPHIC BEHAVIOR OF AROMATIC ACIDS AND DERIVATIVES

Compound	K	Concn., millimoles per liter	i_d/C	$-E_{1/2}$ (S.C.E.)
Benzoic	6.3×10^{-5}	0.9–6	1.90	1.56–1.72
Mandelic	1.4×10^{-4}	2–13	2.20	1.70–1.78
Gallic	3.8×10^{-5}	2–6	2.33	1.71–1.73
Salicylic	1.01×10^{-3}	1–9	2.80	1.66–1.83
Acetylsalicylic		1–7	2.00	1.52–1.6
Anthranilic	1.07×10^{-5}	3–20	1.40	1.60–1.52
Sulfanilic	6.02×10^{-4}	2		1.54
Naphthionic	2×10^{-3}	2–10		1.42–1.52
N-Acetylsulfanilamide	4.2×10^{-6}	2.3–13.4		2.16–2.20
Sulfathiazole	7.6×10^{-8}	2.0–10.4		1.66–1.74
N-[p-Aminosulfonyl]-sulfanilamide	1.4×10^{-8}	2.0– 7.9		1.70–1.86
Sulfapyridine		No wave		
Sulfanilamide		No wave		

TABLE XLI-4

POLAROGRAPHIC BEHAVIOR OF BENZOIC ACID AND DERIVATIVES

Compound	0.05 *N* tetraethylammonium iodide–75% dioxane		0.1 *M* tetrabutylammonium iodide–50% dioxane	
	$-E_{1/2}$ (S.C.E.)	i_d/C	$-E_{1/2}$ (S.C.E.)	i_d/C
Benzoic acid	1.94	1.10	1.87	2.12
	2.46	1.7		
Tetraethylammonium benzoate	1.98	1.1–0.9[a]		
	2.38	2.9–2.7[a]		
Benzonitrile	2.26	3.0–2.5[a]		
	2.37	3.4–1.8[a]		
Ethyl benzoate	2.14	3.5–2.7[a]	2.18	7.62
Benzamide			1.82	9.2

[a] Concentration dependent.

addition of hydrochloric acid, the first wave increases rapidly to a maximum after which the second wave increases progressively and merges with the hydrogen wave. Addition of multivalent cations (barium or calcium) to unbuffered or poorly buffered solutions over a certain pH range may cause the appearance of three waves.

The experiments in buffered solutions were carried out by Furman and

[7] N. H. Furman and C. E. Bricker, *J. Am. Chem. Soc.*, **64**, 660 (1942).

Bricker in acetic acid-sodium acetate buffers in a pH range of 3.6–4.7 and in citric acid-sodium phosphate buffers below pH 3.6. Maxima were suppressed by the addition of methyl cellulose.

Working with 30 ml. of a 0.00333 *N* potassium biphthalate solution in a 0.333 *N* buffer solution with pH 3.6, successive additions of 0.1 ml. of 0.01 *M* barium chloride solution were found to decrease the first (and only) wave observed in the absence of barium. With an increase in barium concentration a second wave was observed. Qualitatively calcium and lanthanum have the same effect as barium. This second wave is probably due to the reduction of a complex between barium and phthalate ions. Conductance data indicate that there is some tendency for such complex formation.

The effect of the acidity was investigated between pH values of 2.80 and 4.21. In the absence of barium choride, only one wave was found which declines in height as the pH increases, as follows: pH 3.46, i_d = 23 microamp.; pH 3.75, i_d = 8.6 microamp.; pH 4.06, i_d = 1.5 microamp.; pH 4.21, i_d = 0.77 microamp.; pH 4.7, no wave. The height of this wave was found to be proportional to the concentration of the biphthalate at pH 3.6 or 4.1. This wave is also considerably larger than the first wave obtained with barium chloride at the same pH.

The height of the first wave in buffered solutions between pH 2.6 and 4.1 containing an excess of barium is proportional to the concentration of the undissociated acid. From this behavior it has been concluded that the first wave is due to a reduction of undissociated molecules of phthalic acid. This conclusion is not entirely justified since the concentrations of undissociated acid were calculated in solutions free from barium ions, whereas barium ions affect the pH of phthalic acid or biphthalate solutions and form complexes.

The following half-wave potentials (*vs.* S.C.E.) in buffered solutions containing an excess of barium ions are reported:

pH	1.5	2.9	3.6	4.1
$E_{1/2}$ (first wave)	−1.19	−1.28	−1.33	−1.39
$E_{1/2}$ (second wave)			−1.49	−1.53

A plot of log $[i/(i_d - i)]$ against the potential yielded a value for $0.059/n$ of 0.033 in phthalic acid solutions at pH 1.5, but of 0.057 at pH 3.6 (absence of barium ions). In the presence of barium ions at pH 3.6, the plot did not give a straight line. The sum of the heights of the two waves formed in the presence of barium ions corresponds to a 2-electron transfer per molecule of phthalic acid or biphthalate ion.

A possible reduction product at the dropping electrode is *trans*-δ-3,5-dihydrophthalic acid, but no conclusive evidence is given that this acid is formed.

It seems possible that the involved relations could be interpreted on the assumption that in the absence of Ba ions the wave corresponds to reduction of undissociated phthalic acid, and that the height of the wave is determined in part by the rate of transformation of biphthalate ion to phthalic acid at the surface of the electrode. The presence of Ba ions may decrease this rate (complex formation), but may promote the reduction of the complex at the dropping electrode. More work should be done to determine whether the currents are partly kinetic.

According to Korshunov[1] phthalic acid shows the following behavior in 0.05 *N* tetramethylammonium iodide solution:

Concn., millimoles/liter	i_d/C	$-E_{1/2}$ (S.C.E.)
1–11	3.22	1.60–1.85

3. UNSATURATED ACIDS

Acids containing a double bond conjugated with the carboxyl group are reducible at the dropping mercury electrode.

Fumaric and Maleic Acids. The most extensive investigations in this series of acids have been made with fumaric and maleic acids.[8] In most of these cases careful control of the factors affecting polarographic reduction were not maintained. The behavior of maleic acid in solutions 0.9 *M* in buffer* are given in Table XLI-5.

The behavior of fumaric acid is slightly different from that of maleic acid in that only one wave is obtained. A summary of the data is given in Table XLI-6.

The ethyl esters show only one wave in all buffers with the exception of those of pH 9.7. The results are given in Tables XLI-7 and XLI-8. The second wave observed at pH 9.70 is apparently due to maleic acid formed by saponification of the ester. The second wave in a buffer of pH 9.63 may be due to partial saponification of the ester in this alkaline buffer.

Buffering capacity was found to be very important in maintaining a constant pH and in affecting the constancy of the half-wave potential. Data for maleic acid in 0.1 *M* concentrations of measured buffer constituent,

[8] For a list of these see reference 1 in the paper by P. J. Elving and C. Teitelbaum, *J. Am. Chem. Soc.*, **71**, 3916 (1949).

* Main constituent of the buffer is 0.9 *M* in concentration.

TABLE XLI-5

POLAROGRAPHY OF MALEIC ACID (0.5 MILLIMOLAR) IN 0.9 *M* BUFFER SOLUTIONS

pH	No. of waves[a]	$-E_{1/2}$ (S.C.E.)	i_d, microamp.
2.43	2	0.73	4.23
4.00	2	0.89	2.90
4.89	1 or 2	0.96	3.76
5.88	1	1.12	4.01
6.93	1	1.27	3.99
7.80	1	1.34	2.46
8.72	1	1.43	2.88
8.60	1	1.37	2.79
9.63	1	1.42	0.96

[a] At higher pH values, a second wave may have escaped observation due to an irregular diffusion current. $E_{1/2}$ refers to the value for the first wave when two waves are obtained.

TABLE XLI-6

POLAROGRAPHY OF FUMARIC ACID (0.5 MILLIMOLAR) IN 0.9 *M* BUFFER SOLUTIONS

pH	$-E_{1/2}$ (S.C.E.)	i_d, microamp.
2.56	0.79	3.99
4.00	0.94	4.00
4.89	1.06	3.49
5.83	1.21	0.91
6.87	1.48	3.88
7.81	1.55	4.33
8.41	1.58	3.94
8.78	1.71	3.98
9.50	1.58	3.56

TABLE XLI-7

POLAROGRAPHY OF DIETHYL MALEATE (0.5 MILLIMOLAR) IN 0.9 *M* BUFFER SOLUTIONS

pH	$-E_{1/2}$ (S.C.E.)	i_d, microamp.
2.20	0.87	3.36
3.98	0.95	2.40
4.88	0.99	2.89
5.86	1.04	2.74
6.96	1.03	2.82
7.90	1.03	3.13
8.65	1.03	3.02
8.68	1.05	2.25
9.70	1.05, 1.46	1.83, 0.92

for example, give a very irregular relationship when the half-wave potential is plotted against pH, while the relation for the 0.9 *M* buffer is quite regular.

The behavior of maleic acid in a buffer of pH 5.90 is rather unusual. Two waves are obtained when low concentrations of buffer are used; these waves gradually merge as the concentration of buffer is increased or if potassium chloride is added. It has been suggested that the phenomenon observed may be caused by an ionic strength effect and that the appearance of two waves is not an indication of a varying ratio of different dissociated forms.

Comparison of the results for maleic and fumaric acids indicates that

TABLE XLI-8
POLAROGRAPHY OF DIETHYL FUMARATE (0.5 MILLIMOLAR) IN 0.9 *M* BUFFER SOLUTIONS

pH	$-E_{1/2}$ (S.C.E.)	i_d, microamp.
2.16	0.75	1.87
3.97	0.84	2.43
4.86	0.86	2.55
5.90	0.93	2.57
6.94	0.97	3.09
7.62	1.01	1.77
8.62	1.02	1.68
8.66	1.07	0.89
9.63	1.15, 1.37	2.48, 1.32

the fumaric acid wave occurs at more negative potentials. The greater thermodynamic stability of fumaric acid is in agreement with this behavior. The more negative half-wave potentials are not in agreement with the comparison of half-wave potentials for *trans*- and *cis*-stilbene since the *cis* compound is more difficult to reduce.

Analysis of the wave indicates that the potential determining step involves one electron:

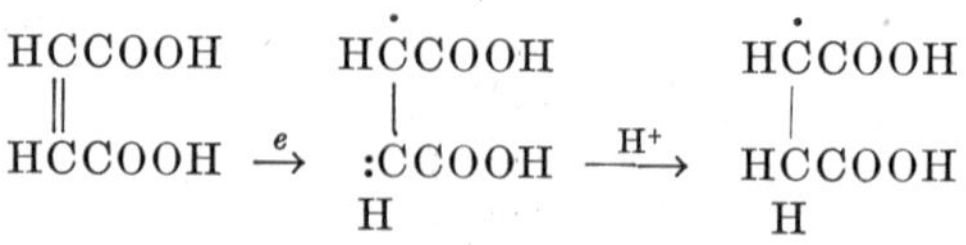

(I)

The pH dependence of the reduction seems to indicate that a proton is added before the second reduction step which then proceeds as follows:

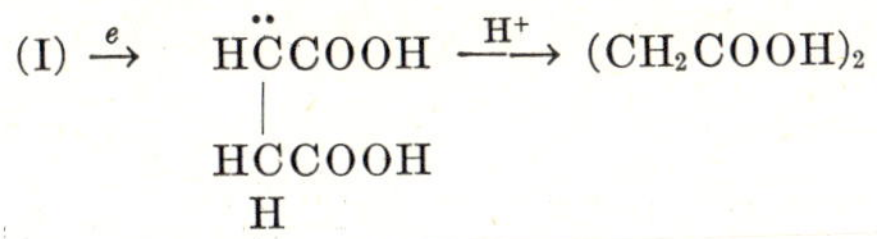

The esters behave differently from the acids. Diethyl fumarate has more positive half-wave potentials than diethyl maleate up to a pH of 8.4. Diethyl maleate shows a pH-dependent mechanism at low pH values and a pH-independent mechanism at high pH values. Diethyl fumarate shows a pH-dependent mechanism at all pH values. The reduction of the latter is probably similar to that of the acids.[9] The irreversibility of the reduction has been demonstrated using the oscillographic method.[10]

Starting with the equation for the reduction of the acid:

$$R_0 + 2H^+ + 2e \rightarrow RH_2$$

a relationship between the hydrogen ion concentration and the half-wave potential may be derived. In well-buffered solutions the potential will be determined by:

$$E = \text{constant} + \frac{RT}{2F_y} \ln \frac{[H^+]^2[R_0]}{[RH_2]} \qquad (1)$$

since:

$$i = K[RH_2]$$

and

$$E = \text{constant} + \frac{RT}{2F_y} \ln [H^+]^2 \frac{K[R_0]}{i}$$

$$= \text{constant} + \frac{RT}{2F_y} \ln [H^+]^2 K \frac{(i_d - i)}{i}$$

$$= E_{1/2} + \frac{RT}{2F_y} \ln \frac{i_d - i}{i} + \frac{RT}{F_y} \ln [H^+] \qquad (2)$$

In this equation R_0 represents the concentration of the undissociated acid in the immediate neighborhood of the electrode. Assuming that dissociation equilibrium of the acid is established instantaneously at the electrode surface, R_0 becomes a function of the pH.[11]

[9] P. J. Elving and C. Teitelbaum, *J. Am. Chem. Soc.*, **71,** 3916 (1949).

[10] M. Kalousek, *Collection Czechoslov. Chem. Communs.*, **13,** 105 (1948).

[11] P. Herasymenko, *Collection Czechoslov. Chem. Communs.*, **9,** 104 (1937).

The dissociation equilibrium of a dibasic acid is expressed by the following:

$$R \rightleftarrows R^- + H^+ \qquad \frac{[H^+][R^-]}{[R]} = K_1 \tag{3}$$

$$R^- \rightleftarrows R^{--} + H^+ \qquad \frac{[H^+][R^{--}]}{[R^-]} = K_2 \tag{4}$$

From eqs. 3 and 4 it is seen that:

$$[R^-] = [R]\, K_1 \frac{1}{[H^+]}$$

$$[R^{--}] = [R] \frac{K_1 K_2}{[H^+]^2}$$

the total concentration ΣR of the acid is:

$$\Sigma R = [R] + [R^-] + [R^{--}] = [R]\left(1 + \frac{K_1}{[H^+]} + \frac{K_1 K_2}{[H^+]^2}\right) \tag{5}$$

and:

$$\Sigma R_0 = [R_0]\left(1 + \frac{K_1}{[H^+]} + \frac{K_1 K_2}{[H^+]^2}\right)$$

or:

$$[R_0] = (\Sigma R_0) \frac{[H^+]^2}{[H^+]^2 + K_1[H^+] + K_1 K_2} \tag{6}$$

Introducing this into eq. 2 we find that:

$$E = E_{1/2} + \frac{RT}{2F_y} \ln \frac{i_d - i}{i} + \frac{RT}{2F_y} \frac{\ln [H^+]^4}{[H^+]^2 + K_1[H^+] + K_1 K_2} \tag{7}$$

In strongly acid medium $K_1[H^+]$ and K_1K_2 become very small with regard to $[H^+]^2$ and eq. (7) reduces to:

$$E = E_{1/2} + \frac{RT}{2F_y} \ln \frac{i_d - i}{i} + \frac{RT}{F_y} \ln [H^+] \tag{8}$$

Equation (8) is obeyed at pH values between 0 and 2 (see Table XLI-8). Equation (7) is not expected to hold at pH values greater than 6 since at such pH values the concentration of the undissociated acid becomes negligibly small and the uni- or divalent anions may be reduced directly. Also, the current may become partly kinetically controlled.

Half-wave potentials calculated for fumaric and maleic acids in the pH range from 0 to 5 are given in Table XLI-9; the half-wave potential at pH 0 was taken equal to −0.58 v. The calculated results are compared with the experimental values given by Elving[9] and Vopicka.[12]

For the quantitative determination of a mixture of the barium salts of maleic and fumaric acids, the use of an ammonium hydroxide-ammonium chloride (1 *M*) buffer at a pH of approximately 8 as the supporting electrolyte is recommended.[13] Under these conditions, maleic and fumaric acids give separate and distinct waves which are suitable for the quantitative analysis of mixtures of the two acids. The previously suggested

TABLE XLI-9

CALCULATED VALUES OF $E_{1/2}$ OF FUMARIC[a] AND MALEIC[b] ACIDS USING EQ. 7

pH	Fumaric acid			Maleic acid		
	$-E_{1/2}$ (calcd.)	$-E_{1/2}$ (exptl.) Vopicka	$-E_{1/2}$ (exptl.) Elving	$-E_{1/2}$ (calcd.)	$-E_{1/2}$ (exptl.) Vopicka	$-E_{1/2}$ (exptl.) Elving
0	0.58	0.58		(0.58)	0.58	
1	0.64	0.64		0.64	0.64	
2	0.71	0.705	0.68	0.71	0.705	0.66
3	0.775	0.775	0.77	0.80	0.775	0.75
4	0.85	0.98	0.90	0.88	0.93	0.85
5	0.95	1.15	1.04	0.97	1.035	0.94

[a] $K_1 = 1.0 \times 10^{-3}$, $K_2 = 3.0 \times 10^{-5}$.
[b] $K_1 = 1.4 \times 10^{-2}$, $K_2 = 2.6 \times 10^{-7}$.

method of using a medium of either 0.5 *N* ammonium chloride,[14] 0.1 *N* lithium chloride or 0.1 *N* potassium chloride and 0.5 to 1 *N* sodium acetate (Vopicka) could not be duplicated by Warshowsky *et al.*[13]

Procedure. Aliquots of the stock standard maleate and fumarate solutions are pipetted into a 100-ml. volumetric flask, 70 ml. of the ammonium hydroxide-ammonium chloride supporting electrolyte solution is added, and the solution is diluted to volume with distilled water. The concentration range investigated is approximately 7.5×10^{-5} to 3.0×10^{-4} gram per milliliter (6.5×10^{-4} to 2.6×10^{-3} *M*) for the acids alone or in mixtures of each other. The solution is electrolyzed in the polarographic cell, using a quiet mercury pool as the anode.

Effect of pH. Polarographic curves over a range of concentration were

[12] E. Vopicka, *Collection Czechoslov. Chem. Communs.*, **8,** 349 (1936).
[13] B. Warshowsky, P. J. Elving, and J. Mandel, *Anal. Chem.*, **19,** 161 (1947).
[14] G. Semerano and I. S. Rao, *Mikrochemie*, **23,** 9 (1937). See also G. Semerano and G. Bettinelli, *Gazz. chim. ital.*, **66,** 744 (1936). G. Semerano, *Mikrochemie*, **24,** 10 (1938).

obtained at pH values of 6.8, 8.2, and 8.9. The diffusion currents of the barium salts of the acids showed a linear relation with concentration in the range investigated at the three pH values selected. This relationship was also found for mixtures of the two substances.

Effect of Concentration. The diffusion current was found to be proportional to the concentration of the acids up to about $2.6 \times 10^{-3}\ M$ (3.0×10^{-4} gram per milliliter). When the concentration exceeds the latter value at pH 8.2, the readings become unsteady and erratic. Moreover, at this upper concentration level, with mixtures of the acids, there is an overlapping of the polarographic waves. It is, therefore, advisable to work with total concentrations no greater than $1.5 \times 10^{-3}\ M$.

Effect of Temperature. The temperature coefficient for the maleate ion was found to be 2.0 to 2.7 per cent, and that of fumarate ion approximately 1 per cent per degree centigrade.

Half-Wave Potentials. It was found that at a pH of 8.2 the half-wave potentials were constant over a temperature range of 20 to 30° C. and a concentration range of 0.65×10^{-3} to $2.60 \times 10^{-3}\ M$ at -1.36 v. for maleate ion and -1.60 v. for fumarate ion. At a pH of 6.8 the half-wave potentials were shifted to the slightly less negative values of -1.31 v. for maleate and -1.55 v. for fumarate ion.

The polarographic method has been used to determine: (a) amounts of fumaric acid formed in molten malic acid,[15] (b) amounts of maleic acid in succinic acid manufactured by the oxidation of benzene,[16] and (c) amounts of unchanged maleic anhydride in styrene-maleic anhydride and vinyl acetate-maleic anhydride polymerizations. The free maleic anhydride hydrolyzes in the presence of water to maleic acid.[17]

Indirectly the method has been used for determining aspartic acid in acid-hydrolyzed proteins. The conversion of this amino acid to a mixture of fumaric and maleic acids is accomplished by the following series of reactions:[18]

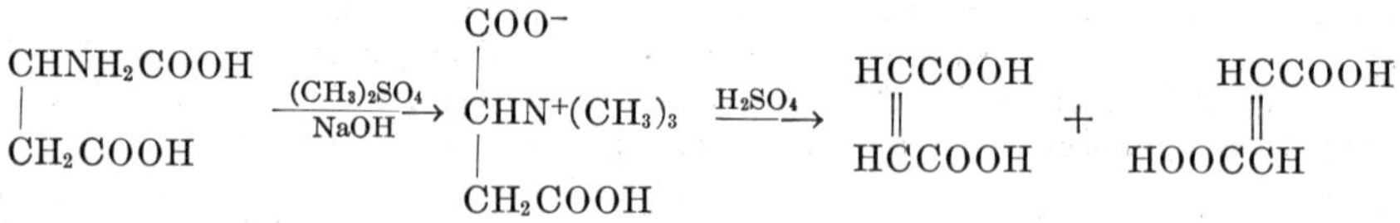

[15] P. Herasymenko and Z. Tyvonuk, *Collection Czechoslov. Chem. Communs.*, **2**, 77 (1930).

[16] L. Silverman, *Chemist-Analyst*, **36**, 57 (1947).

[17] G. C. Whitnack, *Anal. Chem.*, **20**, 658 (1948).

[18] B. Warshowsky, *Anal. Chem.*, **20**, 341 (1948).

Values reported for other unsaturated acids and their derivatives are given in Table XLI-10.

The polarographic method has been used to study the polymerization of methyl methacrylate in the presence of acetone peroxide, diethyl peroxide, and benzoyl peroxide (Neiman, *2*) and to determine α,β-unsaturated esters in the presence of their β,γ-isomers which are not reducible.[19]

TABLE XLI-10
REDUCTION POTENTIAL (45° TANGENT METHOD) (N.C.E.) AND HALF-WAVE POTENTIALS OF UNSATURATED ORGANIC ACIDS AND THEIR DERIVATIVES

Compound	$-E_{1/2}$	i_d/C	Solvent	Ref.
Acrylonitrile	1.94 (S.C.E.) (easily polymerized)	2.0	0.05 *N* $(C_2H_5)_4NI$, 75% dioxane	*1*
Crotonic acid	1.94 (S.C.E.)	0.8	0.05 *N* $(C_2H_5)_4NI$, 75% dioxane	*1*
Ethyl methacrylate	1.7	—	0.1 *N* LiCl	*2*
	2.6	—	0.1 *N* $(CH_3)_4NI$, 25% ethanol	
Acetylene dicarboxylic acid	0.22[a]	—	1 *N* HCl	*3*
Aconitic acid	0.49[a]	—	0.1 *N* HCl	*3*
Mesaconic, citraconic acids	0.51[a]	—	1 *N* HCl	*3*
Sorbic acid	2.01[a]	—	Alkali	*3*
Cinnamic acid	0.96[a]	—	1 *N* HCl	*3*
	1.46[b]	—	0.1 *N* NH_4Cl, 50% ethanol	*4*

[a] Reduction potential (45° tangent method).
[b] Molar reduction potential; the reduction involves two electrons and yields phenylpropionic acid.
(*1*) M. von Stackelberg and W. Stracke, *Z. Electrochem.*, **53,** 118 (1949).
(*2*) M. B. Neiman and M. A. Shubenko, *Zavodskaya Lab.*, **14,** 394 (1948); *Chem. Abstracts*, **43,** 1289 (1949).
(*3*) L. Schwaer, *Collection Czechoslov. Chem. Communs.*, **7,** 326 (1935).
(*4*) G. Semerano and A. Chisini, *Gazz. chim. ital.*, **66,** 510 (1936).

In a medium 0.102 *M* in tetramethylammonium chloride and 76 per cent in ethanol, the diffusion currents of ethyl isopropylidenemalonate, ethyl 1-methylpropylidenemalonate, and ethyl cyclopentylidenemalonate are proportional to concentration and half-wave potentials were found between −2.03 and −2.06 v.

The average and maximum deviations between the calculated and found ratios of the pair of isomers were 1.8 and 5 per cent. The polarographic method is of value in cases where the pure α,β-unsaturated forms can be isolated and supplements the iodometric and bromometric methods de-

[19] A. C. Cope and E. M. Hardy, *J. Am. Chem. Soc.*, **62,** 3319 (1940).

veloped by Linstead,[20] which are most accurate in cases where the pure β,γ-unsaturated isomers can be isolated.

The method has also been used to determine fumaric acid in the presence of aconitic and citraconic acids, to follow the depolymerization of aconitic acid,[21] and to analyze mixtures of *cis*- and *trans*-aconitic acids.[22]

4. HYDROXY ACIDS AND LACTONES

Freshly prepared solutions of hydroxy acids produce only hydrogen waves in 0.1 *N* tetraethylammonium chloride solution. On standing, these acids which can give lactones show waves at values which are dependent upon the structure.

Glycolic, lactic, and glyceric acids produce only hydrogen waves. Lactide formed by heating lactic acid produces waves with inflection points at −2.03 and −2.22 v. which, due to hydrolysis, disappear on standing with the appearance of a hydrogen wave.

The lactones from higher hydroxy acids produce waves presumably caused by their reduction to the corresponding sugar at potentials (0.1 N.C.E.) ranging from −2.2 to −2.7 v.[23]

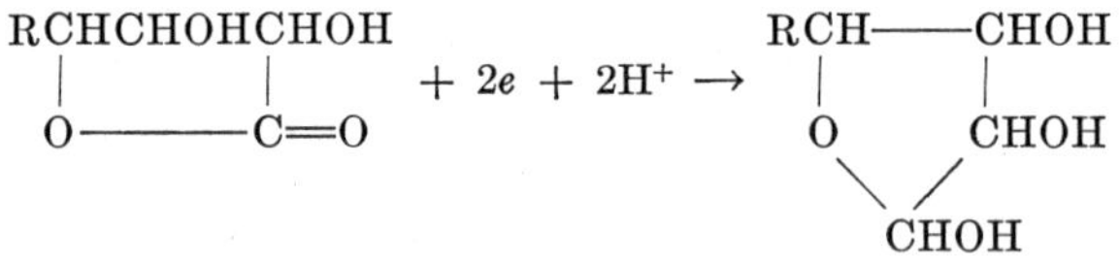

With the exception of the wave for D-glucono-γ-lactone which occurs at −2.2 v., the waves of 17 other γ-lactones occur at potentials ranging from −2.4 to −2.7 v. In some cases two waves appear which suggest a stepwise reduction or the progressive reduction of the lactone. The δ-lactones of D-gluconic, D-mannonic, and L-rhamnonic acids appear to be reduced at slightly lower potentials than the corresponding γ-lactones. The δ-lactone of lactobionic acid is reduced at approximately the same potential as the δ-lactone of D-gluconic acid. Thus, a galactoside group attached to carbon 4 of D-gluconolactone does not materially alter the reduction.

Among the γ-lactones configuration has an important effect upon the reduction potentials. Lactones in which the OH of carbon 2 and the terminal CH_2OH group lie on the same side of the ring (*cis*) give reduction

[20] R. P. Linstead *et al.*, *J. Chem. Soc.*, **1927,** 355, 2565; **1931,** 723.

[21] I. D. S. Rao, *J. Univ. Bombay*, **10.** Pt. 3, 56 (1941); *Chem. Abstracts*, **36,** 3786 (1942).

[22] G. Semerano and L. Sartori, *Mikrochemie*, **24,** 130 (1938).

[23] H. Matheson, H. S. Isbell, and E. R. Smith, *J. Research Natl. Bur. Standards*, **28,** 95 (1942).

potentials in the range from −2.6 to −2.7 v., whereas lactones in which these groups lie on opposite sides of the ring (*trans*) give reduction potentials in the range from −2.3 to −2.55 v., when tetraethylammonium chloride is used as supporting electrolyte.

The polarographic method is very useful for studying the rates of formation and hydrolysis of lactones. Thus, a fresh solution of D-glucono-δ-lactone does not produce a hydrogen wave, but after the solution stands for only a few minutes a hydrogen wave appears and increases in height, while the δ-lactone wave decreases until after 3.3 hours it is not visible. It has not been investigated whether the lactone and lactide waves are of quantitative analytical importance. Since their diffusion currents are found at extremely negative potentials, difficulties will be encountered in their quantitative measurement.

In summary, the following substances have been studied by Matheson[23] *et al.*: glycolic acid, lactic acid, lactyllactic acid, lactide, glyceric acid, D-erythrono-γ-lactone, L-threono-γ-lactone, L-arabono-γ-lactone, L-ribono-γ-lactone, D-xylono-γ-lactone, D-lyxono-γ-lactone, L-rhamnono-lactones, D-gluconic acid and its lactones, D-galactonic acid and its γ-lactone, D-mannono-lactones, D-talonic acid and its γ-lactone, D-gulono-γ-lactone, L-allono-γ-lactone, D-α-galaheptono-γ-lactone, D-α-glucoheptono-γ-lactone, D-α-mannoheptono-γ-lactone, D-β-mannoheptonic acid and its γ-lactone, D-α-guloheptonic acid and its γ-lactone, D-β-guloheptonic acid, and lactobiono-δ-lactone.

Use of the suppression of the oxygen maximum is made in the estimation of polycondensates of lactic acid.[24]

Phthalides. Phthalides are γ-lactones involving an aromatic acid grouping. The simple ones like phthalide (I) or 3-phenylphthalide (II) show one reduction wave which is independent of pH.[25, 26]

Compound	$-E_{1/2}$(S.C.E.)		
	0.1 *M* tetrabutylammonium iodide, 50% dioxane	Buffer I[a] 50% Dioxane	Buffer II[a] 50% Dioxane
Phthalide	2.03	2.05	2.05
3-Phenylphthalide	1.89	—	—
3,3-4′-hydroxydiphenyl) phthalide	1.88	—	—

[a] These buffers differed by about 2 pH units.

The exact course of the reduction is not known.

[24] Z. K. Jelinek, *Chem. Obzor*, **23,** 61 (1948); *Chem. Abstracts*, **42,** 6274 (1948).

[25] S. Wawzonek and J. H. Fossum, *J. Electrochem. Soc.*, **96,** 234 (1949).

[26] S. Wawzonek, H. A. Laitinen, and S. J. Kwiatkowski, *J. Am. Chem. Soc.*, **66,** 827 (1944).

3,3-Diphenylphthalides containing hydroxyl groups in the para position (III)—a structure present in fluorescein and phenolphthalein—show more complicated behavior since ring opening may be brought about very easily by alkali.

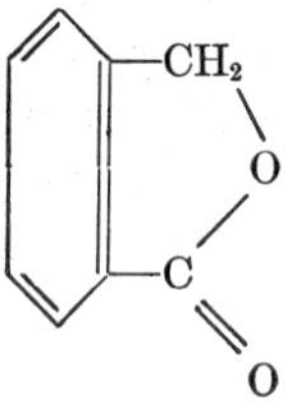

(I)

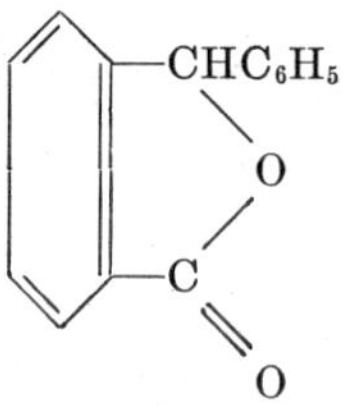

(II)

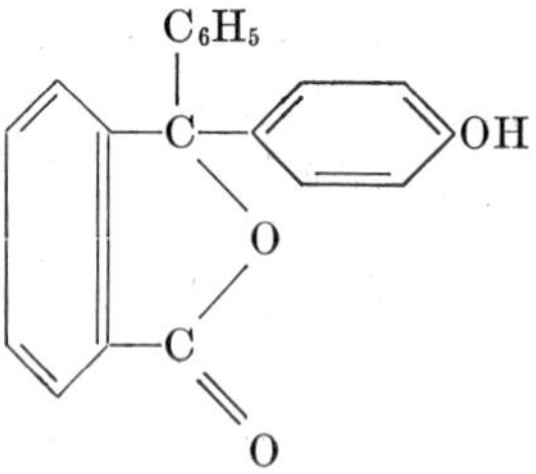

(III)

Phenolphthalein. Phenolphthalein,[27] because of its slight solubility, has been studied in a medium of twenty-five per cent ethanol in buffered solutions. The pH values refer to those of the buffers in water in the absence of alcohol. Basic fuchsin in a concentration of 0.001 per cent served to eliminate maxima. Polarographic waves recorded in 25 per cent ethanol in a pH region between 3.5 and 7.4 (phosphate buffer) are shown in Fig. XLI-1. Up to a pH of 8.5 only one wave is observed with a well-defined diffusion current which is proportional to the concentration of phenolphthalein. Hence this substance can be determined polarographically. The polarogram at a pH of 7.4 has quite an abnormal appearance (see Fig. XLI-1, curve III). The diffusion current remains practically constant in a potential range between -1.0 and -1.2 v., it decreases very slightly between -1.2 and -1.4 v., and then drops suddenly to attain a minimum value at a potential of -1.6 v. This minimum value remains constant up to a potential of -1.75 v., whereupon the wave of the supporting electrolyte appears. The height of the minimum as compared to i_d was found to increase from 0.07 at 25° C. to 0.56 at 75° C. At 85° C. no dip in the diffusion current was observed. At a pH of 4.7 a decrease in the diffusion current with increasing negative potential is also observed, but the hydrogen wave appears before the flat minimum is attained. In still more acid medium (pH 3.5) no decrease in the diffusion current is observed, as the hydrogen wave appears before the decrease in the diffusion current. The dip and the minimum observed at a pH of 7.4 seem to be

[27] I. M. Kolthoff and D. J. Lehmicke, *J. Am. Chem. Soc.*, **70**, 1879 (1948).

related to the electrolyte content of the solution. For example, in a buffer of pyridine and pyridinium hydrochloride with a total chloride concentration of 2.4×10^{-4} *M*, a pyridine concentration of 2×10^{-3} *M*, and a phenolphthalein concentration of 1×10^{-3} *M* (pH 6.85), a normal wave was observed. The diffusion current remained constant until the wave of the supporting electrolyte appeared. The dip became evident upon addition of potassium chloride and became more pronounced with increasing concentration of this salt.

In the pH range corresponding roughly to the color change interval of phenolphthalein two waves are observed. At a pH of about 9 the first diffusion current is about equal to the total diffusion current found at

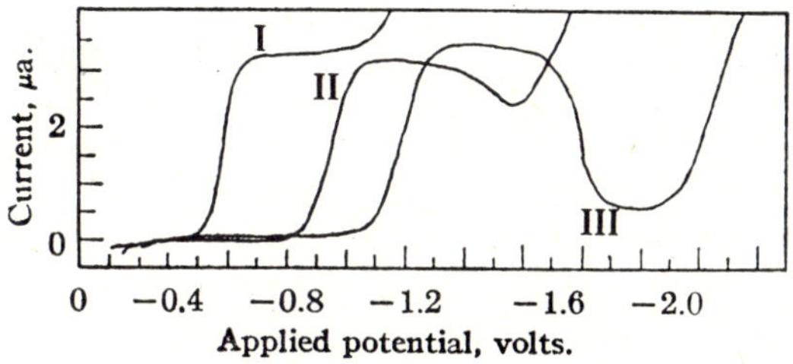

Fig. XLI-1. Current–voltage curves of millimolar phenolphthalein in 25 per cent ethanol: (I) solution is 3.8 *N* in HCl and 0.001 per cent in basic fuchsine; (II) biphthalate buffer, pH 4.73; (III) phosphate buffer, pH 7.40. Abscissa scales for curves (II) and (III) start at 0.1 and 0.2 v., respectively, to the right of curve (I).

lower pH values, the second wave comprising only a few per cent of the total current. In a more alkaline medium the second wave increases at the cost of the first one, the total current remaining the same. At a pH of 10.25 and at higher pH values two waves of equal height are found. No decrease in the total diffusion current with increasing negative potential is found at a pH greater than 9.

In solutions containing free sodium hydroxide the diffusion current decreases with time since the red form of phenolphthalein is changed into the colorless carbinol form which is not reducible at the dropping electrode. The polarographic method was found quite suitable for following the rate of color fading of phenolphthalein in an alkaline medium. Phenolphthalin, the reduced form of phenolphthalein, was not found to be reducible in the pH range between 0 and 13.

Polarograms were made of 0.001 *M* phenolphthalein in a phosphate buffer of pH 7 in the presence of 0.1 *N* potassium chloride at various alcohol concentrations. Up to about 30 per cent ethanol the diffusion current had a normal value, but at higher alcohol contents it decreased considerably. In 70 per cent alcohol its value was only 3 per cent of that in 25 per cent ethanol.

Half-wave potentials at 25° C. in 25 per cent ethanol as a function of pH are reported in Table XLI-11. In the range between pH 0 and 10 the half-wave potential changes according to $E_{1/2} = -0.54 - 0.046$ pH.

TABLE XLI-11

HALF-WAVE POTENTIALS (S.C.E.) OF PHENOLPHTHALEIN AS FUNCTION OF pH

pH	−0.06	0	1	2.5	3.5	4.7	5.8
$-E_{1/2}$ (1)	0.49	0.53	0.60	0.67	0.71	0.75	0.80
$-E_{1/2}$ (2)	—	—	—	—	—	—	—
pH	6.9	7.4	8.2	9.6	10.06	12	13
$-E_{1/2}$ (1)	0.86	0.89	0.93	0.98	1.01	(0.84)	(0.82)
$-E_{1/2}$ (2)	—	—	—	1.35	1.33	1.33	1.23

The one-step reduction of phenolphthalein in an acid medium involves two electrons, phenolphthalin being the reduction product. From the height of the diffusion current a diffusion coefficient of 5.2×10^{-6} cm.^{2} sec.$^{-1}$ at 25° C. was calculated for phenolphthalein in 25 per cent ethanol.

The exact mechanism of the polarographic reduction of phenolphthalein is not yet completely understood. On the basis of further work carried out at the University of Minnesota, the interpretation of the decrease of the apparent diffusion current of phenolphthalein at higher alcohol concentrations is different from that presented in the paper by Kolthoff and Lehmicke.

In neutral and acid media phenolphthalein is mainly present in the form of the colorless lactone (I) which is in equilibrium with quinoid form (II). Apparently, the concentration of the quinoid form is extremely small. Assuming that this form, and not the lactone, is reduced at the dropping electrode, the rate of transformation of (I) into (II) must be very great in order to yield the normal diffusion current in 25 per cent ethanol. The decrease in this current with increasing alcohol concentration might be attributed to a decreasing rate of ionization of (I) due to the decreasing dielectric constant of the medium. The very small current found in 70

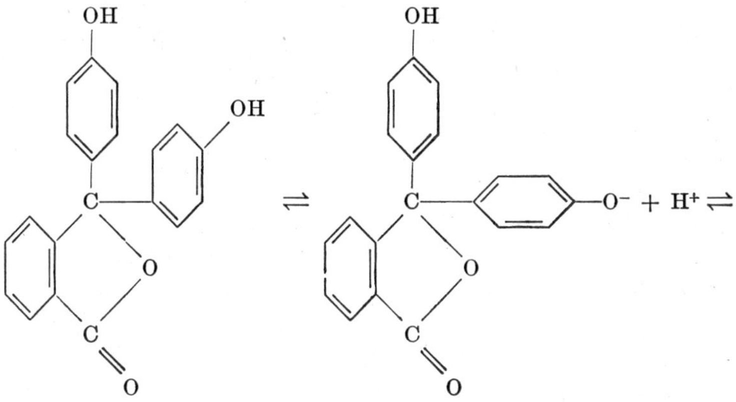

(I)

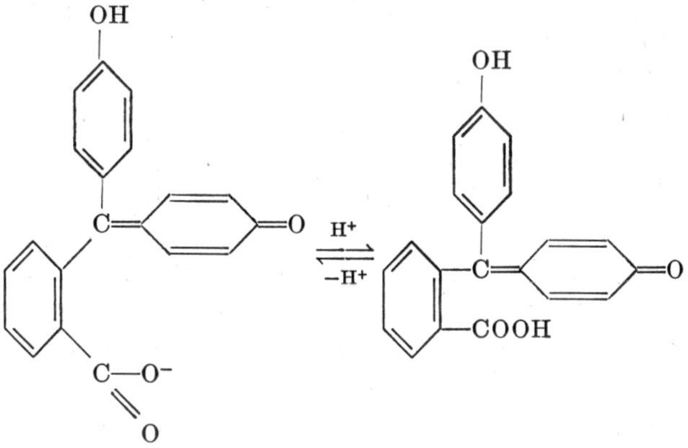

(II)

per cent ethanol should be entirely rate and not diffusion controlled. In agreement with this postulate Nachtsheim (University of Minnesota) found that the height of the wave in 70 per cent ethanol was independent of the height of the mercury in the reservoir (see Chapter XV). The decrease in diffusion current was also observed with solvents like methanol and acetone instead of ethanol.

The lactone form is probably not reducible at the dropping electrode at such positive potentials (see 3,3-(4′-hydroxydiphenyl)phthalide). It resembles in structure the colorless carbinol formed in an alkaline medium which is not reduced at a pH of 7. All the present evidence indicates that the reduction of phenolphthalein at a pH of 7 or less is given by the equation:

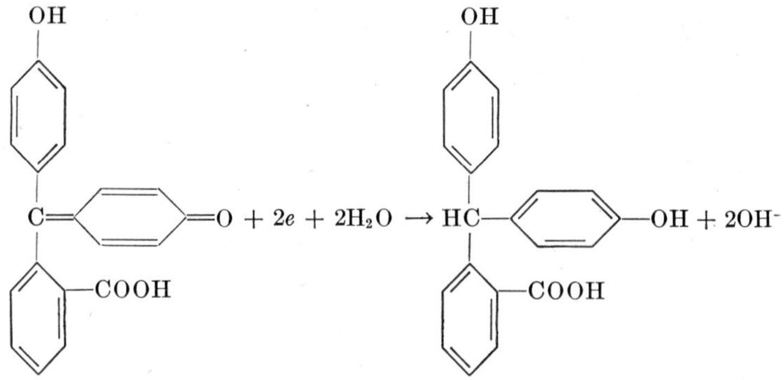

At higher pH values at which the red form appears the second wave is due to the reduction of the colored form (III):

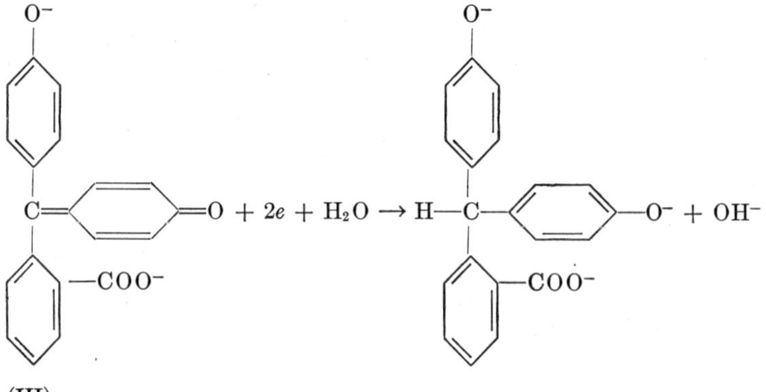

(III)

Fluorescein. The polarographic behavior of fluorescein has been reported by Delahay.[28] In order to prevent the precipitation of fluorescein at pH values less than 5 the solutions were made 6 N in acetic acid. Ordinary buffers were used at pH values greater than 5. The half-wave potentials (S.C.E.) measured at 25° are as follows:

pH	2.0	3.8	5.0	6.6	7.1	7.85	8.5	9.0	9.6	10.1	10.8
$-E_{1/2}$ (1)	0.48	0.57	0.65	0.74	0.80	0.92	0.97	1.07	1.14	1.18	1.20
$-E_{1/2}$ (2)	—	—	—	—	—	—	—	1.38	1.40	1.44	—

[28] P. Delahay, *Bull. soc. chim. France*, **1948**, 348.

Since absorption spectrum data indicate that fluorescein is present in solution as the quinoid form[29] the wave up to pH 6.5 apparently represents reduction of the quinoid form:

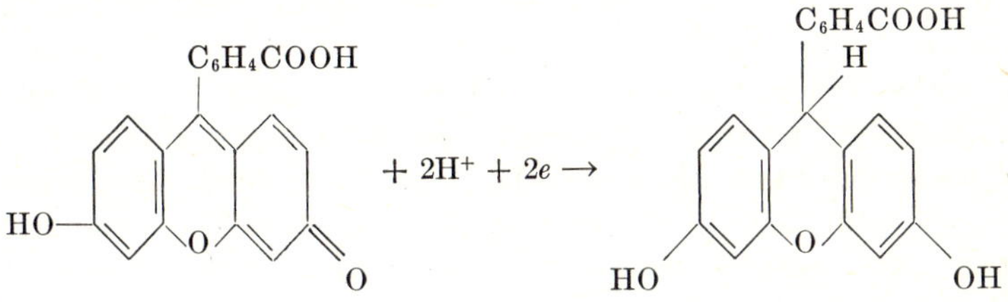

(IV)

At pH values greater than 6.5 the anion form of (IV) appears and is reduced at slightly more negative values. At pH values greater than 9 a third form (V) is possible and would account for the second wave observed in alkaline buffers:

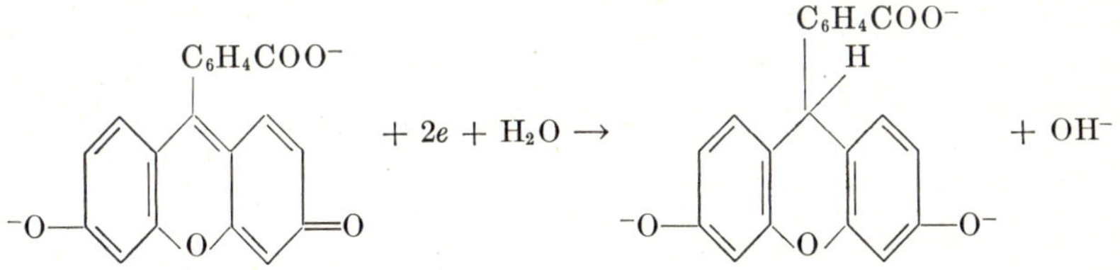

(V)

Plant heart poisons which are naturally occurring lactones are reported to give polarographic waves.[30]

Ascorbic Acid[31] (Vitamin C). Ascorbic acid is an example of an unsaturated hydroxylactone. Only one mention is made in the literature, however, that it yields a cathodic wave. The observation was made in 2 per cent metaphosphoric acid solution and a wave was reported at −1.7 v.[31]

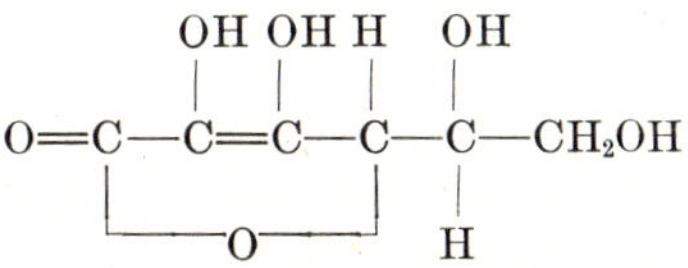

Vitamin C

[29] W. R. Orndorff, R. C. Gibbs, and C. V. Shapiro, *J. Am. Chem. Soc.*, **50,** 819 (1928).
[30] E. B. Hershberg, J. K. Wolfe, and L. F. Fieser, *J. Am. Chem. Soc.*, **62,** 3516 (1940).
[31] M. M. Kirk, *Ind. Eng. Chem., Anal. Ed.*, **13,** 625 (1941).

Because of the ene-diol system, ascorbic acid is a strong reducing agent and produces an anodic wave which shifts with pH.[32] A number of studies[33–35] have been made of the shift of the half-wave potential with pH, and the values are in agreement with each other. Some of the values[35] are given in Table XLI-12.

The anodic oxidation is not completely reversible; the polarographic values are about 0.21 v. more positive than those obtained from potentiometric measurements. Moreover, the oxidation product, dehydroascorbic acid, is not reducible at the dropping mercury electrode. The irreversibility has also been demonstrated by oscillographic measurements.[36]

For analytical purposes a potassium biphthalate buffer (pH 6.2) containing 0.25 per cent oxalic acid or a phosphate buffer (pH 8.0) containing 1.5 per cent metaphosphoric acid have been suggested. Oxalic and metaphosphoric acids are the only two extracting substances for vitamin C which do not interfere with the anodic wave and which do not affect the vitamin. Values obtained in these buffers are given in Table XLI-13. The method has an accuracy of 3 to 4 per cent, depending upon the amounts and the electrolyte used, and checks well with visual and photometric procedures.[37]

TABLE XLI-12

pH	$E_{1/2}$ (N.C.E.)	pH	$E_{1/2}$ (N.C.E.)
0.63	+0.23[a]	4.85	+0.02
1.50	+0.21	5.60	−0.02
2.20	+0.18	6.65	−0.06
2.75	+0.15	7.30	−0.08
3.25	+0.12	8.30	−0.11
3.90	+0.08	9.69	−0.16[a]
4.45	+0.05		

[a] Values of Cattaneo and Sartori.[34]

Better defined anodic diffusion currents are reported if the potential

32 E. Kodicek and K. Wenig, *Nature*, **142**, 38 (1938).

33 O. H. Müller and R. A. Phillips, *Am. J. Physiol.*, **129**, 426 (1940). See also O. H. Müller, *Ind. Eng. Chem., Anal. Ed.*, **14**, 99 (1942).

34 C. Cattaneo and G. Sartori, *Gazz. chim. ital.*, **72**, 351 (1942).

35 Z. Vavrin, *Collection Czechoslov. Chem. Communs.*, **14**, 367 (1949).

36 M. Kalouzek, *Collection Czechoslov. Chem. Communs.*, **13**, 105 (1948).

37 W. S. Gilliam, *Ind. Eng. Chem., Anal. Ed.*, **17**, 217 (1945).

range of the polarograph is 0–1.0 v. instead of 0–3.0 v.[38] Citric acid or potassium citrate may be substituted for the metaphosphoric acid.[39]

Procedure.[37] A given weight of the material to be analyzed is whipped in a Waring Blendor with a known volume of metaphosphoric or oxalic acid for 2 to 5 minutes. The extract is then filtered through a dry fluted filter paper. A 2- to 5-ml.

TABLE XLI-13

POLAROGRAPHIC BEHAVIOR OF ASCORBIC ACID

Supporting electrolyte	$E_{1/2}$ (S.C.E.)	Diffusion coefficient, cm.^{2} sec.$^{-1}$	Diffusion current constant, $i_d/Cm^{2/3}t^{1/6}$
Biphthalate buffer–0.25% oxalic acid......	+0.10	5.7×10^{-6}	3.00
Phosphate buffer–1.5% metaphosphoric acid; pH 3.41..........................	+0.17	6.2×10^{-6}	3.04
Biphthalate buffer–0.25% oxalic acid–50% alcohol................................	+0.06	3.6×10^{-6}	2.34
Biphthalate buffer–3% HPO_3; pH = 2.71..	+0.24	5.3×10^{-6}	2.88

aliquot of the buffer solution is added, and the well-mixed solution is polarographed between −0.1 and +0.3 v. The diffusion current is measured, corrected for the residual current of the supporting electrolyte, and the corresponding concentration is read off the calibration graph.

A procedure for the polarographic determination of free and combined ascorbic acid in plants has been given by Hasselbach.[40]

5. KETO AND ALDEHYDO ACIDS

(a.) Aliphatic Acids

Pyruvic Acid. $CH_3COCOOH$. Scattered data on the reduction of pyruvic acid at the d.e. are reported in the literature.[41] A more systematic study of the plarographic reduction has been made in buffered solutions

[38] J. E. Page and J. G. Waller, *Analyst*, **71**, 65 (1946).

[39] I. Okada, *J. Agr. Chem. Soc. Japan*, **19**, 749, 955 (1943); **20**, 325 (1944); *Chem. Abstracts*, **43**, 1454 (1949).

[40] H. Hasselbach, *Z. Pflanzenernähr. Düngung u. Bodenk.*, **39**, 27 (1947); *Chem. Abstracts*, **42**, 4496 (1948).

[41] (*a*) L. Schwaer, *Collection Czechoslov. Chem. Communs.*, **7**, 326 (1935). (*b*) M. Shikata and K. Shoji, *Bull. Agr. Chem. Soc. Japan*, **4**, 96 (1928). (*c*) H. Adkins and F. W. Cox, *J. Am. Chem. Soc.*, **60**, 1151 (1938). (*d*) A. Winkel and G. Proske, *Ber.*, **69**, 1917 (1936).

and is given below by Müller and Baumberger:[42]

pH	0	2	3	4	5	6	7	8
$-E_{1/2}$ (1) (S.C.E.)	0.65	0.75	0.86	1.09	1.18	1.24	1.30	
$-E_{1/2}$ (2)	—	—	—	—	1.64	1.60	1.57	1.54

At pH values smaller than 4.1 only one wave occurs. At pH values between 4 and 8, two waves are observed, the magnitude of the first one decreasing and that of the second one increasing with increasing pH, with the sum of the two remaining constant. The ratio of the first to the second wave was found to be 1 at pH 5.8, 10 at pH 4.8, and 0.1 at pH 6.8. At a pH of 8 or greater than 8 only one wave is obtained. (Müller and Baumberger wrongly attributed the waves to reduction of the keto form and enol form, respectively.) The first wave corresponds to the reduction of the free acid and the second wave to the reduction of the pyruvate ion.[43] When two waves occur, the height of the first wave is much greater, and that of the second wave much smaller than what corresponds to the concentration of undissociated and dissociated acid in the bulk of the solution. The rate of transformation of the acid into its anion must be considered, as has been discussed on page 269.

The kinetic character of the waves can also be demonstrated with phenylglyoxylic acid ($C_6H_5COCOOH$) which cannot enolize but which behaves like pyruvic acid. Ethyl pyruvate and ethyl phenylglyoxylate show only one wave.

A wave reported by Schwaer[41] at a pH of zero starting at −0.2 v. (N.C.E.) is attributed by him to a polymer of pyruvic acid (trimer) since the height of the wave was found to increase with the time of standing. This was confirmed by Müller and Baumberger.[42] This polymerization has to be considered in the determination of pyruvic acid; at a pH of 8 the polymerization tendency is the smallest.

Pyruvic acid is reduced to lactic acid at the dropping mercury electrode. The concentration of the anion form is increased by the addition of amino acids (glycine) and sodium fluoride, but not by sodium chloride, bromide, or iodide.[44]

Mesoxalic Acid. $CO(COOH)_2$. Mesoxalic acid does not produce a reduction wave because the carbonyl group is hydrated. The hydrazone

[42] O. H. Müller and J. P. Baumberger, *J. Am. Chem. Soc.*, **61,** 590 (1939).
[43] R. Brdicka, *Collection Czechoslov. Chem. Communs.*, **12,** 212 (1947).
[44] A. M. Kuzin and I. E. Elpiner, *Biokhimiya,* **12,** 509 (1947); *Chem. Abstracts*, **42,** 6752 (1948).

and oxime, however, are reduced and produce waves involving two electrons. Products proposed are the substituted hydrazine and hydroxylamine. The half-wave potentials obtained are given below.

Oxime		Hydrazone	
pH	$-E_{1/2}$	pH	$-E_{1/2}$
2.75	0.45	2.60	0.45
3.20	0.47	3.10	0.48
3.92	0.51	4.46	0.54
5.62	0.60	5.25	0.59

Tartronic acid, the reduction product of mesoxalic acid, does not produce an oxidation wave.

The method was used to study the kinetics of the formation of the oxime and hydrazone of mesoxalic acid.[45]

Methyl 4-Keto-2-pentenoate.[46] $CH_3COCH{=}CHCOOCH_3$. This ester shows the following behavior in buffered solutions:

pH	$-E_{1/2}$ (S.C.E.)	i_d/C
1.85	0.37	5.08
2.85	0.65	4.98
5.53	0.68, 0.82	3.69, 1.49
9.02	0.71, 0.97	3.91, 0.89
11.00	1.22	2.77

The wave height corresponds to a 2-electron reduction and to the formation of ethyl levulinate.

The **oxime,** $HON{=}CHCOCH{=}CHCOOCH_3$ (ethyl 4,5-diketo-2-pentenoate-5-oxime), gives two waves of four and two electrons each in a buffer of pH 5.4. The values for other buffers are given below:[46]

	$-E_{1/2}$ (S.C.E.)			Wave height		
pH	1st	2nd	3rd	1st	2nd	3rd
1.6	0.06	0.61	0.90	8.2	2.7	1.8
2.9	0.38	1.02	1.35	8.0	1.6	1.6
5.4	0.44	1.65	—	8.0	4.6	—
9.1	0.46	0.87	1.53	4.5	1.4	2.3
12.3	0.92	1.46	—	6.2	2.3	—

[45] G. Sartori and A. Gaudiano, *Gazz. chim. ital.*, **78,** 77, (1948).

[46] E. D. Hartnell and C. E. Bricker, *J. Am. Chem. Soc.*, **70,** 3385 (1948).

Only the waves in 0.5 *N* NaOH resemble those obtained for methyl 4-keto-2-pentenoate. The actual mechanism of reduction is not known. The method was used to follow the decomposition of this oxime.

(b.) AROMATIC KETO ACIDS

Extensive investigations on the behavior of γ-keto and aldehydo acid derivatives at the dropping mercury electrode have shown that polarographic measurements can be used to assign structures to the compounds in these series.

The behavior of various esters, anhydrides, and amides of phthalaldehydic acid and of 2-benzoylbenzoic acid, their reduction products, and related compounds was studied in both unbuffered and buffered media. A summary of the observed half-wave potentials and diffusion current constants in the various solutions is given in Table XLI-14, XLI-15A, and XLI-15B.[47] Buffer I was composed of a mixture which was 0.172 *M* in tetramethylammonium hydroxide, 0.075 *M* in tetramethylammonium iodide, 0.102 *M* in phosphoric acid, and 50 per cent in dioxane by volume. Buffer II was a mixture which was 0.196 *M* in tetramethylammonium hydroxide, 0.075 *M* in tetramethylammonium iodide, 0.092 *M* in phosphoric acid, and 50 per cent in dioxane by volume.

In general, all of the compounds gave well-defined reduction waves. Maxima could be suppressed in most cases by means of gelatin. It was also noticed that addition of gelatin caused a catalytic hydrogen wave which decreased the voltage range of the buffers and prevented evaluation of waves at more negative potentials.

All the esters and anhydrides of phthalaldehydic acid and 2-benzoylbenzoic acid are reduced to phthalide and 3-phenylphthalide, respectively, since they show a second wave in 0.1 *M* tetrabutylammonium iodide with a half-wave potential which is characteristic of these substances. The reduction proceeds in the same manner in both series; the cyclic derivatives are reduced directly to the phthalide and an alcohol or acid:

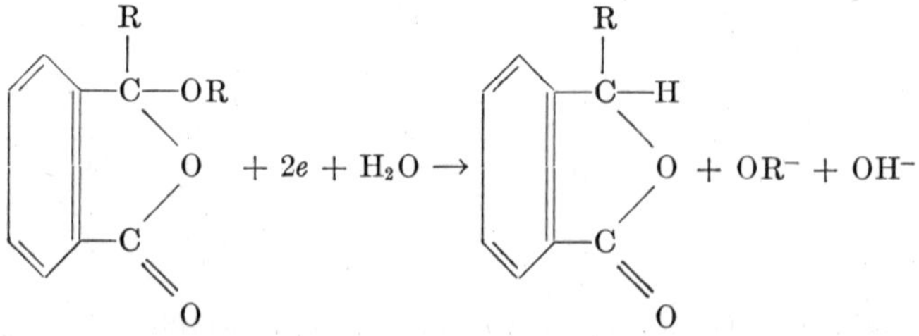

[47] S. Wawzonek and J. H. Fossum, *J. Electrochem. Soc.*, **96**, 234 (1949).

TABLE XLI-14

HALF-WAVE POTENTIALS AND DIFFUSION CURRENT CONSTANTS FOR PHTHALALDEHYDIC ACID DERIVATIVES AND RELATED COMPOUNDS IN VARIOUS SOLUTIONS AT 25° IN 0.1 *M* TETRABUTYLAMMONIUM IODIDE–50 PER CENT DIOXANE

Compound	$-E_{1/2}$ (S.C.E.), v.	i_d, microamp.	C, mmoles/liter	I_d (diffusion current constants)
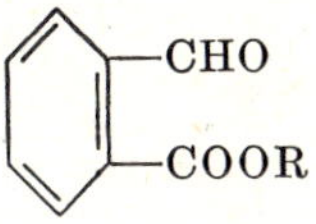				
Acid (R = H)	1.08	2.16	1.15	1.25
	1.55	2.02	—	1.17
	2.11	0.58	—	0.34
n-Methyl ester (R = CH_3)	1.35	3.32	1.15	1.92
	2.07	4.18	—	2.42
n-Ethyl ester R = CH_2CH_3	1.37	2.55	1.06	1.60
	2.05	4.59	—	2.88
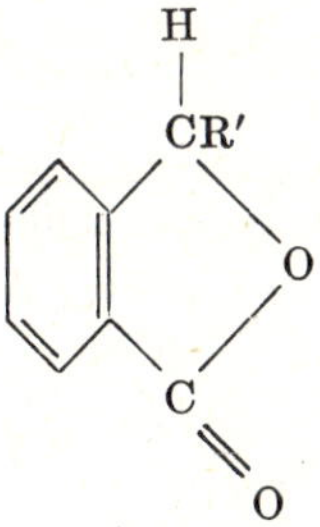				
Phthalide (R′ = H)	2.03	8.05	1.37	3.91
ψ-Methyl ester (R′ = OCH_3)	1.89	10.97	2.39	3.27
	2.02	5.42	—	1.52
ψ-Ethyl ester (R′ = OC_2H_5)	1.89	8.74	2.05	2.84
	2.00	4.67	—	1.51
ψ-Phenyl ester (R′ = OC_6H_5)	1.66	3.44	0.98	2.34
	2.05	4.48	—	3.05
Acetoxyphthalide (R′ = $OCOCH_3$)	1.53	4.49	1.37	2.18
	2.06	5.11	—	2.48
Diphthalyl ether (R′ = O—CH 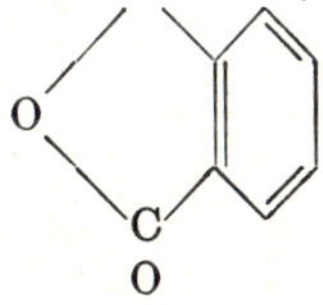)	1.59	6.24	1.04	4.00
	2.04	4.22	—	2.70
Anil (R′ = NHC_6H_5)	1.10	1.34	0.86	1.04
	1.56	1.75	—	1.36
	2.10	0.27	—	0.21

TABLE XLI-14—*Continued*

Compound	$-E_{1/2}$ (S.C.E.), v.	i_d, microamp.	C, mmoles/liter	I_d (diffusion current constants)
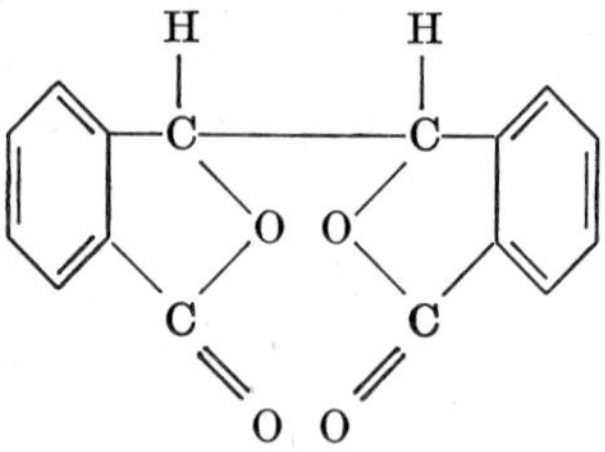				
Dihydrophthallyl	1.91	6.98	1.22	3.81
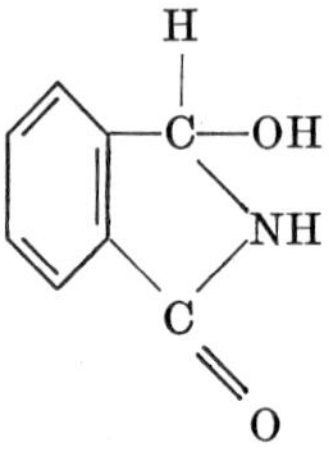				
Amide	2.01	7.19	1.85	2.58
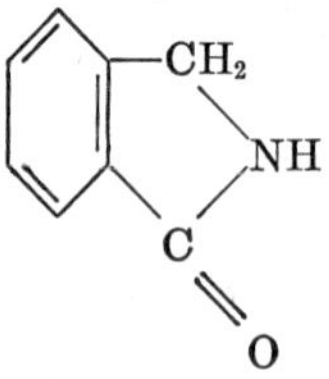				
Phthalimidine	Max.			
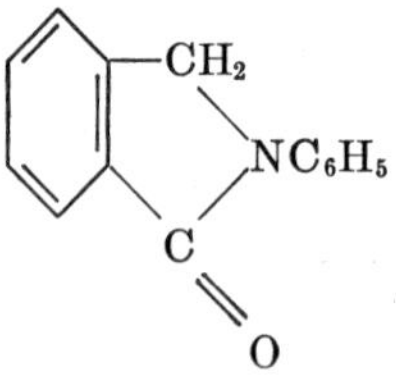				
N-Phenylphthalimidine	2.20	4.00	1.02	2.61
Benzaldehyde	1.54	9.77	5.42	1.20
Buffer I				
Acid	1.38	4.96	1.37	2.41
n-Methyl ester	1.21	3.48	1.04	2.23
n-Ethyl ester	1.22	2.38	0.83	1.92

TABLE XLI-14—*Continued*

Compound	$-E_{1/2}$ (S.C.E.), v.	i_d, microamp.	C, mmoles/liter	I_d (diffusion current constants)
Phthalide	2.05	8.24	1.29	4.25
ψ-Methyl ester	1.90	3.37	1.00	2.25
ψ-Ethyl ester	1.92	2.77	1.01	1.83
ψ-Phenyl ester	1.73	2.78	0.90	2.06
Acetoxyphthalide	1.58	2.77	0.90	2.06
Diphthalyl ether	1.69	6.66	1.30	3.41
Anil	1.16	2.09	0.88	1.58
	1.40	0.60	—	0.45
Dihydrophthallyl	1.92	4.03	1.06	2.53
Amide	2.01	5.34	1.09	3.26
Phthalimidine	Max.	—	—	—
N-Phenylphthalimidine	2.14	11.59	1.00	7.74
Benzaldehyde	1.46	4.66	1.46	2.12
Buffer II				
Acid	1.49	4.74	1.45	2.18
n-Methyl ester	1.46	2.76	1.04	1.77
n-Ethyl ester	1.47	1.82	0.83	1.46
Phthalide	2.05	7.61	1.29	3.92
ψ-Methyl ester	1.88	3.98	1.00	2.66
ψ-Ethyl ester	1.91	3.10	1.01	2.02
ψ-Phenyl ester	1.71	2.48	0.90	1.83
Acetoxyphthalide	1.51	2.88	0.90	2.14
Diphthalyl ether	1.47	10.44	1.51	4.60
Anil	1.15	0.45	0.89	0.34
	1.45	2.49	—	1.86
Dihydrophthallyl	1.52	0.59	0.66	0.60
	1.71	0.56	—	0.57
Amide	2.00	3.87	1.09	2.36
Phthalimidine	Max.	—	—	—
N-Phenylphthalimidine	2.15	17.24	1.00	11.50
Benzaldehyde	1.53	4.45	1.11	2.67

while the normal derivatives must be first reduced to the benzyl alcohol and then lose alcohol or acid to form the phthalide:

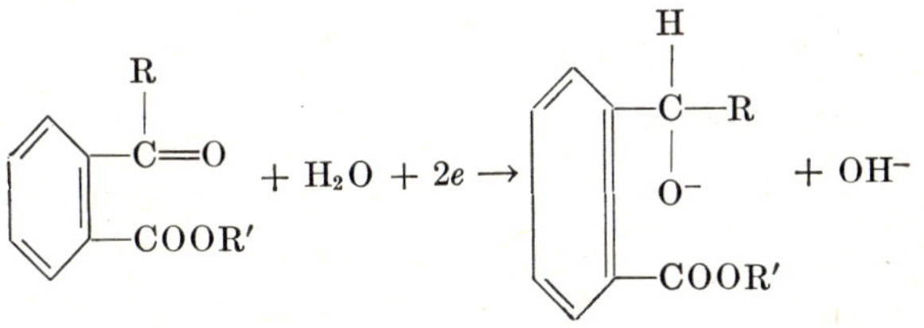

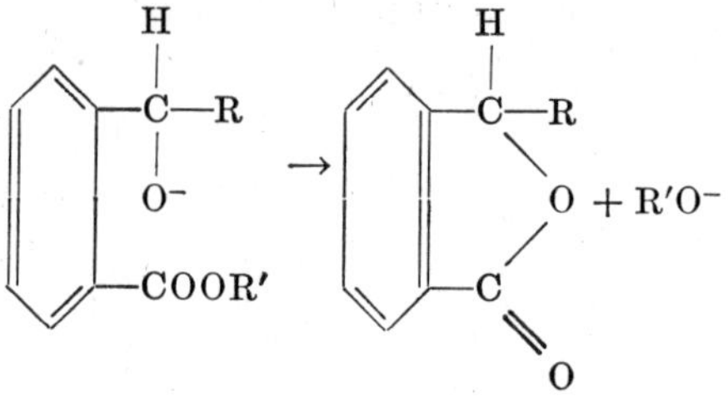

$$R'O^- + H_2O \rightleftharpoons R'OH + HO^-$$

TABLE XLI-15A

POLAROGRAPHY OF 2-BENZOYLBENZOIC ACID DERIVATIVES AND RELATED COMPOUNDS IN VARIOUS SOLUTIONS

Compound	$-E_{1/2}$ (S.C.E.), v.	i_d, microamp.	C, mmoles/liter	i_d/C
In 0.1 *M* tetrabutylammonium iodide–50% dioxane				
Esters				
n-Methyl	1.51	3.51	0.87	4.03
	1.94	3.44	—	3.95
ψ-Methyl	1.85	5.33	0.83	6.53
n-Ethyl	1.49	4.95	1.34	3.86
	1.92	4.49	—	3.51
ψ-Ethyl	1.85	8.58	1.65	5.22
Phenyl (m.p. 83°)	1.30	3.69	1.07	3.45
	1.92	4.09	—	3.82
Phenyl (m.p. 162–163°)	1.53	2.85	0.77	3.70
	1.91	2.75	—	3.57
Thiophenyl (m.p. 112–113°)	1.17	3.75	1.15	3.36
	1.67	1.47	—	1.32
	1.92	2.92	—	2.62
Thiophenyl (m.p. 119–121°)	1.41	4.46	1.36	3.27
	1.91	4.16	—	3.05
β-Naphthyl (m.p. 103–105°)	1.25	3.02	1.00	3.02
	1.94	3.82	—	3.82
β-Naphthyl (m.p. 198–199°)	1.27	3.95	1.08	3.66
4-Bromophenyl (m.p. 92–94°)	1.22	4.15	1.05	3.95
	Max.			
4-Bromophenyl (m.p. 170–172°)	1.47	4.29	1.05	4.09
	1.93	4.39	—	4.18
3-Phenylphthalide	1.89	3.90	1.00	3.90
Benzophenone	1.48	5.50	1.38	3.98
4′-Hydroxydiphenylphthalide	1.88	5.49	1.74	3.16
Ethyl benzoate	2.18	7.70	1.01	7.62
2-Benzoylbenzoic acid	1.11	1.87	1.00	1.87
Benzoic acid	1.87	2.34	1.10	2.12
Acetic 2-benzoylbenzoic anhydride	1.48	3.90	1.00	3.90
	1.92	2.99	—	2.99
2-Benzoylbenzoic anhydride	1.28	2.73	1.00	2.73
	1.58	3.05	—	3.05
	1.90	2.60	—	2.60
3,3′-Diphenyldiphthalyl sulfide	1.35	2.92	0.82	3.56
	1.56	1.20	—	1.46
	1.91	1.90	—	2.32
In 0.1 *M* hydrochloric acid–50% dioxane				
2-Benzoylbenzoic acid	0.95	5.89	1.15	5.12

TABLE XLI-15B

POLAROGRAPHY OF 2-BENZOYLBENZOIC ACID DERIVATIVES IN BUFFERED MEDIA

Compound	$-E_{1/2}$ (S.C.E.), v.	i_d, microamp.	C, mmoles/liter	i_d/C
Buffer I				
Anthraquinone	0.63	2.61	1.06	2.50
Esters				
n-Methyl	1.33	4.71	1.19	3.96
ψ-Methyl	1.82	6.22	1.10	5.16
n-Ethyl	1.39	8.04	1.99	4.02
ψ-Ethyl	1.87	5.43	1.24	4.37
Phenyl (83°)	1.27	4.61	1.23	3.75
Phenyl (162–163°)	1.59	4.50	1.31	3.45
Thiophenyl (112–113°)	1.22	6.16	1.66	3.73
	1.47	2.38	—	1.43
	1.63	3.37	—	2.90
Thiophenyl (119–121°)	1.43	2.89	1.17	2.48
β-Naphthyl (103–105°)	1.30	3.49	1.07	3.26
β-Naphthyl (198–199°)	1.29	2.34	1.05	2.23
4-Bromophenyl (92–94°)	1.29	4.59	1.14	4.03
4-Bromophenyl (170–172°)	1.55	4.84	1.28	3.78
Benzophenone	1.33	6.28	1.38	4.55
Acetic 2-benzoylbenzoic anhydride	1.44	6.43	1.89	3.38
2-Benzoylbenzoic anhydride	1.20	5.57	0.82	6.75
3,3′-Diphenyldiphthalyl sulfide	1.15	3.71	0.82	4.53
2-Benzoylbenzoic acid	1.44	3.61	0.89	4.06
Buffer II				
Anthraquinone	0.76	2.65	1.06	2.50
Esters				
n-Methyl	1.51	4.19	1.13	3.75
ψ-Methyl	1.84	7.08	1.43	4.95
n-Ethyl	1.48	7.77	2.12	3.67
ψ-Ethyl	1.85	12.12	2.38	5.10
Phenyl (83°)	1.28	3.08	1.10	2.80
	1.53	0.79	—	0.72
	1.91	3.42	—	3.11
Phenyl (162–163°)	1.59	4.66	1.30	3.58
	1.93	1.28	—	0.99
Thiophenyl (112–113°)	1.15	4.12	1.29	3.20
	1.46	0.91	—	0.71
	1.69	1.32	—	1.02
Thiophenyl (119–121°)	1.42	4.02	1.36	2.96
	1.94	3.95	—	2.91
β-Naphthyl (103–105°)	1.29	4.99	1.65	3.03
	1.98	5.67	—	3.44
β-Naphthyl (198–199°)	1.30	4.19	1.18	3.55
	2.10	3.18	—	2.70
4-Bromophenyl (92–94°)	1.26	5.85	1.55	3.77
4-Bromophenyl (170–172°)	1.55	3.35	0.95	3.54
Benzophenone	1.45	9.91	2.40	4.13
Acetic 2-benzoylbenzoic anhydride	1.52	5.17	1.50	3.42
	1.91	1.72	—	1.14
2-Benzoylbenzoic anhydride	1.27	1.61	0.72	2.25
	1.55	3.06	—	4.28
	1.89	1.06	—	1.48
Di-2-benzoylbenzoyl sulfide	1.28	2.88	0.69	4.16
	1.48	1.03	—	1.48
	1.91	1.57	—	2.26
2-Benzoylbenzoic acid	1.53	5.23	1.31	4.00

Evidence for this mechanism is the fact that compounds which reduce by the first mechanism do so with an ease which increases with increasing ionization constant of the alcohol, phenol, or acid concerned. If members of this class are stable in buffer II, they produce half-wave potentials which are independent of pH.

The second class, consisting of the open-chain compounds, is more easily hydrolyzed and produces half-wave potentials which vary with pH similar to benzophenone and benzaldehyde. The acids show more complex behavior because of their ability to ionize.

The polarographic method has likewise proved successful in distinguishing between the amides of phthalaldehydic acid and 2-benzoylbenzoic acid which can exist in two isomeric forms and exhibit ring-chain tautomerism of the following types:[48]

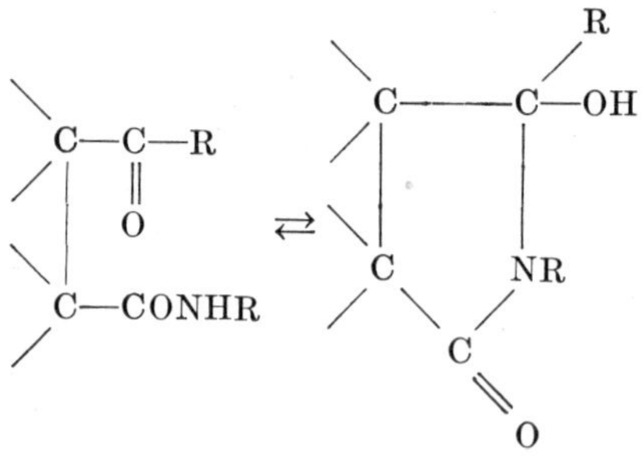

(I) (II)

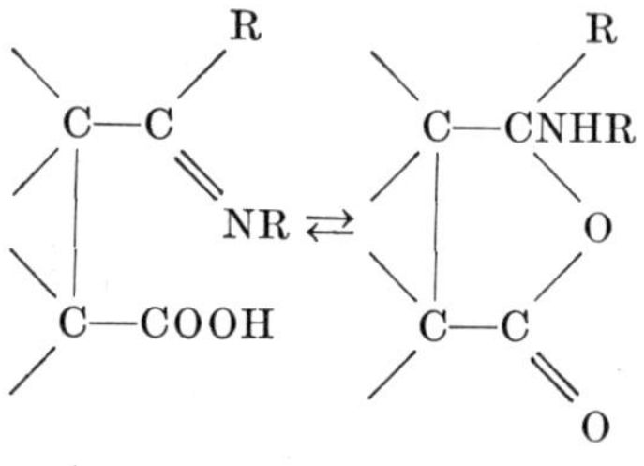

(III) (IV)

A summary of the observed half-wave potentials and individual diffusion current constants in the various solutions is given in Table XLI-16.

Forms (I) and (II) apparently both yield the same reduction products, the 1-ketoisoindole, through mechanisms similar in nature to those pro-

[48] S. Wawzonek, H. A. Laitinen, and S. J. Kwiatkowski, *J. Am. Chem. Soc.*, **66**, 830 (1944).

TABLE XLI-16

POLAROGRAPHY OF THE AMIDES OF 2-BENZOYLBENZOIC ACID, THEIR DERIVATIVES, AND RELATED COMPOUNDS IN VARIOUS SOLUTIONS

Compound	$-E_{1/2}$ (S.C.E.), v.	i_d, microamp.	C, mmoles/liter	i_d/C
In 0.1 *M* tetrabutylammonium iodide–50% dioxane				
n-Anilide (m.p. 195°)	1.48 Max. 2.20	3.12 0.91 3.59	1.00 — —	3.12 0.91 3.59
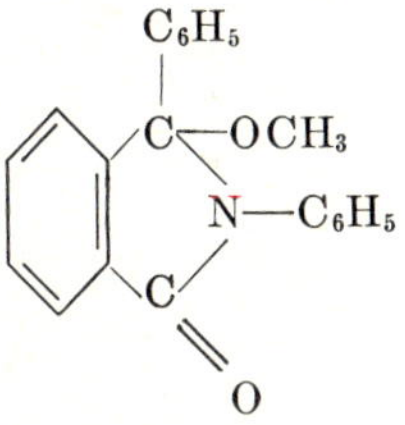	1.88 2.20	4.62 5.56	1.33 —	3.48 4.19
Amide	1.52 1.94 2.26	1.36 2.53 4.42	1.00 — —	1.36 2.53 4.42
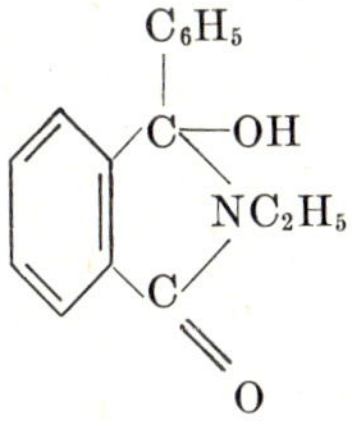	2.06 2.22	4.42 5.74	1.25 —	3.56 4.54
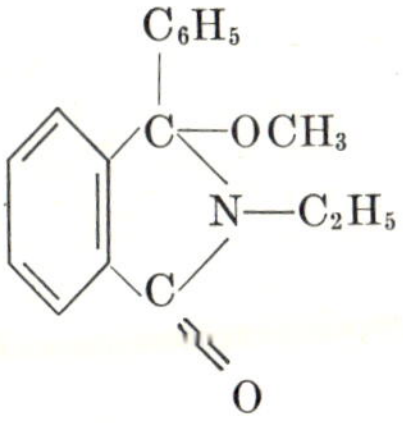	1.95 2.38	3.59 5.19	0.90 —	4.00 5.76
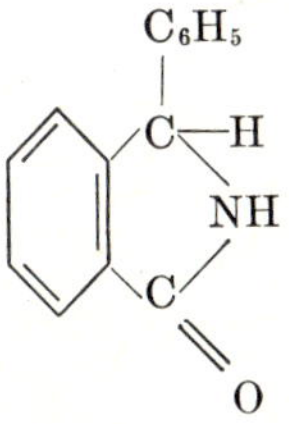	2.25	4.16	1.00	4.16

TABLE XLI-16—*Continued*

Compound	$-E_{1/2}$ (S.C.E.), v.	i_d, microamp.	C, mmoles/liter	i_d/C
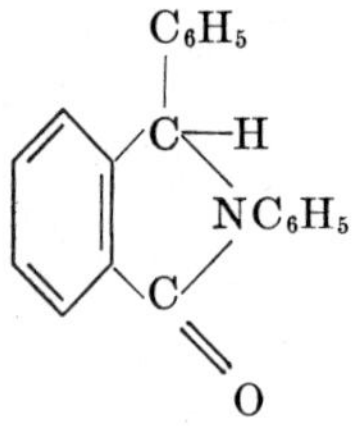	2.16	4.42	1.00	4.42
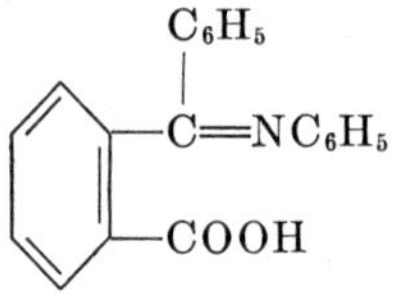	0.96 1.70	1.74 2.21	1.00 —	1.74 2.21
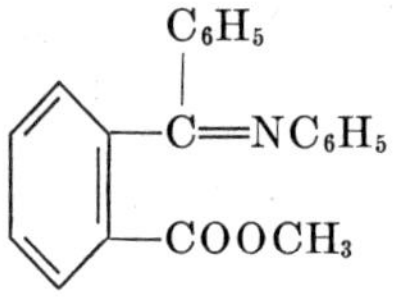	1.42 2.11	3.77 5.85	1.00 —	3.77 5.85
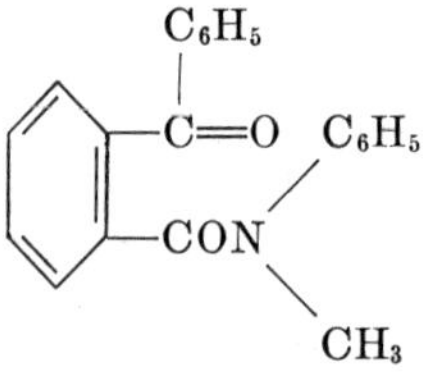	1.46 1.97	3.64 6.37	1.00 —	3.64 6.37
Benzamide	1.82	10.14	1.10	9.21
Buffer I				
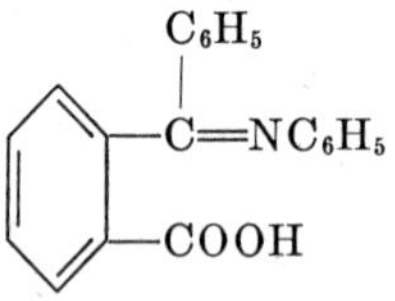	1.15	2.24	0.82	2.72
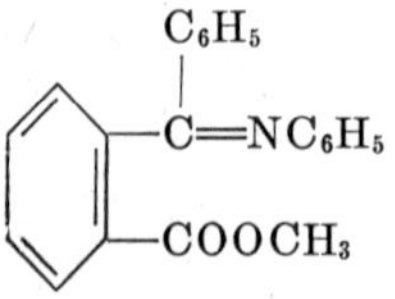	1.32	3.14	0.92	3.43

TABLE XLI-16—*Continued*

Compound	$-E_{1/2}$ (S.C.E.), v.	i_d, microamp.	C, mmoles/liter	i_d/C
Buffer I—*Continued*				
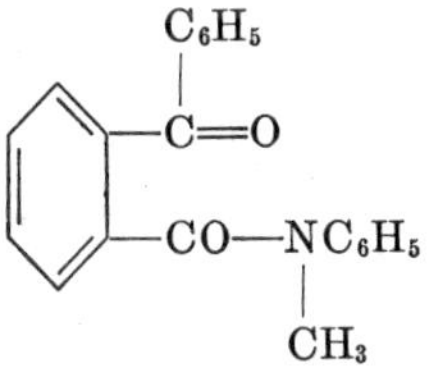	1.30	4.92	1.16	4.24
Buffer II				
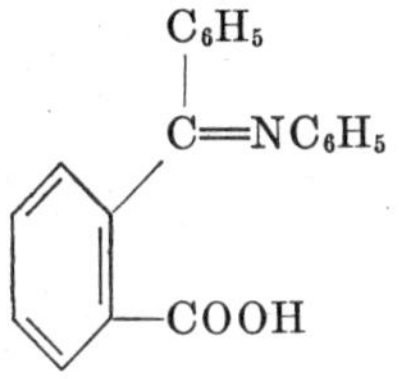	1.61	4.36	1.01	4.33
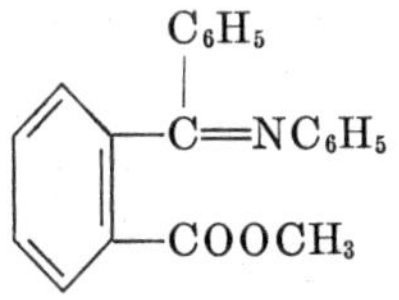	1.45	3.83	0.85	4.52
	2.15	5.02	—	5.94
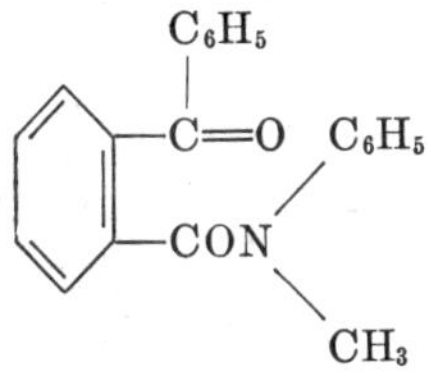	1.48	3.61	1.12	3.22
	2.05	4.38	—	3.91

posed for the production of phthalides from normal and cyclic esters. Forms (III) and (IV) seem to resemble the parent acid in the series. Where equilibrium structures existed no attempt was made to determine whether any of the currents were kinetic in nature.

The polarographic method also proved successful in distinguishing between the *cis-trans* and cyclic derivatives of β-(*p*-bromobenzoyl)crotonic acid.[49] The experiments were carried out in 0.1 *M* tetrabutylammonium

[49] S. Wawzonek, R. C. Reck, W. W. Vaught, Jr., and J. W. Fan, *J. Am. Chem. Soc.*, **67**, 1300 (1945).

iodide solution in 50 per cent dioxane, and in the same medium containing 0.052 *M* tetrabutylammonium hydroxide. The results are summarized in Table XLI-17.

TABLE XLI-17
POLAROGRAPHY OF β-(*p*-BROMOBENZOYL)CROTONIC ACID

Compound	$-E_{1/2}$ (S.C.E.), v.	i_d, microamp.	C, mmoles/liter	I_d (diffusion current constant)
$CH_3CCOC_6H_4Br$ ‖ HCCOR R = OH	0.91	2.08	1.89	0.56
	1.36	1.34	—	0.36
	1.55	5.84	—	1.57
	1.80	5.96	—	1.60
	1.49[a]	10.86[a]	1.56[a]	3.52[a]
	1.77[a]	4.69[a]	—	1.52[a]
R = OCH_3	1.19	4.45	1.07	2.10
	1.57	4.40	—	2.08
	1.73	2.10	—	0.99
R = $N(CH_3)_2$	1.31	4.41	1.14	1.97
	1.59	3.44	—	1.54
	1.74	2.20	—	0.97
	1.30[a]	3.45[a]	1.08[a]	1.62[a]
	1.55[a]	3.68[a]	—	1.73[a]
	1.79[a]	1.88[a]	—	0.88[a]
R = $N(CH_3)C_6H_5$	1.23	6.30	1.58	2.02
	1.59	4.35	—	1.40
	1.76	4.33	—	1.39
$CH_3CCOC_6H_4Br$ ‖ RCOCH R = OH	0.96	2.21	1.15	0.98
	1.51	3.39	—	1.50
	1.79	1.47	—	0.65
	2.32	2.54	—	1.13
	1.39[a]	4.10[a]	1.10[a]	1.91[a]
	1.60[a]	3.15[a]	—	1.45[a]
	1.79[a]	2.18[a]	—	1.30[a]
R = OCH_3	1.03	4.99	1.25	2.02
	1.55	3.95	—	1.60
	1.74	2.58	—	1.04
R = NHC_6H_5	1.06	4.42	1.09	2.06
	1.53	3.04	—	1.41
	1.67	2.48	—	1.16
R = $N(CH_3)C_6H_5$	1.13	5.30	1.26	2.13
	1.59	4.06	—	1.63
	1.74	3.48	—	1.40
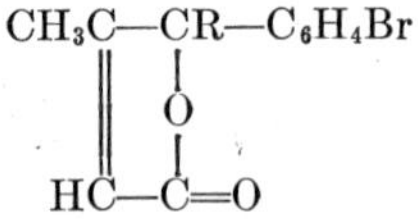				
R = OCH_3	1.71	4.85	1.18	2.08
R = $N(CH_3)C_6H_5$	1.01	3.02	0.83	1.86
	0.96[a]	0.67[a]	0.98[a]	0.35[a]
	1.50[a]	4.69[a]	—	2.43[a]
	1.79[a]	1.98[a]	—	1.02[a]

TABLE XLI-17—*Continued*

Compound	$-E_{1/2}$ (S.C.E.), v.	i_d, microamp.	C, mmoles/liter	I_d (diffusion current constant)
R = NHC_6H_5	0.99	2.11	1.17	0.91
	1.63	2.24	—	0.97
	2.34	4.58	—	1.99
β-(*p*-Bromobenzoyl)butyric acid......	1.34	2.82	1.54	0.93
	1.68	8.86	—	2.92
	1.55[a]	6.40[a]	1.27[a]	2.55[a]
	1.74[a]	3.22[a]	—	1.29[a]

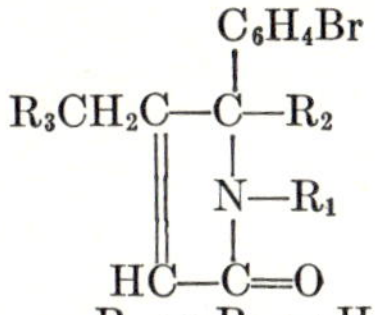

Compound	$-E_{1/2}$ (S.C.E.), v.	i_d, microamp.	C, mmoles/liter	I_d (diffusion current constant)
$R_1 = R_3 = H$; $R_2 = OH$.........	2.02	10.53	1.38	3.87
	2.44	4.76	—	1.75
	2.10[a]	5.88[a]	1.09[a]	2.72[a]
	2.35[a]	0.91[a]	—	0.42[a]
$R_1 = R_3 = H$; $R_2 = OCH_3$.......	2.01	11.63	1.47	4.01
	2.50	3.83	—	1.35
$R_1 = R_2 = R_3 = H$..............	2.07	6.95	1.60	2.20
	2.40	7.10	—	2.26
$R_1 = H$, $R_2 = OH$...............	1.85	6.29	1.26	2.53
$R_3 = Cl$	2.27	4.68	—	1.89
$R_1 = H$, $R_2 = OH$...............	1.77	4.51	0.88	6.95
$R_3 = Cl$	2.23	4.68	—	2.68
$R_1 = CH_3$, $R_2 = OH$.............	2.23	15.41	1.12	6.95
$R_3 = H$	—	—	—	—
$R_1 = CH_3$, $R_2 = OCH_3$..........	1.94	5.36	1.50	1.76
$R_3 = H$	Max.	14.21	—	4.78
$R_1 = CH_3$, $R_2 = R_3 = H$........	2.30	13.7	1.55	4.47
$R_1 = CH_3$, $R_2 = OH$.............	2.13	14.65	1.04	7.14
$R_3 = Cl$	—	—	—	—
$R_1 = CH_3$, $R_2 = OCH_3$..........	1.78	5.79	1.16	2.53
$R_3 = Cl$	2.22	12.10	—	5.28
$R_1 = C_6H_5$, $R_2 = OH$............	1.89	3.88	1.52	1.30
$R_3 = H$	2.23	6.85	—	2.28
$R_1 = C_6H_5$, $R_2 = OCH_3$..........	1.78	8.24	1.44	2.89
$R_3 = H$	2.16	7.36	—	2.59
$R_1 = C_6H_5$, $R_2 = R_3 = H$.......	2.23	7.30	0.99	3.75

[a] In 0.1 *M* tetrabutylammonium iodide–0.052 *M* tetrabutylammonium hydroxide–50 per cent dioxane.

The *cis* and *trans* acid derivatives produce half-wave potentials which do not change as the solution becomes alkaline. All are reduced first at the double bond for they show two final waves characteristic of the ketone group. The *trans* form is reduced at more positive potentials than the *cis* form in all instances.

The cyclic forms which have a furanone structure give only one wave corresponding to a 2-electron reduction to a dihydrofuranone. The anilide having this structure resembles the anil in the 2-benzoylbenzoic acid series. Amides having a hydroxypyrroline structure are reduced to the pyrrolidones.

The acids, because of their ability to ionize, show a more complex behavior.

6. HALOGENATED ACIDS

The most complete polarographic study of a halogenated acid has been carried out with α-bromopropionic acid.[50] The values obtained in buffers of varying pH are given in Table XLI-18. No wave was observed in the buffer range pH 5.5 to 6.9.

TABLE XLI-18

POLAROGRAPHY OF α-BROMOPROPIONIC ACID

	$-E_{1/2}$ (N.H.E.)		i_d (0.00125 *M*), microamp.	
pH	25°	0°	25°	0°
0.8	0.26	—	9.3	—
2.2	0.244	0.29	9.2	5.3
2.6	0.249	—	9.0	—
3.04	0.300	—	8.8	—
3.53	0.364	—	8.6	—
4.4	0.57	—	8.3	—
5.5	0.74	0.92	7.0	5.5
6.0	0.90	—	6.8	—
7.1	1.20	—	7.6	—
7.5	1.20	1.31	6.9	4.1
8.1	1.21	—	7.0	—

The height of the wave corresponds to a 2-electron reduction and points to the following behavior in acid medium:

$$CH_3CH(Br)COOH + H^+ + 2e \rightarrow CH_3CH_2COOH + Br^- \quad \text{(acid)}$$

and to the following reaction in alkaline medium:

$$CH_3CH(Br)COO^- + H_2O + 2e \rightarrow CH_3CH_2COO^- + Br^- + HO^- \quad \text{(alkaline)}$$

An explanation is offered for the variation of half-wave potentials with

[50] E. Saito, *Bull. soc. chim. France*, **1948**, 404.

pH but no reason is given for the absence of a wave between pH 5.5 and 6.9.

Other halogenated acids and derivatives studied are given in Table XLI-19.

TABLE XLI-19
POLAROGRAPHIC DATA FOR HALOGENATED ACIDS AND THEIR DERIVATIVES

Compound	$-E_{1/2}$	i_d/C	Electrolyte
Chloroacetic acid (*1*)	1.70 (S.C.E)	1.0	0.05 *M* tetrabutylammonium bromide–75% dioxane
Chloroacetamide (*2*)	1.6 (*2*)	—	0.1 *N* NaOH
Bromoacetic acid (*3*)	1.4	—	0.1 *N* KCl
Iodoacetic acid (*4*)	0.4	—	0.1 *N* KCl
Dichloroacetic acid (*1*)	1.19 (S.C.E.) 1.64 (S.C.E.)	0.9 1.1	0.05 *M* tetrabutylammonium bromide–75% dioxane
	1.4 (*2*)		0.1 *N* NaOH
Trichloroacetic acid (*1*)	0.80 *ca.* (S.C.E.) 1.66 (S.C.E.)	1.7 1.0	0.05 *M* tetrabutylammonium bromide–75% dioxane
	0.4 (*2*)	—	0.1 *N* NaOH

(*1*) M. von Stackelberg and W. Stracke, *Z. Electrochem.*, **53,** 118 (1949).
(*2*) M. B. Neiman, A. V. Ryabov, and E. M. Sheyanova, *Doklady Akad. Nauk S. S. S. R.*, **68,** 1065 (1949); *Chem. Abstracts*, **44,** 1360 (1950). These investigators used, as an electrode, a fritted glass disc through which mercury passed.
(*3*) M. B. Neiman, T. A. Petukhovskaya, A. V. Ryabov, and E. M. Sheyanova, *Zavodskaya Lab.*, **15,** 1277 (1949); *Chem. Abstracts*, **44,** 3817 (1950). Used fritted glass disc electrode.
(*4*) R. Brdicka, *J. Gen. Physiol.*, **19,** 843 (1936).

CHAPTER XLII

Nitro Compounds, Their Reduction Products, and Related Compounds

1. NITRO COMPOUNDS

Nitro compounds show reduction waves at the dropping mercury electrode which are dependent upon the pH of the solution and the structure of the compound.

(a) Aliphatic Nitro Compounds

The polarographic behavior of nitromethane, nitroethane, 1-nitropropane, and 1-nitrobutane has been studied in buffered solutions.[1] In an acid medium only one wave is found, but at pH 4.5 a second wave (starting at about -1.5 v.) is observed, which is obscured by the hydrogen wave in a more strongly acid medium. The second wave disappears in a more strongly alkaline medium (pH greater than 10).

The half-wave potential of the first wave is about the same for nitromethane, -ethane, -propane, and -butane, and it changes only slightly with pH. The variation of the half-wave potential of nitromethane with pH is given below:

pH	1.8	3.3	4.6	5.7	6.8	8.0–11.9
$-E_{1/2}$ (S.C.E.)	0.75	0.83	0.85	0.87	0.89	0.90

The diffusion current is proportional to the concentration in the investigated range between 2×10^{-4} and 10^{-3} M. The ratio of the diffusion currents of the four nitroparaffins is 1:0.83:0.67:0.65. At pH 7 or greater the diffusion current of the freshly prepared solution decreases on standing until an equilibrium value is attained. This decrease is due to a slow transformation of the nitro form into the *aci* form, which is not reducible. The rate of this tautomeric reaction can be found polarographically.

The reduction, which is irreversible, involves four electrons and produces the hydroxylamine. The second wave is due to the irreversible reduction of the hydroxylammonium ion to the corresponding alkylammonium ion:

$$CH_3NO_2 \xrightarrow[H_2O]{4e \; 4H^+} CH_3NHOH \xrightarrow[H_2O]{2e \; 3H^+} CH_3NH_3^+$$

[1] F. Petru, *Collection Czechoslov. Chem. Communs.*, **12**, 620 (1947).

This mechanism was verified by a study of N-methylhydroxylamine and ordinary hydroxylamine. In both cases curves were obtained which were similar to the second curve observed in the reduction of the nitroparaffins. The following half-wave potentials (S.C.E.) for hydroxylamine are reported: pH 4.6, −1.46 v; pH 6.8, −1.52 v; pH 8.0, −1.58 v; pH 9.2, −1.69 v; and pH 10.4, −1.75 v.

Values (N.C.E.) obtained for tetranitromethane are: pH 1, −0.59 v; pH 12, −0.409 v; and pH 13.2, −0.423 v. The wave at pH 12 is suitable for concentration determination but overlaps that of nitrobenzene if this compound is present. The tetranitromethane is removed from such solutions by adding methanol and vacuum distilling at 70°.[2]

For quantitative estimation of nitroparaffins an acid solution (0.05 *M* sulfuric acid) is recommended as an electrolyte since the diffusion current is proportional to the concentration. In a medium of 0.05 *M* sodium sulfate (unbuffered) the diffusion current does not increase linearly with the concentration due to formation of hydroxyl ions at the electrode which cause a shift in the equilibrium between the normal and *aci* forms in favor of the *aci* form. The diffusion current has a temperature coefficient of 1.5 per cent per degree. The half-wave potentials (S.C.E.) in 0.05 *M* sulfuric acid and 0.05 *M* sodium sulfate (unbuffered) are given in Table XLII-1. The half-wave potentials shift to slightly less negative potentials with increasing concentration.

TABLE XLII-1

$-E_{1/2}$ (0.0005 *M* SOLUTIONS *vs.* S.C.E.)

Substance	Medium	
	0.05 *M* H_2SO_4	0.05 *M* Na_2SO_4
Nitromethane	0.685	0.89
Nitroethane	0.663	0.90
1-Nitropropane	0.617	0.89
1-Nitrobutane	0.565	0.84
2-Nitropropane	0.677	0.94
2-Nitrobutane	0.645	0.93

The polarographic method has been used to determine rate of enolization of a nitroparaffin by introducing the compound into a buffered solution of a given pH and following the decrease in diffusion current with time;[3] it has also been used to determine the amount of nitromethane in air.[4]

[2] A. K. Cruse and R. Haul, *Z. Electrochem.*, **53,** 115 (1949).

[3] E. W. Miller, A. P. Arnold, and M. J. Astle, *J. Am. Chem. Soc.*, **70,** 3971 (1948).

[4] H. N. Wilson and W. Hutchinson, *Analyst*, **72,** 432 (1947).

Recovery of nitromethane in air by passage through gas absorption bottles containing 0.2 *N* sulfuric acid was unsatisfactory and was abandoned in favor of a technique using an evacuated winchester.

Procedure. Evacuate a winchester of known volume to about 30 mm. of mercury pressure, measuring the pressure with a manometer. Then open it in the atmosphere to be tested, taking the temperature at the same time. Introduce 50 ml. of 0.2 *N* sulfuric acid by cooling the winchester slightly and allowing the acid to be drawn in. Shake well and allow to stand for 30 minutes. Polarograph the sulfuric acid solution, without deoxygenating, at a suitable sensitivity, over the voltage range from −0.3 to −1.3 v. against the saturated calomel electrode. The method has been proved satisfactory for determining 0.2 to 14 g. of nitromethane per liter.

The polarographic method has also been used to determine nitroethane in blood and air using 1 per çent hydrochloric acid as the electrolyte.[5]

(b) Aromatic Nitro Compounds

The reduction of aromatic nitro compounds at the dropping mercury electrode differs from the classical reduction mechanism proposed by Haber in that, in general, the aromatic nitro compounds are reduced in a 4-electron step to a phenylhydroxylamine. In an acid medium the phenylhydroxylamine, or better the phenylhydroxylammonium ion, is reduced further to the aniline in a second separate wave. In an alkaline medium, depending upon the strength of the base, the phenylhydroxylamine is stable to further reduction but may, under the influence of alkali, form condensation products which are electroreducible.

Exceptions to this generalization are *o*- and *p*-nitrophenols, *p*-nitroaniline, and *p*-dinitrobenzene, which form phenylhydroxylamines that can be converted to reducible quinoid forms. This type is reduced to the substituted aniline in neutral and alkaline media:

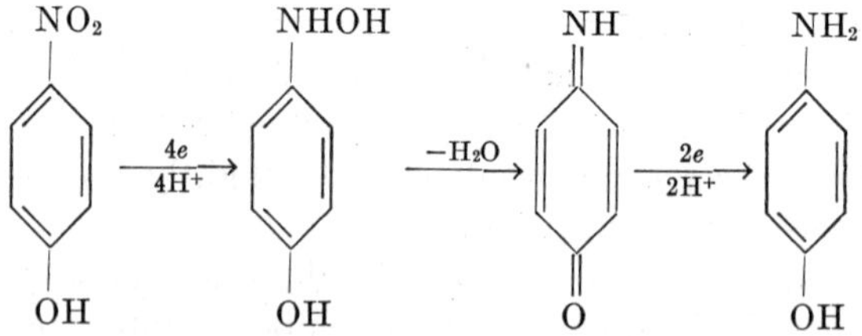

Nitrobenzene. Nitrobenzene gives two waves of 4 and 2 electrons, respectively, at a pH less than 5. Values obtained in buffers made from sodium acetate and hydrochloric acid, potassium acid phthalate and hydro-

[5] E. W. Scott, *J. Ind. Hyg. Toxicol.*, **25**, 20 (1943).

chloric acid or sodium hydroxide, and borax, and containing 8 per cent ethanol by volume, are listed in Table XLII-2. The half-wave potential of the second wave, which corresponds to a further 2-electron reduction, is not given. The half-wave potentials are independent of and proportional to the concentration in the range of 0.04 to 0.6 millimole.

TABLE XLII-2

HALF-WAVE POTENTIALS OF NITROBENZENE

pH	$-E_{1/2}$ (S.C.E.) first wave ($n = 4$)	pH	$-E_{1/2}$ (S.C.E.) first wave ($n = 4$)
0.5	0.16	4.1	0.42
1.7	0.26	5.7	0.48
2.5	0.30	7.4	0.58
3.0	0.34	9.2	0.70
3.8	0.38		

TABLE XLII-3

POLAROGRAPHY OF NITROBENZENE

	First step		Second step	
pH	$-E_{1/2}$ (S.C.E.)	I_d	$-E_{1/2}$ (S.C.E.)	I_d
1	0.22	5.90	0.675	1.95
2	0.30	5.65	0.745	2.40
3	0.34	5.35	0.860	2.15
4	0.415	5.50	0.975	0.95
6	0.535	5.35		
8	0.645	5.05		
10	0.74	5.80		
12	0.83	5.75		

Plots of the half-wave potential against log $[i/(i_d - i)]$ give a straight line with a slope of 0.075 at pH 2.5 and 0.088 at a pH of 9.2.[6]

Similar results are obtained in buffers made from sodium chloride and hydrochloric acid, sodium phosphate and citric acid or sodium hydroxide, and glycine and sodium hydroxide. The results obtained in a 10 per cent ethanol solution are listed in Table XLII-3.[7] Gelatin (0.01 per cent) is used to suppress maxima but is reported to shift the half-wave potential and decrease the diffusion current.[8]

[6] J. Pearson, *Trans. Faraday Soc.*, **44,** 683 (1948).
[7] J. E. Page, J. W. Smith, and J. G. Waller, *J. Phys. & Colloid Chem.*, **53,** 545 (1949).
[8] I. A. Korshunov and A. S. Kirillova, *J. Gen. Chem. U. S. S. R.*, **18,** 785 (1948).

The ratio of the heights of the two waves in an acid medium is 2:1 only for concentrations less than 10^{-4} M. While the concentration dependence of the height of the first wave is strictly linear, the height of the second wave increases less rapidly with concentration. The temperature coefficient is less for the second wave than the first (1.5 per cent per degree). The height of the second wave decreases more rapidly with decreasing height of mercury than is expected of a diffusion current. On addition of gelatin the ratio of the heights of the two waves approaches the theoretical value 2:1. These facts point to a reaction which competes with the electrode process and may be the following:[9]

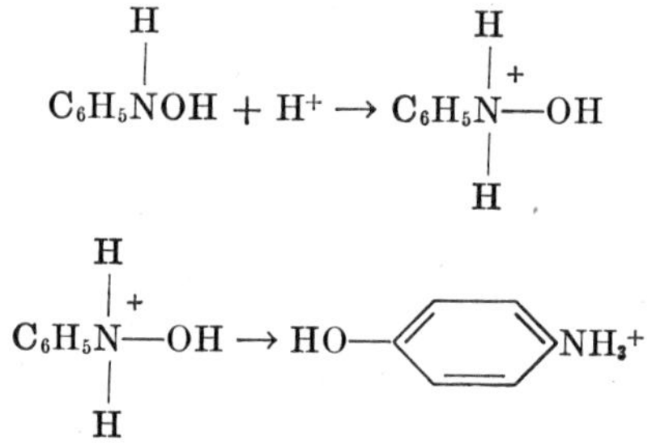

(c) Nitroalkylbenzenes

The nitrotoluenes behave in a manner which is similar to that of nitrobenzene. The only difference is that the two waves are observed over a greater pH range. The values obtained in 8 per cent ethanol are listed in Table XLII-4.[6] Values reported in 50 per cent ethanol for the nitrotoluenes are approximately 0.35 v. more negative.[10]

The half-wave potentials change very little in 80 per cent dioxane solution. Substitution of another methyl group in the ring has little effect upon the half-wave potentials unless the groups are both ortho to the nitro group as in nitromesitylene and nitrodurene. The values are given in Table XLII-5. Maxima are eliminated with 0.01 per cent gelatin.[11]

Nitromesitylene and nitrodurene resemble the nitroparaffins in their ease of reduction. They give a reduction wave equivalent to the formation of the phenylhydroxylamine in an acid medium at a more negative potential than the other aromatic nitro compounds. The second reduction to the amine occurs at too negative a value to be observed in the buffers used.

[9] D. Stocesova, *Collection Czechoslov. Chem. Communs.*, **14**, 615 (1949).
[10] S. F. Dennis, A. S. Powell, and M. J. Astle, *J. Am. Chem. Soc.*, **71**, 1484 (1949).
[11] M. Fields, C. Valle, Jr., and M. Kane, *J. Am. Chem. Soc.*, **71**, 421 (1949).

In more alkaline solutions interaction of the nitro compounds with alkali are responsible for their behavior. In a more strongly alkaline medium

TABLE XLII-4

HALF-WAVE POTENTIALS OF NITROTOLUENES IN 8 PER CENT ETHANOL

	$o\text{-}NO_2C_6H_4CH_3$				$m\text{-}NO_2C_6H_4CH_3$				$p\text{-}NO_2C_6H_4CH_3$			
	First wave		Second wave		First wave		Second wave		First wave		Second wave	
pH	$-E_{1/2}$ (S.C.E.)	n	$-E_{1/2}$	n	$-E_{1/2}$	n	$-E_{1/2}$	n	$-E_{1/2}$	n	$-E_{1/2}$	n
0.5	0.20	4	?[a]	6	0.15	4	?	6	0.10	4	?	6
1.7	0.32	4	?	6	0.25	4	?	6	0.22	4	?	6
2.5	0.35	4	?	6	0.30	4	?	6	0.30	4	?	6
3.0	0.44	4	?	6	0.37	4	?	6	0.35	4	?	6
3.8	0.45	4	?	6	0.41	4	?	5	0.43	4	?	6
4.1	0.50	4	?	5	0.46	4	?	?	0.48	4	?	?
5.7	0.55	4	—	—	0.51	4	—	—	0.50	4	—	—
7.4	0.62	4	—	—	0.59	4	—	—	0.58	4	—	—
9.2	0.75	4	—	—	0.71	4	—	—	0.69	4	—	—

[a] ? indicates half-wave potentials not measured.

TABLE XLII-5

HALF-WAVE POTENTIALS (S.C.E.) OF NITROALKYLBENZENES

	pH 1.0		pH 7.2	pH 12.0
Compound	$-E_{1/2}$ (1)	$-E_{1/2}$ (2)	$-E_{1/2}$ (1)	$-E_{1/2}$ (1)
Nitrobenzene	0.24	0.84	0.62	0.83
o-Nitrotoluene	0.26	0.64	0.66	0.84
m-Nitrotoluene	0.22	0.71	0.60	0.82
p-Nitrotoluene	0.24	0.71	0.62	0.91[a]
2-Nitro-1,4-dimethylbenzene	0.25	0.58	0.64	0.82
3-Nitro-1,2-dimethylbenzene	0.25	0.59	0.65	0.82
Nitromesitylene	0.38	—	0.75	0.88[b], –1.36[c]
Nitrodurene	0.38	—	0.75	0.87[b], –1.42[c]

[a] A slight inflection was observed; since the "break" between the two steps is only barely perceptible, the value above refers to the total wave.
[b] First wave.
[c] Second wave.

the poor definition of the total wave heights makes it difficult to estimate reliable values for the diffusion current constants. The values obtained are listed below together with those of 2-nitro-1,4-dimethylbenzene for comparison:

Compound	pH 1.0		pH 7.2	pH 12	
	I_d (1)	I_d (2)	I_d (1)	I_d (1)	I_d (2)
2-Nitro-1,4-dimethylbenzene	7.3	9.3	4.9	5.3	—
Nitromesitylene	7.7	—	4.6	1.2	2.8
Nitrodurene	7.6	—	5.1	1.9	5.2

The more negative half-wave potentials for nitromesitylene and nitrodurene point to steric inhibition of the conjugation of the nitro group with the aromatic ring (resonance) by the two *o*-methyl groups:

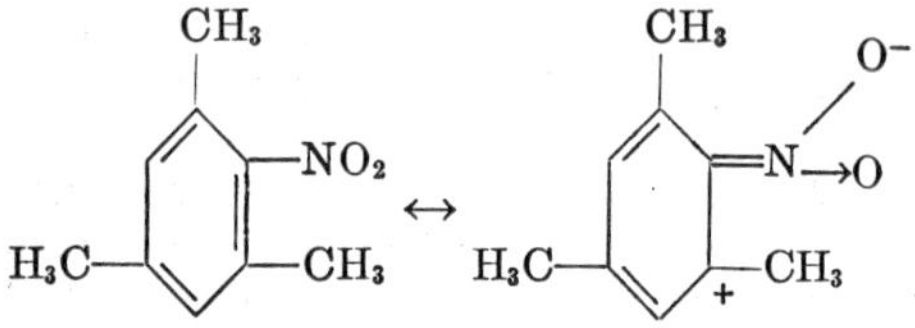

The nitro group, therefore, behaves more like an isolated or aliphatic nitro group.

Introduction of more nitro groups into nitrobenzene or the nitrotoluenes results in a behavior more complicated than that of the mononitro compounds since two or more reducible groups are present which are reduced stepwise. The extent of the reduction is dependent upon the pH and the relative positions of the nitro groups. The results are listed[12] in Table XLII-6.

Similar results for *m*-dinitrobenzene and *sym*-trinitrobenzene are obtained in ammonium chloride solutions containing varying amounts of hydrochloric acid and sodium hydroxide.[13] Values for the dinitrobenzenes in 50 per cent ethanol are reported to be approximately 0.3 v. more negative[14] than the half-wave potentials in 8 per cent ethanol.

Nitrobenzoic Acids.[15] Substitution of a carboxyl group into nitrobenzene results in a reduction behavior which resembles that of pyruvic acid.

The esters are normal and produce two waves in acid solution and one in neutral and alkaline media. The wave observed in buffers of pH 12 is for the anion of the acid formed by saponification of the esters. The

[12] J. Pearson, *Trans. Faraday Soc.*, **44,** 683 (1948).
[13] A. K. Cruse and R. Haul, *Z. Elektrochem.*, **53,** 115, (1949).
[14] S. F. Dennis, A. S. Powell, and M. J. Astle, *J. Am. Chem. Soc.*, **71,** 1484 (1949).
[15] J. E. Page, J. W. Smith, and J. G. Waller, *J. Phys. & Colloid Chem.*, **53,** 545 (1949).

TABLE XLII-6

POLAROGRAPHIC BEHAVIOR OF NITROBENZENES AND NITROTOLUENES (*vs.* S.C.E.) IN 8 PER CENT ETHANOL

Compound	pH 0.5		pH 1.7		pH 2.5		pH 3.0		pH 3.8		pH 4.1		pH 5.7		pH 7.4		pH 9.2	
	$-E_{1/2}$	n	$-E_{1/2}$	n	$-E_{1/2}$	n	$-E_{1/2}$	n	$-E_{1/2}$	n	$-E_{1/2}$	n	$-E_{1/2}$	n	$-E_{1/2}$	n	$-E_{1/2}$	n
o-$C_6H_4(NO_2)_2$	0.01	4	0.10	4	0.12	4	0.18	4	0.22	4	0.25	4	0.28	4	0.33	4	0.38	4
	0.16	12	0.27	12	0.32	12	0.41	12	0.45	11	0.47	10	0.54	10	0.62	10	0.74	10
m-$C_6H_4(NO_2)_2$	0.03	4	0.12	4	0.17	4	0.20	4	0.23	4	0.25	4	0.31	4	0.39	4	0.46	4
	0.12	8	0.26	8	0.29	8	0.35	8	0.38	8	0.42	8	0.47	8	0.58	8	0.68	8
	?	12	?	12	?	12	?	10	?	9	—	—	—	—	—	—	—	—
p-$C_6H_4(NO_2)_2$	0.01	4	0.08	4	0.12	4	0.14	4	0.19	4	0.21	4	0.28	4	0.31	4	0.34	4
	0.18	12	0.30	12	0.33	12	0.40	12	0.45	12	0.50	12	0.55	10	0.69	10	0.80	8
																	?	12
1,3,5-$C_6H_3(NO_2)_3$	0.04	4	0.06	4	0.08	4	0.12	4	0.16	4	0.20	4	0.24	4	0.31	4	0.34	4
	0.01	8	0.12	8	0.15	8	0.20	8	0.24	8	0.29	8	0.34	8	0.44	8	0.48	8
	0.06	12	0.20	12	0.24	12	0.30	12	0.34	12	0.42	12	0.47	12	0.61	12	0.65	12
	?	14	?	14	?	14	?	14	?	?	?	?	—	—	—	—	—	—
2,4-Dinitrotoluene	0.05	4	0.13	4	0.18	4	0.22	4	0.26	4	0.29	4	0.35	4	0.40	4	0.50	4
	0.14	8	0.25	8	0.30	8	0.36	8	0.40	8	0.43	8	0.48	8	0.58	8	0.70	8
	?	12	?	12	?	12	?	12	?	?	?	?	—	—	—	—	—	—
2,6-Dinitrotoluene	0.10	4	0.19	4	0.24	4	0.29	4	0.32	4	0.37	4	0.41	4	0.46	4	0.55	4
	0.19	8	0.34	8	0.40	8	0.44	8	0.48	8	0.54	8	0.57	8	0.64	8	0.74	8
	?	12	?	12	?	12	?	12	?	12	—	—	—	—	—	—	—	—
2,3-Dinitrotoluene	0.03	4	0.13	4	0.16	4	0.23	4	0.28	4	0.30	4	0.34	4	0.40	4	0.45	4
	0.22	12	0.34	12	0.41	12	0.50	12	0.56	11	0.58	10	0.63	10	0.69	10	0.81	10
3,4-Dinitrotoluene	0.03	4	0.13	4	0.16	4	0.22	4	0.26	4	0.28	4	0.31	4	0.36	4	0.42	4
	0.21	12	0.32	12	0.46	12	0.46	12	0.50	11	0.53	10	0.60	10	0.69	10	0.79	10
Trinitrotoluene	0.01	4	0.07	4	0.10	4	0.14	4	0.17	4	0.19	4	0.25	4	0.31	4	0.40	4
	0.08	8	0.16	8	0.20	8	0.24	8	0.28	8	0.31	8	0.38	8	0.48	8	0.55	8
	0.14	12	0.24	12	0.31	12	0.36	12	0.41	12	0.45	12	0.53	12	0.62	12	0.73	12
	?	?	?	?	?	?	?	?	?	?	—	—	—	—	—	—	—	—

(?) indicates that the wave is ill defined.

TABLE XLII-7

EFFECT OF pH ON THE REDUCTION OF 5×10^{-4} *M* *o*-, *m*-, AND *p*-NITROBENZOIC ACIDS AND THEIR METHYL ESTERS IN 10 PER CENT ETHANOL

pH	o-$NO_2C_6H_4COOH$				m-$NO_2C_6H_4COOH$				p-$NO_2C_6H_4COOH$			
	$-E_{1/2}$ (S.C.E.)*	i_d	$-E_{1/2}$ (2)	i_d	$-E_{1/2}$ (S.C.E.)*	i_d	$-E_{1/2}$ (2)	i_d	$-E_{1/2}$*	i_d	$-E_{1/2}$ (2)	i_d
1	0.170	5.95	—	0	0.150	6.05	0.570	2.15	0.110	5.65	0.645	1.80
2	0.225	5.10	0.725	1.80	0.200	5.40	0.695	2.15	0.170	5.50	0.740	1.85
3	0.320	5.10	—	—	0.275	5.10	0.845	1.95	0.235	4.95	—	0
4	0.380	5.15	—	—	0.335	5.10	—	0	0.300	4.95	—	0
6	0.530	5.00	—	—	0.515	5.10	—	0	0.470	5.10	—	0
8	0.780	4.20	0.835	1.10	0.730	3.80	1.155	3.55	0.655	4.05	1.110	1.25
10	0.855	2.40	1.045	2.75	0.770	1.90	1.070	5.10	0.705	2.65	0.990	2.55
12	0.845	1.05	1.075	3.35	0.760	1.50	1.070	4.00	0.705	1.45	1.010	3.80
pH	o-$NO_2C_6H_4COOCH_3$				m-$NO_2C_6H_4COOCH_3$				p-$NO_2C_6H_4COOCH_3$			
1	0.180	5.95	0.530	2.70	0.170	6.45	0.510	2.50	0.125	6.20	0.555	2.75
2	0.245	5.75	0.740	2.55	0.240	6.00	0.675	2.50	0.200	5.95	0.725	2.45
3	0.295	5.50	0.915	2.65	0.285	5.70	0.805	2.40	0.250	5.70	0.825	2.15
4	0.355	5.50	—	0	0.340	5.90	0	0	0.300	5.75	—	0
6	0.470	5.65	—	0	0.455	5.90	—	0	0.400	5.90	—	0
8	0.590	5.80	—	0	0.570	5.80	—	0	0.510	6.00	—	0
10	0.655	5.80	—	0	0.650	5.80	—	0	0.610	5.80	—	0
12	0.675[a]	0.45	1.10	5.00	0.730[a]	1.60	1.070	3.90	0.700[a]	1.50	1.020	4.00

* Values reported by Dennis, Powell, and Astle[14] are more negative by 0.25 to 0.3 v.

[a] Saponified.

results obtained are summarized in Table XLII-7. Similar values at low pH values are reported for *p*-nitrobenzoic acid by other investigators.[16]

Chloronitrobenzenes. Values reported by Dennis *et al.*[14] for the chloronitrobenzenes in 50 per cent ethanol are listed below:

	Ortho		Meta		Para	
pH	$-E_{1/2}$ (S.C.E.)	i_d	$-E_{1/2}$ (S.C.E.)	i_d	$-E_{1/2}$ (S.C.E.)	i_d
2.7	0.64	12.2	0.53	11.7	0.59	12.4
4.6	0.75	11.5	0.65	11.5	0.69	12.4
7.4	0.91	11.7	0.86	11.5	0.87	12.2
10.0	1.01	11.8	0.98	11.5	1.04	12.4
11.7	1.07	11.7	1.05	11.7	1.15	13.0

***p*-Nitrobenzenesulfonamide.** This compound (I) is soluble in water and gives two waves in an acid medium[17] (0.1 *N* hydrochloric acid); the height of the first wave corresponds to a transfer of 4 electrons, and that of the total wave to a transfer of 6 electrons. In an alkaline medium only one wave, corresponding to an electron transfer of 4, is found. The half-

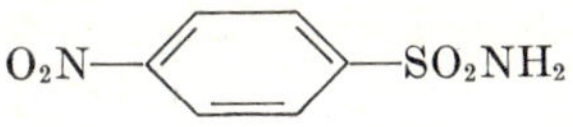

(I)

wave potentials (*vs.* S.C.E.) at 25° are −0.05 v., −0.48 v., and −0.65 v. at pH values of 1, 7.9, and 13.0, respectively.

Nitrobenzaldehydes. The nitrobenzaldehydes show two reduction waves for the ortho and meta isomers and three reduction waves for the para compound. The results in dilute alcohol are given in Tables XLII-8 and XLII-9. Greater amounts of alcohol shift the half-wave potentials.[18]

The first wave represents the reduction of the nitro group to the hydroxylamine since the half-wave potentials are similar to those obtained with nitrobenzene. The second wave in acid solution represents a combination of the reduction of the phenylhydroxylamine to the amine and the reduction of the aldehyde group apparently to the carbinol since a total of four electrons is involved in each case. In buffers of pH greater than 5, the behavior is only normal with the meta isomer; the second wave represents the reduction of the aldehyde group to the carbinol. Abnor-

[16] H. Eyring, L. Marker, and T. C. Kwoh, *J. Phys. & Colloid Chem.*, **53,** 1453 (1949).

[17] N. I. Levitan, I. M. Kolthoff, W. G. Clark, and D. J. Tenenberg, *J. Am. Chem. Soc.*, **65,** 2265 (1943).

[18] A. Korshunov and L. N. Sazanova, *Zhur. Fiz. Khim.*, **23,** 1299 (1949).

TABLE XLII-8

POLAROGRAPHIC BEHAVIOR OF *o*- AND *m*-NITROBENZALDEHYDES IN DILUTE ALCOHOL

	o-Nitrobenzaldehyde				*m*-Nitrobenzaldehyde				
pH	$-E_{1/2}(1)$ (S.C.E.)	i_d, microamp.	$-E_{1/2}$ (2)	i_d	pH*	$-E_{1/2}(1)$	i_d	$-E_{1/2}$ (2)	i_d
1.31	0.14	12.2	0.90	11.2		0.21	13.2	1.01	13.2
2.00	0.15	12.7	0.95	12.2		0.28	11.6	1.20	10.0
2.39	0.17	10.2	1.00	10.2		0.42	11.3	1.30	10.9
2.75	0.24	12.7	1.10	12.0		—	—	—	—
3.64	0.31	10.2	1.24	10.3	3.52	0.34	11.6	1.24	10.9
4.01	0.34	11.6	1.26	10.3		0.50	11.6	1.41	10.2
4.57	0.39	10.7	1.32	7.8	5.25	0.56	9.9	1.48	5.64
5.45	0.40	11.1	1.36	7.8	6.37	0.60	11.6	1.30	6.4
7.52	0.41	11.1	1.40	8.8		0.66	12.0	1.45	6.0
8.26	0.42	10.1	1.40	8.5		0.74	12.3	1.53	5.64
8.98	0.44	11.1	1.41	9.1		0.68	12.3	1.54	5.64
9.79	0.54	10.6	1.56	9.1		0.77	12.8	1.58	6.00
11.95	0.49	10.1	1.61	8.8		0.78	13.2	1.56	5.64
12.00	0.49	10.1	1.43	8.8		—	—	—	—

* Same as for *o*-nitrobenzaldehyde except where indicated.

TABLE XLII-9

POLAROGRAPHIC BEHAVIOR OF *p*-NITROBENZALDEHYDE IN DILUTE ETHANOL SOLUTION

pH	$-E_{1/2}(1)$ (S.C.E.)	i_d	$-E_{1/2}$ (2)	i_d	$-E_{1/2}$ (3)	i_d
1.62	0.19	10.9	1.06	11.3	—	—
2.08	0.24	11.3	1.09	13.2	—	—
3.61	0.26	11.7	1.15	12.0	—	—
4.68	0.44	12.4	1.35	10.5	—	—
6.53	0.46	11.7	1.29	5.3	1.62	4.0
7.36	0.46	11.3	1.29	4.9	1.66	4.5
8.67	0.47	11.3	1.29	5.3	1.63	4.9
8.98	0.48	11.3	1.30	5.0	1.66	4.0
9.79	0.52	10.9	1.34	4.5	1.68	3.9
10.60	0.60	10.9	1.33	4.3	1.68	4.71
11.74	0.56	10.5	1.37	4.9	1.68	4.1
11.95	0.53	10.9	1.40	4.9	1.69	4.1

mally high diffusion currents for the ortho isomer and two waves for the para isomer suggest secondary reactions between the groups in the molecule and reduction of the resulting products. One possibility that could occur with the ortho compound is the formation of anthranil:

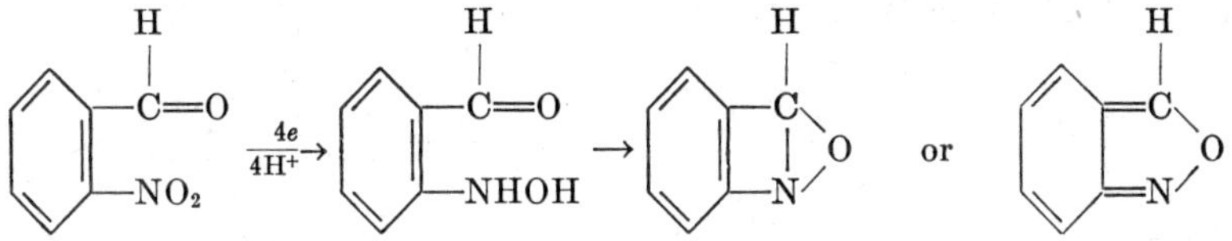

The more positive half-wave potentials obtained for the second waves are in agreement with such a postulation.

Nitrophenols. *o*- and *p*-Nitrophenols differ in their behavior from that of nitrobenzene in that reduction waves involving six electrons are ob-

TABLE XLII-10

POLARGRAPHIC BEHAVIOR OF NITROPHENOLS IN 8 PER CENT ETHANOL

	pH 2.0		pH 4.0		pH 6.0		pH 8.0		pH 10.0		pH 11.9	
Compound	$-E_{1/2}$ (S.C.E.)	*n*	$-E_{1/2}$	*n*	$-E_{1/2}$	*n*	$-E_{1/2}$	*n*	$-E_{1/2}$	*n*	$-E_{1/2}$	*n*
o-$HOC_6H_4NO_2$	0.23	4	0.37	4	0.50	5.2	0.65	5.5	0.80	6	0.91	6
	?	6	?	6	—	—	—	—	—	—	—	—
p-$HOC_6H_4NO_2$	0.35	4	0.46	4	0.64	6	0.82	(5)	0.92	2	0.96	2
	?	6	?	6	—	—	1.28	6	1.33	6	1.67	6
m-$HOC_6H_4NO_2$	0.25	4	0.37	4	0.53	4	0.62	4	0.76	4	0.80	(1)
	0.84	5.5	—	—	—	—	—	—	—	—	1.27	4
	1.10	6	—	—	—	—	—	—	—	—	—	—

tained in alkaline buffers. The meta compound behaves in a fashion similar to that of nitrobenzene.

Values obtained in 8 per cent ethanol are given in Table XLII-10.[19] Similar behavior has been reported in 10 per cent ethanol for 5×10^{-4} *M* buffered solutions[20] (Table XLII-11).

Results reported by Astle and McConnell[21] for *o*-nitrophenol could not be verified by Stocesova.[22]

o-Nitrophenol differs from *p*-nitrophenol in behavior in the neutral region in that apparently a normal 4-electron reduction is obtained at a pH of 6. In more acid and alkaline regions this reduction becomes a 6-electron step. The reason for this behavior, according to Stocesova, is that the conversion of *o*-hydroxyphenylhydroxylamine to the quinoneimine

[19] J. Pearson, *Trans. Faraday Soc.*, **44**, 692 (1948).

[20] J. E. Page, J. W. Smith, and J. G. Waller, *J. Phys. & Colloid Chem.*, **53**, 545 (1949).

[21] M. J. Astle and W. V. McConnell, *J. Am. Chem. Soc.*, **65**, 35 (1943).

[22] D. Stocesova, *Collection Czechoslov. Chem. Communs.*, **14**, 615 (1949).

TABLE XLII-11

POLAROGRAPHIC BEHAVIOR OF NITROPHENOLS IN 10 PER CENT ETHANOL

	o-Nitrophenol				*m*-Nitrophenol				*p*-Nitrophenol			
pH	$-E_{1/2}(1)$(S.C.E.)	i_d	$-E_{1/2}(2)$	i_d	$-E_{1/2}(1)$	i_d	$-E_{1/2}(2)$	i_d	$-E_{1/2}(1)$	i_d	$-E_{1/2}(2)$	i_d
1	0.185	7.25	0.635	0.90	0.190	5.58	0.690	1.60	0.26	6.95	0.755	1.15
2	0.240	6.65	0.670	1.25	0.250	5.45	0.765	2.10	0.34	6.70	0.740	1.55
3	0.305	6.15	0.770	1.55	0.310	5.15	0.885	1.65	0.405	6.40	0.855	1.50
4	0.355	5.70	0.940	1.85	0.365	5.15	—	0	0.470	6.20	0.975	1.53
6	0.475	5.95	—	0	0.495	5.25	—	0	0.615	6.60	—	0
8	0.640	7.20	—	0	0.630	4.95	—	0	0.810	6.70	1.260	0.75
10	0.760	7.35	—	0	0.800	5.05	—	0	0.985	4.65	1.280	2.85
12	0.900	7.20	—	0	0.940	1.60	1.145	3.30	0.975	1.80	1.565	5.20

TABLE XLII-12

REDUCTION OF 5×10^{-4} *M* *o*-, *m*-, AND *p*-NITROANISOLES IN 10 PER CENT ETHANOL

	o-Nitroanisole				*m*-Nitroanisole				*p*-Nitroanisole			
pH	$-E_{1/2}(1)$ (S.C.E.)	i_d	$-E_{1/2}(2)$	i_d	$-E_{1/2}(1)$	i_d	$-E_{1/2}(2)$	i_d	$-E_{1/2}(1)$	i_d	$-E_{1/2}(2)$	i_d
1	0.210	8.10	—	0	0.190	6.50	0.590	1.85	0.260	7.75	0.600	1.0
2	0.285	7.35	0.575	6.10	0.275	6.10	0.690	2.50	0.350	7.25	0.635	1.55
3	0.380	6.50	0.745	2.20	0.325	5.85	0.820	2.65	0.395	6.75	0.695	1.80
4	0.460	5.70	—	0	0.405	5.70	0.925	1.85	0.460	6.12	0.840	2.20
6	0.565	5.65	—	0	0.510	5.80	—	0	0.590	5.80	—	0
8	0.675	5.60	—	0	0.630	5.80	—	0	0.715	5.80	—	0
10	0.760	5.75	—	0	0.705	5.85	—	0	0.800	5.65	—	0
12	0.795	5.55	—	0	0.755	5.80	—	0	0.855	5.90	—	0

which is necessary for a 6-electron step, is acid-base catalyzed:

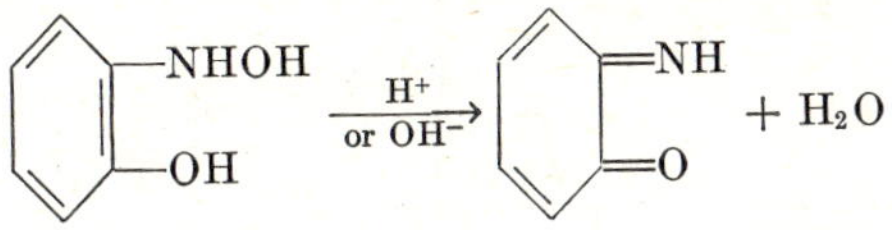

Methylation of the nitrophenols to the corresponding nitroanisoles prevents the quinoidation possibility and results in behavior similar to that of nitrobenzene. The results obtained by Page *et al.*[20] are listed in Table XLII-12.

2,4-Dinitrophenol and 2,6-dinitrophenol behave in a manner somewhat similar to that of the analogous nitrobenzenes in that complete reduction occurs at low pH values and partial reduction at high pH values. One of the nitro groups is reduced to the amine at all hydrogen ion concentrations. This process takes place in two stages at high pH values with the 2,4-isomer.

Picric acid gives waves which are not sufficiently well defined to give measurable half-wave potentials. The final diffusion current corresponds approximately to 16 electrons. Pearson[19] gives the following values for the dinitrophenols in 8 per cent ethanol:

Dinitrophenol compound	pH 2.0		pH 4.0		pH 6.0		pH 8.0		pH 10.0		pH 11.9	
	$-E_{1/2}$	n	$-E_{1/2}$	n	$-E_{1/2}$	n	$-E_{1/2}$	n	$-E_{1/2}$	n	$-E_{1/2}$	n
2,4-	0.14	4	0.25	4	0.42	4	0.59	4	0.66	4	0.83	4
	0.32	8	0.48	9	0.69	10	0.91	10	0.96	8	0.99	8
	?	12	?	11	—	—	—	—	?	10	1.63	10
2,6-	0.09	4	0.23	4	0.40	4	0.51	4	0.62	4	0.83	4
	0.24	10	0.36	10	0.54	10	0.69	10	0.80	10	0.97	10
	?	12	—	—	—	—	—	—	—	—	—	—

The reduction of the nitrocresols depends upon the relative position of the phenolic and nitro groups. 5-Nitro-*o*-cresol and 6-nitro-*m*-cresol, 3-nitro-*o*-cresol and 4-nitro-*m*-cresol, in which the hydroxyl group is ortho to the nitro group, behave like *o*-nitrophenol; 2-nitro-*p*-cresol and 4-nitro-*o*-cresol, in which the hydroxyl group is meta to the nitro group, behave like *m*-nitrophenol, as shown in Table XLII-13.

The compounds of Class I produce waves of equal heights in acid and alkaline media corresponding to an electron transfer of 6. The half-wave potentials of these compounds are appreciably more negative than those of the other two classes. Compounds of Class II produce waves in an acid

TABLE XLII-13

HALF-WAVE POTENTIALS (*vs.* S.C.E.) AT 25° C. OF NITROCRESOLS IN WATER AS FUNCTION OF pH[23]

Cresol	Class	pH 2.2	pH 3.6	pH 5.4	pH 6.4	pH 8.0	pH 9.0	pH 10.1	pH 11.3
6-Nitro-*m*-	I	−0.35	−0.42	−0.54	−0.58	−0.68	−0.81	−0.91	−0.93
5-Nitro-*o*-	I	−0.30	−0.35	−0.45	−0.52	−0.61	−0.72	−0.82	−0.78
4-Nitro-*m*-	II	−0.20	−0.24	−0.37	−0.42	−0.52	−0.65	−0.71	−0.68
3-Nitro-*o*-	II	−0.20	−0.22	−0.31	−0.37	−0.51	−0.58	−0.69	−0.71
3-Nitro-*p*-	II	−0.21	−0.27	−0.36	−0.42	−0.51	−0.63	−0.69	−0.67
2-Nitro-*p*-	III	−0.28	−0.32	−0.44	−0.50	−0.57	−0.66	−0.75	−0.76
4-Nitro-*o*-	III	−0.23	−0.30	−0.42	−0.48	−0.56	−0.65	−0.73	−0.74

TABLE XLII-14

HALF-WAVE POTENTIALS (*vs.* S.C.E.) AT 25° C. OF NITRODIHYDROXYBENZENE IN WATER AS FUNCTION OF pH[24]

pH	2-Nitrohydroquinone	3-Nitrocatechol	4-Nitrocatechol	2-Nitroresorcinol	4-Nitroresorcinol
2.1	−0.20	−0.16	−0.34	−0.17	−0.26
3.6	−0.25	−0.24	−0.45	−0.25	−0.36
5.2	−0.32	−0.33	−0.52	−0.32	−0.47
6.4	−0.38	−0.37	−0.57	−0.36	−0.54
8.0	−0.42	−0.49	−0.65	−0.44	−0.68
9.0	−0.51	−0.74	−0.82[a]	−0.49	−0.80
			−1.32[a]	—	—
9.5	−0.55	−0.72	−0.84[a]	−0.57	−0.85
			−1.32[a]	—	—
10.1	−0.63	−0.73	−0.84[a]	−0.66	−0.97
			−1.32[a]	—	—
11.3	−0.78	−0.73	−0.93[a]	−0.93	−1.27
			−1.37[a]	—	—

[a] Two waves.

medium corresponding to an electron transfer of 4 and in alkaline medium (pH greater than 9.5) corresponding to an electron transfer of 6. Compounds of Class III are reduced in acid solution in two steps, first to the hydroxylamine (4 electrons), then to the amine. In alkaline solutions the reduction stops at the hydroxylamine stage.[23]

TABLE XLII-15

POLAROGRAPHIC BEHAVIOR OF NITRORESORCINOLS IN 8 PER CENT ETHANOL

	pH 2.0		pH 4.0		pH 6.0		pH 8.0		pH 9.2		pH 10.0	
Compound	$-E_{1/2}$ (S.C.E.)	n	$-E_{1/2}$	n	$-E_{1/2}$	n	$-E_{1/2}$	n	$-E_{1/2}$	n	$-E_{1/2}$	n
2-Nitroresorcinol.	0.21	6	0.32	6	0.41	6	0.50	6	0.61	6	0.72	6
2,4-Dinitroresorcinol....	0.14	6	0.21	6	0.30	6	0.55	6	0.86	6	0.97	6
	0.26	12	0.35	12	0.50	12	0.80	12	1.14	12	1.20	12
4,6-Dinitroresorcinol....	0.09	6	0.19	6	0.28	6	0.40	6	0.63	6	0.73	6
	0.27	12	0.40	12	0.52	12	0.74	12	0.95	12	1.04	12
Styphnic acid....	0.07	6	0.15	6	0.25	6	0.38	6	0.57	6	0.70	6
	0.12	12	0.24	12	0.38	12	0.66	12	0.89	12	1.05	12
	0.23	17	0.48	17	0.55	17	0.82	17	1.17	17	1.30	17

The reduction of nitrodihydroxybenzenes, with the exception of 4-nitrocatechol, resembles that of *p*-nitrophenol in that a 6-electron wave is obtained. The half-wave potentials are presented in Table XLII-14. 4-Nitrocatechol produces two waves in buffers with an alkalinity of pH 9 or greater. The first wave involves 2 electrons and the second wave 4 electrons. It has been suggested that the first reduction is to the nitroso compound and the second to the amine.[24] Substitution of two or more nitro groups into resorcinol results in two waves for the dinitro compounds and three waves for the trinitro (styphnic acid). The results obtained by Pearson[25] in 8 per cent ethanol are listed in Table XLII-15.

Nitroanilines. *p*-Nitroaniline produces a wave involving 6 electrons at

[23] M. J. Astle and W. P. Cropper, *J. Am. Chem. Soc.*, **65,** 2395 (1943).
[24] M. J. Astle and S. P. Stephenson, *J. Am. Chem. Soc.*, **65,** 2399 (1943).
[25] J. Pearson, *Trans. Faraday Soc.*, **45,** 199 (1949).

all pH values studied. In water solutions the half-wave potential (S.C.E.) varies with pH according to the following equation: $E_{1/2} = -0.24 - 0.061$ pH in the range of pH from 1 to 9. Addition of alcohol shifts the half-wave potential to more negative values. Values in this mixture of solvents are given for 7.95 *M* ethanol by: $E_{1/2} = -0.293 - 0.068$ pH and for 12.21 *M* ethanol by $E_{1/2} = -0.332 - 0.074$ pH. Diffusion currents decrease with an increase in ethanol concentration and parallel the change in viscosity of these solutions.[26]

o-Nitroaniline and *m*-nitroaniline have been studied but only reduction potentials (N.C.E. tangent method) are reported.[27]

Comparison of the half-wave potentials for the various substituted nitrobenzenes indicates that reduction becomes more difficult with the substituents in the following order: $SO_2NH_2 > NO_2 > COOH > COOCH_3 > CHO > Cl > H > CH_3 > OCH_3 > OH > NH_2$ for the para derivatives, and $NO_2 > COOH > COOCH_3 > CHO > OH > OCH_3 > H > CH_3$ for the meta compounds. The ortho series follows the order $NO_2 > CHO > COOCH_3 > COOH > OH > H > Cl > OCH_3 > CH_3$. This order is in fairly good agreement with the electronegativity rule of Shikata and Tachi,[28] and also with the electron-donating and electron-attracting effects of the various groups, with one exception. The hydroxyl group which is an electron-donating group (+T) lowers the half-wave potential of the nitro group if it is present in the ortho position. This abnormal effect is due to hydrogen bonding which interferes with the normal resonance of the nitro group and makes the reduction easier. The hydroxyl group in the meta position is normal and has little effect on the reduction but, when present in the para position, shifts the half-wave potential to more negative values due to its (+T) effect. If the hydroxyl group is replaced by methoxyl as in the anisoles the behavior of the ortho derivative is normal.

Uses. The polarographic method has been used to demonstrate the presence of hydrogen bonding in N-phenyl-N′-*o*-nitrophenylacetamidine:

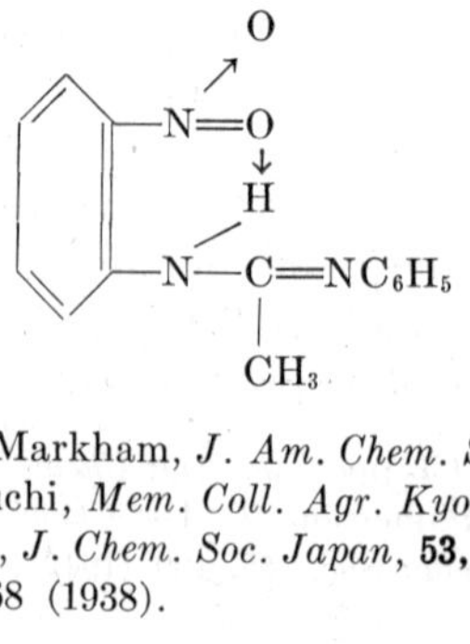

[26] O. D. Shreve and E. C. Markham, *J. Am. Chem. Soc.*, **71**, 2993 (1949).

[27] M. Shikata and E. Taguchi, *Mem. Coll. Agr. Kyoto Imp. Univ.*, **29**, 1 (1934).

[28] M. Shikata and I. Tachi, *J. Chem. Soc. Japan*, **53**, 834 (1932); *Collection Czechoslov. Chem. Communs.*, **10**, 368 (1938).

and consequently the stabilization of one tautomer.

All three possible isomeric nitrophenylacetamidines were studied in 50 per cent ethanol at varying pH values and found to behave in a manner similar to that of nitrobenzene; two waves are obtained in an acid medium up to a pH of 8 and one wave at a pH greater than 8. The presence of hydrogen bonding in the ortho compound was indicated by the half-wave potential which was about 0.1 v. more positive than that of the para isomer in all buffers studied. This behavior parallels that observed for *o*-nitrophenol. The half-wave potentials of the three isomers at varying pH values in 50 per cent ethanol are given in Table XLII-16. Buffer solutions were used and the pH values refer to the aqueous buffers before dilution with ethanol.[29]

TABLE XLII-16

POLAROGRAPHIC BEHAVIOR (*vs.* S.C.E.) OF NITROPHENYLACETAMIDINE

pH	Ortho-		Meta-		Para-	
	First wave	Second wave	First wave	Second wave	First wave	Second wave
2.2	−0.50	−1.14	−0.60	−1.13	−0.58	−1.18
3.0	−0.52	−1.28	−0.60	−1.19	−0.58	−1.26
4.0	−0.54	−1.30	−0.65	−1.28	−0.59	−1.36
5.0	−0.58	−1.34	−0.67	−1.43	−0.61	−1.41
6.0	−0.59	−1.34	−0.68	−1.46	−0.63	−1.45
7.0	−0.61	−1.36	−0.70	−1.48	−0.67	−1.51
8.0	−0.67	−1.38	−0.75	−1.52	−0.74	−1.58
9.4	−0.67	—	−0.76	—	−0.74	—
10.9	−0.72	—	−0.83	—	−0.75	—
12.1	−0.76	—	−0.90	—	−0.78	—
14.5	−0.77	—	−0.87	—	−0.76	—

The method has been used to determine traces of nitrobenzene in aniline.[30]

Procedure: 4 volumes of aniline are mixed with 1 volume of concentrated hydrochloric acid and the mixture is brought to temperature in the thermostat, the diffusion current of nitrobenzene is then determined at a potential of *ca.* —0.45 v. In routine analyses it suffices to determine the part of the c.v. curve between 0.2 and 0.6 v. Except at extremely small nitrobenzene concentrations the diffusion current is proportional to the concentration of this constituent. The concentration of nitrobenzene is found by adding a known amount of nitrobenzene to practically nitrobenzene-free aniline. Assuming proportionality between i_d and the nitrobenzene concentration, the nitrobenzene content of the original samples is calculated from the increase in i_d. In this way the nitrobenzene content of aniline can be determined in the range between 0.001 and 1 per cent.

[29] M. E. Runner, M. L. Kilpatrick, and E. C. Wagner, *J. Am. Chem. Soc.*, **69,** 1406 (1947).

[30] J. V. A. Novak, *Collection Czechoslov. Chem. Communs.*, **11,** 573 (1939).

Maxima can best be suppressed by adding migrosine hydrochloride.[31]

The same technique can be applied for determining nitrobenzene in benzidine and dinitrobenzene.[32]

α-Nitronaphthalene has been determined in the presence of α-naphthylamine in a 0.1 *N* hydrochloric acid solution containing 50 per cent methanol. Other nitro compounds must be absent since the half-wave potential (−0.26 v. *vs.* N.C.E.) is in the same region. The diffusion current is proportional in the range of 20–200 mg. per liter. Gelatin suppresses maxima.[33]

The polarographic determination of *m*-dinitrobenzene in 90 per cent ethanol containing 0.1 *N* ammonium chloride plus enough ammonia to turn phenol red barely red has been used to study its action as a retarder in the emulsion polymerization of styrene.[34] The diffusion current is measured at −1.40 v. (S.C.E.).

The determination of *m*-dinitrobenzene is used indirectly to determine benzene alone or in the presence of toluene and xylene. Benzene is nitrated to dinitrobenzene and is separated in this form from the nitrated toluenes and xylenes by oxidizing the latter to the acids and extracting with petroleum ether. The polarographic analysis is carried out in 50 per cent ethanol containing 2.0 g. of metol, 63 g. of anhydrous sodium sulfite, and 12.6 g. of anhydrous potassium carbonate per liter with bromothymol blue as a maxima suppressor. This procedure has been used to determine benzene in the atmosphere[35] and in biological fluids[36] by aspirating air from these sources into the nitrating acid. A similar base solution is used to determine *m*-dinitrobenzene in blood serum and plasma.[37] The analysis for nitrobenzene in blood can be carried out in a simpler manner by adding the sample to 20 per cent alcohol.[38]

Picric acid can be determined in concentrations of 1×10^{-4} to 6×10^{-4} *M* in buffers below pH 4 under which condition all three waves coincide. Phenol even in large amounts does not interfere.[39]

---

[31] J. Haslam and L. H. Cross, *J. Soc. Chem. Ind.*, **63**, 94 (1944).

[32] I. A. Korshunov, A. V. Ryabov, L. N. Sazanova, and A. S. Kirillova, *Zavodskaya Lab.*, **14**, 519 (1948); *Chem. Abstracts*, **43**, 4976 (1949).

[33] Y. I. Vainshtein, *Zavodskaya Lab.*, **14**, 517 (1948); *Chem. Abstracts*, **43**, 4978 (1949).

[34] I. M. Kolthoff and F. A. Bovey, *J. Am. Chem. Soc.*, **70**, 794 (1948).

[35] A. S. Landry, *Anal. Chem.*, **21**, 675 (1949).

[36] S. Skramovsky and J. Teisinger, *Arch. maladies profess., méd. travail et sécurité sociale*, **8**, No. 1, 22 (1947); *Chimie & industrie*, **59**, 48 (1948); *Chem. Abstracts*, **42**, 7816 (1948).

[37] J. Roubal, K. Tuhy, and F. Pokorny, *Casopis Lekaru Ceskych*, **85**, 1002 (1946); *Chem. Abstracts*, **42**, 3454 (1948).

[38] J. Teisinger, *Mikrochemie ver. Mikrochim. Acta*, **25**, 151 (1938).

[39] M. B. Neiman, L. I. Kuznetsov, I. B. Rabinovich, and A. V. Ryabov, *Zavodskaya Lab.*, **15**, 1280 (1949); *Chem. Abstracts*, **44**, 3845 (1950).

2. NITROSO COMPOUNDS

No simple aromatic nitroso compound has been studied at the dropping mercury electrode. The behavior would no doubt parallel that of nitrobenzene except that two electrons less would be involved.

***p*-Nitrosophenol.** This compound produces a wave involving 4 electrons and is reduced to *p*-aminophenol similar to *p*-nitrophenol. The half-wave potential at pH 4 (−0.20 v. *vs.* S.C.E.) occurs at a more positive value than that for the corresponding nitro compound (−0.68 v.).[40]

α-Nitroso-β-Naphthol.[41] This compound behaves like *p*-nitrosophenol and produces the following half-wave potentials (S.C.E.):

pH	4.1	5.9	7.8	9.9
$-E_{1/2}$	0.01	0.11	0.21	0.33

This reduction is used as the basis for the amperometric titration of cobalt and copper.

3. PHENYLHYDROXYLAMINE DERIVATIVES

Phenylhydroxylamine. This compound is reduced in acid medium in a 2-electron step to aniline. The wave at a pH less than 2 occurs at the same point as the second wave of nitrobenzene.[42] Beyond this pH value the wave is drawn out.

At higher pH values (6.3) phenylhydroxylamine produces no cathodic wave but produces an anodic wave with a half-wave potential of −0.02 v. (S.C.E.). In more alkaline solutions phenylhydroxylamine is converted into azoxybenzene which is reducible.[43] A similar reaction is probably the explanation for the reduction reported for *p*-hydroxylaminobenzenesulfonamide in 0.1 *N* sodium hydroxide.[44]

N-Nitrosophenylhydroxylamine, $C_6H_5N(NO)OH$. The ammonium salt of this acid, called cupferron, was investigated[45] in buffer solutions of various pH values at 25° C. The free acid decomposes rapidly at a pH of 3 or less, the rate increasing with decreasing pH. Only one wave is observed in an acid medium up to a pH of 6. The diffusion or limiting current decreases with increasing pH, the change becoming very pronounced when the pH becomes greater than 6. At a pH between 3 and 6 the limiting

---

40 M. J. Astle and W. V. McConnell, *J. Am. Chem. Soc.*, **65,** 35 (1943).
41 I. M. Kolthoff and A. Langer, *J. Am. Chem. Soc.*, **62,** 3172 (1940).
42 D. Stocesova, *Collection Czechoslov. Chem. Communs.*, **14,** 615 (1949).
43 D. Lester and L. A. Greenberg, *J. Am. Chem. Soc.*, **66,** 496 (1944).
44 N. I. Levitan, I. M. Kolthoff, W. G. Clark, and D. J. Tenenberg, *J. Am. Chem. Soc.*, **65,** 2265 (1943).
45 I. M. Kolthoff and A. Liberti, *J. Am. Chem. Soc.*, **70,** 1884 (1948).

current (corrected for i_r) is proportional to the concentration of the cupferron in the buffers. In the pH range between 7 and 9 two waves are observed, the first wave being poorly defined. The second wave attains a constant value at a pH greater than 8.

The half-wave potential (S.C.E.) of the first wave varies with the pH according to the expression: $E_{1/2} = -0.58 - 0.128$ pH. The half-wave potential of the second wave is practically independent of pH, its value being -1.80 v.

The height of the wave in an acid medium (pH 1) corresponds to a reduction involving 6 electrons and indicates reduction to phenylhydrazine. The reduction in an alkaline medium involves 4 electrons and corresponds to a reauction of the anion to $C_6H_5(NH_2)NO^-$. N-Nitrosophenylhydroxylamine and its anion form a mesomeric system. In an acid medium at a pH value between 3 and 6 the height of the wave is much greater than that corresponding to the concentration of the undissociated acid. The polarographic behavior of the nitroso compound is very similar to that of pyruvic acid and of phenylglyoxalic acid[46] (see p. 729). The ionization constant of nitrosophenylhydroxylamine is estimated to be 5×10^{-5}. Hence, at a pH value greater than 6 the equilibrium concentration of the acid form becomes negligibly small and the limiting current becomes a typical kinetic current, its value being determined by the rate of formation of the acid form, HA, at the surface of the electrode: $H^+ + A^- \rightarrow HA$. At a pH of about 8.2, where the kinetic current is very small as compared to the diffusion current, the height of the former becomes independent of the height of the mercury in the reservoir, while the second current is diffusion controlled.

Cupferron precipitates a great number of metal ions in an acid medium and can be used in the amperometric titration of iron and copper (see p. 930).

4. AZOXY COMPOUNDS

Azoxybenzene. Azoxybenzene produces a half-wave potential at -0.63 v. (S.C.E.) in a phosphate buffer (pH 6.3) in 20 per cent ethanol. Apparently 4 electrons are involved in its reduction to hydrazobenzene and water.[43]

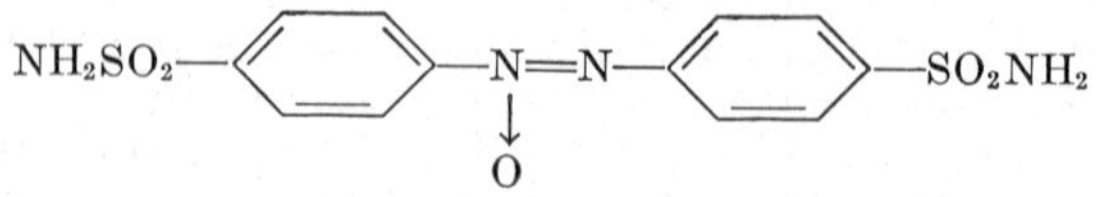

(I)

[46] R. Brdicka and K. Wiesner, *Collection Czechoslov. Chem. Communs.*, **12,** 138 (1947).

***p,p'*Azoxybenzenesulfonamide.** (I) is reducible in both acid and alkaline solutions containing 50 per cent methanol. The half-wave potentials (S.C.E.) at 25° are −0.05, −0.30, −0.45, and −0.70 v. at pH values of 1.5, 4.2, 7.2, and 12.5, respectively.[44]

5. AZO COMPOUNDS

Azobenzene and its derivatives are reduced in neutral, acid, and alkaline solutions at the dropping mercury electrode and give a wave involving 2 electrons.[47] The wave represents a reduction of the azo compound to the hydrazobenzene:

$$\text{ArN{=}NAr} + 2\text{H}^+ + 2e \rightarrow \text{ArNHNHAr}$$

Studies using platinized platinum electrodes indicate that the azobenzene–hydrazobenzene system is reversible.[48] Polarographic work has not been extensive enough to decide whether this is true at the dropping electrode.

Azobenzene. From some scattered data and a graph presented by Tachi[49] it may be inferred that the half-wave potential of azobenzene in 40 per cent ethanol changes between pH 0 and pH 11 according to the equation:

$$E_{1/2} = +0.03 - 0.069 \text{ pH (N.C.E., 25° C.)}$$

The results are not dependable since buffer solutions containing varying amounts of alcohol were used without taking into account the effect of alcohol upon the pH.

The half-wave potential of azobenzene remains constant with concentration in the range of 4.72×10^{-4} to 1.70×10^{-3} M in 0.104 N LiCl dissolved in 79.3 per cent alcohol. The value reported is −0.908 v. (N.C.E.).[50]

The *cis* isomer produces a reduction wave with a half-wave potential of −0.80 v. (S.C.E.?) in 70 per cent dioxane containing 0.1 N tetraethylammonium chloride and the *trans* isomer gives a value of −0.97 v. (S.C.E.?).[51]

Substituted Azobenzenes. Half-wave potentials for a number of azo dyes are reported in a series of buffers. The half-wave potential (S.C.E.)

---

[47] M. Shikata and I. Tachi, *Mem. Coll. Agr. Kyoto Imp. Univ.*, **17**, 45 (1931). I. Tachi, *ibid.*, **40**, 1, 11 (1937); **42**, 1 (1938).

[48] J. B. Conant and M. F. Pratt, *J. Am. Chem. Soc.*, **48**, 2468 (1926).

[49] I. Tachi, *Mem. Coll. Agr. Kyoto Imp. Univ.*, **42**, 36 (1938).

[50] A. Volpi, *Gazz. chim. ital.*, **77**, 473 (1947).

[51] F. Wessely and J. Wratil, *Mikrochemie ver. Mikrochim. Acta*, **33**, 248 (1947).

was found to vary less with concentration in buffers than in an unbuffered medium[52] (0.1 *N* HCl, 0.1 *N* KCl, and 0.1 *N* KOH). Values obtained for Orange II and Metanil yellow (II) are listed in Table XLII-17.

TABLE XLII-17
HALF-WAVE POTENTIALS OF AZO DYES

pH	Orange II $-E_{1/2}$	Metanil yellow $-E_{1/2}$
1.93	—	0.150
2.97	0.273	0.165
3.95	—	0.189
4.96	—	0.237
5.91	—	0.295
6.98	0.483	0.402
8.04	0.561	0.493
9.94	0.667	0.585
11.04	0.682	0.682
12.32	0.708	0.723

TABLE XLII-18

Compound	pH	$-E_{1/2}$ (S.C.E.) in various concentrations (moles/liter)						
		10^{-5}	2×10^{-5}	4×10^{-5}	10^{-4}	2×10^{-4}	5×10^{-4}	10^{-3}
Metanil yellow	2.97	0.244	0.244	0.240	0.173	0.173	0.202	0.272
Orange II	6.97	0.479	0.468	0.458	0.475	0.527	0.538	0.568
Orange IV	6.97	0.414	0.404	0.393	0.404	0.419	0.434	—
Auravante	2.97	0.229	0.229	0.229	0.231	0.242	0.272	0.282
Chrysoine	6.97	0.432	0.419	0.394	0.364	0.368	0.389	0.429
Nacarat	11.04	0.657	0.657	0.652	0.635	0.616	0.619	0.625

Orange II: 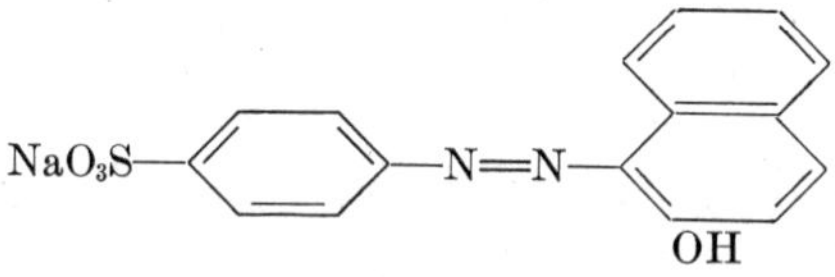

(1/6000 *N*). Metanil yellow:

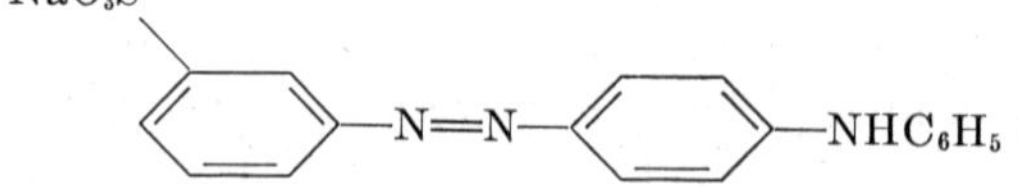

(1/6000 *N*).

The effect of concentration upon the half-wave potential is shown in Table XLII-18.

[52] N. T. Hoang, *J. chim. phys.*, **36,** 164 (1939); *Compt. rend.*, **208,** 1979 (1939).

Reduction potentials are reported for *p*-aminoazobenzene, *p*-dimethylaminoazobenzene,[53] methyl red,[54] and methyl orange.[55]

The polarographic method has been used to show (in agreement with x-ray studies) that the green and red *β*-*p*-azophenols are not *cis-trans* isomers.[56] Identical cathodic waves were obtained at pH 5.75 and similar anodic waves at pH 9.2. *Cis-trans* isomers would give different half-wave potentials.

The method has been used to follow the combination of methyl orange with serum albumin and gelatin. The reduction wave of the dye is lowered until the serum albumin concentration is 0.5 per cent for 10^{-4} *M* methyl orange. From the diffusion coefficient of the complex and the decrease in

TABLE XLII-19

POLAROGRAPHIC HALF-WAVE REDUCTION POTENTIALS OF QUATERNARY AMMONIUM IONS

Positive ion	Negative ion	$-E_{1/2}$ (S.C.E.)
NH_4^+	Cl^-	2.21
$N(CH_3)_3H^+$	Cl^-	2.23
$N(C_2H_4OH)_3C_2H_5^+$	Br^-	2.38
$N(C_3H_7)_4^+$	I^-	2.52
$N(C_4H_9)_4^+$	I^-	2.57
$N(CH_3)_3C_2H_4OCOCH_3^+$	Cl^-	2.64
$N(C_2H_5)_4^+$	Br^-	2.67
$N(CH_3)_3 \cdot C_2H_4OH^+$	Cl^-	2.72
$N(CH_3)_4^+$	Br^-	2.93 (tangent potential)

i_d at various concentrations of the protein, the molecular weight and the binding constant, respectively, can be calculated. No effect of the serum albumin on the diffusion current is noticed in alkali.[57]

6. AMINES

Aliphatic amines are apparently not reduced at the dropping mercury electrode but produce catalytic hydrogen waves since well-defined diffusion currents are not obtained.[58]

---

53 I. Tachi, *Mem. Coll. Agr. Kyoto Imp. Univ.*, **29,** 1 (1934).

54 A. Pittoni, *Ricerca sci. e ricostruz.*, **17,** 1396 (1947); *Chem. Abstracts*, **43,** 7368 (1949).

55 A. Pittoni, *Atti soc. med. chir. Padova*, **25,** 125 (1947); *Chem. Abstracts*, **43,** 7835 (1949).

56 W. M. Lauer, H. P. Klug, and S. A. Harrison, *J. Am. Chem. Soc.*, **61,** 2776 (1939).

57 W. Stricks and I. M. Kolthoff, *J. Am. Chem. Soc.*, **71,** 1519 (1949).

58 J. Pech, *Collection Czechoslov. Chem. Communs.*, **6,** 126 (1934).

Salts of the amines, especially the quaternary salts are reducible at the dropping mercury electrode. Values obtained for a number of these[59] are listed in Table XLII-19. The ease of reduction in this series is apparently dependent upon the purity of the salts. Von Stackelberg and Stracke[60] report that tetraethylammonium bromide is reduced at a more negative potential than tetramethylammonium bromide. Catalytic hydrogen waves are suggested for these waves and not reduction of the cation.

These quaternary salts, because of their very negative reductions, are useful as supporting electrolytes in polarography whenever the reduction reactions occur at potentials more negative than those corresponding to the reduction of the ions of the alkali metals.

Preparation of $(n\text{-}C_4H_9)_4NI$.[61, 62] Tri-*n*-butylamine (200 ml.) purified by fractional distillation (b.p. 213–215°) and *n*-butyl iodide (100 ml.) are heated together on a steam bath for 65 hours. The resulting solid is ground up, triturated with 300 ml. of anhydrous ethyl acetate, filtered, and then dissolved in the least amount of cold ethanol. The resulting solution is mixed with an equal volume of 10 per cent potassium hydroxide in ethanol and poured into water. Removal of the alcohol under reduced pressure gives a crystalline precipitate of tetrabutylammonium iodide which after drying and three recrystallizations from anhydrous thiophene-free benzene melts at 141–142°.

Directions for the preparation of tetramethylammonium chloride are given by Baret and Leveque.[63] This salt has the disadvantage of being hygroscopic.

7. AMINO ACIDS AND THEIR DERIVATIVES

The effect of amino acids on the catalytic hydrogen wave in ammoniacal cobalt solutions is discussed in the chapter on biological applications of polarography (Chap. XLVI). Amino acids do not produce a reduction wave at the dropping mercury electrode unless a reducible group is present, *e.g.*, the iodine in thyroxine. Certain esters and amides which are used as local anesthetics apparently produce catalytic hydrogen waves.[64] The reported results are listed in Table XLII-20.

Amino acids in the aromatic series and their derivatives, as illustrated by *p*-aminobenzoic acid, sulfanilamide, and related compounds, are surface active and lower the oxygen maximum in varying degrees. The extent of

[59] P. Van Rysselberghe and J. M. McGee, *J. Am. Chem. Soc.*, **67**, 1038 (1945).
[60] M. von Stackelberg and W. Stracke, *Z. Electrochem.*, **53**, 118 (1949).
[61] N. L. Cox, C. A. Kraus, and R. N. Fuoss, *Trans. Faraday Soc.*, **31**, 749 (1935).
[62] H. A. Laitinen and S. Wawzonek, *J. Am. Chem. Soc.*, **64**, 1765 (1942).
[63] C. Baret and P. Leveque, *Bull. soc. chim. France*, **1949**, 832.
[64] A. R. McIntyre and R. E. King, *Federation Proc.*, **1**, 160 (1942).

lowering is proportional to the physiological activity in the sulfanilamide series.[65]

Amino acids can be determined indirectly by forming a metal complex and determining this substance polarographically. The most common complex used is the copper complex which is formed by adding the amino acid to a solution of copper phosphate in disodium phosphate solution. The wave appears between −0.7 to −1.0 v. and is related to the amino acid content of the solution. For 5 micromoles of amino acids in the same volume of electrolyte currents of the following magnitudes are obtained: glycine, 0.136 microamp.; leucine, 0.112 microamp.; α,γ-diaminobutyric acid, 0.225 microamp.; and lysine, 0.09 microamp. For glycine the current

TABLE XLII-20

POLAROGRAPHIC BEHAVIOR OF SOME LOCAL ANESTHETICS

Compounds	pH	$-E_{1/2}$	i_d, mm. of deflection
Protocain, p-$C_4H_9NHC_6H_4COOCH_2CH_2N(CH_3)_2$	6.91	1.65	5500
Procain, p-$NH_2C_6H_4COOCH_2CH_2N(C_2H_5)_2 \cdot HCl$	6.91	1.68	1800
Intracain	6.91	1.75	400
Nupercain[a]	6.34	1.0	857
	—	1.5	2571

a 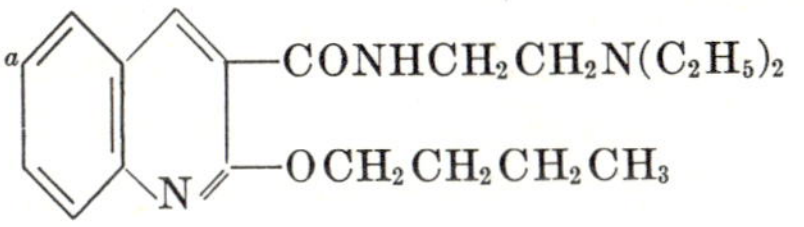

is proportional to the concentration over a range of 80–260 micromoles per liter. The half-wave potentials do not appear to differ significantly for the various complexes.[66]

This method has been applied to the quantitative determination of amino acids separated by paper chromatography. The paper chromatogram is treated overnight with a definite amount of the electrolyte and the resulting solution is examined polarographically. Maxima are suppressed by a starch solution.[67]

The copper complex of ethylenediamine tetraacetic acid is used for the polarographic determination of this acid in the absence of copper ions. Uncomplexed copper is removed by the addition of magnesium oxide.

[65] H. Veldstra and E. Havinga, *Rec. trav. chim.*, **66**, 273 (1947).

[66] T. S. G. Jones, *Biochem. J.*, **42**, ix (1947).

[67] A. J. P. Martin and R. Mittelmann, *Biochem. J.*, **43**, 353 (1948).

The method has been applied to the determination of ethylenediamine tetraacetic acid in the commercial sodium salt.[68]

Thyroxine. Thyroxine gives three waves at −1.20 v., −1.42 v., and −1.70 v. (S.C.E.) involving a total of 8 electrons in a solution containing 0.5 *N* sodium carbonate and either 1 per cent tetramethylammonium iodide[69] or 1 per cent tetramethylammonium bromide in 40 per cent ethanol or 20 per cent isopropyl alcohol.[70] The concentration was found to be a linear function of the diffusion current in the range 0.00006–0.0012 *M* (0.05–1.0 mg. per milliliter). The half-wave potentials remain constant in this range. Gelatin (0.05 per cent) was used to suppress maxima.

Out of 23 related substances which were studied, only 3,5-diiodo-4-(4′-methoxyphenoxy)benzoic acid and 3,5-diiodothyronine interfere in the determination of thyroxine. The data obtained are given in Table XLII-21.

3,5-Diiodotyrosine, which is often obtained along with thyroxine in the hydrolysis of proteins, does not interfere if the molar ratio is not greater than 10. In the presence of this amino acid, the first wave of thyroxine is used for the determination since the two waves of diiodotyrosine at −1.50 v. and −1.70 v. (S.C.E.) are in the same region as the second and third waves of thyroxine. If the ratio of diiodotyrosine to thyroxine is greater than 10, the thyroxine is eluted with 1-butanol from a solution of pH 3.5 and separated from the diiodotyrosine by extracting the latter with a mixture of sodium hydroxide and sodium bicarbonate.

The method has been used to determine thyroxine in Ardein (ground nut protein), thyroid powders, casein,[71] and iodinated casein.[70] The smallest quantity that can be detected is about 0.1 per cent of thyroxine in 1.0 g. of iodinated protein.

Procedure. Hydrolyze 1.0 g. of iodinated casein (containing about 5 mg. of thyroxine) with 40 per cent baryta for 20 hours. Adjust the pH value of the resulting 100 ml. of solution to 3.5 by means of dilute hydrochloric acid and extract the solution with one 50-ml. portion followed by two 25-ml. portions of 1-butanol. Wash the butanol extract with two 50-ml. portions of a solution containing 16 per cent of sodium hydroxide and 5 per cent of sodium bicarbonate, and evaporate to dryness under reduced pressure. Dissolve the residue in 10 ml. of 1.0 *N* sodium carbonate, transfer to a 20-ml. graduated flask, add 2.0 ml. of 10 per cent tetramethylammonium bromide and 4.0 ml. of isopropyl alcohol, and make the solution up to 20 ml. with distilled water. Polarograph a 3.0-ml. oxygen-free portion of the final solution over the po-

---

[68] W. Furness, P. Crawshaw, and W. C. Davies, *Analyst*, **74,** 629 (1949).

[69] G. K. Simpson and D. Traill, *Biochem. J.*, **40,** 116 (1946).

[70] E. T. Borrows, B. A. Hems, and J. E. Page, *J. Chem. Soc.*, **1949,** Suppl. Issue No. 1, p. 204.

[71] G. K. Simpson, A. G. Johnston, and D. Traill, *Biochem. J.*, **41,** 181 (1947).

TABLE XLII-21

HALF-WAVE POTENTIALS (S.C.E.) AND DIFFUSION CURRENTS (IN MICROAMP.) FOR THYROXINE AND RELATED COMPOUNDS

Substance	Molar concn. × 10^4	1st step $-E_{1/2}$ (S.C.E.)	1st step i_d	2nd step $-E_{1/2}$ (S.C.E.)	2nd step i_d	3rd step $-E_{1/2}$ (S.C.E.)	3rd step i_d
C_6H_5COOH	8.0	—	—	—	—	—	—
o-IC_6H_4COOH	20.0	—	—	1.50	8.0	—	—
m-IC_6H_4COOH	20.0	—	—	1.44	7.5	—	—
p-IC_6H_4COOH	20.0	—	—	1.44	8.4	—	—
p-HOC_6H_4COOH	18.0	—	—	—	—	—	—
3,5-I_2-2-HOC_6H_2COOH	12.8	1.25	4.5	1.50	3.2	—	—
3,5-I_2-4-HOC_6H_2COOH	6.2	—	—	1.55	1.96	1.72	1.66
3,5-I_2-4-$AcOC_6H_2COOH$	5.8	1.08	1.63	1.25	1.79	1.63	0.48
3,5-I_2-4-$CH_3OC_6H_2COOH$	12.5	1.20	3.80	1.50	3.5	—	—
p-$HOC_6H_4CH_2CH(NH_2)COOH$	14.0	—	—	—	—	—	—
p-$HOC_6H_4CH_2CH(NHCOC_6H_5)COOC_2H_5$	3.2	—	—	—	—	—	—
4-HO-3,5-I_2-$C_6H_2CH_2CH(NH_2)COOH$	11.4	—	—	1.51	3.7	1.72	3.9
4-HO-3,5-$I_2C_6H_2CH_2CH(NHCOCH_3)COOH$	10.5	—	—	1.47	2.85	1.67	3.35
4-AcO-3,5-$I_2C_6H_2CH_2CH(NHAc)COOH$	4.8	—	—	1.48	1.15	1.68	1.91
4-C_4H_9O-3,5-$I_2C_6H_2CH_2CH(NHAc)COOH$	9.5	1.16	3.28	1.42	2.70	—	—
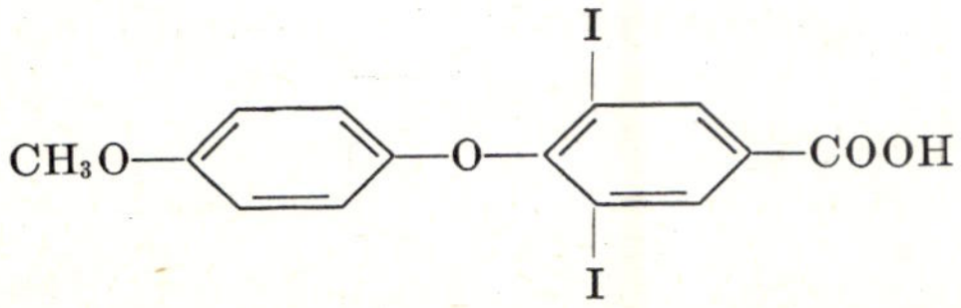	5.0	1.11	1.30	1.30	1.40	—	—

Table Continued

TABLE XLII-21—*Continued*

Substance	Molar concn. × 10^4	1st step $-E_{1/2}$ (S.C.E.)	1st step i_d	2nd step $-E_{1/2}$ (S.C.E.)	2nd step i_d	3rd step $-E_{1/2}$ (S.C.E.)	3rd step i_d
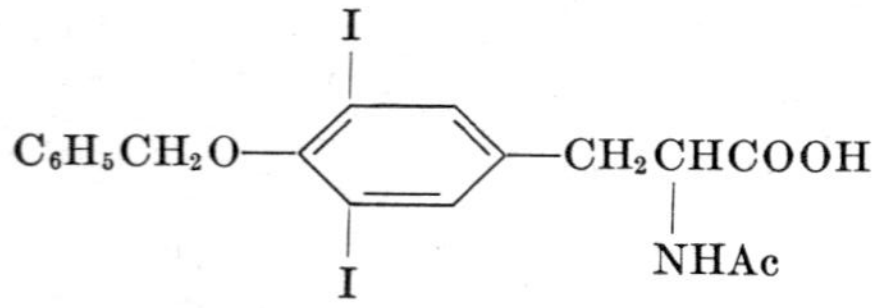	8.9	1.09	2.55	1.33	2.73	—	—
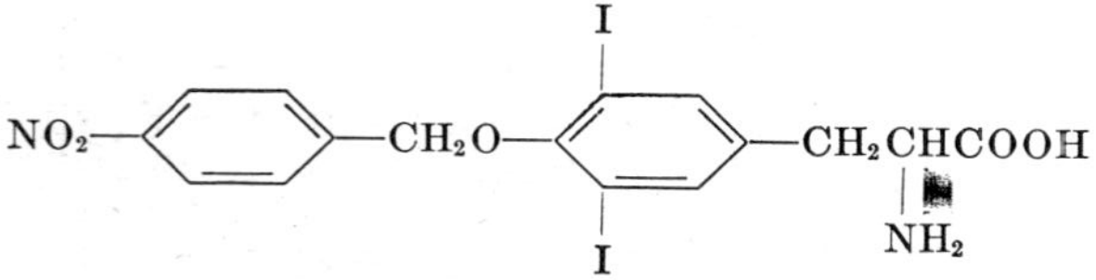	8.2	0.78	5.40	0.94	2.45	1.34	2.75
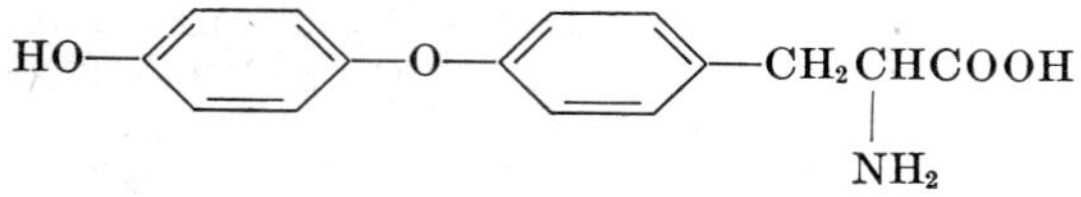	8.9	—	—	—	—	—	—
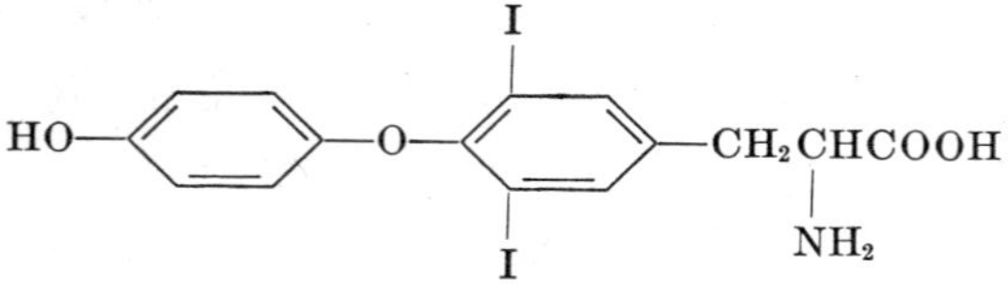	9.5	1.18	2.71	1.37	2.56	—	—
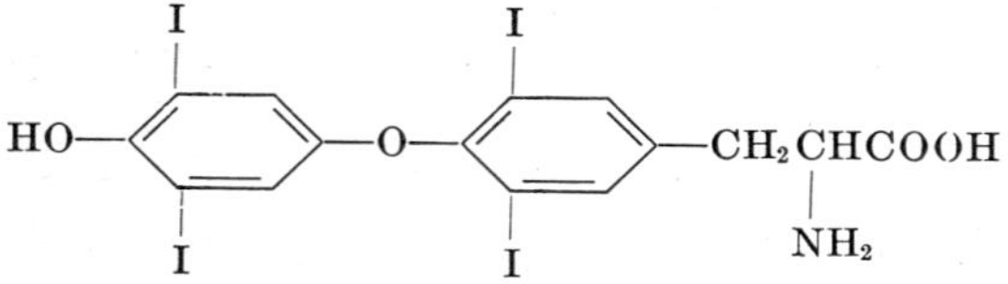	3.1	1.12	1.02	1.30	1.60	1.51	0.90

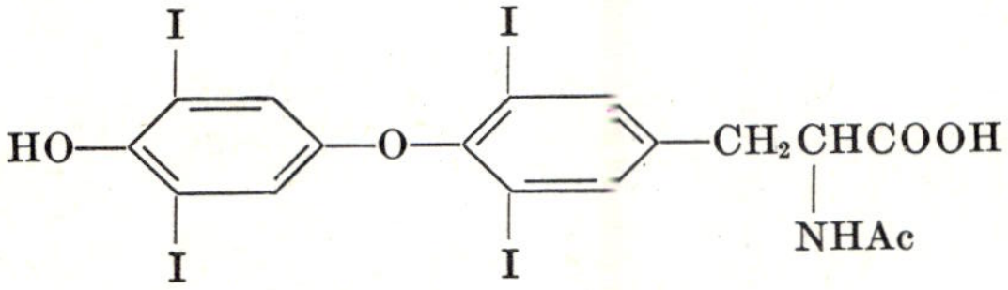	3.0	1.10	0.90	1.29	1.40	1.50	0.60
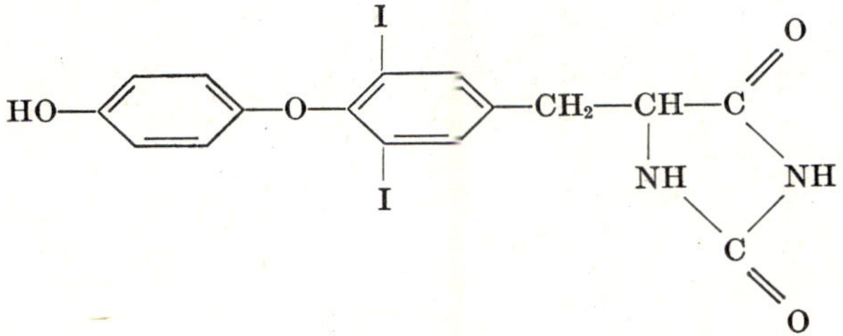	9.5	1.19	3.10	1.39	2.50	—	—
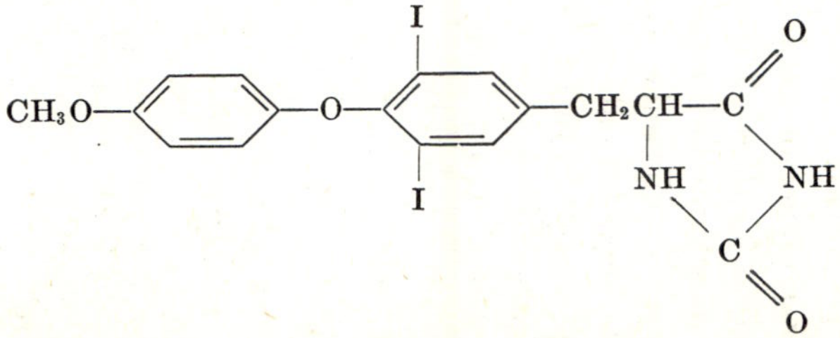	9.0	1.10	2.56	1.30	2.66	—	—

tential range of −0.6 to −2.0 v. Measure the heights of the first thyroxine and the combined thyroxine and diiodotyrosine steps.

For preparations containing more than 0.5 per cent thyroxine the accuracy of the method is about ±10 per cent.

The polarographic behavior of kynurenine resembles that of acetophenone, *o*-aminobenzaldehyde, and lobeline and favors the structure (I) for this compound.[72]

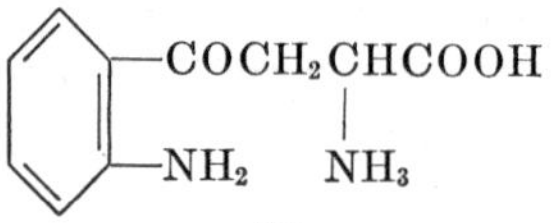

(I)

$C_6H_5N(CH_3)_2$
↓
O

(II)

8. AMINE OXIDES

Dimethylaniline N-Oxide. This compound (II) gives a reduction potential of −0.7047 v. (S.C.E.) at a pH of 3.5.[73] No information is given on the mechanism of reduction. The chemical reduction product is the amine. The behavior of amine oxides of pyridine and quinoline is described in Chapter XLV.

9. DIAZONIUM SALTS

Diazotized aniline yields two well-defined waves in buffer solutions in the presence of tetramethylammonium chloride. The first wave does not vary with pH while the second becomes more negative with increasing pH.[74]

Substituted diazotized anilines behave in a manner similar to that of the parent compound. The results obtained are summarized in Table XLII-22. The reported half-wave potentials have only qualitative significance since the shapes of the waves are irregular and the polarograms are difficult to measure.[75]

The total diffusion current is roughly independent of pH and is equivalent to a 4-electron reduction. The product in all cases must be a phenylhydrazine:

$$ArN_2^+ + 4e + 3H^+ \rightarrow ArNHNH_2$$

---

[72] F. Santavy, *Experimentia*, **5**, 70 (1949).

[73] E. Ochiai, *Proc. Imp. Acad.* (*Tokyo*), **19**, 307 (1943); *Chem. Abstracts*, **41**, 5880 (1947).

[74] E. R. Atkinson and H. H. Warren, Report before the American Chemical Society Meeting, September, 1947.

[75] R. M. Elofson, R. L. Edsberg, and P. A. Mecherly, *J. Electrochem. Soc.*, **97**, 166 (1950).

The first step of the reduction is about one-half to one-third as large as the second wave and is marked by decreases in height as the pH increases. The independence of the half-wave potential with change in pH suggests a reduction of the following type for this step:

$$ArN_2^+ + e \rightarrow ArN_2$$

TABLE XLII-22

HALF-WAVE POTENTIALS OF DIAZOTIZED AROMATIC AMINES AS A FUNCTION OF pH

Compound	pH	$E_{1/2}$ (S.C.E.)	
		First wave	Second wave
p-Anisidine	1 *N* HCl	−0.083, −0.184	−0.434
	0.1 *N* HCl	−0.086	−0.524
	4.2	−0.056	−0.575
	7.8	−0.079	−0.919
	9.75	−0.160	−0.930
Aniline	0.1 *N* HCl	−0.18	−0.67
	4.02	+0.029, −0.057	−0.690
	6.02	+0.051, −0.050	−0.920
	8.00	+0.020, −0.154	−0.990
	9.92	−0.202	−0.970
Dichloroaniline	0.1 *N* HCl	−0.106	−0.554, −0.832
	3.92	−0.022	−0.525, −0.875
	5.90	−0.029	−0.802
	7.60	−0.112	−0.820
Sulfanilic acid	0.1 *N* HCl	−0.326	−0.555, −0.831
	4.02	−0.095, −0.260	−0.626, −0.970
	6.08	−0.112, −0.273	−0.557, −0.885
	7.9	−0.300	−0.606, −0.983
	10.0	−0.233, −0.332	−0.961

Reduction of the diazotized aromatic amine depends on the amine and is not observed above a pH of 8–10. The change takes place at the same point at which the absorption spectrum changes abruptly for the diazonium salt; it occurs at a lower pH for negatively substituted diazotized amines than when electron-donating groups are present. In this pH range the diazonium ion begins to disappear and the diazotates and isodiazotates, which are not reducible, are formed.

The first wave when measured at a potential of −0.3 v. (S.C.E.) is proportional to the concentration and is used for quantitative estimation of diazonium salts. At concentrations greater than 10^{-4} *M* maxima are sometimes observed and are suppressed by Orthner's compound, a commercial surface-active agent.* Gelatin is only of limited value.

* General Aniline and Film Co. Chemical Number 147.

The method has been used to study the rate of coupling of diazonium salts with phenols and related compounds, diazotization rates, decomposition rates of diazonium salts, and the formation and hydrolysis of diazoamino compounds. The latter compounds are reduced at a much higher voltage than that of the first wave of the diazonium salts.

The polarographic determination of diazonium salts is the basis for the amperometric titration of couplers such as pyrazolone and naphthols.[76]

---

[76] R. M. Elofson and P. A. Mecherly, *Anal. Chem.*, **21**, 565 (1949).

CHAPTER XLIII

Sulfur Compounds, the Cysteine–Cystine System, Peroxides, and Miscellaneous Organic Compounds

1. SULFUR COMPOUNDS

Organic compounds containing the thiol group (—SH) produce anodic waves and also give rise to catalytic hydrogen waves in ammoniacal cobalt solutions at −1.8 v. if present on the α-carbon atom of an acid or on the β-carbon atom of an amino acid (see p. 849). In solutions containing 1 *M* ammonia-ammonium chloride buffer (pH 9.8) and 10^{-3} *M* cobalt chloride a maximum occurs at approximately −1.8 to −1.9 v. (S.C.E.), varying heights are obtained with mercaptoacetic, α-mercaptopropionic, α-mercaptosuccinic, α-mercaptoglutaric, α-mercaptoadipic, and α,α′-dimercaptoadipic acids. β-Mercaptopropionic and mercaptomethylsuccinic acids show only a slight inflexion at −1.8 v. No thiol inflection is given by BAL, $CH_2OHCHSHCH_2SH$, or thioethyleneglycol $HSCH_2CH_2OH$.[1]

Ethyl mercaptan and butyl mercaptan produce two anodic curves in 85–90 per cent alcohol in the presence of 0.025 *N* sulfuric acid and in buffered solutions of acetic acid and sodium acetate. The first curve is used for quantitative purposes while the second wave is not reproducible.[2]

Disulfides (cystine, see p. 784) and aromatic thiocyanates are the only types of sulfur groupings which are reported to be reducible at the dropping mercury electrode. Sulfoxides, as exemplified by dibenzyl sulfoxide, are not reducible in the range up to −2.0 v. (S.C.E.)[3] The disulfide group may be reduced even if attached to a thiocarbonyl group as in tetramethylthiuram disulfide (I). This compound gives a well-defined wave in hydrochloric acid, potassium chloride, and ammonium hydroxide.

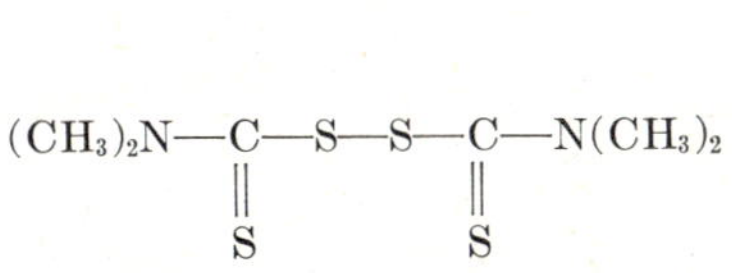

(I)

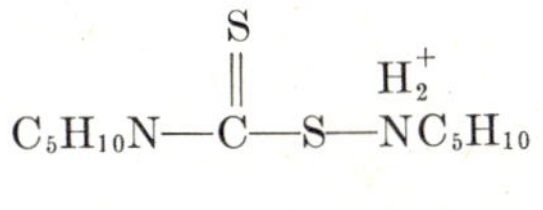

(II)

[1] J. B. Fraser, L. N. Owen, and G. Shaw, *Biochem. J.*, **41,** 328 (1947).

[2] M. I. Gerber, *Zhur. Anal. Khim.*, **2,** 265 (1947); *Chem. Abstracts*, **43,** 6943 (1949).

[3] K. G. Stone, *J. Am. Chem. Soc.*, **69,** 1832 (1947).

The group involved in the reduction is not exactly known since the piperidine salt of pentamethylene dithiocarbonic acid (II) is also reduced.[4] The waves are suitable for quantitative estimation.

The polarographic method has been used to follow the decomposition of allicin (III) with time in a 0.5 *M* phosphate buffer of pH 6.5. The nature of the reduction in this compound is not known.[5]

$$CH_2{=}CH{-}CH_2{-}S{-}\underset{\substack{\downarrow \\ O}}{S}{-}CH_2CH{=}CH_2$$

(III)

Aromatic Thiocyanides. *p*-Aminophenyl thiocyanide and *p*-dimethylaminophenyl thiocyanide are reported[6] to be reduced in 80 per cent ethanol and aqueous solutions containing 0.1 *N* tetraethylammonium iodide as an electrolyte. The half-wave potential in aqueous solution of varying pH were found to be constant at approximately −1.56 v. (N.C.E.) in the range of pH from 1.8 to 6.7. At pH 7.3, 8.5, and 10.1 the values were −1.44 v., −1.48 v., and −1.33 v., respectively. The dimethylamino compound gives a half-wave potential of −1.58 v. at pH 6.3. In 80 per cent alcohol the amino compound gives a half-wave potential at −1.84 v. and the dimethylamino compound one at −1.90 v.

The wave height at pH 8.5 for the amino compound was three times as high as that for the same amount of oxygen and suggested a 6-electron reduction of the thiocyanide group to the thiophenol:

$$NH_2C_6H_4SC{\equiv}N + 6e + 6H_2O \rightarrow NH_2C_6H_4SH + CH_3NH_2 + 6OH^-$$

Since the wave height decreases with an increase in pH, the possibility that these waves represent the reduction of the anilinium ion instead of the thiocyanide group cannot be discounted.[6]

The Cysteine–Cystine System. The oxidation potential of the biologically very important cysteine–cystine system has been the subject of many investigations.[7]

---

[4] G. Proske, *Angew. Chem.*, **53,** 550 (1940).

[5] C. J. Cavallito, J. S. Buck, and C. M. Suter, *J. Am. Chem. Soc.*, **66,** 1952 (1944).

[6] K. Schwabe, *Z. Electrochem.*, **53,** 347 (1949).

[7] M. Dixon and J. H. Quastel, *J. Chem. Soc.*, **123,** 2943 (1923). E. C. Kendall and F. F. Nord, *J. Biol. Chem.*, **69,** 295 (1926). M. Dixon and H. E. Tunnicliffe, *Biochem. J.*, **21,** 844 (1927). E. C. Kendall and D. F. Loewen, *ibid.*, **22,** 649, 669 (1928). M. Dixon, *Proc. Roy. Soc. London*, **B101,** 57 (1927). D. C. Harrison and J. H. Quastel, *Biochem. J.*, **22,** 683 (1928). L. Michaelis and L. B. Flexner, *J. Biol. Chem.*, **79,** 689 (1928). J. W. Williams and E. M. Drissen, *ibid.*, **87,** 441 (1930). E. K. Fischer, *ibid.*,

Cysteine, HOOC—CH(NH_2)—CH_2SH, which in the following will be referred to as RSH, can be oxidized to cystine by anodic electrolysis at a platinum electrode. Cystine, HOOC—CH(NH_2)—CH_2—S—S—CH_2-—CH(NH_2)—COOH, designated in the following as RSSR, can be reduced to cysteine by suitable reducing agents or by electrolytic reduction. From potential measurements carried out in the classical way it is apparent that the reaction:

$$\mathrm{RSSR} + 2\mathrm{H}^+ + 2e \rightarrow 2\mathrm{RSH} \qquad (1)$$

is not strictly reversible. When noble metals are used in the measurement of the oxidation potential it is found that the concentration of cystine does not affect the potential, but that the latter does depend on the hydrogen ion concentration, and on the concentration of cysteine, according to the equation:

$$E = E_0 + \frac{RT}{F_y} \ln [\mathrm{H}^+] - \frac{RT}{F_y} \ln [\mathrm{RSH}]$$

Barron, Flexner, and Michaelis[7a] showed that mercury is not an "indifferent electrode" to cysteine. Upon air oxidation of cysteine at a mercury surface they found that mercuric cysteinate Hg(RS)$_2$ was formed. Hence it is not to be expected that reaction (1) will be reversible at the dropping mercury electrode. Extensive voltammetric studies on the system cysteine–cystine have been carried out by Kolthoff and Barnum.

Anodic Waves of Cysteine. Kolthoff and Barnum[8] found an anodic wave of cysteine at the dropping electrode with a diffusion current that was proportional to the concentration when the electrolysis was carried out in 0.1 *N* perchloric acid. A typical wave is shown by curve 1 in Fig. XLIII-1. A diffusion coefficient of cysteine of 7.0×10^{-6} cm.^{2} sec.$^{-1}$ at 25° C. was calculated with the aid of the Ilkovic equation. In a medium of 0.1 *N* perchloric acid the half-wave potential was found to be −0.05 v. (*vs.* S.C.E.) at 25° C., and independent of the cysteine concentration. Hence in a medium of 0.1 *N* perchloric acid cysteine can be detected and determined quantitatively with the dropping electrode. Hydrochloric acid should not be used in the solution as the anodic wave of chloride interferes with the cysteine wave. From an analysis of the waves,

89, 753 (1930). J. C. Ghosh, S. N. Raychadhuri, and S. C. Ganguli, *J. Indian Chem. Soc.*, **9,** 43 (1932). J. C. Ghosh and S. C. Ganguli, *Biochem. J.*, **28,** 381 (1934). D. E. Green, *ibid.*, **27,** 678 (1933). J. S. Fruton and H. T. Clarke, *J. Biol. Chem.*, **106,** 667 (1934). H. Borsook, E. L. Ellis, and H. M. Huffman, *ibid.*, **117,** 281 (1937).

[7a] E. S. G. Barron, L. B. Flexner, and L. Michaelis, *J. Biol. Chem.*, **81,** 743 (1929).

[8] I. M. Kolthoff and C. Barnum, *J. Am. Chem. Soc.*, **63,** 3061 (1940).

and from anodic electrolysis experiments in a quiet pool of mercury, Kolthoff and Barnum obtained evidence that the anodic wave of cysteine does not correspond to an oxidation to cystine but to the reversible reaction:

$$RSH + Hg \rightleftharpoons HgRS + H^+ + e \qquad (2)$$

which yields the following equation of the polarographic wave:

$$E = E_{1/2} + \frac{RT}{F_y} \ln \frac{i}{i_d - i} \qquad \text{(at constant pH)}$$

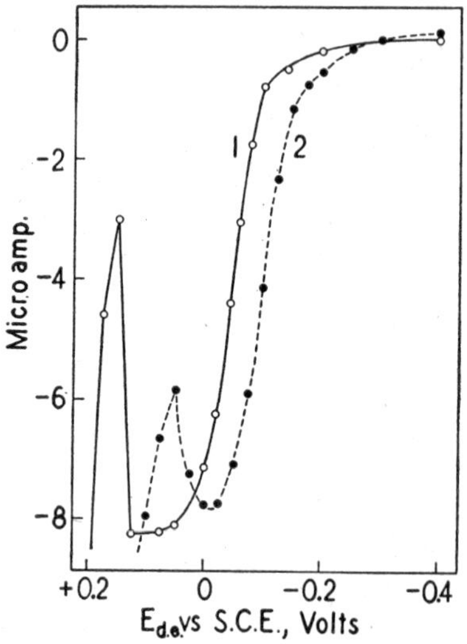

Fig. XLIII-1. Anodic waves of cysteine: (*1*) 0.002 *M* cysteine in 0.1 *N* perchloric acid; and (*2*) 0.002 *M* cysteine in Clark and Lubs buffer of pH = 2.0.

This study for the first time gives evidence that at extremely small concentrations (of the order of 10^{-20} *M*) mercurous ions in solution are mainly present as Hg^+ and not as Hg_2^{++}. Undoubtedly there is an equilibrium $Hg_2^{++} \rightleftharpoons 2Hg^+$, which shifts to the right with decreasing mercurous ion concentration.

At a platinum wire microelectrode reaction (2) cannot occur. Voltammetric experiments at a quiet platinum wire microelectrode did not yield well-defined diffusion currents. The striking difference between the curves obtained at the platinum and the dropping mercury electrodes is that the wave at the platinum electrode starts at a potential about 0.6 v. more positive than the wave at the dropping electrode in solutions of the same pH. Typical anodic c.v. curves obtained with both types of electrodes at a pH of 2 are given in Fig. XLIII-2. Apparently, the reaction occurring at the platinum electrode is an irreversible oxidation of cysteine to cystine, while at the mercury electrode mercurous cysteinate is formed.

This mercurous cysteinate is soluble in strongly acid and strongly alkaline media, but is quite insoluble at intermediate pH values. It separates as a film on the dropping electrode during the electrolysis and thereby causes irregularities on the c.v. curves. This is even noticeable in 0.01 *N* perchloric acid, in which case no well-defined diffusion current is obtained The situation becomes worse with increasing pH, as shown by the curves in Fig. XLIII-3, which have been obtained in buffer solutions with a pH of

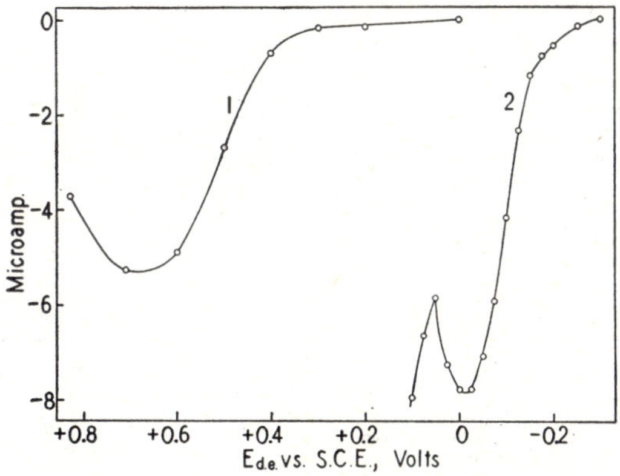

Fig. XLIII-2. Anodic cysteine waves at pH = 2 with a platinum microelectrode (curve *1*) and with the dropping mercury electrode (curve *2*); concentration of cysteine was 0.002 *M*.

4.73 (acetate) and of 6.0 (phosphate). It is seen that as the pH increases the waves start at a more negative potential, but they soon level off to give a constant current (false diffusion current) which is much smaller than the true diffusion current. The current finally increases at fairly positive potentials, but no diffusion current is observed as the wave coincides with the dissolution curve of mercury. In agreement with the postulate that the abnormalities are due to film formation, Kolthoff and Barnum found that the constant current (in their experiments about 1 microamp.) was independent of the cysteine concentration, presumably because the levelling off of the current occurs at a value at which a film of a certain critical thickness is formed. When the cysteine concentration was so small that the diffusion current was 1 microamp. or less a true diffusion current was found and the abnormalities disappeared.

At a pH of 9.2 (borax) a discontinuity was still found in the c.v. curve

(Fig. XLIII-4), but a well-defined diffusion current was obtained, while in 0.1 *N* sodium hydroxide a normal wave was found until the diffusion current was reached. Slight irregularities were observed, however, in 0.1 *N* sodium hydroxide when the potential became more positive than −0.4 v.

Cathodic Waves of Cystine. A few measurements of cathodic waves of cystine have been reported by Brdicka,[9] who found that in a buffer solution which was 0.1 *N* in ammonia and 0.1 *N* in ammonium chloride the diffusion

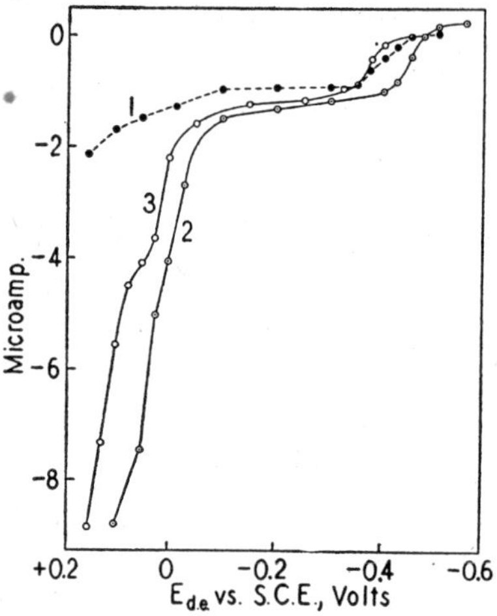

Fig. XLIII-3. Anodic waves of cysteine: (*1*) 0.00025 *M* cysteine in phosphate buffer of pH = 6.0; (*2*) 0.002 *M* cysteine in phosphate buffer of pH = 6.0; and (*3*) 0.002 *M* cysteine in acetate buffer of pH = 4.7.

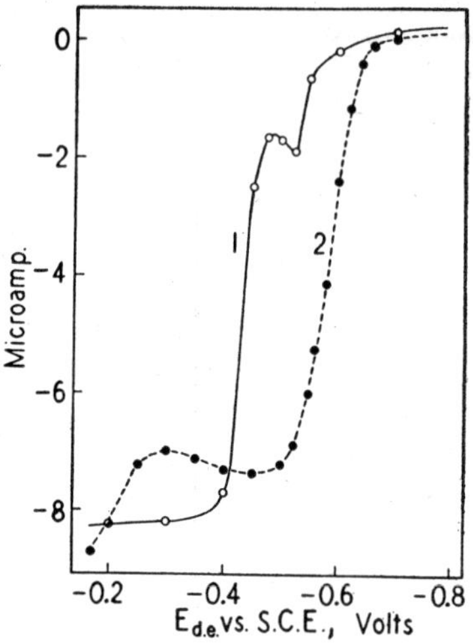

Fig. XLIII-4. Anodic waves of cysteine in alkaline medium: (*1*) 0.002 *M* cysteine in borax buffer of pH = 9.2; and (*2*) 0.002 *M* cysteine in 0.1 *N* sodium hydroxide.

current was proportional to the concentration. Brdicka also measured the depolarization potential of 0.005 *M* cystine solution in 1 *N* hydrochloric acid, and in 1 *N* sodium hydroxide, and found the values −0.37 v. and −1.15 v. (*vs.* N.C.E.). From this dependence of the reduction potential on the pH Brdicka concluded that the reduction of cystine to cysteine at the dropping electrode occurs in a reversible way. This is not correct as is shown below. Roncato[10] studied the polarographic reduction of cystine in unbuffered neutral, acid, and alkaline media. Since the reduction of

[9] R. Brdicka, *Collection Czechoslov. Chem. Communs.*, **5**, 238 (1933).
[10] A. Roncato, *Arch. sci. biol. Italy*, **20**, 146 (1934).

cystine is dependent on the pH the waves in unbuffered solutions are not subject to exact analysis.[11]

The interpretation of the cystine waves is quite involved as is apparent from a systematic study carried out by Kolthoff and Barnum.[12] Before discussing the interpretation of the waves the various experimental characteristics found by these authors are given first. Fig. XLIII-5 shows the c.v. curve of 0.001 *M* cystine solution in 0.1 *N* hydrochloric acid. A pronounced maximum is noticeable, but at a potential of −0.7 v. (S.C.E.)

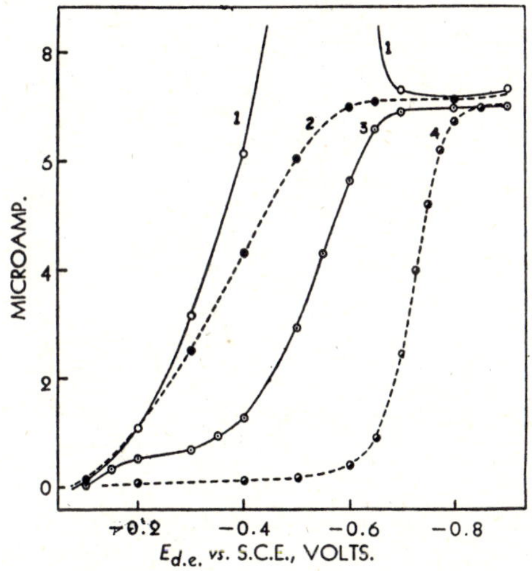

Fig. XLIII-5. Effect of thymol on the cathodic wave of 0.001 *M* cystine in 0.1 *N* hydrochloric acid; (*1*) no thymol; maximum reached at 20.4 microamp.; (*2*) 2.4×10^{-5} *M* thymol; (*3*) 1.2×10^{-4} *M* thymol; and (*4*) 4.8×10^{-4} *M* thymol.

the diffusion current is reached. This i_d in the presence and in the absence of maximum suppressors is proportional to the concentration of cystine, and hence the latter can be determined polarographically. The diffusion coefficient of cystine in 0.1 *N* hydrochloric acid calculated from i_d with the aid of the Ilkovic equation was found to be 5.3×10^{-6} cm.^{2} sec.$^{-1}$ at 25° C.

Neither polyvalent cations (lanthanum) nor polyvalent anions (ferrocyanide) were found to suppress the maximum. The latter, however, is suppressed by addition of capillary-active substances, such as phenol, resorcinol, thymol, gelatin, methylene blue, methyl red, or camphor. The

[11] O. H. Müller, *Chem. Revs.*, **24**, 95 (1939).

[12] I. M. Kolthoff and C. Barnum, *J. Am. Chem. Soc.*, **63**, 520 (1941).

effect of these substances is very peculiar, because in addition to eliminating the maxima *they also shift the waves to more negative potentials.* From a practical viewpoint thymol was found to be most suitable for the elimination of the cystine maximum. It is seen from Fig. XLIII-5 that the maximum of the 0.001 *M* cystine solution in 0.1 *N* hydrochloric acid was completely repressed after the solution had been made 2.4×10^{-5} *M* in thymol. The shift of the waves to more negative potentials with increasing concentration of thymol is also noticed in Fig. XLIII-5. The effect of varying concentrations of camphor on the cystine wave is most unusual

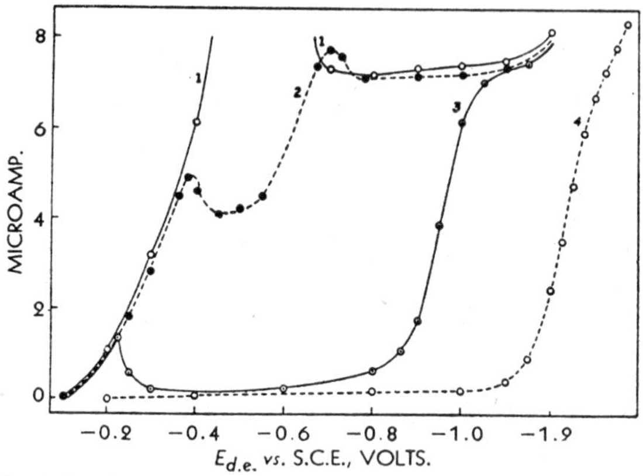

Fig. XLIII-6. Influence of camphor on the cathodic wave of cystine in 0.1 *N* hydrochloric acid: (*1*) no camphor; maximum reached at 20.4 microamp.; (*2*) 3.5×10^{-4} *M* camphor; (*3*) 10^{-3} *M* camphor; and (*4*) solution saturated with camphor.

as is evident from Fig. XLIII-6. The slope of the wave becomes steeper with increasing camphor concentration. It is also seen that the wave is displaced by almost 1 v. to more negative potentials in a solution which is saturated with camphor.

In 1 *N* hydrochloric acid Kolthoff and Barnum observed a wave which almost coincided with that obtained in 0.1 *N* hydrochloric acid. In the more strongly acid medium the diffusion current appeared to be a few per cent smaller than in 0.1 *N* hydrochloric acid. This difference, undoubtedly, is due to the fact that the diffusion coefficient of the cystine cation is smaller than that of uncharged cystine. At pH values between 3 and 7 cystine is very slightly soluble and c.v. curves were determined with saturated solutions, the concentrations of which varied between 7 and 4×10^{-4} *M*. No maxima were found at pH values above 2. The character of the waves at pH values between 3 and 9 is quite different from that at a pH of 1. It is seen from Fig. XLIII-7 that at the former pH values two-step waves are found, the first step being fairly steep while the second step has a flat slope. The "depolarization potential" at the beginning of the first wave was found to shift 0.056 v. to more negative potentials with a unit increase in pH. In 0.1 N sodium hydroxide only one wave with a very flat slope is found, the latter being typical of an irreversible reaction.

Again, it was found that at the various pH values camphor displaced the waves to more negative potentials, at the same time making the slope much steeper. From Fig. XLIII-8 it is seen that saturation with camphor shifts

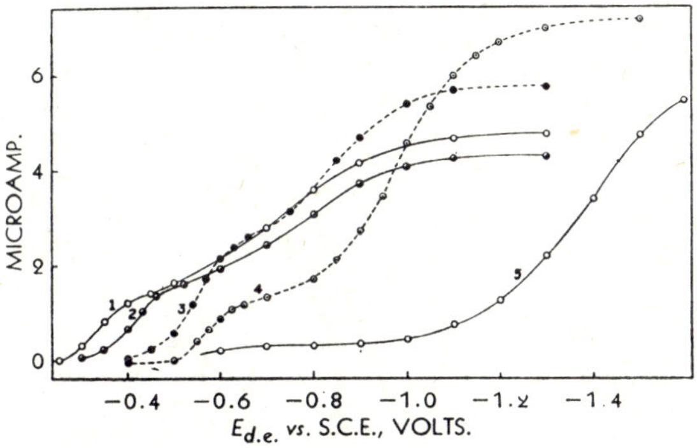

Fig. XLIII-7. Effect of pH on the cathodic cystine wave: (*1*) acetate buffer of pH = 4.7 saturated with cystine; (*2*) phosphate buffer of pH = 6.0 saturated with cystine; (*3*) phosphate buffer of pH = 8.0 saturated with cystine; (*4*) 0.001 M cystine in borax buffer of pH = 9.2; and (*5*) 0.001 M cystine in 0.1 N sodium hydroxide.

the wave of cystine at pH values of 1 and 2 to exactly the same potentials. At pH values between 4 and 9.2 saturation with camphor shifts the waves to still more negative potentials, all the waves starting at about −1.35 v. The wave in 0.1 N sodium hydroxide is also shifted to −1.35 v. after saturation with camphor, but has become more a typical two-step wave (Fig. XLIII-8).

In order to explain the shift of the cystine waves to more negative potentials Kolthoff and Barnum postulate that the cystine has to be oriented (adsorbed) in a favorable position at the surface of the dropping mercury electrode to be reduced. Apparently capillary-active substances counteract or prevent the orientation (adsorption) of cystine at the interface and thus displace the waves to more negative potentials. The effect of camphor is particularly striking (Fig. XLIII-6). When the concentration of camphor is very small the initial portion of the wave almost coincides with the wave (showing the pronounced maximum) obtained in the absence of camphor. Apparently, at these small concentrations the camphor is unable to prevent the orientation of cystine completely. With increasing negative potentials the adsorption of camphor is apparently favored, and dependent on the concentration of camphor, the current suddenly decreases

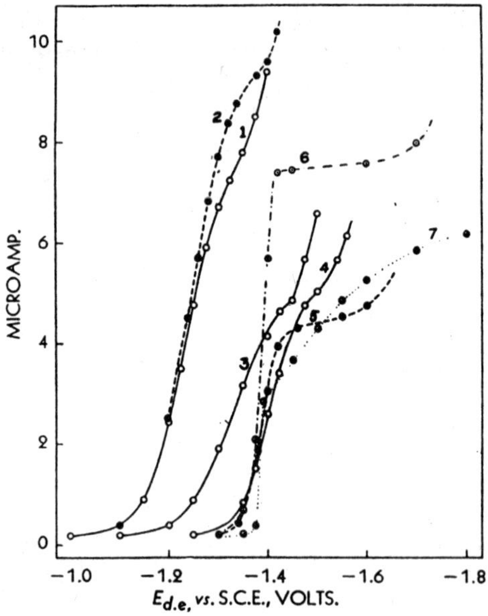

Fig. XLIII-8. Effect of camphor on cystine waves at various pH values: (*1*) 0.001 *M* cystine in 0.1 *N* HCl; (*2*) buffer solution of pH = 2.0 saturated with cystine; (*3*) buffer solution of pH = 3.0 saturated with cystine; (*4*) buffer solution of pH = 4.0 saturated with cystine; (*5*) buffer solution of pH = 6.0 saturated with cystine; (*6*) 0.001 *M* cystine in borax buffer of pH = 9.2; and (*7*) 0.001 *M* cystine in 0.1 *N* NaOH. All solutions were saturated with camphor.

and the remainder of the wave is displaced to more negative potentials. A quantitative interpretation is not possible since the relative adsorbabilities of cystine and camphor at various potentials of the mercury drop, and the relative speeds of adsorption and orientation are not known.

The above interpretation is substantiated by the electrocapillary behavior of camphor as shown in Fig. XLIII-9. These curves show the drop time of mercury in 0.1 *N* hydrochloric acid at various potentials, and in the same medium saturated with camphor. The two electrocapillary curves coincide at a potential of about −1.28 v. (*vs.* S.C.E.). At this potential the camphor becomes completely desorbed, and hence can no longer interfere with the reduction of cystine. In agreement herewith the beginning of the cystine waves in solutions with a pH of 1 to 2 which are saturated with camphor were found to be displaced to a potential of about −1.2 v. At pH values greater than 4 the electrocapillary curve of camphor was found to coincide with the normal curve at a potential of −1.5 v., this being the potential at which the cystine waves start at these pH values in solutions saturated with camphor. Incidentally, it may be said that the desorption of camphor is also evident from the appearance of a small wave in solutions free of reducible matter which are saturated with camphor. The apparent wave is not a reduction wave but indicates the sudden change of

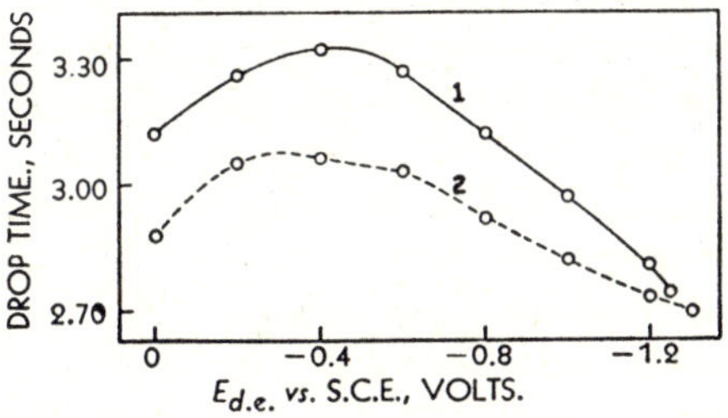

Fig. XLIII-9. Influence of camphor on the electrocapillary curve in 0.1 *N* hydrochloric acid: (*1*) no camphor; and (*2*) solution saturated with camphor.

the residual current at the potential where the camphor is desorbed and where the capacity of the double layer changes markedly.

The double wave in a 0.1 *N* sodium hydroxide solution which is saturated with camphor (Fig. XLIII-8) is also easily accounted for. The slope of the wave in the absence of camphor (Fig. XLIII-7) is flat; in the presence of camphor, however, the beginning of the wave is displaced to −1.35 v. where the camphor is desorbed. Between −1.35 v. and −1.45 v. the desorption of camphor becomes complete and the wave of cystine is fairly steep in this potential region. At potentials more negative than −1.45 v. the wave in the presence of camphor coincides with that in the absence of camphor.

If the reduction of cystine at the dropping electrode were reversible:

$$\text{RSSR} + 2\text{H}^+ + 2e \rightleftharpoons 2\text{RSH}$$

it is easily derived that:

$$E = E^0 + 0.0295 \log \frac{i_d - i}{i^2} - 0.0591 \text{ pH } (25° \text{ C.})$$

Plotting E against log $(i_d - i/i^2)$ Kolthoff and Barnum found straight lines, which, however, did not have the theoretical slope. For example, in the analysis of the wave in 0.1 N hydrochloric acid which was 2.4×10^{-5} M in thymol (maximum eliminated) the slope was found to be 0.132 instead of 0.0295. With increasing thymol concentrations the slope became closer to the theoretical value but did not become equal to it. With 4.8×10^{-4} M thymol a slope of 0.046 was found. Similar relations hold when camphor is added instead of thymol. The straight-line relation also applied in the presence of gelatin, but the deviation from the theoretical value became greater with increasing gelatin concentrations. For example, in the presence of 0.01 per cent of gelatin a slope of 0.150 was found at a pH of 2. From the above it is evident that the reduction of cystine at the dropping electrode is not thermodynamically reversible.

Since the plot of E against log $(i_d - i/i^2)$ yields a straight line, it can be inferred that the half-wave potential should change with the concentration of cystine, other conditions remaining the same. Indeed Kolthoff and Barnum found a slight shift of the half-wave potential to more negative values with increasing cystine concentrations, which was in agreement with the shift calculated from the above equation.

Finally, the two-step waves found at pH values between 3 and 9.2 need interpretation. The first and steeper part of these double waves is caused by the facts that the reduction of cystine is irreversible and that part of the cysteine formed at potentials on the first step of the waves reacts anodically with mercury with the formation of mercurous cysteinate (see first section of this chapter). The rather complex relations are clarified by the diagram in Fig. XLIII-10, which gives the observed reduction wave of cystine and the anodic cysteine wave at a pH of 6. The dotted line A is an extrapolation of the second step of the cystine wave and represents the hypothetical "true reduction wave." At point C on the experimental reduction wave the current is zero, and the potential corresponds to that of the cystine solution measured at the dropping electrode. This is a typical "mixed potential" in the sense of Kolthoff and Miller.[13] At this potential a reduction of cystine occurs, the cathodic current being equal to CA. The cysteine formed reacts immediately with mercury at this potential to yield mercurous cysteinate (HgRS) and the anodic current CB is equal to the cathodic current CA. Similarly, at a potential D the actual current DG is equal to the difference between the hypothetical cathodic current DE and the anodic current DF. At the potential I cysteine no longer yields an anodic current and the experimental current IH now

[13] I. M. Kolthoff and S. C. Miller, *J. Am. Chem. Soc.*, **62**, 2171 (1940).

corresponds to the *true* reduction current of cystine. The same, of course, is true at potentials more negative than *I*, and the experimental curve *HK* is also the true reduction wave. It may be mentioned that potentials of the cystine-cysteine system measured with the dropping electrode do not correspond to the reversible potentials of the system $RSSR + 2H^+ + 2e \rightleftarrows 2RSH$, but represent typical mixed potentials.

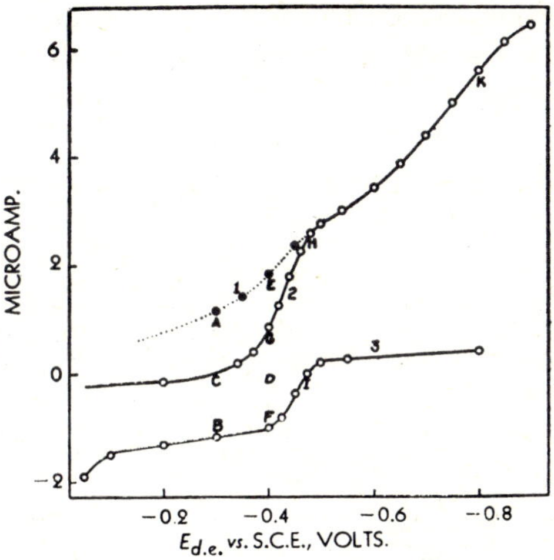

Fig. XLIII-10. (*1*) Hypothetical reduction wave of cystine at pH = 6. (*2*) Experimental curve of 0.001 *M* cystine at pH = 6. (*3*) Anodic wave of 0.002 *M* cysteine at pH = 6.

2. ORGANIC PEROXIDES

Organic peroxides have reduction potentials which are dependent upon their structures. The waves obtained vary in appearance and are drawn out and flat for methyl hydroperoxide (CH_3OOH) and ethyl hydroperoxide (CH_3CH_2OOH) and more normal for *tert*-butyl hydroperoxide ($(CH_3)_3COOH$). Because of their reactivity side reactions with the electrolyte used may take place and interfere with the analysis. Benzoyl peroxide, $(C_6H_5COO)_2$, for example, produces a wave in lithium chloride and methyl sulfuric acid solutions but is decomposed in lithium methoxide. An outside reference electrode should be used instead of the pool of mercury. "Reduction potentials" obtained for a number of simple peroxides are listed in Table XLIII-1.

From the quantitative point of view 0.1 *N* lithium chloride is the best supporting electrolyte since in this medium the diffusion current is proportional to the concentration of the peroxide. This medium has been used for the simultaneous polarographic determination of hydroperoxides and aliphatic aldehydes, such as formaldehyde and acetaldehyde. The peroxide is reduced first and then the aldehyde and both yield distinct diffusion currents. Satisfactory results are also observed in 0.2 *N* hydrochloric acid (Neiman and Gerber, ref. (*2*); Table XLIII-1).

For the polarographic study of higher molecular weight peroxides, solutions of equal volumes of methanol and benzene containing 0.2 *M* methyl hydrogen sulfate, 0.3 *M* lithium chloride, or 0.1 *M* lithium methoxide were used. These solutions are also suitable for simple compounds like ethyl,

TABLE XLIII-1

REDUCTION POTENTIALS (VOLTS) OF PEROXIDES

Substance	0.01 *N* HCl	0.01 *N* NaOH	0.1 *N* NaOH	0.1 *N* LiCl	0.01 *N* LiOH	Ref.
Methyl hydroperoxide, CH_3OOH	−0.25	−0.25	−0.75	−0.75	−0.3	*1, 2*
Ethyl hydroperoxide, C_2H_5OOH.	−0.25	−0.25	−0.60	−0.25	−0.25	*1, 2*
Ethyl ether peroxide, $(C_2H_5)_2O_2$.	−0.55	−0.5	—	−0.25	−0.5	*1, 2*
Formaldehyde peroxide, $(CH_2OH)_2O_2$.	−0.4	−0.4	−0.60	−0.35	−0.4	*1*
Hydrogen peroxide.	−0.75	−0.8	−0.8	−0.75	−0.8	*1*
Acetone triperoxide.	−0.6[a]	—	—	—	—	*2*
Cyclohexene peroxide.	−0.05[a]	—	—	—	—	—

[a] In 0.2 *N* HCl.

(*1*) V. Shtern and S. Pollyak, *Acta Physicochim. U.R.S.S.*, **11,** 797 (1939); *J. Gen. Chem. U.S.S.R.*, **10,** 21 (1940). See also N. M. Kharlov and M. B. Neiman, *Zavodskaya Lab.*, **9,** 166 (1940); *Chem. Abstracts,* **34,** 5788 (1940).

(*2*) M. B. Neiman and M. I. Gerber, *Zhur. Anal. Khim.*, **1,** 211 (1946); *Chem. Abstracts,* **41,** 4096 (1947).

isopropyl, or *n*-butyl ether peroxides. The polarographic behavior of these compounds is not alike in the 3 solutions. Ether peroxides and gasoline peroxide produce complete waves in the lithium chloride and lithium methoxide solutions but not in the methyl hydrogen sulfate solution. Benzoyl peroxide, on the other hand, produces a wave in lithium chloride and methyl hydrogen sulfate and is destroyed in lithium methoxide.[14]

The waves obtained are not regular or well defined enough to assign half-wave potentials. They serve, however, for the quantitative estimation of the peroxides. The method permits detection of 1 milliequivalent of fat peroxide per kilogram of fat and 0.1 milliequivalent of gasoline or ether peroxide per liter. Lithium chloride is the best supporting electrolyte since no hydrogen wave is obtained and no maximum formation or peroxide destruction occurs.[14]

[14] W. R. Lewis, F. W. Quackenbush, and T. DeVries, *Anal. Chem.*, **21,** 762 (1949).

The polarographic method has been used to demonstrate that ether peroxide is not acetaldehyde peroxide $(CH_3CH(OH))_2O_2$, vinyl ether, or a solution of hydrogen peroxide in ether.[15]

In the polarographic determination of ether peroxide, the solution should be saturated with ether since it decreases the catalytic splitting of ether peroxide by mercury. The following procedure has been devised for acetaldehyde and ether peroxide:[15]

Ether containing peroxide (5.00 ml.) is mixed with an equal volume of aldehyde-free ethyl alcohol and $N/25$ tetramethylammonium hydroxide to a total volume of 20 ml. First, a polarographic determination of acetaldehyde is performed on this mixture prior to the removal of the oxygen. Subsequently, the oxygen is driven off with ether-saturated nitrogen and the height of the peroxide wave is measured. The peroxide concentration is calculated by comparing the wave height with standards determined by titration.

The method has also been used to establish the structure of the styrene-oxygen copolymer as a polymeric styrene peroxide (I):[16]

$$[O—O—CH_2—CH(C_6H_5)—O—O—CH_2CH(C_6H_5)]_n$$

(I)

Solutions of the polymer in a mixture of 58 per cent benzene, 37 per cent ethanol, and 5 per cent water which is 0.05 M in ammonium acetate give waves resembling those found for diethyl peroxide up to potentials of −1.5 v. (S.C.E.) but differing from these in that a drop of diffusion current occurs beyond this point. *tert*-Butyl hydroperoxide in the same medium gives a normal curve with a small maximum. The half-wave potential is close to −1.0 v. (S.C.E.). It has been suggested by Bovey and Kolthoff[16] that the dip in the diffusion current for the polymer is due to the desorption of the compound at potentials more negative than −1.5 v.

The polarographic method has been used to follow the autoxidation of lard and to demonstrate the presence of three types of peroxides in autoxidized methyl oleate, methyl linoleate, and lard. All three waves disappear when the lard is treated with potassium iodide.[17]

---

15 F. Reimers, *Quart. J. Pharm. Pharmacol.*, **19**, 473 (1946).
16 F. A. Bovey and I. M. Kolthoff, *J. Am. Chem. Soc.*, **69**, 2143 (1947).
17 W. R. Lewis and F. W. Quackenbush, *J. Am. Oil Chemists' Soc.*, **26**, 53 (1949).

MISCELLANEOUS

Organometallic Compounds. Monochlorotriethyllead, $(C_2H_5)_3PbCl$,[18] and dichlorodiethyltin, $(C_2H_5)_2SnCl_2$,[19] are reducible in 1 *N* potassium chloride at the dropping mercury electrode. The monochlorotriethyllead wave gives a half-wave potential of −0.723 v. (N.C.E.) and involves a 1-electron change. The dichlorodiethyltin wave has a half-wave potential of −0.574 v. (S.C.E.) and involves a 2-electron change. The products formed are triethyllead, $(C_2H_5)_2Pb$ and diethyltin, $(C_2H_5)_3Sn$, and may be stable under the conditions used. Fuchsin is used to suppress maxima.

Tetraethyllead is not reducible at the dropping mercury electrode but can be determined indirectly by decomposition with nitric acid and polarographic determination of the lead.[20]

Phenylarsonic Acids ($C_6H_5AsO_3H$). A rather incomplete polarographic study of these acids has been carried out using 0.01 *M* solutions of the acid in 0.1 *N* lithium chloride and in a Sörensen citrate buffer of pH 2.97.[21] The values of the diffusion current were not reported, nor was the proportionality between the concentration and i_d tested.

Hymatomelanic Acid. This acid yields a wave in 0.1 *N* lithium hydroxide which starts at −1.4 v. (N.C.E.).[22] The reduction potential varies with the concentration of hymatomelanate. It is −1.461 v. in 0.011 per cent solution, −1.39 v. in 0.056 per cent, −1.37 v. in 0.11 per cent, and −1.35 v. in 0.2 per cent solutions. The diffusion current was found to be well defined and proportional to the concentration, when the latter was less than 0.085 per cent. Similar waves were found in calcium and barium hydroxide solutions.

Humic Acid. Under the foregoing conditions, this compound is not reduced. Therefore, the above authors[22] applied the polarographic technique to the determination of hymatomelanic acid in peat.

---

[18] L. Riccoboni, *Gazz. chim. ital.*, **72,** 47 (1942).

[19] L. Riccoboni and P. Popoff, *Atti reale ist. Veneto sci.*, **107,** Pt. II, 123 (1949); *Chem. Abstracts*, **44,** 6752 (1950).

[20] M. Ya. Khlopin and N. S. Litvinova, *Zavodskaya Lab.*, **15,** 677 (1949); *Chem. Abstracts*, **44,** 487 (1950).

[21] B. Breyer, *Ber.*, **71,** 163 (1938); *Biochem. Z.*, **301,** 65 (1939).

[22] M. Tokuoka and J. Ruzicka, *Z. Pflanzenernähr. Düngung. u. Bodenk*, **A35,** 79 (1934).

CHAPTER XLIV

Heterocyclic Compounds Containing Oxygen

The furan and pyran ring systems are not reducible at the dropping mercury electrode. Functional groups attached to these rings or incorporated in the ring are found to behave normally.

1. FURAN DERIVATIVES

Furfural. Compound (I) exhibits a behavior at the dropping mercury electrode which resembles that of benzaldehyde. In acid and alkaline

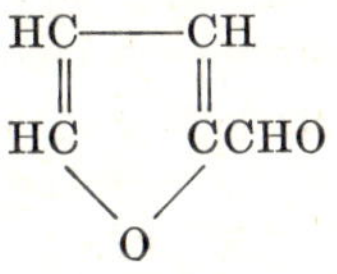

(I)

solutions a single wave is obtained; in buffers of pH 4–7 a two-step curve is obtained. The height of the first step decreases with increase in pH while the second wave increases, the total height remaining constant. The half-wave potentials (N.C.E.) and diffusion currents obtained at 14° C. are listed in Table XLIV-1.[1]

TABLE XLIV-1

HALF-WAVE POTENTIALS (N.C.E.) AND DIFFUSION CURRENTS FOR FURFURAL AT VARIOUS pH VALUES

	First wave		Second wave	
pH	$-E_{1/2}$	i_d	$-E_{1/2}$	i_d
4	1.24	53	—	—
5.5	1.36	46.2	1.78	10.2
5.82	1.42	28.2	1.74	22.8
6.07	1.45	13.7	1.74	37.3
6.10	1.45	12.3	1.73	39.7
6.15	1.47	9.2	1.73	42.6
6.50	1.52	3.0	1.72	55.6
7.25	—	—	1.70	56.0

[1] I. A. Korshunov and S. A. Ermolayeva, *J. Gen. Chem. U. S. S. R.*, **17,** 181 (1947).

These results are at variance with the values reported by Tachi:[2]

pH	1.1	2.0	3.0	4.0	4.5	6.0	7.0
$-E_{1/2}$ (N.C.E.)	0.93	1.0	1.07	1.13	1.17	1.30	1.35
pH	8.0	9.0	10.0	11.0	12.0	13.0	—
$-E_{1/2}$ (N.C.E.)	1.38	1.41	1.45	1.47	1.47	1.47	—

for a 0.0001 *M* solution of furfural at 25° C. and the results reported by MacKinney and Temmer:[3]

pH	2.8	3.9	7.1	7.6
$-E_{1/2}$ (S.C.E.)	0.99	1.06	1.32	1.38

Substitution of a hydroxymethyl group in the 5-position changes the half-wave potential very little to −1.36 v. (N.C.E.) at pH 7.6.[4] For analytical purposes either 0.1 *N* lithium chloride,[5] or a 10 per cent alcohol solution which is 0.05 *N* in sodium acetate[6] are recommended. The method has been used to detect furfural in darkened apricot extracts[3] and to determine furfural in the presence of formaldehyde and in extracts from resins.[6]

Only one example of a furan with a functional group incorporated in the ring has been studied. Aucubin hexaacetate, a naturally occurring furan, for which the structure (II) has been postulated,

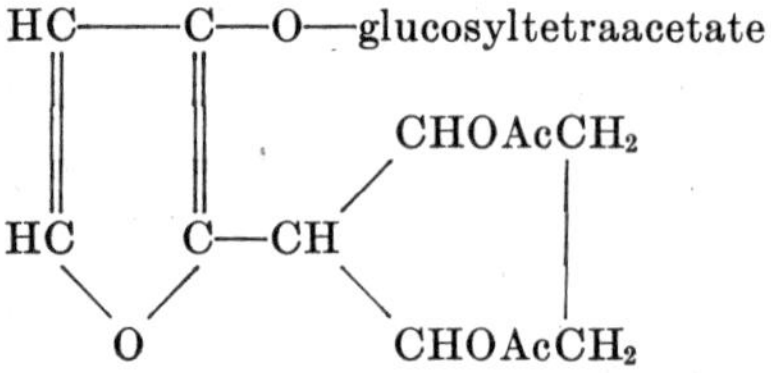

(II)

is reported to give two waves with half-wave potentials of −0.074 v. and −0.974 v. (N.C.E.) in 80 per cent ethanol containing 0.01 *N* tetramethylammonium bromide.[7] If this structure is correct, it is difficult to interpret the results.

[2] I. Tachi, *Bull. Agr. Chem. Soc. Japan,* **14,** 1371 (1938).

[3] G. MacKenney and O. Temmer, *J. Am. Chem. Soc.,* **70,** 3586 (1948).

[4] S. M. Cantor and Q. P. Peniston, *J. Am. Chem. Soc.,* **62,** 2113 (1940).

[5] M. Tokuoka and J. Ruzicka, *Z. Pflanzenernähr. Düngung u. Bodenk,* **A35,** 79 (1934).

[6] N. I. Malyingima and I. A. Korshunov, *Zhur. Anal. Khim.,* **2,** 341 (1947); *Chem. Abstracts,* **43,** 7377 (1949).

[7] P Karrer and H. Schmid, *Helv. Chim. Acta,* **29,** 529 (1946).

2. REDUCED FURANS

Streptomycin. The aldehyde group is probably the group involved in the reduction of streptomycin (III) at the dropping mercury electrode; a half-wave potential of −1.45 v. against the mercury anode is observed in 3 per cent tetramethylammonium hydroxide solution. The diffusion coefficient is of the order to be expected, 3×10^{-6} cm.^{2} sec.$^{-1}$, but the possibility of a catalytic hydrogen wave is not excluded.[8]

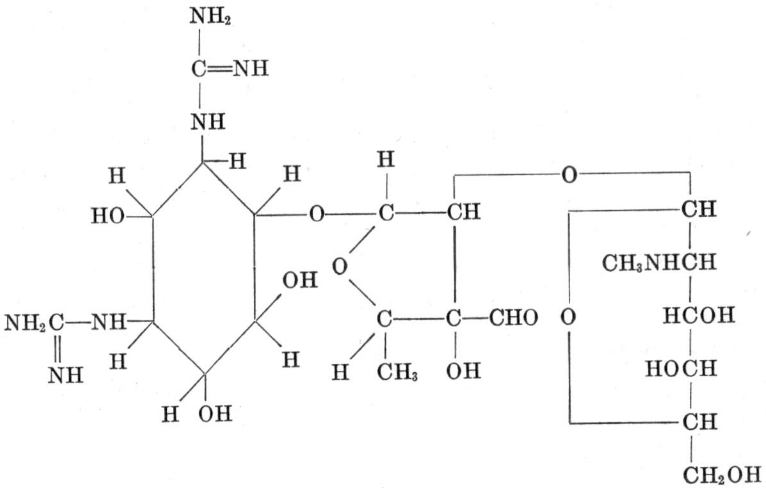

(III)

The diffusion current at 13.6° in aqueous 3 per cent tetramethylammonium hydroxide is proportional to the concentration in the range 100–1000 mg. per liter. The polarographic method compares favorably with the microbiological method. For twenty-five consecutive samples treated in both ways, the average deviation of the polarographic values from those obtained by the microbiological method was ±6 per cent which is well within the accuracy of the latter method.

3. PYRAN DERIVATIVES

In the simple pyran series, the ketones are the only members which have been studied at the dropping mercury electrode.

γ-Pyrone. Compound (IV) gives half-wave potentials of −1.69 v. and

[8] G. B. Levy, P. Schwed and J. W. Sackett, *J. Am. Chem. Soc.*, **68,** 528 (1946).

−1.91 v. in 0.1 *N* ammonium chloride and 0.2 *N* tetramethylammonium hydroxide, respectively, against a mercury pool anode.[9]

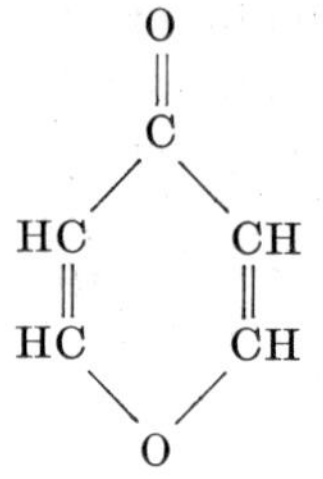

(IV)

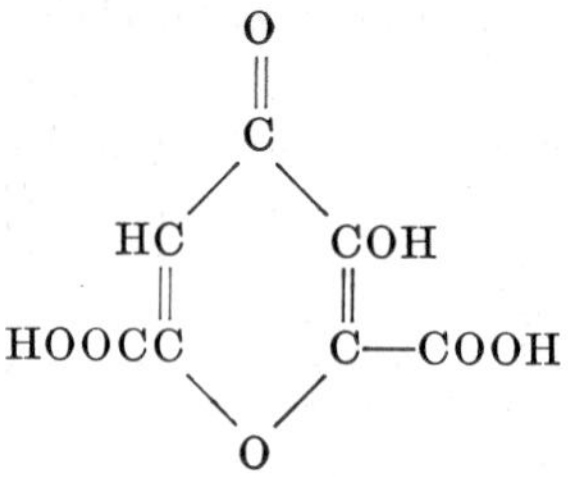

(V)

Meconic Acid. Compound (V) gives two coalesced waves at pH 3 in acetic acid; the first wave is about seven times as high as the second wave.[10] At higher pH values only one wave is obtained which becomes smaller and ill defined with increase in pH. This behavior suggests that a combined reduction of hydrogen ions and possibly the carbonyl group is involved. At pH 3, the diffusion current is proportional to the concentration in the range 0.3 to 10 $\times$ 10^{-3} *M*. The half-wave potential of the large wave is −1.2 v. and is independent of concentration up to 1 $\times$ 10^{-3} *M*.

In the determination of meconic acid in opium the acid is separated as the lead salt. This is decomposed with hydrogen sulfide and the acid is then determined in 0.025 *M* acetic acid.

Citrinin. The polarographic behavior of citrinin (VI) resembles that of the fuchsones. One wave with a half-wave potential of −0.81 v. (S.C.E.) is obtained in a buffer of pH 2.05 in 75 per cent ethanol[11] and of −1.14 v. (S.C.E.) in an aqueous buffer of pH 4.63.[12]

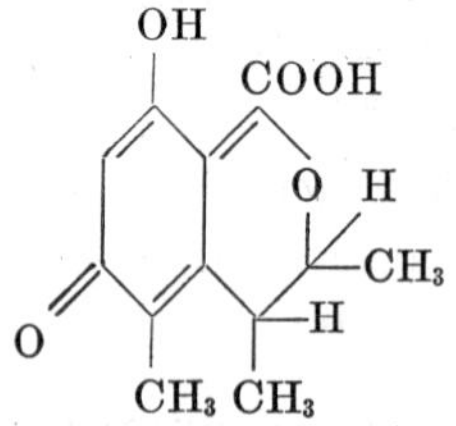

(VI)

[9] H. Adkins and F. W. Cox, *J. Am. Chem. Soc.*, **60,** 1151 (1938).
[10] H. F. Kirkpatrick, *Quart. J. Pharm. Pharmacol.*, **18,** 338 (1945).
[11] H. W. Hirschy and P. M. Ruoff, *J. Am. Chem. Soc.*, **64,** 1490 (1942).
[12] J. E. Page and F. A. Robinson, *J. Chem. Soc.*, **1943,** 133.

4. BENZOPYRONES AND BENZOPYRANS

Flavones. Flavones (VII) as a rule give two reduction waves at the dropping mercury electrode. With polyhydroxyflavones or derivatives of

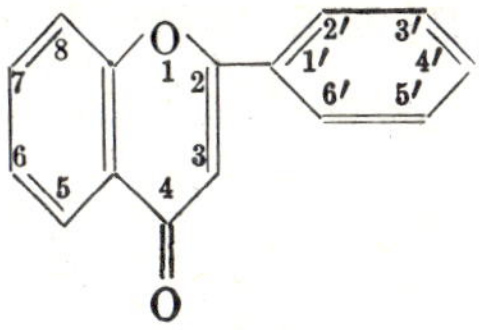

(VII)

these, the two waves coalesce and give the appearance of one wave. The half-wave potentials (S.C.E.) and diffusion currents for a number of examples are listed in Tables XLIV-2 and XLIV-3.

TABLE XLIV-2

POLAROGRAPHIC BEHAVIOR OF FLAVONES[13]

Flavone	pH 6.1		pH 7.5		pH 8.6		pH 9.6	
	$-E_{1/2}$	i_d/C	$-E_{1/2}$	i_d/C	$-E_{1/2}$	i_d/C	$-E_{1/2}$	i_d/C
Unsubstituted	1.26	1.53	1.26	1.99	1.41	—	1.42	—
	1.38	—	1.44	—	1.75	—	1.75[a]	—
4-Hydroxy	1.25	1.62	1.35	1.37	1.53	—	1.54	—
	1.49	2.34	1.50	1.29	1.53	—	—	—
5-Hydroxy	1.28	1.58	1.33	—	1.37[b]	—	1.37	—
	1.41	—	1.45	—	1.51	—	1.51	—
					1.65	—	1.65	—
7-Hydroxy	1.26	1.23	1.36[b]	—	1.59	—	1.60	—
	1.51	2.95	1.50	1.26	—	—	—	—
4′-Acetoxy	1.14	1.48	1.23	1.47	1.37	—	1.37	—
	1.37	2.16	1.39	1.65	1.75[a]	—	1.73[a]	—
5-Acetoxy	1.15	1.56	—	—	1.35	—	—	—
	1.31	1.60	—	—	—	—	—	—
7-Acetoxy	1.14	1.24	—	—	1.38	—	—	—
	1.41	1.88	—	—	1.72	—	—	—
4′-Methoxy	1.20	1.09	—	—	1.47	—	—	—
	1.37	—	—	—	—	—	—	—

[a] Very small wave.
[b] Triple wave with small breaks.

In Table XLIV-2 i_d/C values are reported only for those curves which were clearly separated from adjoining waves. In some cases the compounds were only partially soluble in the medium used and the concentration was not known.[13]

The heights observed for the flavones point to a 1-electron step for each of the waves. Wave analysis indicates that the potential-determining step involves 1 electron. The data suggest that the flavones are reduced in the following manner:

First Wave:

(1)

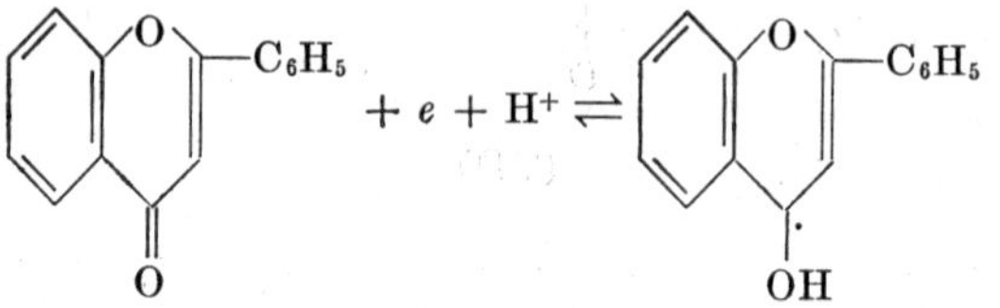

(2)

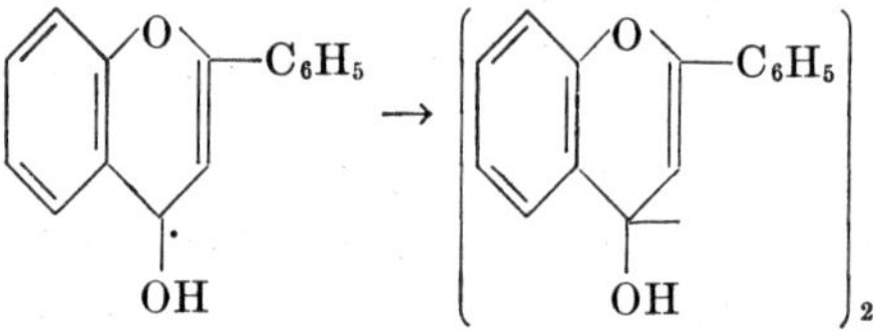

Second Wave:

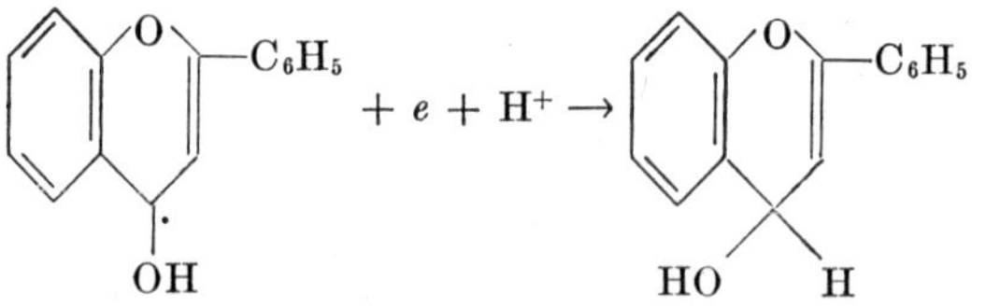

The two-step wave is possible because of the stabilization of the free radical (by resonance) toward further reduction at the more positive potentials which allows dimerization to a pinacol:

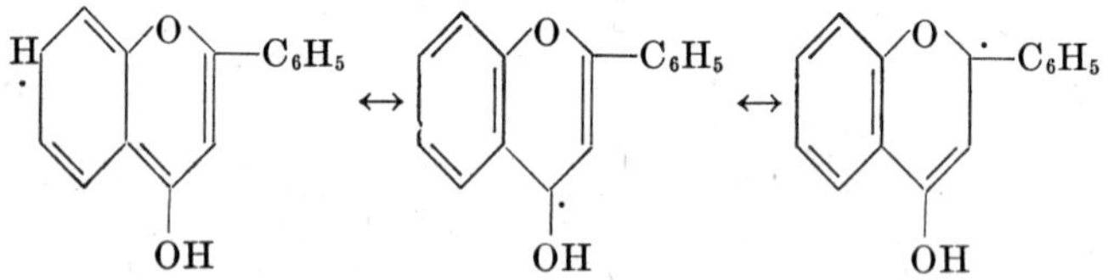

At more negative potentials further reduction occurs before dimerization can take place. The suggestion that the free radical persists at the electrode and is reduced further is made without chemical evidence.

The substitution of a hydroxyl group decreases the ease of reduction of

[13] T. A. Geissman and S. L. Friess, *J. Am. Chem. Soc.*, **71**, 3893 (1949).

the flavones ($E_{1/2}$ becomes more negative) if the group is conjugated with the carbonyl group. The extent varies according to the acidities of the hydroxyl groups in this series and is more pronounced at higher pH values where the formation of anions takes place. The difficulty in reduction follows the order: 7—OH > 4′—OH > 5–OH. The behavior of the 5-hydroxyl group is an exception in the buffer of pH 6.2 because of the strong

TABLE XLIV-3

HALF-WAVE POTENTIALS (S.C.E.) AND DIFFUSION CURRENT CONSTANTS FOR FLAVONES IN 50 PER CENT ISOPROPYL ALCOHOL[14]

Compound	pH 5.6		pH 7.7		pH 10.4	
	$-E_{1/2}$	I_d	$-E_{1/2}$	I_d	$-E_{1/2}$	I_d
Apigenin[a] R = R′ = H	1.57	2.9	1.63	2.8	1.65	2.5
Quercitin[b] R = H, R′ = H	1.53	3.0	1.62	2.73	1.81	2.1
Quercitrin[b] R = H, R′ = $C_6H_{11}O_5$	1.46	3.1	1.58	2.9	1.81	3.4
Quercitintetramethyl ether[b] R = CH_3, R′ = H	1.41	1.9	1.47	1.6	1.59	1.82
Quercitinpentamethyl ether[b] R = R′ = CH_3	1.27	1.8	1.43	1.8	1.54	2.0
Quercitinpentaacetate[b] R = R′ = $COCH_3$	1.10	1.8	1.16	1.9	1.72	3.2
Flavonol[c]	1.25	2.08	1.33	2.06	1.45	2.02

a
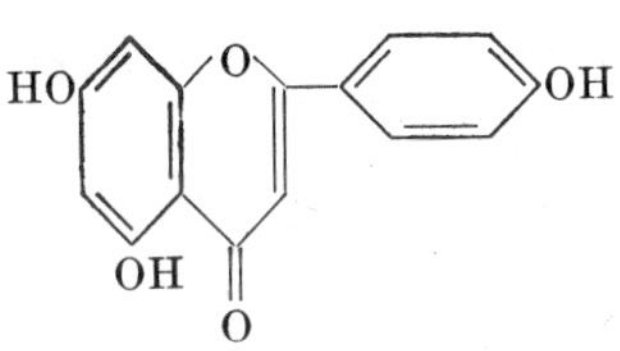

b
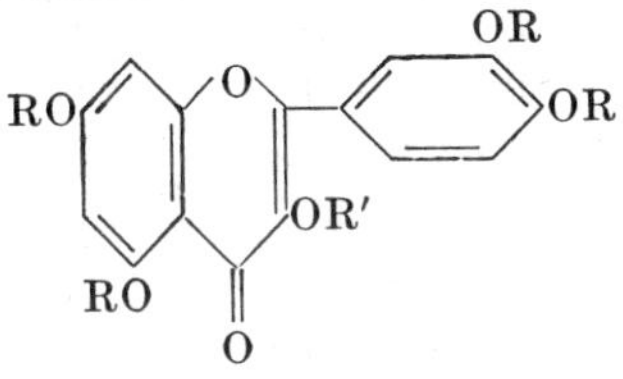

c
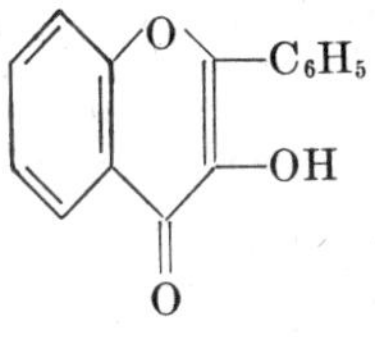

hydrogen bonding between this group and the carbonyl group under these conditions. This type of structure would lend anionic properties to the carbonyl group.[14]

Introduction of a hydroxyl group in the 3-position which is not conjugated with the carbonyl group affects the ease of reduction only slightly. Acetylation of the hydroxyl group reduces its electron-donating properties and results in more positive half-wave potentials. Methylation exerts a similar effect but to a smaller extent. Substitution of an ethyl group for the 2-phenyl as in 2-ethylchromone shifts the half-wave potentials to more

[14] D. W. Engelkemeir, T. A. Geissman, W. R. Crowell, and S. L. Friess, *J. Am. Chem. Soc.*, **69**, 155 (1947).

negative values; half-wave potentials of −1.53 v. and −1.66 v. are reported in 0.1 *N* ammonium chloride and 0.2 *N* tetramethylammonium hydroxide, respectively.[15]

Flavanones. Flavanones produce only one reduction wave at the dropping mercury electrode. The half-wave potentials and diffusion currents for these compounds are listed in Table XLIV-4.

TABLE XLIV-4

POLAROGRAPHIC BEHAVIOR OF FLAVANONES[16]

Compound	pH 6.1		pH 7.5		pH 8.6		pH 9.6	
	$-E_{1/2}$	i_d/C	$-E_{1/2}$	i_d/C	$-E_{1/2}$	i_d/C	$-E_{1/2}$	i_d/C
Unsubstituted	1.29	2.78	1.37	2.62	—	—	1.51	2.43
4′-Hydroxy	1.29	—	1.37	—	1.51	0.62	1.52	0.73
4′,5-Dihydroxy	1.39	2.47	1.45	2.25	1.53	1.72	1.54	1.91
4′,7-Dihydroxy	1.45	2.20	1.57	2.25	1.58	0.83	1.59	0.67
4′,5,7-Trihydroxy	1.57	2.18	1.62	2.24	1.64	0.82	1.66	0.57
3′,4′,5,7-Tetrahydroxy	1.54	2.44	1.61	1.98	1.65	0.73	1.66	0.70
3′-Methoxy-4′,5,7-trihydroxy	1.55	2.49	1.59	1.99	1.65	0.91	1.66	0.60
4′-Methoxy	1.28	2.81	1.34	2.41	1.51	2.19	1.52	—
4′,7-Dimethoxy	1.40	2.28	1.46	—	1.64	2.25	1.64	2.24
4′,5,7-Trimethoxy	1.36	—	1.60	—	—	—	—	—
	—	—	1.43	—	1.68	—	1.69	—
4′,7-Dimethoxy-5-hydroxy	1.50	—	1.53	—	1.65	—	1.67	—
4′-Acetoxy	1.26	—	1.52	—	—	—	1.49	—
4′,5-Diacetoxy	1.22	—	1.27	—	—	—	1.46	1.91
4′,5,7-Triacetoxy	1.19	—	1.25	—	—	—	1.38	—

Analysis of the curves points to a 1-electron potential-determining step. The height of the diffusion current in most cases is equivalent to a 2-electron change and suggest that the flavanones are reduced in the following manner:

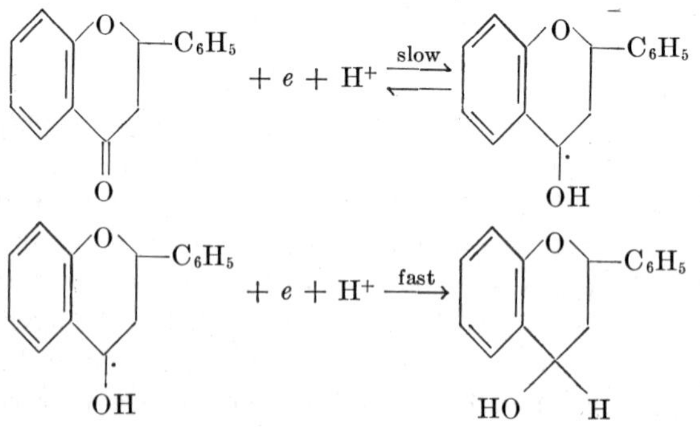

[15] H. Adkins and F. W. Cox, *J. Am. Chem. Soc.*, **60,** 1151 (1938).

[16] T. A. Geissmann and S. L. Friess, *J. Am. Chem. Soc.*, **71,** 3893 (1949).

A comparison of the half-wave potentials indicates that introduction of various groups into the 2-phenyl group has no effect on the ease of reduction. This behavior is to be expected since the phenyl group is not conjugated with the carbonyl group as in the flavones. Groups in the 5- and 7-positions which are conjugated with the carbonyl group show an effect which is dependent upon their ability to release electrons. The difficulty of reduction decreases in the following manner: HO > CH_3O > H > CH_3COO.[16]

Flavanones in the naphthalene series behave in a manner similar to that of the parent compound. Values obtained for several compounds in 40 per cent isopropyl alcohol containing 0.4 *N* ammonium sulfate are listed below:[17]

Compound	$-E_{1/2}$, v.
Flavanone	1.80
2′-Hydroxy-5,6-benzoflavanone	1.74
2′-Hydroxy-5′-bromo-5,6-benzoflavanone	1.74

In 0.1 *N* sodium hydroxide–50 per cent isopropyl alcohol flavanone forms the chalcone and produces waves which are characteristic for this substance.[17]

The polarographic method has been used to follow the sodium amalgam reduction of a substituted chromanone.[18]

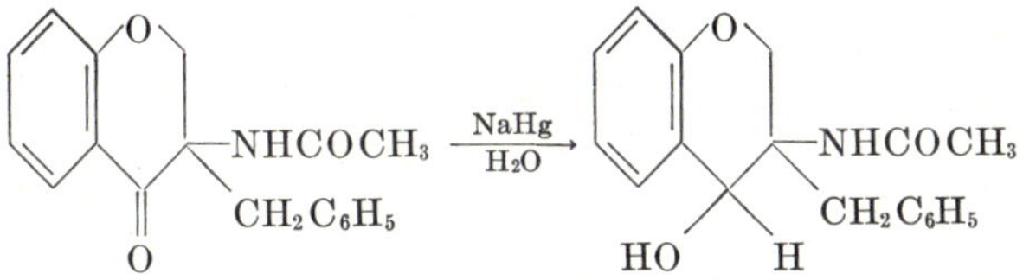

6-Hydroxychromans and 5-Hydroxycoumarans. 6-Hydroxychromans and 5-hydroxycoumarans with three alkyl groups on the benzenoid ring give anodic waves at the dropping mercury electrode. The results obtained in a 0.1 *M* anilinium perchlorate–0.1 *M* aniline buffer containing 50 per cent methanol are listed in Table XLIV-5.[19]

The diffusion currents are proportional to—and the half-wave potentials

[17] E. Schraufstatter, *Experientia*, **4,** 192 (1948).

[18] P. Pfeiffer and H. Epler, *Ann.* **545,** 263 (1940).

[19] L. I. Smith, I. M. Kolthoff, S. Wawzonek, and P. M. Ruoff, *J. Am. Chem. Soc.*, **63,** 1018 (1941).

TABLE XLIV-5

ANODIC WAVES OF 6-HYDROXYCHROMANS AND 5-HYDROXYCOUMARANS IN A BUFFER OF pH 3.6 (50 PER CENT METHANOL)

Compound	$E_{1/2}$ (S.C.E.)	i_d/C
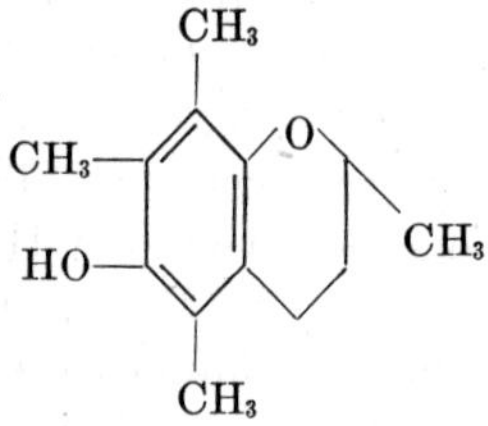	+0.230	3.72
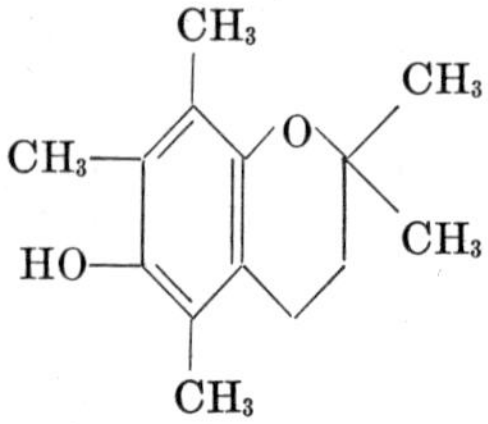	+0.232	3.55
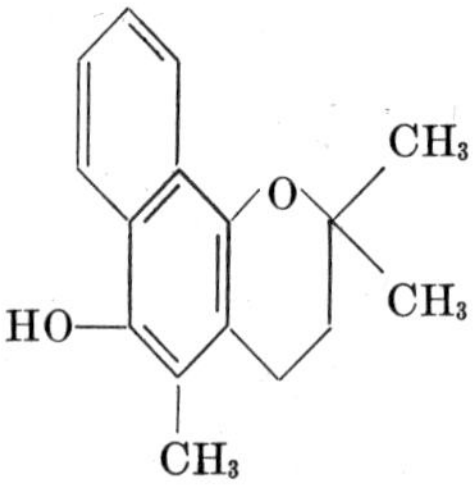	+0.114	4.39
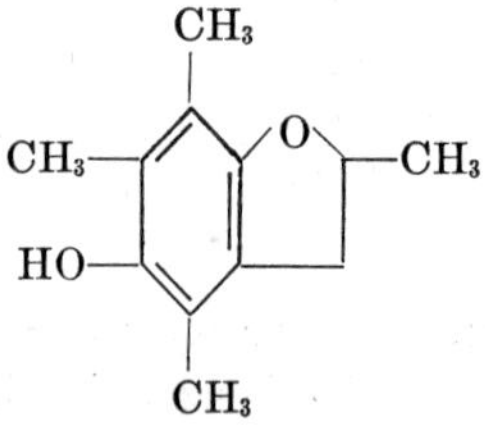	+0.219	3.57
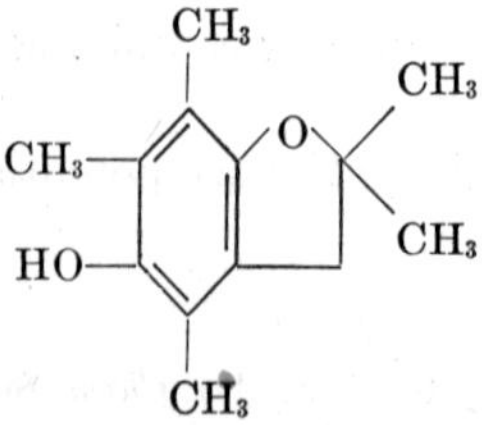	+0.219	3.68

TABLE XLIV-5—*Continued*

Compound	$E_{1/2}$ (S.C.E.)	i_d/C
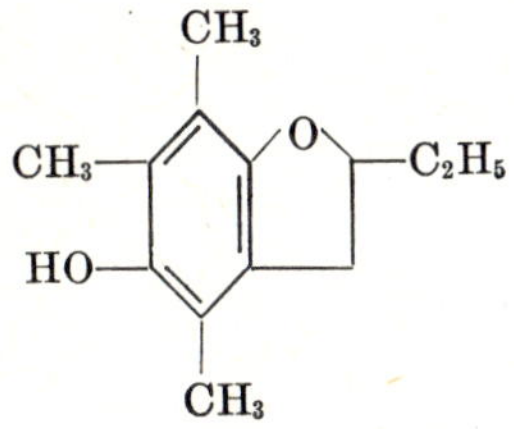	+0.214	3.64
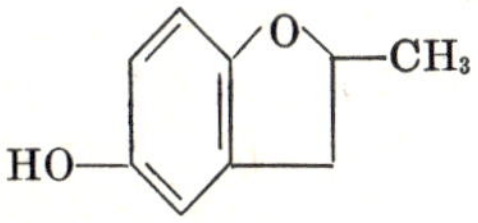	No wave	

independent of—the concentration in the range studied (0.0005–0.001 *M*).[19]

Analysis of the wave for 2,2,5,7,8-pentamethyl-6-hydroxychroman gives a slope of 0.0324 and points to the following reaction at the mercury drop:

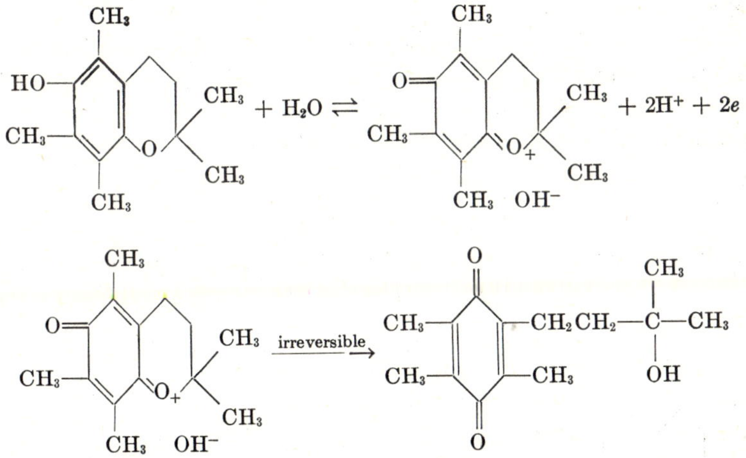

Further evidence for this mechanism is the fact that the difference between the half-wave potentials of the following two hydroxychromans was found to be equal to the difference between the half-wave potentials of the related quinones:

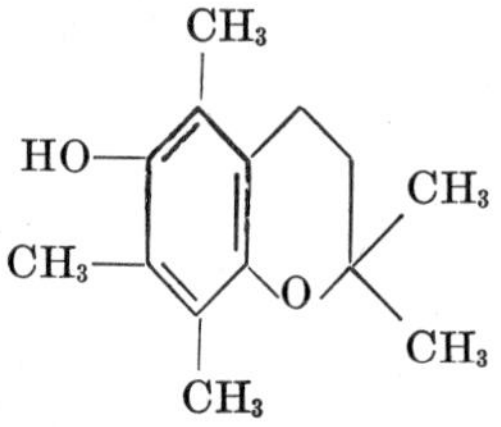

$E_{1/2} = +0.232$

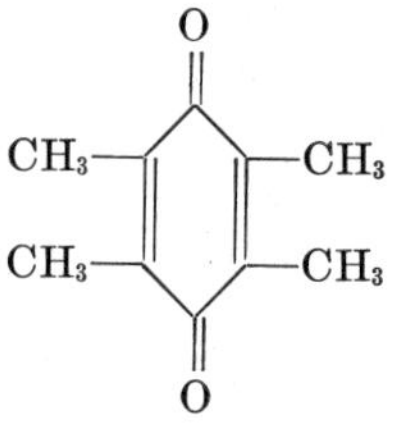

$E_{1/2} = -0.093$

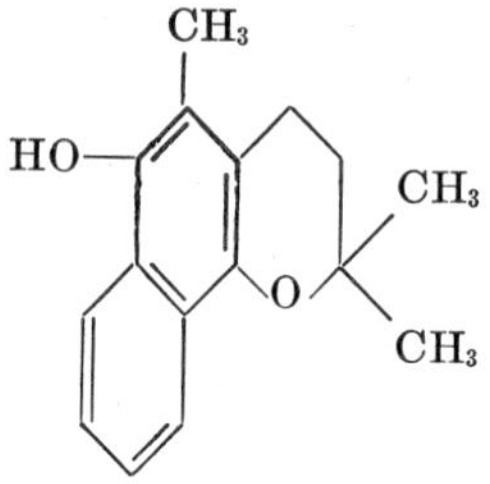

$E_{1/2} = +0.114$
$\Delta = 0.12$

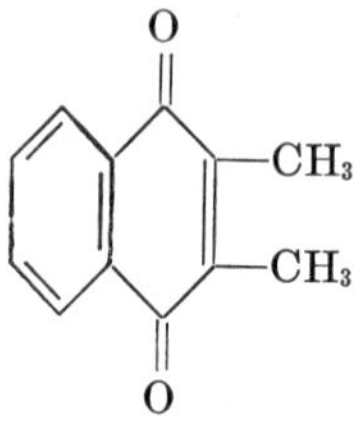

$E_{1/2} = -0.216$
$\Delta = 0.12$

This behavior points to a common rate constant for the oxidations of the two chromans.

The difference of 10 millivolts between the half-wave potentials of 6-hydroxychromans and 5-hydroxycoumarans is a significant and characteristic value and offers a method for distinguishing between these two types of compounds. This is the only physical method that is capable of doing this.

TABLE XLIV-6

ANODIC WAVES OF 6-HYDROXYCHROMANS AND 5-HYDROXYCOUMANANS IN A BUFFER OF pH 4.02 (75 PER CENT ETHANOL)

Compound	$E_{1/2}$ (S.C.E.)	i_d/C
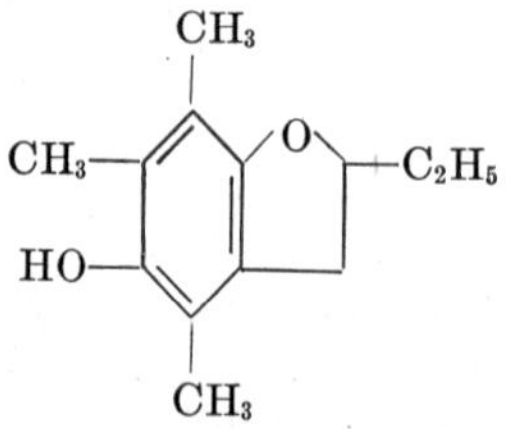	+0.267	3.08

TABLE XLIV-6—*Continued*

Compound	$E_{1/2}$ (S.C.E.)	i_d/C
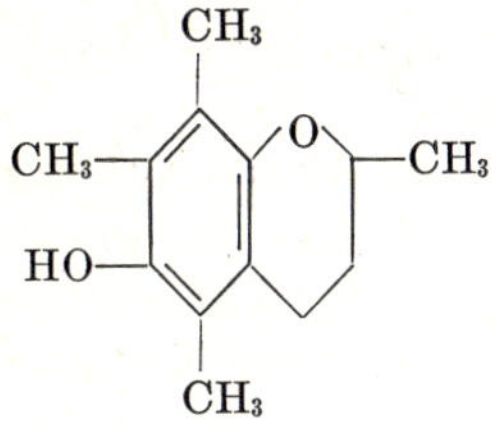	+0.278	2.73
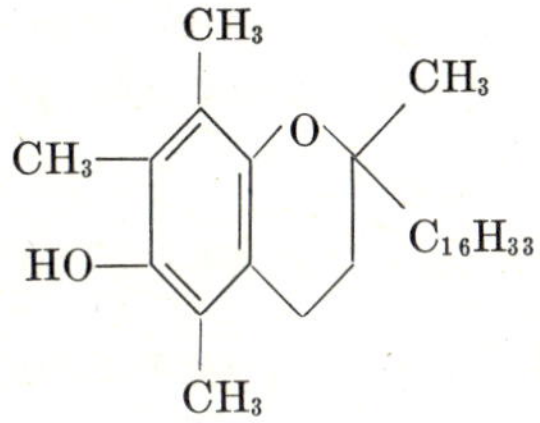 (vitamin E)	+0.284	2.21
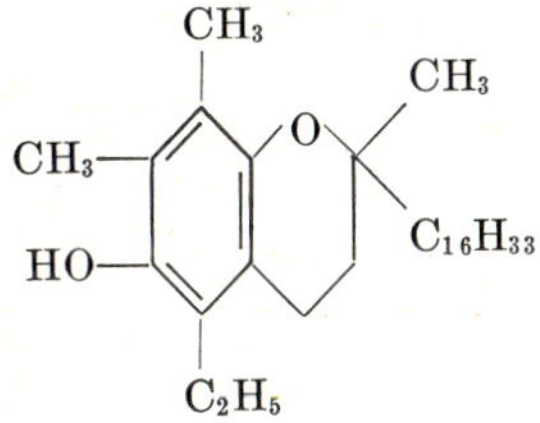	+0.286	1.96
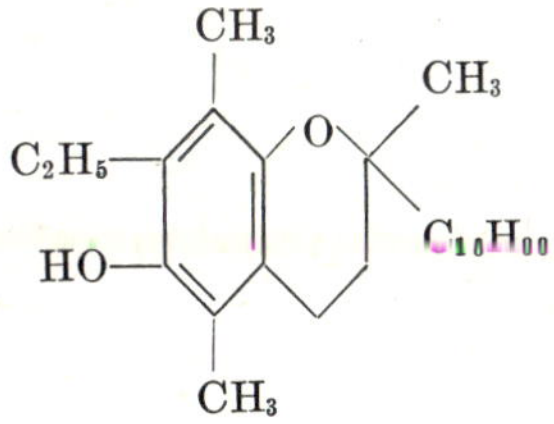	+0.291	1.85
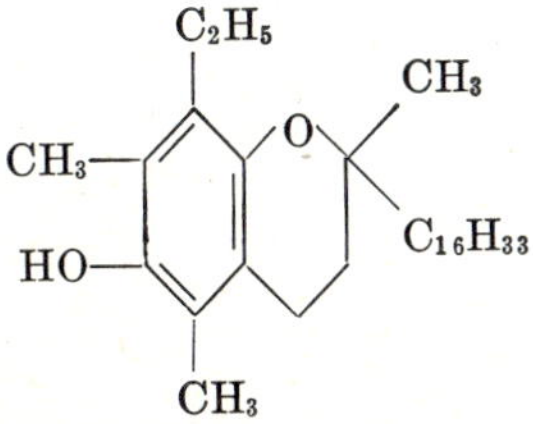	+0.291	2.06

The polarographic method has been used to confirm the chroman structure of α-tocopherol (vitamin E). The results obtained in a 0.1 *N* anilinium perchlorate, 0.1 *N* aniline buffer (pH 4.02) at 25° in 75 per cent ethanol are listed in Table XLIV-6. The half-wave potential for vitamin E is in the range obtained for the simple chroman.

β- and γ-Tocopherols do not produce anodic waves under the conditions used.[20] The use of the polarographic method to determine α-tocopherol quantitatively is subject to an error of 3 to 5 per cent in 0.001 *M* solutions. Oils if present to the extent of 2.5 to 5 per cent and cholesterol in amounts of 0.15 per cent make this method useless for determining α-tocopherol in natural substances.[21] α-Tocopherol (vitamin E) may be determined by an amperometric titration with gold chloride. The excess of gold chloride is determined polarographically.[22]

[20] L. I. Smith, L. J. Spillane, and I. M. Kolthoff, *J. Am. Chem. Soc.*, **64,** 447 (1942).

[21] J. J. Beaver and H. Kaunitz, *J. Biol. Chem.*, **152,** 363 (1944).

[22] L. I. Smith, I. M. Kolthoff, and L. J. Spillane, *J. Am. Chem. Soc.*, **64,** 646 (1942).

CHAPTER XLV

Heterocyclic Compounds Containing Nitrogen

The polarographic behavior of heterocyclic compounds containing nitrogen is complicated by the fact that many of them give catalytic hydrogen waves in addition to the normal type of reduction waves.

Catalytic Waves. This type of wave is attributed to a catalytic discharge of hydrogen and is subject to great variations depending upon the conditions. The height and shape of the catalytic wave and the tendency to yield maxima are dependent upon the following factors: (1) The pH of the solution is important and therefore differences between buffered solutions are often observed. (2) The wave may change entirely in shape on changing the composition of a given buffer, keeping the pH constant. (3) The wave shifts to more negative values with increasing concentration of the compound. (4) The rate of increase of potential especially in an acid medium affects the height of the steps. Reduction of the speed reduces the height of the wave.

The catalytic waves differ from normal reduction waves in that the height produced is considerably higher than corresponds to a diffusion current (sometimes in the range of twenty to one[1]).

1. PYRROLES, INDOLES, AND CARBAZOLES

Substituted pyrroles have been studied in 0.1 *N* ammonium chloride in 50 per cent ethanol[2] and in 10 per cent ethanol.[3] The reduction potentials for these compounds are listed in Tables XLV-1 and XLV-2.

Tho following pyrroles, as well as pyrrole itself, are reported as nonreducible at the dropping mercury electrode: 1-methyl, 2-methyl, 1-allyl, 2,4-dimethyl, 2,5-dimethyl, 3-methyl-4-ethyl, 2,3,5-trimethyl, 2,5-dimethyl-1-ethyl, 2,4-dimethyl-3-ethyl, 2,4-dimethyl-3-propyl, 2,4-dimethyl-3-ethyl-1-carbethoxy, 2,4-dimethyl-3,5-dicarbethoxy.[4]

Indole is apparently not reducible at the dropping mercury electrode but

[1] H. F. W. Kirkpatrick, *Quart. J. Pharm. Pharmacol.*, **18,** 245 (1945).

[2] G. B. Bonino and G. Scaramelli, *Ber.*, **75,** 1948 (1942).

[3] M. Dezelic, *Rad. Hrvat. Akad.*, **271,** 21 (1941); *Chem. Abstracts*, **42,** 5899 (1948).

[4] G. Scaramelli, *Boll. sci. facoltà chim. ind. Univ. Bologna*, **3,** 205 (1942); *Chem. Zentr.*, *II*, **1943,** 1180.

TABLE XLV-1

REDUCTION POTENTIALS (N.C.E.) OF SUBSTITUTED PYRROLES IN 0.1 *N* AMMONIUM CHLORIDE–50 PER CENT ETHANOL

-pyrrole	Reduction potential, volts (1.5×10^{-3} *M*)
2-Formyl	−1.452
2,3,4-Trimethyl-5-formyl	−1.534
2-Methyl-3,4-diethyl-5-formyl	−1.536
2,4-Dimethyl-3-ethyl-5-formyl	−1.538
2,4-Dicarbethoxy-3-methyl-5-formyl	−0.935
2-Carbethoxy-3,5-dimethyl-4-formyl	−1.524
1-Methyl-2-formyl	−1.420
2-Acetyl	−1.646
1-Acetyl	No reduction
2,3,5-Trimethyl-4-acetyl	No reduction
2,4-Dimethyl-5-carbethoxy-3-acetyl	No reduction
1-Methyl-2-acetyl	−1.700
2,5-Diacetyl	−1.031, −1.650
1-Methyl-2,5-diacetyl	−1.053, −1.685
Di-[3-methyl-5-ethylpyrryl]ketone	−1.618
Di-[5-methyl-3-ethylpyrryl]ketone	−1.640
Di-[3,5-dimethyl-4-propyl]ketone	−1.680

TABLE XLV-2

REDUCTION POTENTIALS OF SUBSTITUTED PYRROLES IN 0.1 *N* AMMONIUM CHLORIDE–10 PER CENT ETHANOL

-pyrrole	Reduction potential, v. (1×10^{-3} *M*)
2,3,4,5-Tetramethyl	−0.95, −1.40
2-Formyl	−1.47
4-Methyl-3-ethyl-2-formyl	−1.48
4-Methyl-3-ethyl-5-formyl	−1.50
2,4-Dimethyl-3-formyl-5-carbethoxy	−1.52
2,4-Dimethyl-5-formyl-3-carbethoxy	−1.44
2-Formyl-4-methyl-3-ethyl-5-carboxy	−0.99, −1.50
3-Carbethoxy-2-formyl-4-methyl-5-carboxy	−0.80, −1.35
5-Formyl-2,4-dimethyl-3-[β-carboxyethyl]	−1.35, −1.71
5-Bromo-4-methyl-3-ethyl-2-formyl	−1.15, −1.49
2-Bromo-4-methyl-3-ethyl-5-formyl	−1.26, −1.53
3-Methyl-4-ethyl-2,5-diformyl	−0.84, −1.51
Methylethylmaleimide	−0.97

several derivatives of indole are.[5] The exact nature of the reduction of these compounds is not known.

[5] J. Pech, *Collection Czechoslov. Chem. Communs.*, **6**, 126 (1934).

Isatin. The behavior of isatin (I) at the dropping mercury electrode is complicated and is dependent upon the buffer and the period of time

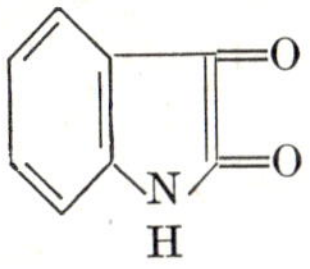

(I)

that the solution is allowed to stand before the analysis is carried out. The results[6] are listed in Table XLV-3.

TABLE XLV-3

HALF-WAVE POTENTIAL ($-E_{1/2}$) (S.C.E.) OF ISATIN AT VARIOUS pH VALUES

pH	Molecular variety				
	A	A′	B	C	D
2.88	0.22	0.42	—	—	—
3.01	0.23	0.43	0.54	—	—
3.40	0.32	—	—	—	—
4.00	0.28	0.46	0.63	—	—
4.35	0.30	0.49	0.74	—	—
5.43	—	—	0.80	1.02	—
5.83	—	—	—	1.04	—
7.10	0.39	0.65	—	1.06	1.35
8.00	0.44	0.68	—	1.08	1.36
9.75	—	—	—	—	1.44
11.10	—	—	—	—	1.47

The first two waves A and A′ must be associated with the reduction of the β-ketone group in isatin since the solutions exhibiting waves A and A′ are predominantly yellow in color. These waves may represent reduction of isatin to the pinacol and alcohol, respectively. It has been suggested that wave A′ represents the reduction of the hydrate of isatin since its height decreases with the addition of alcohol.

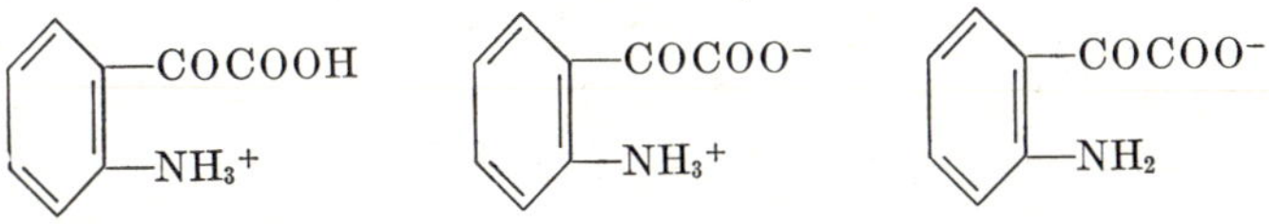

(I) (II) (III)

[6] W. C. Sumpter, Y. L. Williams, P. H. Wilken, and B. L. Willoughby, *J. Org. Chem.*, **14**, 713 (1949).

Wave B is probably due to the reduction of the carbonyl in the salt of isatinic acid (I) since the wave is at its maximum when the sodium salt is acidified. Wave B disappears slowly upon standing and produces waves A and A′, *i.e.*, the isatinic acid under the acid conditions slowly cyclizes and forms isatin. Waves C and D are probably due to the reduction of the carbonyl in the dipolar molecule (II) and in the anion of isatinic acid (III) since they appear in neutral and alkaline buffers. Wave heights are in agreement with such postulations.

Dihydroxyindole (IV). This reduction product of isatin produces a wave with a half-wave potential of −0.29 v. at pH 2.64.[7] Above pH 4 it apparently undergoes reactions with the buffer since the waves obtained vary with time.

Indigo Carmine (V). According to Korshunov *et al.*[7] this dye produces waves which do not vary with time but change with concentration; at pH 2.64 a half-wave potential of −0.24 v. is obtained at a concentration of 0.004 *M* and −0.4 v. at 0.030 *M*.

Phthalimide (VI). Phthalimide (according to Korshunov) produces (in 0.2 *N* HCl at 25°) a wave with a half-wave potential of −0.7 v.

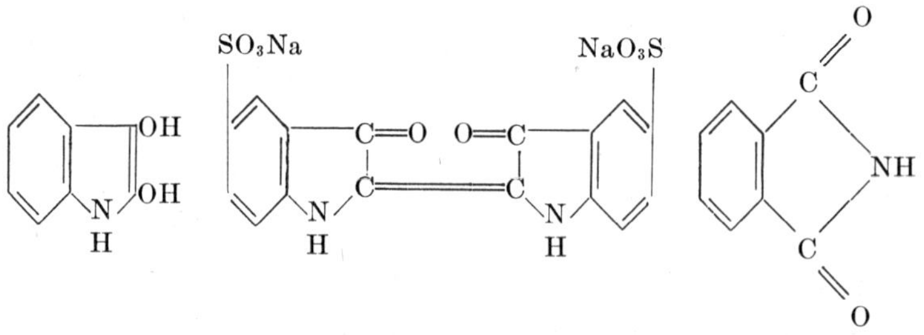

(IV) (V) (VI)

N-Acetylcarbazole. This compound is not reducible at the dropping mercury electrode.[8]

2. PYRIDINES, QUINOLINES, AND ACRIDINES

(a) Pyridines

Pyridine (I). The pyridinium ion produces a catalytic hydrogen wave at the dropping mercury electrode:

---

[7] I. A. Korshunov, L. N. Sazanova, M. K. Shchennikova, and O. P. Malkova, *Zavodskaya Lab.*, **15**, 1287 (1949); *Chem. Abstracts*, **44**, 3846 (1950).

[8] L. Palasciano, *Boll. sci. facoltà chim. ind. Univ. Bologna*, **2**, 83 (1941); *Chem. Zentr.*, *I*, **1942**, 1363.

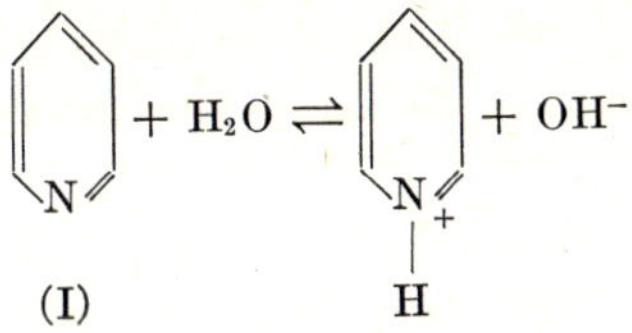

(I)

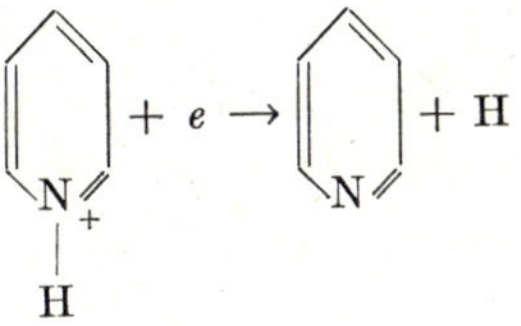

$$2H \rightarrow H_2$$

Since this ion is in equilibrium with the free base, the height of the wave is dependent upon pH. Thus no wave is formed at pH 12 or in 0.1 *N* sodium hydroxide solution.

The half-wave potential becomes more negative with an increase in pH; a value of $E_{1/2}$ of -1.49 v. is reported[9,10] in a solution containing 0.001 *M* hydrochloric acid and 0.1 *N* potassium chloride (pH 2.9) and one of -1.8 v. in a buffer of pH 8.7

At a pH greater than 6, the wave is a reduction wave. The current is proportional to the concentration of pyridine within ten per cent between 0.001 *M* and 0.01 *M* pyridine. For quantitative applications, a well-buffered solution (0.1 *N* phosphate) between pH 6 and 8 are advised.

Picolinic Acid. Substitution of a carboxylic group into the 2-position of pyridine shifts the half-wave potentials to more positive values than those of pyridine with the result that picolinic acid (II) produces a small wave even at pH 12. No wave is found in 0.1 *N* sodium hydroxide. The half-wave potential (S.C.E.) in 0.1 *N* hydrochloric acid is -0.93 v., -1.17 v. at pH 5 and -1.5 v. at pH 9.[11] The diffusion current for 0.001 *M* picolinic acid in 0.1 *N* hydrochloric acid is 5.9 microamp., at pH 7.2 it is 5.98. At pH 12, the limiting current is 0.43. The diffusion current is proportional to the concentration of the acid in a given buffer.

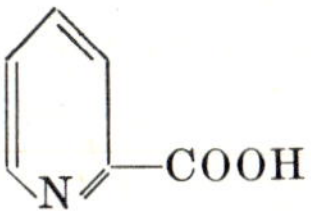

(II)

[9] P. C. Tompkins and C. L. A. Schmidt, *J. Biol. Chem.*, **143,** 643 (1942).

[10] E. Knobloch, *Collection Czechoslov. Chem. Communs.*, **12,** 407 (1947).

[11] P. C. Tompkins and C. L. A. Schmidt, *Univ. California Pub. Physiol.*, **8,** 229 (1944).

Isonicotinic Acid (III). This acid is reducible in the same pH range as picolinic acid. The half-wave potentials at various pH values are about 0.1 v. more positive.[11]

Nicotinic Acid. Nicotinic acid (IV) produces a catalytic hydrogen wave which appears in the buffer range pH 1–10. The best waves appear in sodium borate buffers between pH 8 and 9. The current under these circumstances is not proportional to the concentration unless the buffer concentration is at least 0.6 *M*. At a pH 8.7 the half-wave potential is −1.66 v. (S.C.E.) and changes with pH in phosphate solutions by 90 millivolts per pH unit.

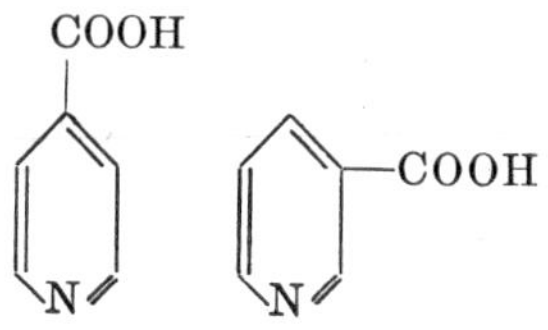

(III) (IV)

The wave decreases in height as the pH increases and disappears when the nicotinic acid is all in the form of its potassium salt, pH 9.2. For 0.0001 *M* solutions of nicotinic acid in 0.4 *M* hydrochloric acid the limiting current is 5.9 microamp., in 0.1 *M* hydrochloric acid 5.1 microamp., and in 0.6 *M* sodium borate (pH 8.7) 1.0 microamp.

Since both the pH and the buffer capacity exert a great effect on the diffusion current, extrême care must be used in the quantitative determination; 0.4 *M* and 0.1 *M* hydrochloric acid have been recommended as electrolytes.[12, 13]

Nicotinamide (Niacin). The polarographic behavior of nicotinamide (V) differs from that of pyridine and the various pyridinecarboxylic acids in that two waves occur.[14] The first wave is well defined and is found in buffers of pH 8 or greater. The diffusion current of this wave does not vary with pH and is proportional to the concentration. The reduction corresponds to a 2-electron change and suggests the following electrode reaction:

---

[12] P. C. Tompkins and C. L. A. Schmidt, *J. Biol. Chem.*, **143,** 643 (1942).

[13] J. J. Lingane and O. L. Davis, *J. Biol. Chem.*, **137,** 567 (1941).

[14] P. C. Tompkins and C. L. A. Schmidt, *Univ. California Pub. Physiol.*, **8,** 237, 247 (1944).

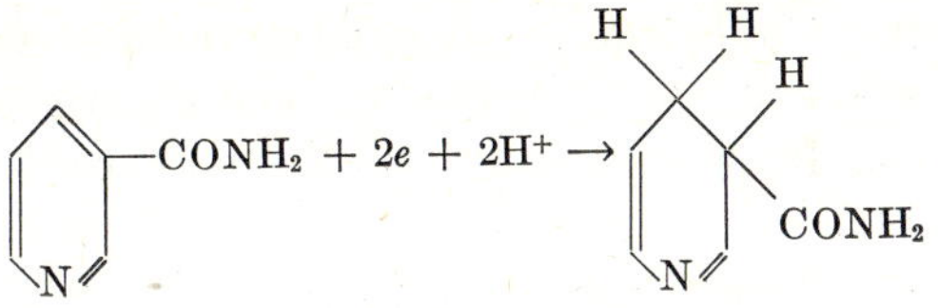

(V) (VI)

The half-wave potential is independent of the concentration and changes from −1.56 v. at pH 8.7 (borate buffer) to −1.75 v. (S.C.E.) at pH 13 (0.1 *N* sodium hydroxide).

The second wave occurs at more negative potentials than the first; it is a catalytic wave brought about by reduction product (VI). This wave has a maximum and is even observed at a concentration of 5×10^{-5} *M*. Coming from the alkaline side, the catalyzed wave appears at around pH 7.24 and increases as the solution is made more acidic. The maximum disappears at pH 4.78. The wave coalesces at pH 4.56 with the hydrogen wave.[15]

For the quantitative determination of nicotinamide, the normal reduction wave is used in 0.1 *N* potassium chloride, 0.1 *N* sodium hydroxide, 0.1 *N* sodium carbonate[16] or in a buffered solution.[14]

(b) Pyridinium Compounds

Pyridinium compounds exhibit a polarographic behavior which is dependent upon the substituents in the pyridine ring.

Methylpyridinium Hydroxide (I) and Trigonelline (II). These compounds exhibit similar polarographic behavior.[14] Both compounds produce a diffusion current in buffered solutions more alkaline than pH 10 which is independent of pH. A second wave which becomes more apparent in

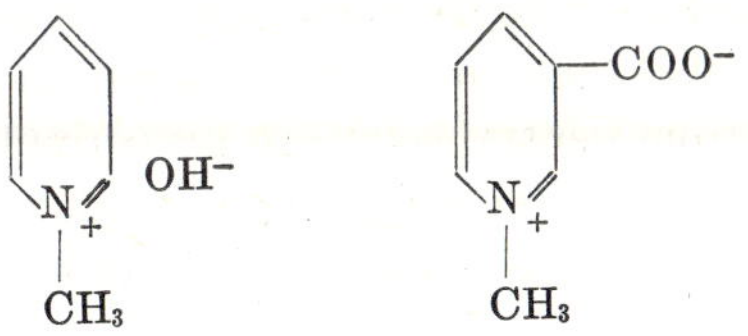

(I) (II)

more acid buffers obscures the diffusion current in buffered solutions more

[15] E. Knobloch, *Collection Czechoslov. Chem. Communs.*, **12,** 407 (1947).

[16] K. Wenig and M. Kopecky, *Casopis Ceskoslov, Lekarnictva,* **56,** 49 (1943); *Chem. Zentr., I,* **1944,** 34.

acid than pH 10. The half-wave potential of the first wave for methylpyridinium hydroxide is −1.47 v. (S.C.E.) and for trigonelline is −1.35 v. The diffusion current is proportional to concentration in the limits 10^{-4} to 10^{-2} *M*.

Analysis of the curve gives a slope of 0.059. The height of the curve points to a 1-electron change and suggests the following electrode reaction:

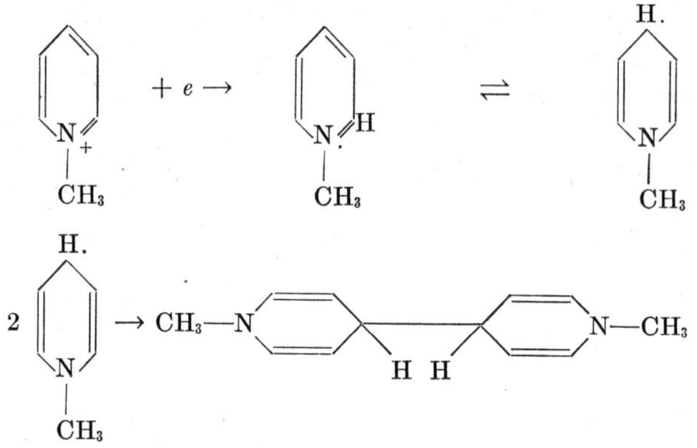

The second wave, which is pH dependent, may represent a reduction to a dihydro derivative:

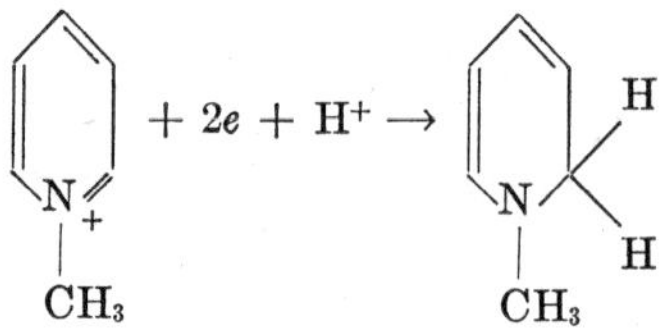

For quantitative determinations of these two substances a sodium hydroxide solution is best. The method cannot be used for the separate determination of the two compounds in mixtures. The methylpyridinium ion can be removed from neutral solutions by permutite while trigonelline is not affected. Gelatin is used to suppress maxima.

Nicotinamide Methochloride (III). This compound exhibits behavior

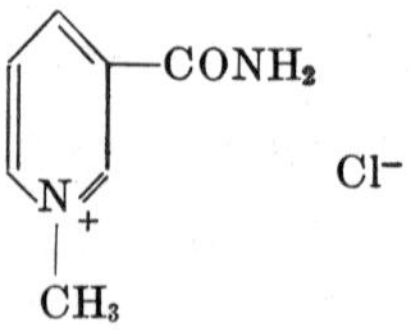

(III)

very similar to that of methylpyridinium hydroxide and trigonelline in that the first wave is independent of the pH and the electrolyte used.[14] The wave is observed in buffers as acid as pH 5.3 and has a half-wave potential of -1.025 ± 0.005 v. (S.C.E.). The second wave is better defined for this compound and varies in half-wave potential and slope with the pH and the type of buffer. This wave is observed in solutions more alkaline than pH 8 (-1.59 v. *vs.* S.C.E.) and is obscured by the hydrogen discharge in more acid solutions.

Each wave involves a 1-electron change and has a slope close to 0.059 v. The mechanism of reduction is similar to that proposed for methylpyridinium hydroxide. Quantiatively this compound can be analyzed for in the presence of nicotinamide in a buffer of pH 7.3 (phosphate) if only the first wave of the quaternary compound is used.

Homarine (IV) and Isonicotinic Acid Methochloride (V). These compounds in which the carboxyl group is conjugated with the quaternary nitrogen, exhibit different polarographic behavior from that of methylpyridinium hydroxide.[16]

Homarine produces well-defined waves in buffers more alkaline than pH 5. At pH 8.7 the single wave separates into a double wave which has half-wave potentials of -1.23 v. (S.C.E.) and -1.53 v. (S.C.E.), respectively. The half-wave potentials and diffusion currents vary with the pH and the buffer used.

Isonicotinic acid methochloride produces one wave in the pH range of 1–12 and two waves in more alkaline buffers; the second wave appears at the expense of the first wave. The half-wave potential is -1.20 v. at pH 7.8 and varies with the pH, the electrolyte, and the concentration.

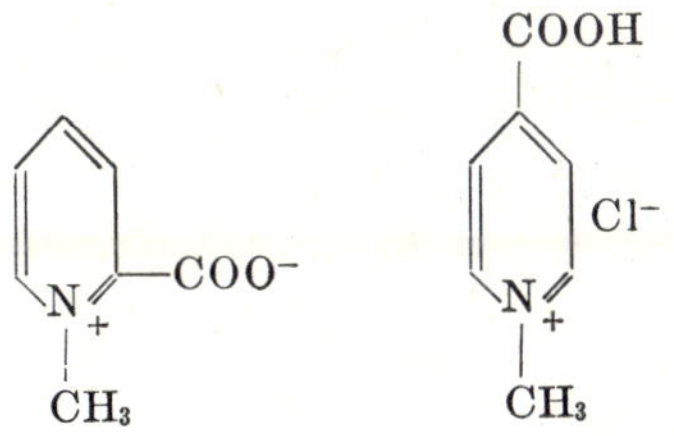

(IV) (V)

The heights of the waves obtained for both compounds point to an electron transfer of 2 electrons in neutral and alkaline media. Analysis of the waves indicates a slope of 1 electron for each of the waves of homarine and to a varying slope for isonicotinic acid methochloride. The former behavior suggests the following electrode reaction:

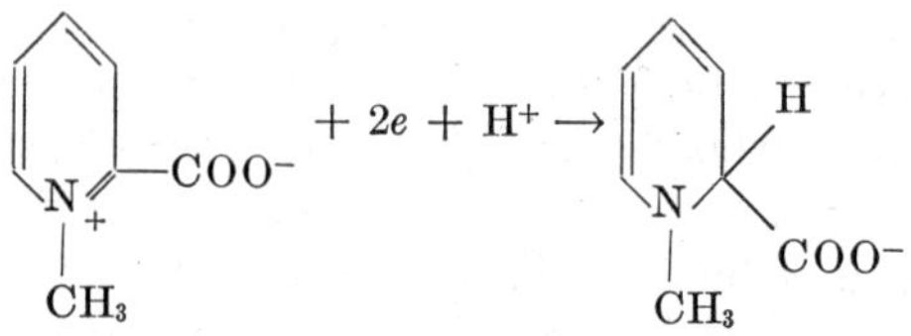

Quantitative determination of homarine is best carried out in a phosphate buffer of pH 7.2, while that of isonicotinic acid methochloride is best in a phosphate buffer of pH 8.3. The diffusion current is proportional to the concentration in the range of 0.28×10^{-3} to 5×10^{-3} *M* for the former, and in the range of 0.18×10^{-3} to 8.4×10^{-3} *M* for the latter.

Miscellaneous Pyridine Derivatives. Methyl nicotinate, nicotinic acid methiodide, γ-aminopyridine, and piperidine are reported to produce catalytic hydrogen waves.[15]

(c) Quinolines

Quinoline. Quinoline (I) produces three waves at the dropping mercury electrode in weakly acid solutions[17] and two waves of about equal height in neutral and alkaline media. In a 0.0013 *M* solution of quinoline in 0.1 *N* ammonium chloride (unbuffered) the first wave starts at −1.06 v. and the second at −1.64 v., in lithium chloride (unbuffered) at −1.28 and −1.74 v., in 0.1 *N* magnesium chloride at −1.28 and −1.68 v.; in 0.02 *N* tetramethylammonium chloride the wave starts at −1.43 and −1.58 v., and in 0.1 *N* lithium hydroxide at −1.37 and −1.57 v. *vs.* the N.C.E.[18] Half-wave potentials of −1.22 v. (N.C.E.) for a 0.001 *M* solution of quinoline in 0.1 *N* ammonium chloride and −1.54 v. in 0.2 *N* tetramethylammonium hydroxide are reported.[19]

Quinoline is also reported to produce a catalytic hydrogen wave.[20]

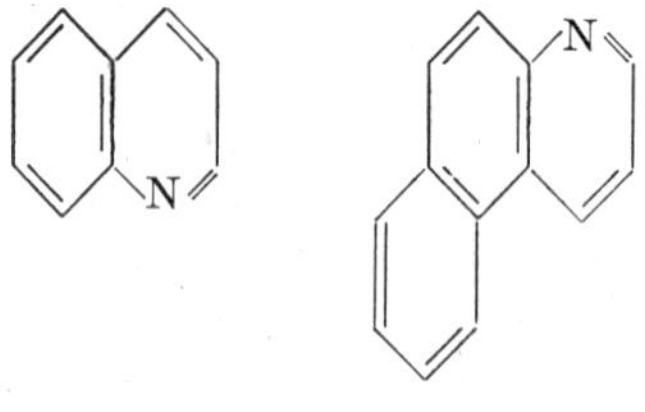

(I) (II)

[17] I. Tachi and H. Kabai, *J. Electrochem. Assoc. Japan*, **3,** 250 (1935); *Chem. Abstracts*, **30,** 2500 (1936).

[18] J. Pech, *Collection Czechoslov. Chem. Communs.*, **6,** 126 (1934).

[19] H. Adkins and F. W. Cox, *J. Am. Chem. Soc.*, **60,** 1151 (1938).

[20] E. Knobloch, *Collection Czechoslov. Chem. Communs.*, **12,** 407 (1947).

β-Naphthoquinoline (5,6-Benzoquinoline) (II). This substituted quinoline produces only three waves. The variation of half-wave potential (S.C.E.) and diffusion current constant with pH are given by Korshunov.[21]

Quinoline Ethiodide (III). This compound produces four waves in buffered solutions. The first two waves no doubt represent reduction of the ion and involve 1 electron. A comparison of the wave heights for this

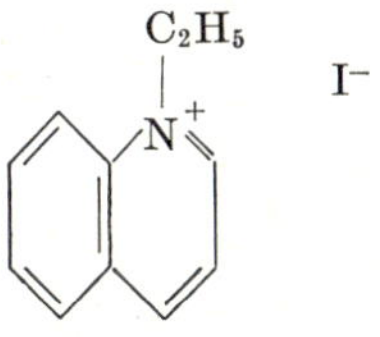

(III)

process with the total wave height in buffers more alkaline than 7 suggests a reduction involving 3 electrons. The actual electrode reaction is not known.[21]

(d) Hydroxyquinolines

2-Hydroxyquinoline (Carbostyril). This compound produces a wave only in a borate buffer of pH 8 and in potassium chloride solution. In both solutions the half-wave potential is −1.9 v. (S.C.E.).[22]

8-Hydroxyquinoline (Oxine). This compound is reducible at the dropping mercury electrode in buffers ranging from 2 to about 12. The behavior is complicated and strongly dependent upon the pH of the medium.

In strongly acid solutions a single poorly defined wave, A, is observed. The foot of this wave suggests the presence of a small ill-defined wave, B, which is visible in the pH range from 2.5 to 7. Between pH 6.3 and 6.7 wave A merges with a succeeding wave and produces a wave about 0.1 v. more negative than that expected of wave A alone. In faintly alkaline solutions this larger wave is no longer visible and wave A acquires a maximum and is followed by a long drawn-out wave, C. Wave C persists up to about pH 10.5 where it merges with the residual current. In strong alkaline solutions wave A diminishes in height. The half-wave potentials and wave heights observed[23] are given in Figs. XLV-1 and XLV-2.

The interpretation of the electrode processes which take place is difficult. Wave B is an absorption wave since its height is independent of pH and

[21] M. K. Shchennikova and I. A. Korshunov, *J. Phys. Chem. U. S. S. R.*, **22**, 503 (1948).

[22] K. G. Stone, *J. Am. Chem. Soc.*, **69**, 1832 (1947).

[23] J. T. Stock, *J. Chem. Soc.*, **1949**, 586.

concentration. Wave A may represent reduction of the 8-hydroxyquinoline to the dihydro compound and wave C a reduction of the anion produced

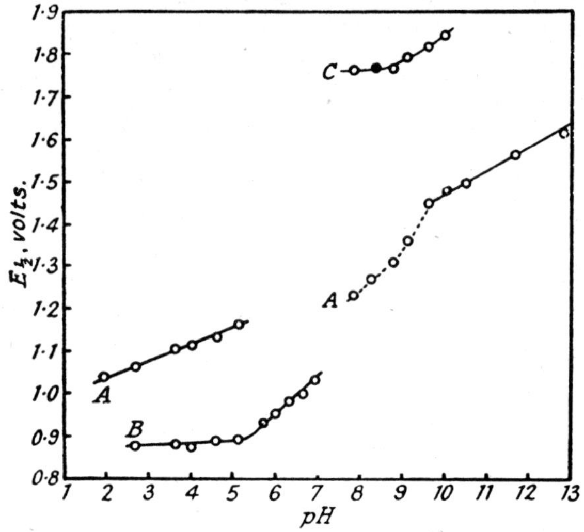

Fig. XLV-1. Half-wave potentials (S.C.E.) of 8-hydroxyquinoline in various buffers.[23]

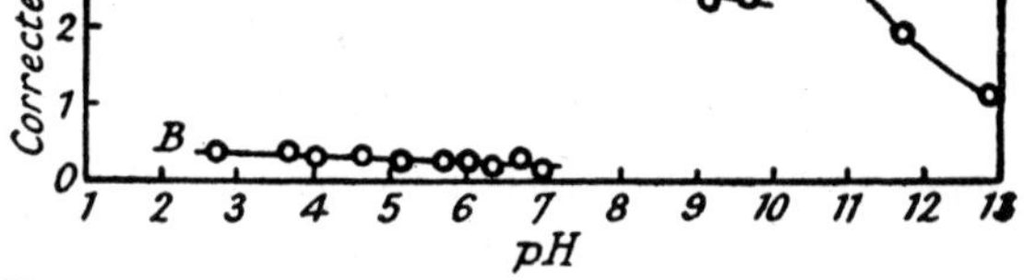

Fig. XLV-2. Wave heights of 8-hydroxyquinoline in various buffers.[23]

by the basic buffers in the same way. Results obtained for 8-hydroxyquinoline by Korshunov[21] in buffered solutions differ from those mentioned above.

The polarographic behavior of 8-hydroxyquinoline is used to determine magnesium and aluminum. An excess of hydroxyquinoline is added and the excess is determined polarographically without removing the precipitate.[24]

For the quantitative determination of 8-hydroxyquinoline a pH range of 9.2–10.5 is recommended since wave A is well formed without a maximum and is proportional to the concentration up to 2.5×10^{-3} *M*. Magnesium is determined in an ammonia-ammonium chloride solution at pH 10 using wave A at −1.39 v. (S.C.E.). The second wave, C, at −1.61 v. (S.C.E.) is not used since it shifts with concentration. Polarographically, the solubility of magnesium 8-hydroxyquinolate has been found to be 1.9×10^{-6} moles per liter.[24]

Procedure. 5 ml. of standard 8-hydroxyquinoline (0.05 per cent in 5 per cent ethanol) and 10 ml. of the buffer are mixed, diluted to 25 ml., and a polarogram is taken. For the precipitation, a given amount of standard magnesium solution or of unknown is added and the mixture is diluted to 25 ml. The polarogram is taken after shaking at frequent intervals for 1 to 2 hours, depending upon the amount of magnesium.

A solution with a pH of 7.09 to 7.2 gives less satisfactory results since three waves are obtained which are not well defined unless the concentration is kept between 0.914 and 0.1148×10^{-3} *M*. The diffusion current is measured at about −1.5 v. and, after correction for residual current, is referred to an empirical calibration curve to obtain the concentration of the magnesium.[25]

The determination of aluminum is carried out in an ammonium chloride-ammonia buffer of pH 9.8.[26]

Procedure. Place 25 ml. of pH 9.8 buffer solution and 20 ml. of 0.05 per cent alcoholic oxine solution in each of four 50-ml. volumetric flasks. To the first three flasks add enough standard aluminum solution to give 0.1, 0.2, and 0.3 mg. of aluminum, respectively. (The standard aluminum solution should be only slightly acid.) Dilute the flasks to the mark with distilled water. Shake at 5 minute intervals, and after 1 hour obtain polarograms on each solution. Measure all the wave heights and, using the aluminum-free solution as a standard, plot differences in wave heights for the other solutions against the amounts of aluminum added. Draw the best straight line through the points.

Prepare samples to be analyzed in such a way that 3 ml. of solution contains between 0.1 and 0.4 mg. of aluminum. Treat the sample solutions in the same manner as the standards, using 3.00 ml. of sample solution, and determine the decrease in wave heights. Read the amount of aluminum in the portion taken from the calibration curve.

---

24 K. G. Stone and N. H. Furman, *Ind. Eng. Chem., Anal. Ed.*, **16**, 596 (1944).
25 G. Carruthers, *Ind. Eng. Chem., Anal., Ed.*, **15**, 412 (1943).
26 T. D. Parks and L. Lykken, *Anal. Chem.*, **20**, 1102 (1948).

(e) Quinoline Carboxylic Acids

Quinaldinic Acid (I). This compound is reducible over the entire pH range examined.[27] The behavior is complicated and dependent upon the pH of the medium. In strongly acid solutions, the first wave, A, is surmounted by a "maximum" and followed by a second smaller wave, B,

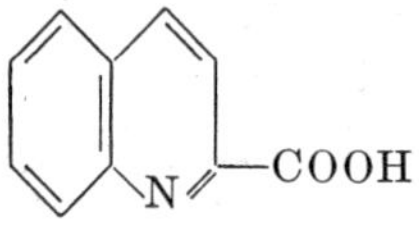

(I)

which immediately precedes the residual current. As the pH of the solution is increased, wave A is displaced to more negative potentials, and wave B decreases in height, vanishing completely at about pH 4. In the neutral region four waves A, C, D, and E appear. Wave C is apparent in the pH range 6.4–8 and decreases in height with increase in pH. Its appearance is accompanied by a marked decrease in the height of wave A. Wave D appears at pH 5.5 and wave E at about pH 7.4. Both increase in height with increase in pH and merge at pH 9. In solutions more alkaline than pH 9, wave A becomes very small; it is still apparent at pH 12.3 but disappears at pH 13. Half-wave potentials (S.C.E.) and diffusion currents are plotted in Figs. XLV-3 and XLV-4.

The wave height of wave A is directly proportional in all buffers to the concentration of quinaldinic acid in the range of 0–2 $\times$ 10^{-3} M. The wave height of A at pH 5 where it is the only wave present corresponds to a 2-electron change. A similar value is obtained for wave E in 0.1 M sodium hydroxide. This behavior suggests that the following electrode reactions occur:

At pH 5:

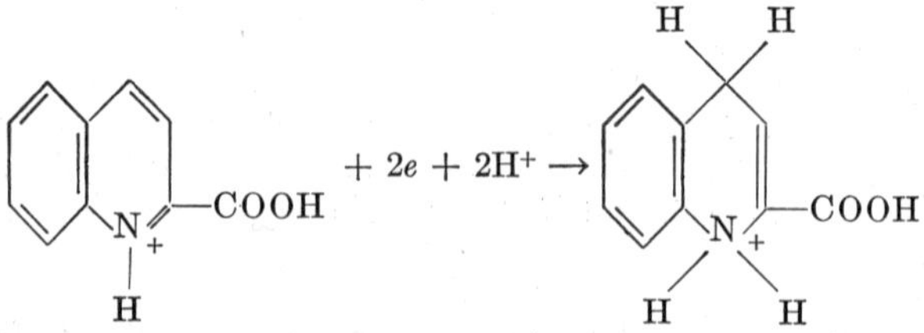

[27] J. T. Stock, *J. Chem. Soc.*, **1944,** 427.

At pH 13:

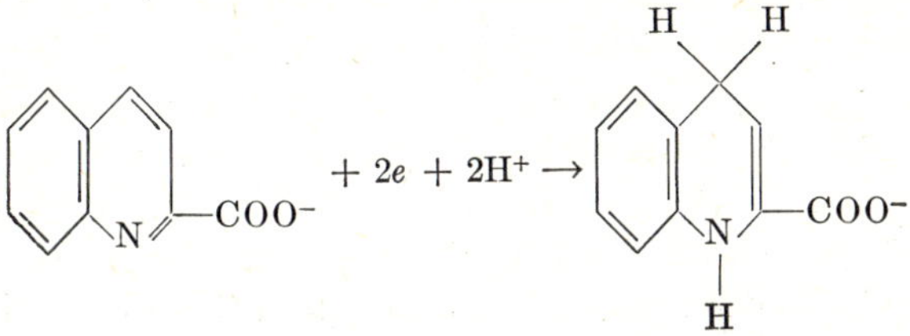

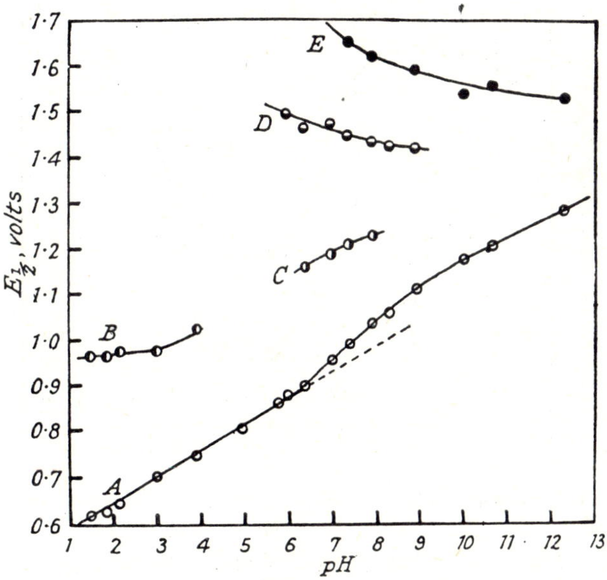

Fig. XLV-3. Half-wave potentials (S.C.E.) of quinaldinic acid in various buffers.[27]

Quinoline-8-carboxylic Acid (II). The polarographic behavior of this acid resembles that of quinaldinic acid in many respects. Cathodic waves are obtained over pH range 1–12. Up to about pH 5.5 wave A is difficult

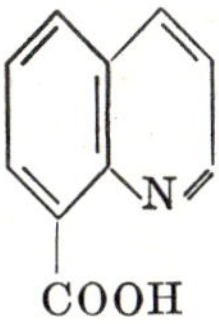

(II)

to distinguish but beyond this value it becomes more apparent. A small prewave is observed up to about pH 6. Between pH 6 and 7.5, a maximum appears beyond A and is followed by another wave, C. At pH 7.53 the maximum disappears, C becomes more distinct and is preceded by small waves D and E. In solutions more alkaline than pH 8, the only waves visible are A and C. The former diminishes in height while the latter increases with the change in pH. The relationship between half-wave potentials (S.C.E.), diffusion currents, and pH are shown in Figs. XLV-5

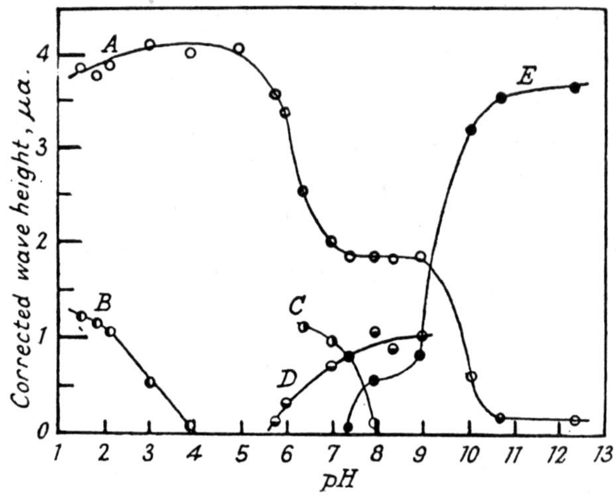

Fig. XLV-4. Wave heights of quinaldinic acid in various buffers.[27]

and XLV-6. The mechanism of reduction is probably similar to that proposed for 8-hydroxyquinoline and quinaldinic acid.[27a]

(f) Pyridine-N-oxides and Quinoline-N-oxides

A number of pyridine-N-oxides and quinoline-N-oxides have been studied by Ochiai[27b] and the reduction potentials (S.C.E.) obtained in buffers are reported in Table XLV-3A. Values for the quinoline derivatives were obtained in a buffer of pH 3.0 while the others are in a buffer of pH 3.5.

Comparison of the values for 4-aminoquinoline-N-oxide and 4-nitroquinoline-N-oxide indicates that the nitro group is more easily reduced than the N-oxide.

[27a] J. T. Stock, *J. Chem. Soc.*, **1949**, 763.
[27b] E. Ochiai, *J. Pharm. Soc. Japan*, **69**, 1 (1949).

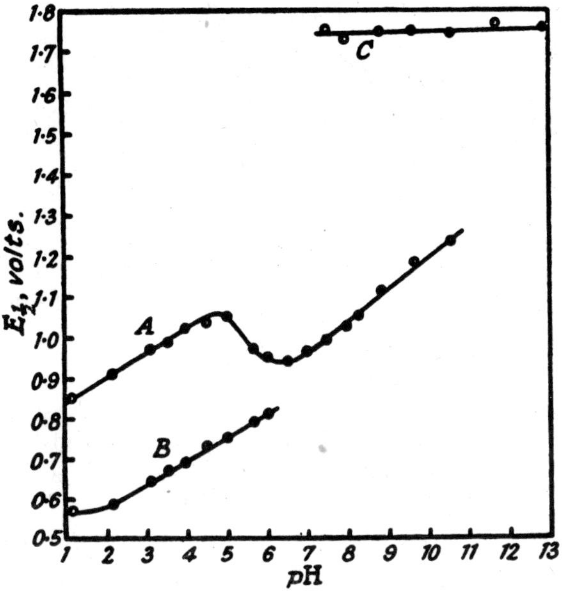

Fig. XLV-5. Half-wave potentials (S.C.E.) of quinoline-8-carboxylic acid in various buffers.[27a]

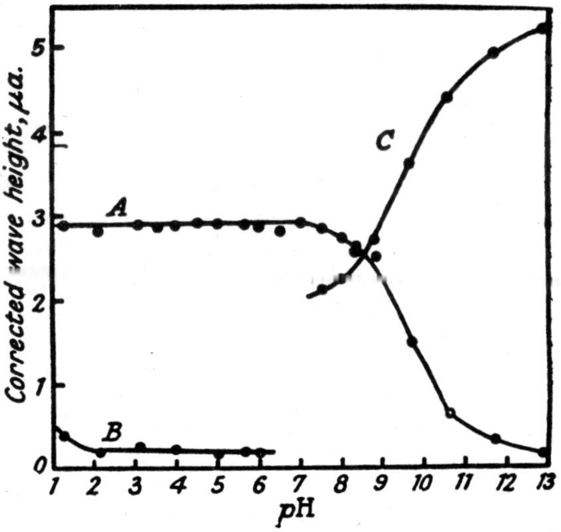

Fig. XLV-6. Wave heights of quinoline-8-carboxylic acid in various buffers.[27a]

TABLE XLV-3A

Compound	Reduction potential (S.C.E.)
Pyridine-N-oxide	1.2786
2,4,6-Collidine-N-oxide	1.3924
4-Phenoxypyridine-N-oxide	1.2820
4-Nitropyridine-N-oxide	0.2890, 1.6622
Quinoline-N-oxide	1.0692
4-Ethoxyquinoline-N-oxide	1.1802
4-Aminoquinoline-N-oxide	1.3822
4-Nitroquinoline-N-oxide	0.2802, 1.1192, 1.3722

(g) Acridines

Acridines (III). These compounds produce two waves at the dropping mercury electrode which shift to more negative potentials with an increase in pH The half-wave potentials (N.H.E.) obtained for a series of acridines are listed in Table XLV-4.

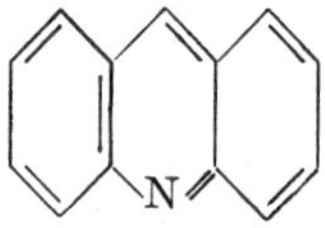

(III)

The diffusion current is proportional to the concentration only in the range of 10^{-3} to 10^{-5} M except for acridine and 4-aminoacridine which show proportionality in the range 10^{-4} to 10^{-5}. In these two cases concentrations above 10^{-4} give three waves. The half-wave potentials change with concentration. The value for acridine, for example, shifts from 0.042 to -0.099 v. at pH 2 and from -0.054 v. to -0.124 v. at pH 4 when the concentration is varied from 10^{-3} to 10^{-4} M. Maxima can be suppressed in some cases by 0.05 per cent gelatin or methyl cellulose.

The actual electrode process is dependent upon the form in which the acridine is present in solution. The first reduction wave must involve an acridinium salt since, if the half-wave potentials are plotted against pH, curves with breaks occurring at the pK values of the acridines are obtained. This type of reduction is indicated by the behavior of acridone which is not basic and produces only one wave corresponding to the second wave of the acridines.[28]

More conclusive evidence for the electrode reaction has been obtained

[28] B. Breyer, C. S. Buchanan, and H. Duewell, *J. Chem. Soc.*, **1944**, 360.

from a polarographic study of 9-phenylacridine and 9-(*o*-iodophenyl)-acridine, and controlled potential electrolysis. The results obtained for these compounds in 50 per cent ethanol are listed below.

Solution	0.1 *N* HCl		0.2 *N* HCl		0.1 *N* $(CH_3)_4NOH$		0.1 *N* KOH 0.5 *N* KOAc
	$E_{1/2}$ (S.C.E.)	*n*	$E_{/2}$	*n*	$E_{1/2}$	*n*	$E_{1/2}$
9-Phenylacridine			−0.65 −0.72	1.3–2	−1.20	2	
9-(*o*-Iodophenyl)-acridine	−0.59	—			−1.21 −1.33	4	−1.32 −1.62

Controlled potential electrolysis of 9-(*o*-iodophenyl)acridine in 0.1 *N* potassium hydroxide–0.5 *N* potassium acetate at −1.39 v. gave 9-(*o*-iodophenyl)-dihydroacridine, and at −1.70 v. gave 9-phenyldihydroacridine. The electrode reaction in alkaline solution is therefore the following:

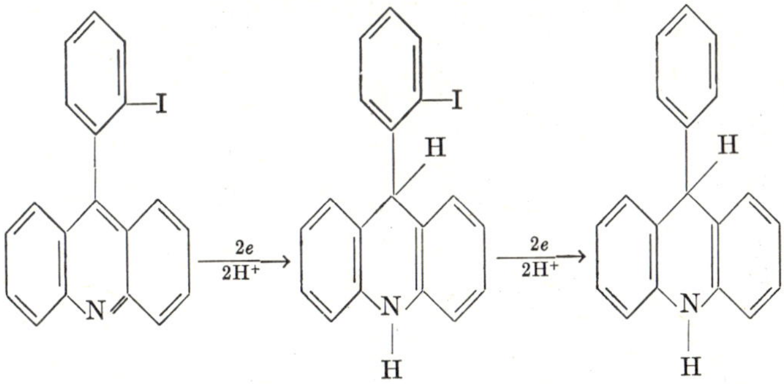

The actual electrode reaction in acid solution was not demonstrated as conclusively. Controlled potential electrolysis at −0.85 v. of 9-phenylacridine in 1 *N* hydrochloric acid–50 per cent ethanol gave a red precipitate which apparently formed 9-phenylacridine and 9-phenyldihydroacridine upon standing or treatment with alcohol.[29]

The half-wave potentials of the aminoacridines have been found to be related to their chemotherapeutic activity. Compounds having a more negative half-wave potential than −0.400 v. (N.C.E.) proved to be the most active antiseptics.[28]

3. ALKALOIDS

Alkaloids in general produce catalytic waves at the dropping mercury electrode to an extent that is dependent upon the ring system involved.

[29] J. J. Lingane, C. G. Swain, and M. Fields, *J. Am. Chem. Soc.*, **65,** 1348 (1943).

TABLE XLV-4

HALF-WAVE POTENTIALS ($-E_{1/2}$ *vs.* N.H.E.) OF ACRIDINES (10^{-4} M) IN VARIOUS BUFFERS

Substance	pH 2		pH 4		pH 5.5		pH 7.3		pH 9		pH 11.8	
	1st	2nd	1st	2nd	1st	2nd	1st	2nd	1st	2nd	1st	2nd
Acridine	0.099	(a)	0.124	(a)	0.157	1.063	0.313	1.057	0.472	1.069	0.731	1.055
1-Aminoacridine	0.081	0.521	0.176	0.614	0.260	0.735	0.394	0.979	0.570	1.030	0.842	1.041
2-Aminoacridine	0.338	0.672	0.396	0.781	0.418	0.875	0.468	1.009	0.549	1.119	0.720	1.134
3-Aminoacridine	0.124	0.566	0.199	0.674	0.235	0.749	0.369	0.896	0.489	1.014	0.753	1.050
4-Aminoacridine	0.166	0.635	0.200	0.729	0.209	0.847	0.301	0.954	0.423	1.005	0.672	1.006
5-Aminoacridine	0.658	(a)	0.732	(a)	0.834	(b)	0.916	(b)	0.941	(b)	1.073	(b)
2,8-Diaminoacridine	0.448	0.597	0.604	0.774	0.690	0.929	0.731	1.059	0.706	1.177	0.830	1.225
2,8-Diamino-10-methylacridinium bromide	0.422	0.588	0.560	0.728	0.634	0.889	0.650	1.055	0.639	1.153	0.627	1.218
2-Chloro-8-aminoacridine	0.264	0.646	0.302	0.767	0.378	0.868	0.420	(b)	0.473	1.077	(c)	(c)
3-Hydroxyacridine	0.162	0.257	0.202	0.318	0.229	1.029	0.369	1.030	0.506	1.050	0.884	1.046
Acridine-3-sulfonic acid	0.095	(a)	0.099	0.932	0.155	0.925	0.281	0.952	0.422	0.962	0.691	0.960
Acridone	(c)	(c)	—	0.325	—	0.847	—	1.191	—	1.265	—	1.391

(a) Hydrogen step overlaps

(b) Maximum formation.

(c) Insufficiently soluble.

The ability to produce catalytic waves follows the order: indole ≈ quinoline ≈ isoquinoline > pyridine ≈ lupinane > tropane > glyoxaline > amine. The more complex the molecule the greater the catalytic effect. Structures, for example, containing two of the above ring systems will give a greater catalytic wave than structures with only one nitrogen atom.[30] If reducible groups are present in the molecule, these produce normal reduction waves. The waves are suitable for the quantitative determination of alkaloids but not for their qualitative detection in mixtures.

(a) Opium Alkaloids

The results obtained for a number of opium alkaloids and their derivatives are listed in Table XLV-5.

TABLE XLV-5
POLAROGRAPHIC BEHAVIOR OF OPIUM ALKALOIDS AND RELATED COMPOUNDS

Alkaloid	pH range in which catalytic wave is observed	Buffer (pH) for quantitative determination	Range in which $i_d = kC$, moles/liter	Range of concn. in calibration, moles/liter
Morphine	5–9	8.0	$0.03-0.13 \times 10^{-3}$	$0.025-0.27 \times 10^{-3}$
Codeine	5–10	8.5	$0.025-0.13 \times 10^{-3}$	$0.025-0.25 \times 10^{-3}$
Ethymorphine	5–10	9.0	$0.020-0.40 \times 10^{-3}$	$0.03-0.40 \times 10^{-3}$
Diamorphine	5–9	7.5	$0.07-0.17 \times 10^{-3}$	$0.03-0.23 \times 10^{-3}$
Apomorphine	5–7 (Two waves)	6.5 (1st wave)	$0.04-0.30 \times 10^{-3}$	—
Narcotine	3–10	8.0	—	$0-0.5 \times 10^{-3}$

Cotarnine, narceine, and papaverine yield normal reduction waves:

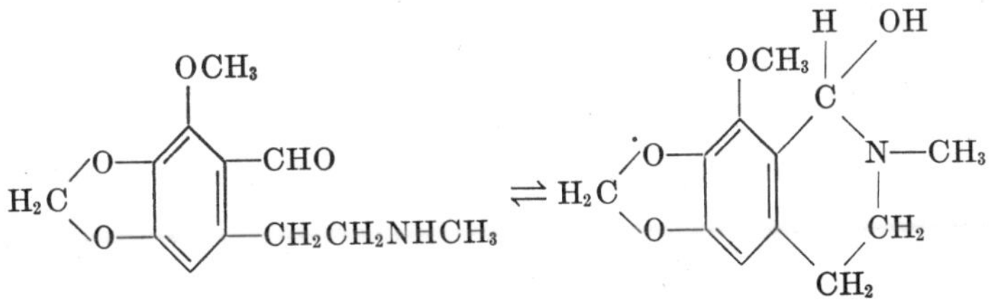

(I)

Cotarnine (I). This compound produces well-defined waves at all pH values investigated. The behavior differs from that of an aromatic aldehyde in that the height of the waves obtained in all buffers corresponds to 2 electrons. The results obtained for a 1×10^{-3} M solution are listed

[30] H. F. W. Kirkpatrick, *Quart. J. Pharm. Pharmacol.*, **20**, 87 (1947).

below:[31]

pH	3	4	5	6	7	8	9	10
$-E_{1/2}$ (S.C.E.)	1.09	1.11	1.04	1.11	1.13	1.10	1.06	0.98
			1.17	1.25	1.30	1.30	1.32	1.33

Narceine. This compound behaves like a typical ketone and produces a wave for which the half-wave potential is given by the equation $E_{1/2}$(S.C.E.) $= -1.02 - 0.06$ pH for pH values between 3 and 8. The diffusion current corresponds to a 2-electron change and at pH 3 is proportional to the concentration in the range of 0.15 to 1.0×10^{-3} M. The catalytic wave is negligibly small.[31]

Papaverine. This compound produces both a catalytic and a reduction wave. The latter is well defined at pH values greater than 5 and corresponds to a 2-electron change. The half-wave potential is about -1.45 v. (S.C.E.) and is independent of pH. The catalytic wave follows the normal wave and makes its measurement difficult. The diffusion current at pH 6.5 is proportional to the concentration within the range 0.07 to 0.14×10^{-3} M.[31] The method is applicable to the determination of morphine and diamorphine in injection solutions and in pills containing a single alkaloid.[32]

Morphine can be determined[33] in the presence of alkaloids which do not form nitroso compounds, such as narcotine, papaverine, and codeine, by converting it to 2-nitrosomorphine with acid and excess nitrite. The procedure consists in dissolving a 5- to 100-mg. sample in 100 ml. N hydrochloric acid and adding 2 ml. of 1 M potassium nitrite solution. After 5 minutes 3 ml. of 20 per cent potassium hydroxide and 1 ml. of 2 per cent methyl cellulose are added and the polarogram is run immediately after removing dissolved air with nitrogen since nitrosomorphine is decomposed by acid. The diffusion current is proportional to the concentration of morphine up to concentrations of about 120 mg. per 100 ml. of solution.

The polarographic method has been used to determine cotarnine in tablets.[34]

---

[31] H. F. W. Kirkpatrick, *Quart. J. Pharm. Pharmacol.*, **18**, 338 (1945).

[32] H. F. W. Kirkpatrick, *Quart. J. Pharm. Pharmacol.*, **20**, 87 (1947).

[33] H. B. Rasmussen, C. Hahn, and K. Ilver, *Dansk Tids. Farm.*, **19**, 41 (1945); *Chem. Abstracts*, **39**, 5405 (1945).

[34] F. Santavy, *Collection Czechoslov. Chem. Communs.*, **14**, 377 (1949).

(b) Alkaloids of Hydrastis and Ipecacuanha

Alkaloids in this series produce both catalytic waves and reduction waves. Results for a number of these are given in Table XLV-6.[35]

TABLE XLV-6

POLAROGRAPHIC BEHAVIOR OF HYDRASTIS AND IPECACUANHA ALKALOIDS

Alkaloid	Buffer range (pH) in which catalytic wave is observed	Buffer (pH) for quantitative estimation	Range in which $i_d = kC$, moles/liter	Range obtained by calibration, moles/liter
Hydrastine	3–10	8.0	—	$0\text{–}0.5 \times 10^{-3}$
Emetine	3–10	3.0	$0.08\text{–}0.25 \times 10^{-3}$	—
		8.0	—	$0.01\text{–}0.1 \times 10^{-3}$
Cephaeline	3–10	3.0	$0.08\text{–}0.25 \times 10^{-3}$	—
		8.0	—	$0.01\text{–}0.1 \times 10^{-3}$

Hydrastinine. The behavior of hydrastinine (I) is similar to that of

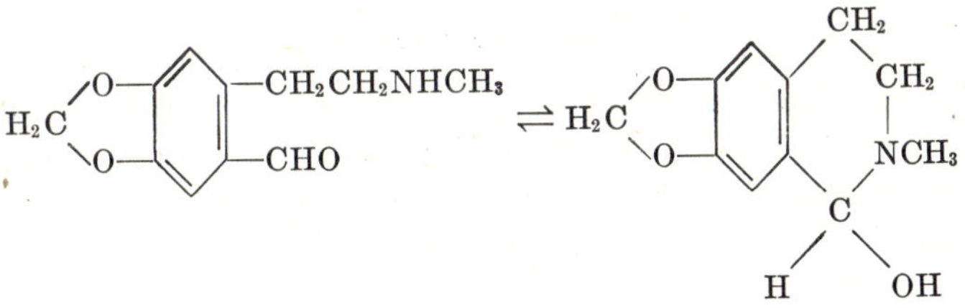

(I)

cotarnine since they differ only by a methoxyl group. The half-wave potentials in various buffers are given below for 1×10^{-3} M solutions.[35]

pH	3	4	5	6	7	8	9	10
$-E_{1/2}$ (S.C.E.)	1.09	1.12	1.15	1.20	1.12	1.04	0.99	(?)
					1.29	1.29	1.31	1.32

The diffusion current at pH 3 is proportional to the concentration in the range 0.1 to 1×10^{-3} M.

Berberine (II). This compound because of its tetrahydroquinoline nucleus produces both a diffusion controlled and a catalytic wave. Only one wave appears at pH 3 which is proportional to the concentration in the range $0.075\text{–}0.15 \times 10^{-3}$ M.[35] At pH 4 and 5 a coalescence of

[35] H. F. W. Kirkpatrick, *Quart. J. Pharm. Pharmacol.*, **19**, 8 (1946).

two waves occurs; these waves separate at higher pH values and are indicative of a 2-electron reduction. The catalytic wave appears at more

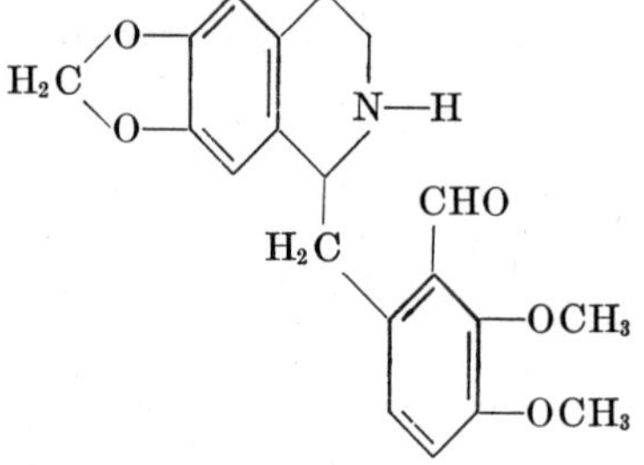

(II)

negative potentials, is at its maximum at pH 6, and disappears at pH 9. The catalytic wave at pH 6 may be used for quantiative purposes in the range of 0.02 to 0.15 $\times 10^{-3}$ M if a calibration curve is used.[34]

(c) Cinchona Alkaloids

Quinine and quinidine show both catalytic and normal waves.[36] The catalytic curve is observed between pH 5 and 8 and can be used for quantitative determination at pH 8 over the range 10^{-5} to 10^{-4} M using a calibration curve.

TABLE XLV-7
HALF-WAVE POTENTIAL ($-E_{1/2}$) OF CINCHONA ALKALOIDS

pH	3	4	5	6	7	8	9	10
Quinine, Quinidine	1.00	1.02	1.05	1.08	1.10	1.10[a] 1.26	1.10[a] 1.30	1.10[a] 1.33
Cinchonine, Cinchonidine	0.90	0.95	1.01	1.05	1.08	1.08[a] 1.27	1.02[a] 1.27	1.02 1.27

[a] Small prewave.

Cinchonine and cinchonidine behave like quinine in acetate and phosphate media but have a less catalytic effect in a borate buffer. The normal waves observed are similar to those obtained for quinoline. The reduction of the vinyl group is eliminated as a possibility by the similar polarographic behavior of hydroquinine. The half-wave potentials (S.C.E.) obtained in various buffers are given in Table XLV-7.

Quantitative determination is possible at pH values greater than 8 for concentrations up to 1 millimolar.

[36] H. F. W. Kirkpatrick, *Quart. J. Pharm. Pharmacol.*, **19**, 127 (1946).

(d) Indole Group

Alkaloids of the indole group produce catalytic waves.[36] The results are listed in Table XLV-8.[35]

The method is suitable for determining strychnine in a nux vomica preparation after separation of the strychnine.[37]

TABLE XLV-8

POLAROGRAPHIC BEHAVIOR OF INDOLE ALKALOIDS

Alkaloid	pH range in which catalytic wave is observed	Buffer (pH) for quantitative estimation	Range in which $i_d = kC$, moles/liter	Range in calibration, moles/liter
Strychnine	5–10	8.0	—	10^{-5} to 10^{-4}
Brucine	5–12	8.0	—	10^{-5} to 10^{-4}
Ergometrine	<9	6.0	$0.02–0.10 \times 10^{-3}$	$0.01–0.15 \times 10^{-3}$
Ergotoxine	<9	6.0	$0.02–0.10 \times 10^{-3}$	$0.01–0.15 \times 10^{-3}$
Physostigmine	5–9	6.0	—	$0.01–0.15 \times 10^{-3}$
Yohimbine	5–9	6.0	—	10^{-5} to 10^{-4}

(e) Tropane Alkaloids

The tropane alkaloids are characterized by catalytic waves only in borate buffers and at high concentrations.[37] Phosphate media have very little effect. The results are listed in Table XLV-9.

TABLE XLV-9

POLAROGRAPHIC BEHAVIOR OF TROPANE ALKALOIDS

Alkaloid	pH range in which catalytic wave is observed	Buffer for quantitative estimation	Range in which $i_d = kC$, moles/liter
Atropine	8–10	8	$0.2–0.6 \times 10^{-3}$
		9	$0.5–1.0 \times 10^{-3}$
		10	$1.0–2.0 \times 10^{-3}$
Hyoscine	8–10	8	$1.0–2.0 \times 10^{-3}$
Homatropine	8–10	10	$1.0–2.0 \times 10^{-3}$
Cocaine	8–9 (Two waves)	8	$1.0–2.0 \times 10^{-3}$

(f) Miscellaneous Alkaloids

Catalytic waves reported for a number of miscellaneous alkaloids are listed in Table XLV-10. The method is reported to detect one part of

[37] H. F. W. Kirkpatrick, *Quart. J. Pharm. Pharmacol.*, **20**, 87 (1947).

nicotine in 10^7 parts of solution in the presence of pyridine and ammonium salts.[38]

TABLE XLV-10

POLAROGRAPHIC BEHAVIOR OF MISCELLANEOUS ALKALOIDS

Alkaloid	pH range in which catalytic wave occurs	Buffer (pH) for quantitative estimation	Range of concn. in calibration, moles/liter
Nicotine	5–10	8	$0.05–0.5 \times 10^{-3}$
Sparteine	5–9	8	$0.02–0.2 \times 10^{-3}$
Pilocarpine	7–9	8	$2.0–4.0 \times 10^{-3}$

Colchicine. This alkaloid has been widely studied polarographically both from the quantitative point of view and also in an attempt to arrive at its structure (see p. 631). Single reduction waves are obtained at pH 3 and 4, two waves between pH 6 and 9, and one at pH values greater than 9. The half-wave potentials obtained in various buffers are listed below:

pH	3	4	5	6	7	8	9	10
$-E_{1/2}$ (S.C.E.)	1.22	1.30	1.30	1.36	1.38	1.40	1.44	1.40
	—	—	—	1.75	1.67	1.61	1.54	—

The wave in acid solution corresponds to a 1-electron change, while the wave in alkali corresponds to a 2-electron process. For quantitative purposes the waves at pH 3 and pH 10 may be used. No catalytic wave is observed.[39]

The polarographic method gives results which are about 6 per cent high when applied to the determination of colchicine in seed extracts or in a tincture. More accurate results are obtained if the extracts are purified chromatographically. The most suitable electrolyte is 0.1 *N* ammonium hydroxide.[40]

Lobeline (I). This compound can be determined polarographically in the form of its hydrochloride in tetramethylammonium iodide as an indifferent electrolyte.[41] The wave of lobeline lies so close to that of acetophenone that lobeline cannot be determined in the presence of this cleavage

[38] G. Semerano, *Giorn. chim. applicata,* **14,** 608 (1932).

[39] H. F. W. Kirkpatrick, *Quart. J. Pharm. Pharmacol.,* **19,** 527 (1946).

[40] G. Santavy, *Pharm. Acta Helv.,* **23,** 380 (1948).

[41] P. Nyman and F. Reimers, *Dansk Tids. Farm.,* **11,** 292 (1941); *Chem. Abstracts,* **36,** 5951 (1942).

product. Sympathol has no influence on the results. In extracts of Lobelia the lobeline content and that of the other alkaloids of this plant can be determined after isolation of the alkaloids with ethyl ether from an alkaline medium.

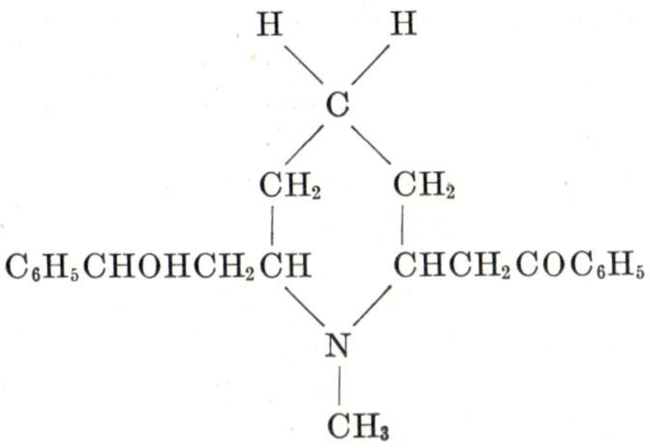

(I)

(g) Catalytic Potentials

Potentials have been measured for the various catalytic waves in a manner similar to that used for reduction potentials (tangent method). These "catalytic potentials" show no simple relation to the pH of the electrolyte. Values obtained under specified conditions[37] are listed in Table XLV-11.

4. MISCELLANEOUS HETEROCYCLIC SYSTEMS

Penicillin (I). This compound can be determined polarographically by using indirect methods. The compound itself does not produce a catalytic wave but, upon treatment with alkali and acid, gives dimethylcysteine which produces a catalytic wave in ammoniacal cobalt solution.

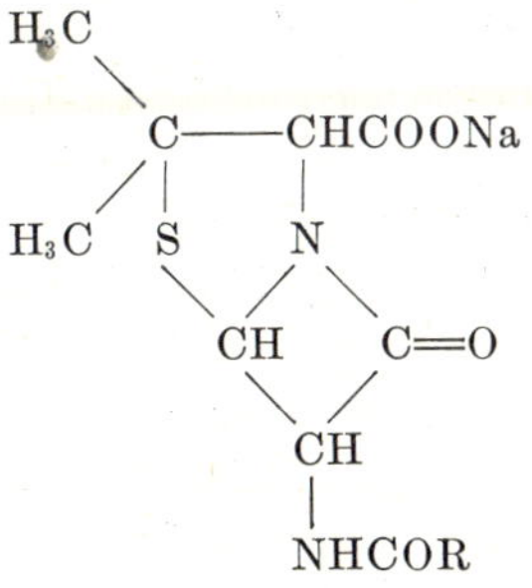

(I)

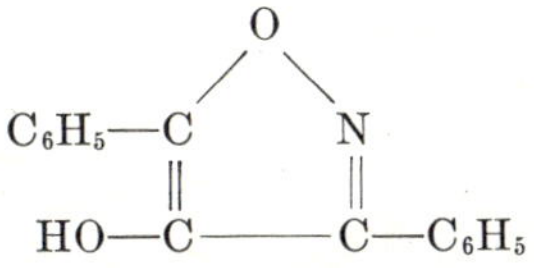

(II)

TABLE XLV-11

CATALYTIC POTENTIALS OF ALKALOIDS (*vs.* S.C.E.)

Alkaloid	Concn., $M \times 10^3$	pH 6.0	pH 8.0
Cephaeline	.1	−1.78	−1.64
Emetine	0.1	−1.76	−1.62
Narcotine	0.1	−1.52	−1.48
Hydrastine	0.1	−1.58	−1.46
Apomorphine	0.1	−1.52	......
Diamorphine	0.1	−1.73	−1.60
Morphine	0.1	−1.80	−1.67
Codeine	0.1	−1.80	−1.67
Ethylmorphine	0.1	−1.78	−1.65
Berberine	0.1	−1.75	−1.67
Papaverine	0.1	−1.62	−1.62
Quinine	0.1	......	−1.63
Quinidine	0.1	......	−1.63
Cinchonine	0.1	......	−1.63
Cinchonidine	0.1	......	−1.63
Strychnine	0.1	−1.64	−1.55
Brucine	0.1	−1.66	−1.58
Yohimbine	0.1	−1.75	−1.65
Ergometrine	0.1	−1.64	−1.60
Ergotoxine	0.1	−1.60	−1.52
Physostigmine	0.1	−1.50	−1.50
Nicotine	0.1	−1.72	−1.68
Sparteine	0.1	−1.70	−1.64
Atropine	0.5	......	−1.80
Hyoscyamine	0.5	......	−1.80
Homatropine	0.5	......	−1.80
Cocaine	0.5	......	−1.75
Hyoscine	1.0	......	−1.68
Pilocarpine	2.0	......	−1.80

It has the disadvantage that it lacks specificity and responds to breakdown products of penicillin and therefore cannot be used for fermentation liquors. The accuracy is similar to that of the plate method of biological assay.[42]

Penicillin may also be determined by studying its action on the oxygen consumption of various bacteria (see p. 555). It is possible to construct an activity *vs.* rate of oxygen consumption curve, which can be used to determine penicillin with an accuracy of about 5 per cent.[43]

3,5-Diphenyl-4-isoxazole (II). This compound produces no reduction waves in alkali or in acetic acid-potassium acetate buffers in 95 per cent

[42] E. L. Smith, *Analyst*, **73**, 197, 214 (1948).

[43] A. Trifonov, *Collection Czechoslov. Chem. Communs.*, **13**, 23 (1948).

ethanol in the range from pH 1.35–6.0. Waves are obtained in aqueous acid buffers with a half-wave potential of −1.24 v. (S.C.E.) at pH 3.55 and −1.05 v. (S.C.E.) in 0.1 *M* hydrochloric acid.[44]

Purine Derivatives. Adenine (III), adenosine, adenylic acid, and an adenine-cytosine dinucleotide are reducible at the dropping mercury elec-

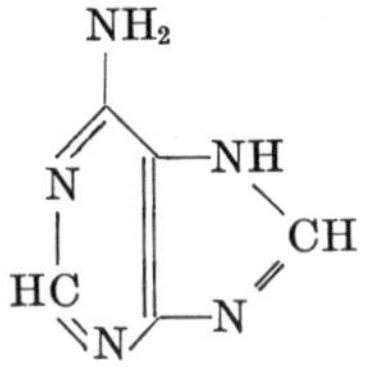

(III)

trode in dilute perchloric acid solution.[45] Guanine, guanosine, guanylic acid, cytidine, cytidylic acid, uracil, and 5-methyluracil are not reducible.

The half-wave potentials vary with increasing concentration and pH. Thus, adenine at pH 1.30 gives a half-wave potential which varies from −1.046 to −1.073 v. (S.C.E.) over a concentration range of 0.50×10^{-4} *M* to 1.96×10^{-4} *M*, and which shifts to −1.129 v. (S.C.E.) at pH 2.24. The wave height corresponds to a 2-electron change. The actual electrode process is not known.

The method can be used to determine adenine in the presence of the other purines with an accuracy of ±2 per cent and has been used on hydrolyzates of ribose and desoxyribose nucleic acids. Caffeine, theobromine, and freshly prepared uric acid solutions are not reducible at the dropping mercury electrode. Uric acid becomes reducible after prolonged exposure to air.[46]

Pterins. The results obtained with a number of pterins and compounds related to vitamin B_c, or pteroylglutamic acid, are listed in Table XLV-12.

The similarity of the behavior of the pterins to that of pteroylglutamic acid is good evidence for the structure of the latter.

The polarographic method has been used for the quantitative estimation of vitamin B_c or pteroylglutamic acid or folic acid.[49] In a buffer of pH 9

---

[44] E. D. Hartnell and C. E. Bricker, *J. Am. Chem. Soc.*, **70,** 3385 (1948).

[45] J. C. Heath, *Nature*, **158,** 23 (1946).

[46] J. Pech, *Collection Czechoslov. Chem. Communs.*, **6,** 126 (1934).

[47] K. J. Carpenter and E. Kodicek, *Biochem. J.*, **43,** ii (1948).

[48] E. L. Rickes, N. R. Trenner, J. B. Conn, and J. C. Keresztesy, *J. Am. Chem. Soc.*, **69,** 2751 (1947).

[49] J. B. Duncan and J. E. Christian, *J. Am. Pharm. Assoc.*, **37,** 507 (1948).

containing boric acid, potassium chloride, and sodium hydroxide the diffusion current is proportional to the concentration in the range from 10^{-3} to 10^{-4} *M*. The half-wave potential in this buffer is −0.87 v. (S.C.E.) and does not change with concentration.

A 1 per cent tetramethylammonium hydroxide solution buffered to pH 9 to 9.5 with ammonium chloride as supporting electrolyte and cadmium

TABLE XLV-12

POLAROGRAPHIC BEHAVIOR OF PTERINS (S.C.E.)

Compound	pH 6.1	pH 8.5	pH 9.12		pH 11.3[a]
	$-E_{1/2}$	$-E_{1/2}$	$-E_{1/2}$	i_d/C	$-E_{1/2}$
Xanthopterin (I)	—	0.8	0.8	27.2	1.12
Leucopterin (II)	—	—	—	—	1.63
Rhizopterin (III)	—	—	0.84	14.2	—
Aporhizopterin (IV)	—	—	0.81	15.1	—
Pteroylglutamic acid (PGA) (V)	0.66	0.8	0.79	8.64	1.0
	1.09	—	—	—	—
	1.3	—	—	—	—
Irradiated PGA (VI)	0.2	0.35	—	—	0.5
	0.47	0.66	—	—	0.82
	0.6	0.79	—	—	1.25
	1.12	1.34	—	—	1.52

[a] A 0.1 *N* K_2HPO_4-NaOH buffer was used for pH 11.3[47] and a 0.1 *M* $Li_2B_4O_7$ buffer for pH 9.12.[48]

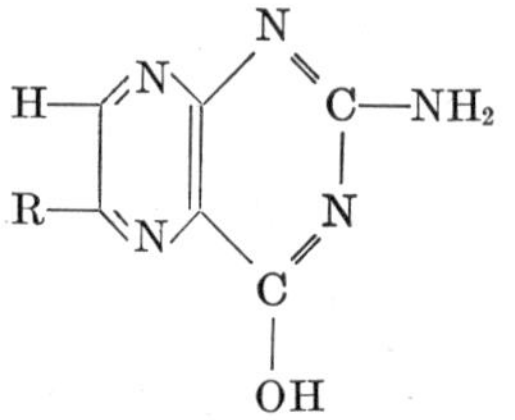

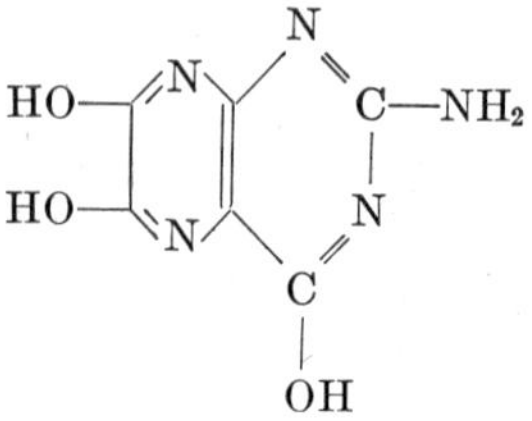

(II)

(I) R = OH

(III) R = 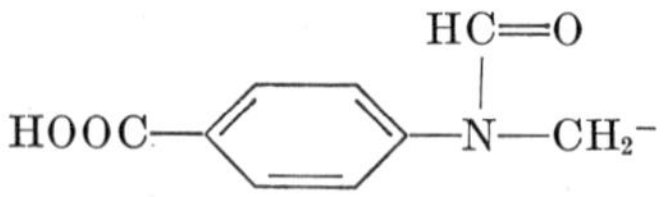

(IV) R = 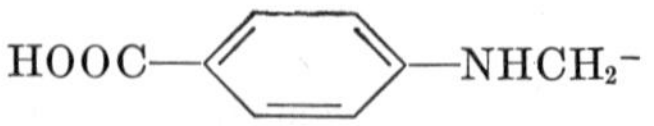

(V) R = 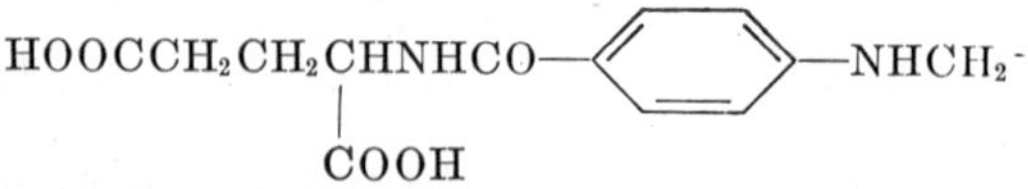

(VI) R = —CHO (?)

chloride as an internal standard permits the quantitative estimation of folic acid with a precision of ±2 per cent.[50] Tablet constituents such as other members of the vitamin B complex do not interfere in the determination. Ferrous sulfate if present interferes and increases the height of the wave.

Saccharin (I). This compound yields well-defined waves in acid, neutral, and alkaline media.[51] In an acid medium the reduction starts at about −1.0 v. and in ammonium chloride at −1.7 v. (N.C.E.). The polarographic method is very suitable for the quantitative determination of saccharin in an electrolyte which is about 0.05 *N* in hydrochloric acid and 0.05 *N* in potassium chloride.

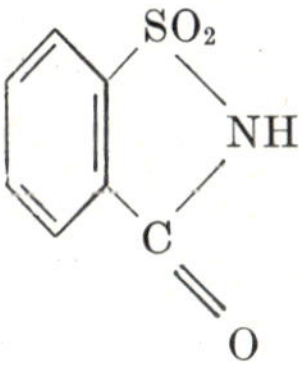

(I)

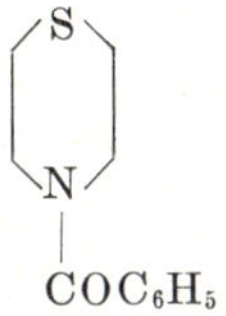

(II)

Benzthiomorpholide (II). This compound is reducible polarographically

TABLE XLV-13

POLAROGRAPHIC BEHAVIOR OF BENZTHIOMORPHOLIDE

Buffer	pH	$-E_{1/2}$ (S.C.E.)	i_d, microamp.
Acetate	4.0	1.16	3.92
Acetate	5.0	1.16	3.68
Acetate	7.0	1.55	3.92
Acetate	8.0	1.57	4.00
Phosphate	8.0	1.64	3.60
Borate	8.0	1.71	0.04
Ammonia	9.0	1.56	3.84
Ammonia	10.0	1.62	3.76
Tartrate	11.0	1.59	2.56[a]
Phosphate	11.0	1.69	2.80
Borate	11.0	1.70	3.12
Phosphate	12.0	1.64	2.48
Potassium chloride	—	1.60	3.28

[a] Wave is poorly defined.

[50] W. J. Mader and H. A. Frediani, *Anal. Chem.*, **20**, 1199 (1948).
[51] J. Pech, *Collection Czechoslov. Chem. Communs.*, **6**, 1269 (1934).

in acid and alkaline media.[52] The results obtained for a 0.48×10^{-3} *M* solution in 2 per cent ethanol are listed in Table XLV-13. The diffusion current is not proportional to the concentration and corresponds approximately to a 4-electron change. The electrode process is difficult to interpret since the oxygen analog, benzmorpholide, is not reducible.

Pyrimidine Derivatives

Alloxan (I) is reported[53] to produce two reduction waves at +0.182 and −0.588 v., while dialuric acid (II), the reduction product, produces an oxidation wave at +0.182 v. Alloxantin (III) produces two waves at +0.182 v. and −0.588 v., and alloxanic acid (IV) one at −0.588 v.

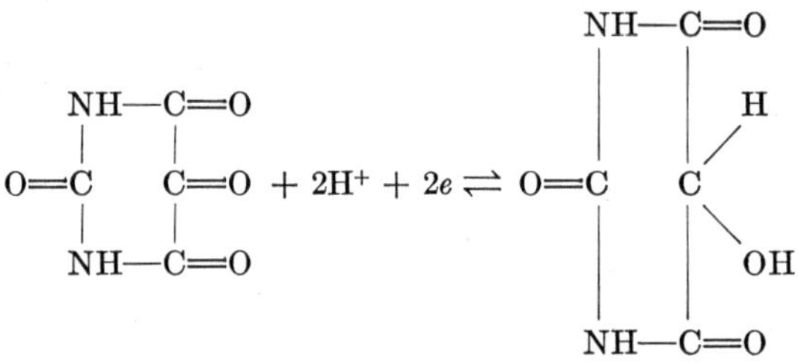

(I) (II)

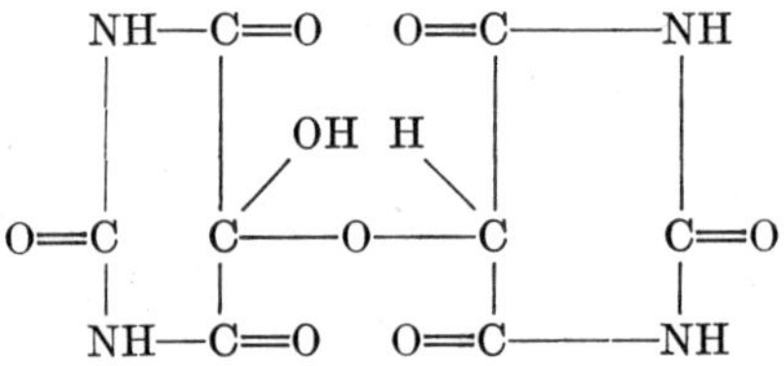

(III)

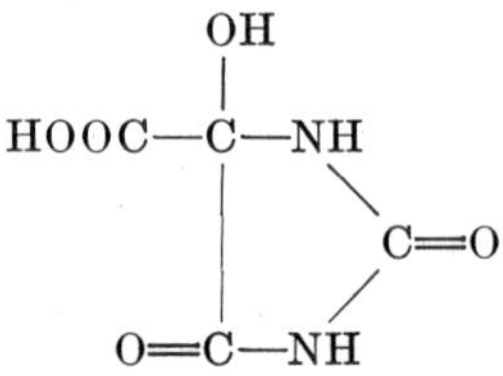

(IV)

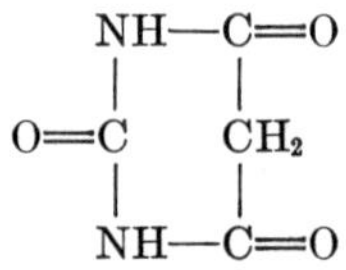

(V)

Barbituric acid (V) is not reducible in 0.1 *M* ammonium chloride, 0.2 *M* tetraethylammonium bromide, or buffers in the pH range from 3 to 10.[54]

The polarographic method is suitable for determining alloxan in blood and in a buffer of pH 5.5–6.2.[55]

---

[52] K. G. Stone, *J. Am. Chem. Soc.*, **69,** 1832 (1947).

[53] G. Sartori and A. Liberti, *Ricerca sci.*, **16,** 313 (1946); *Chem. Abstracts*, **43,** 7835 (1949).

[54] J. B. Duncan and J. E. Christian, *J. Am. Pharm. Assoc.*, **37,** 507 (1948).

[55] R. I. Veksler, *Biokhimiya*, **14,** 341 (1949); *Chem. Abstracts*, **43,** 8979 (1949).

Picrolonic Acid (VI). This compound in unbuffered solutions of potassium or lithium chloride yields two waves when the maximum is suppressed with camphor (gelatin and thymol are less suitable under these conditions).[56]

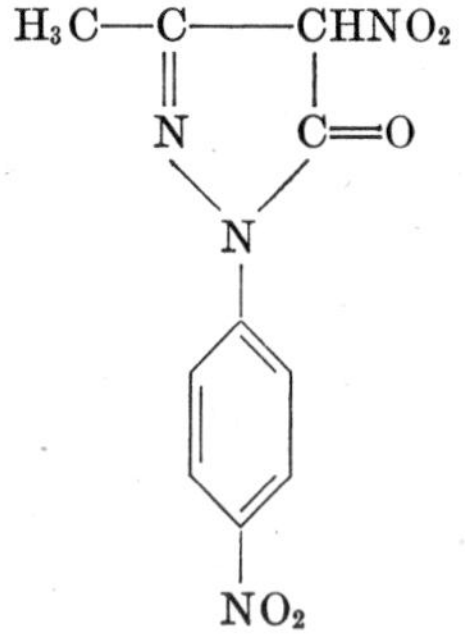

(VI)

The first wave yields a diffusion current which is constant at an applied e.m.f. between −1.0 and −1.3 v., the second one at an e.m.f. more negative than −1.4 v.

In acid solutions, gelatin is a better maximum suppressor than camphor or thymol. In some acetate buffers no maximum suppressor is required. In acetate buffers of pH 3.6 to 3.8, and in the presence of gelatin, two waves were observed; the first one is ill defined especially at higher concentrations of picrolonic acid. The height of the second wave is proportional to the concentration. In the absence of gelatin there is a constant diffusion current. (The total wave is found at an applied e.m.f. between −1.15 and −1.3 v.)

The polarographic determination of picrolonic acid may be used to determine calcium.

Determination of Calcium. Calcium can be determined in a concentration range from 0.001 to 0.01 *M* by precipitation as calcium picrolonate—$Ca(C_{10}H_7O_5N_3)_2 \cdot 2H_2O$—with an excess of standard picrolonic acid solution. The excess of reagent is determined polarographically without filtering.

Picrolonic Acid. A product recrystallized from 33 per cent acetic acid. A 0.01 *M* solution of the acid in water is prepared.

Procedure. Determine the approximate calcium content by noticing the time required for the first appearance of a precipitate of calcium picrolonate when the solution is mixed with the acetate buffer (see above) and picrolonic acid and by comparing with solutions of known calcium content.

Add to a known volume of the calcium solution a quantity of buffer solution (1 *M*

[56] G. Cohn and I. M. Kolthoff, *J. Biol. Chem.*, **147**, 705 (1943).

in acetic acid, 0.125 *M* in lithium acetate, and 1 *M* in lithium chloride) such that it is diluted about 10 times in the final mixture. Add such an amount of standard 0.01 *M* picrolonic acid solution to 5 to 10 ml. of the unknown that the concentration of picrolonic acid after complete precipitation becomes at least 0.001 *M*. The amount of picrolonic acid shall not exceed the 4-fold molar concentration of calcium. Keep the mixture overnight in an icebox when the calcium concentration is equal to or smaller than 5×10^{-3} *M*. When the calcium concentration is greater, keep it at room temperature (not above 20°). Without filtering determine the residual concentration of picrolonic acid polarographically at 20° C. at a cathode potential of −1.1 v. (*vs.* S.C.E.) or, when a mercury pool anode is used, at an applied e.m.f. of −1.2 v.

The method has an accuracy of 1 to 2 per cent in the determination of 0.001 to 0.01 *M* calcium solutions, also when the concentrations of alkali, magnesium sulfate, and phosphate ions are from 10 to 100 times larger than that of the calcium.[56]

2-Mercaptobenzthiazole (I). This compound and its disulfide are reported to give the following depolarization potentials:[57]

Compound	HCl	KCl	NH_3	NaOH
Mercaptabenzthiazole	−0.16	−0.37	−0.41	−0.48
Disulfide	−0.26	−0.44	−0.36	−0.44

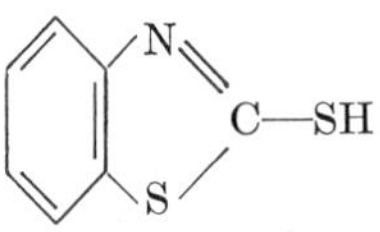

(I)

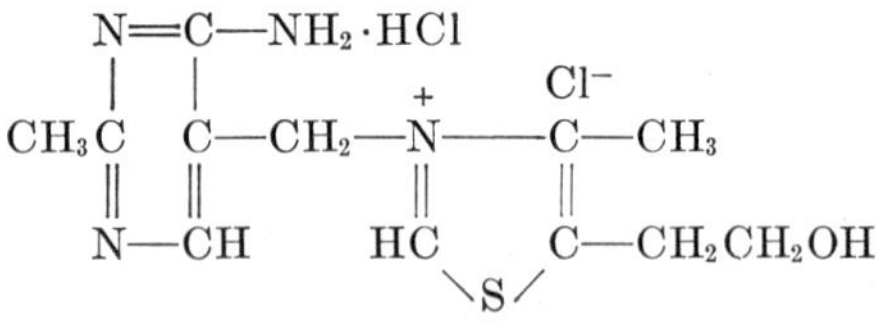

(II)

Vitamin B₁ (Thiamin) (II). The following polarographic behavior is reported[58] for a 10^{-4} *M* thiamin solution at 12°:

pH	$-E_{1/2}$	pH	$-E_{1/2}$
1.81	1.13	5.72	1.37
2.56	1.15	6.37	1.40
3.29	1.18	6.80	1.44
4.10	1.23	7.24	1.46
4.56	1.28	7.96	1.51
5.02	1.33	8.69	1.52

The half-wave potential varies with concentration.

[57] G. Proske, *Angew. Chem.*, **53**, 550 (1940).

Thiamin may be determined in the presence of riboflavin and nicotinic acid, using 0.1 *N* potassium chloride as an electrolyte. The half-wave potential in this solution is −1.25 v. (S.C.E.) and in a buffer of pH 7.2 it is −1.30 v. (S.C.E.).[59]

Phenazines. This system has been studied both by means of the standard classical procedure with a platinum electrode and at the dropping mercury electrode.[60] In exact studies of oxidation-reduction potentials the classical method is preferred.

Neutral Red (I). This compound, which is a reversible oxidation-re-

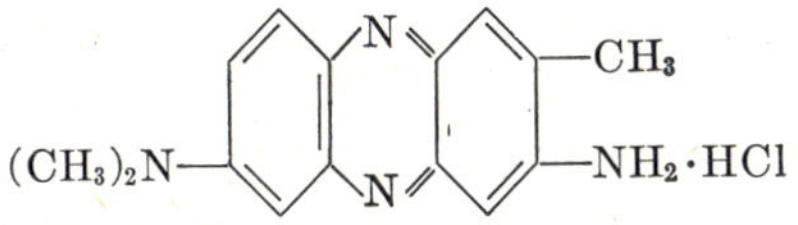

(I)

duction indicator[61] shows the following polarographic behavior at 25° C.:[62]

pH	0.1	1.1	2.0	3.0	4.0	5.0	6.0	7.0	8.0	9.0	10.0
$-E_{1/2}$[a]	0.080	0.165	0.25	0.32	0.40	0.49	0.55	0.61	0.68	0.73	0.76

[a] *vs.* N.C.E.

From the change in the half-wave potential with pH the following values of the ionization constants were calculated: K_b (neutral red) 3.2×10^{-7}, K_b (leuco form) 1.3×10^{-9}, K_a (leuco form) 3.2×10^{-9}.

Rosindulin G (I). This dye yields at least three waves at the dropping mercury electrode.[60]

The wave which occurs at the most negative potential gives a half-wave potential which is in agreement with the E^0, measured by the classical method.[63] The analysis of this wave at a pH of 2.2 indicates the formation of a semiquinone during the reduction. The first anomalous wave observed is an absorption current. The reason for the second wave is not clear.[64]

1-Hydroxyphenazine (II). This compound produces two waves of equal

[58] G. Sartori and C. Cattaneo, *Gazz. chim. ital.*, **74,** 166 (1944).

[59] J. J. Lingane and O. L. Davis, *J. Biol. Chem.*, **137,** 567 (1941).

[60] O. H. Müller and J. P. Baumberger, *Trans. Electrochem. Soc.*, **71,** 181 (1937).

[61] W. M. Clark, *et al.*, "Studies on Oxidation and Reduction," I-X, *Hygienic Laboratory Bulletin*, **1928** (No. 151).

[62] I. Tachi, *Mem. Coll. Agr. Kyoto, Imp. Univ.*, **42,** 27 (1938).

[63] L. Michaelis, *J. Biol. Chem.*, **91,** 369 (1931); **92,** 211 (1931).

[64] O. H. Müller, *Trans. Electrochem. Soc.*, **87,** 441 (1945).

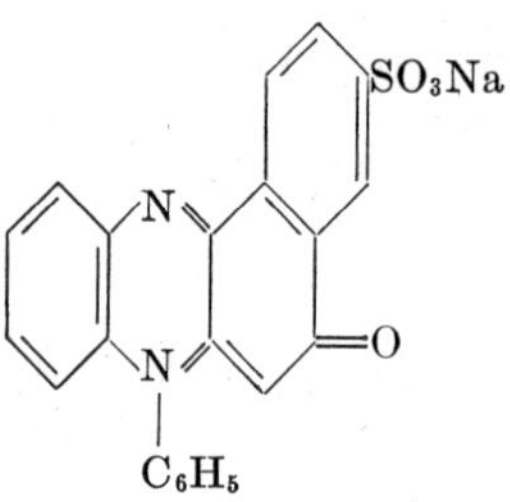

(I)

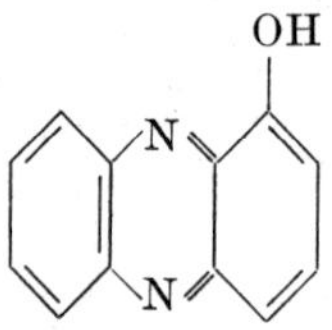

(II)

height at a pH of 2.13.[65] The first wave corresponds to the reduction of the phenazine to a semiquinone and the second to the reduction of the semiquinone to the final reduction product.[66] At higher pH values the two waves overlap and the semiquinone formation is revealed only by a more involved mathematical analysis.[67] The half-wave potentials are in agreement with the values obtained potentiometrically. Peculiarities appear in the current–voltage curves due to an adsorption phenomenon[68] (see Chap. XIV).

Pyocyanine (III). This compound gives half-wave potentials which are in agreement with the potentiometrically determined values.[69] Abnormal waves are interpreted on the basis of adsorption currents (see Chap. XIV).[70]

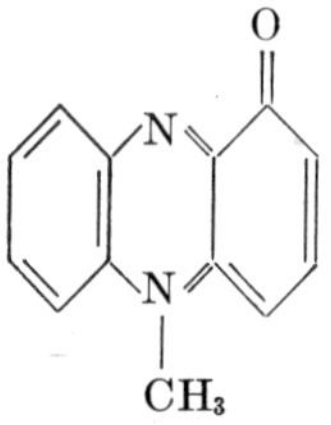

(III)

Riboflavin (Lactoflavin, Vitamin B_2). Riboflavin (IV) produces reduction waves at the dropping mercury electrode which are preceded by a small anomalous reduction wave due to the adsorption of the reduced form on the drop (see Chap. XIV).

[65] O. H. Müller, *Cold Spring Harbor Symp. Quant. Biol.*, **7,** 59 (1939).
[66] L. Michaelis, *Chem. Revs.*, **16,** 243 (1935).
[67] L. Michaelis and M. P. Schubert, *Chem. Revs.*, **22,** 437 (1938).
[68] O. H. Müller, *J. Biol. Chem.*, **145,** 425 (1942).
[69] C. Cattaneo and G. Sartori, *Gazz. chim. ital.*, **72,** 38 (1942).
[70] M. Voriskova, *Collection Czechoslov. Chem. Communs.*, **12,** 607 (1947).

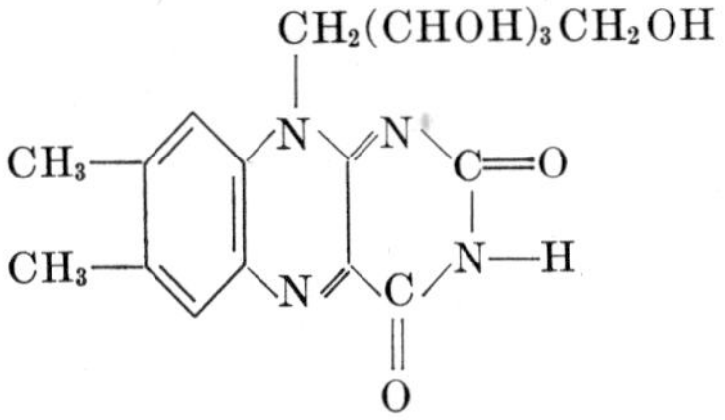

(IV)

The half-wave potential of the main wave is in agreement with the potentiometrically determined potential while that of the anomalous wave occurs at a more positive potential.[71]

The polarographic method is suitable for the quantitative estimation of riboflavin since the diffusion current is proportional to the concentration in the range from 10^{-5} to 10^{-4} M. The method is successful in determining the riboflavin content of yeast[72] and has been used to study the photolysis of riboflavin.[73]

Methylene Blue (V). This compound is reversibly reduced at the dropping mercury electrode.[74, 75] The polarograms are quite involved due to the occurrence of an adsorption current (Chap. XIV).

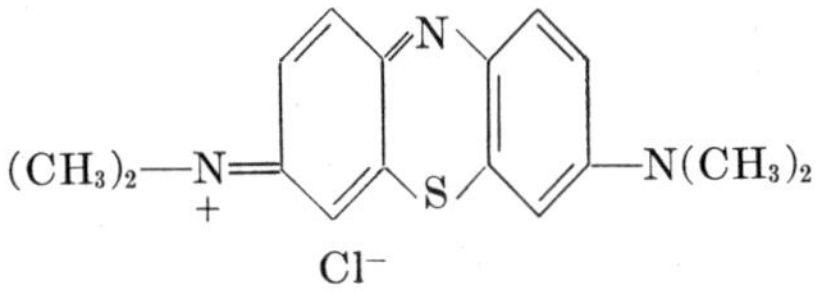

(V)

From a practical point of view it is of interest to note that the total diffusion current is not linearly proportional to the concentration. The value of k in $i_d = kC$ decreases with increasing concentration. Undoubtedly this must be attributed to association of methylene blue in more concentrated solutions.[76] The following values are reported for the diffusion

---

[71] R. Brdicka and E. Knobloch, *Z. Electrochem.*, **47,** 721 (1941). R. Brdicka, *ibid.*, **48,** 686 (1942).

[72] J. J. Lingane and O. L. Davis, *J. Biol. Chem.*, **137,** 567 (1941).

[73] R. Brdicka, *Collection Czechoslov. Chem. Communs.*, **14,** 130 (1949).

[74] M. von Stackelberg, *Z. Electrochem.*, **45,** 466 (1939).

[75] R. Brdicka, *Z. Electrochem.*, **48,** 278 (1942); *Collection Czechoslov. Chem. Communs.*, **12,** 522 (1947).

[76] See, however, O. H. Müller, *Trans. Electrochem. Soc.*, **87,** 441 (1945).

coefficient at varying molar concentrations, C:

$C \times 10^4$	0.61	1.18	1.70	2.19	2.89	3.67	4.50	7.95
$D \times 10^6$	6.2	5.2	4.9	4.6	4.4	4.2	4.1	3.4

Chlorophylls. Chlorophylls are reported to produce reduction waves in tetramethylammonium bromide solutions.[77] The actual electrode process involved is not known.

[77] P. V. Rysselberghe, J. M. McGee, A. H. Gropp, and R. W. Lane, *J. Am. Chem. Soc.*, **69**, 809 (1947).

Part Five

BIOLOGICAL APPLICATIONS OF POLAROGRAPHY

CHAPTER XLVI

Catalytic Waves of Cysteine and Other Mercaptoacids, of Their Disulfides and of Proteins. Applications to the Serological Detection of Cancer

1. INTRODUCTION

Heyrovsky and Babicka[1] discovered a polarographic effect of proteins which consists of a characteristic wave at a potential of -1.6 v. on the current–voltage curve of a solution of 0.1 *N* ammonium chloride and 0.1 *N* ammonia containing traces of proteins. This wave occurs at a potential 0.2 v. more positive than the wave of ammonium ion. A general review on limiting currents of catalyzed hydrogen waves is given in Chapter XIII. Here we shall deal with the special case of catalytic waves which are produced by compounds containing sulfhydryl or disulfide groups.

Heyrovsky and Babicka showed that the wave is not due to reduction of the protein but to the evolution of hydrogen catalyzed by the presence of the protein at the cathode interface. Brdicka[2] found that these catalytic waves also occur in certain buffer solutions in the absence of ammonia. The height of the catalytic wave depends not only on the pH but also on the kind of buffer used and the concentrations of the buffer constituents. Working with the same type of buffer, and coming from the alkaline side, the height of the wave increases with decreasing pH down to a pH of 8. Buffers of different types but with the same pH do not give catalytic waves of the same height. Increasing the total buffer concentration (buffer capacity)—but keeping the pH constant—increases the height of the wave. Undoubtedly we are dealing here with an example of general acid-base catalysis in the Brönsted sense. Brdicka attributes the catalytic wave to a direct reduction of the acid sulfhydryl group (—SH) to hydrogen:

$$\text{—SH} + e \rightarrow \text{—S}^- + \text{H}$$

The —S^- ions react with any acid, A, present to reform —SH:

$$\text{—S}^- + \text{A} \rightarrow \text{—SH} + \text{B}$$

[1] J. Heyrovsky and J. Babicka, *Collection Czechoslov. Chem. Communs.*, **2,** 270 (1930).

[2] R. Brdicka, *Collection Czechoslov. Chem. Communs.*, **8,** 366 (1936).

where B is the conjugate base of A. The net reaction is the discharge of hydrogen from the acid A. One would expect that the catalytic wave would increase with increasing acid strength A, and—when working with the same buffer—with increasing concentration of A. This, indeed, has been observed by Brdicka.[2]

Analytical use of this catalytic wave has been made in the polarographic determination of traces of proteins. Brdicka,[3] in continuing the study of the protein effect, found that it was quite different in character when the ammoniacal solution contained a salt of divalent or trivalent cobalt. In such a medium sulfur-containing proteins cause a "double catalytic wave" as shown in Fig. XLVI-1. Curve *2* shows the catalytic protein wave in an ammonia-ammonium chloride solution (Heyrovsky and Babicka). Curve *3* represents the current–voltage curve of 3×10^{-3} N hexammino-cobaltic chloride in the ammonia buffer. The first wave is due to the reduction of Co^{+++} to Co^{++}, and the second wave, yielding a pronounced maximum, to the further reduction of Co^{++} to cobalt amalgam. The two catalytic protein waves are shown in curve *4*. The trace of protein suppresses the cobalt maximum completely. Brdicka found that the height of these protein waves increases with the protein concentration. No linear relation was found between the height of the catalytic waves and the protein concentration. A limiting height of the protein wave was obtained at a concentration of 0.8 to 1 g. of protein per liter. The concentration at which the limiting value is found varies with the kind of protein used. For example, Tropp *et al.*[4] found the limiting concentration in the case of globulin at 3.2 per cent, and of fibrogen at 0.42 per cent. These authors also found that the shapes of the waves with different proteins were not identical. In general, the height and shape of the catalytic protein waves are greatly dependent on the composition of the medium and the protein concentration. Thus, the double wave produced by a protein solution at a low ammonium hydroxide concentration may be transformed into a single one when a large amount of ammonia is present. In increasing the ammonium chloride concentration from 4.8×10^{-3} N to 3.5×10^{-2} N Brdicka found a linear relation between the salt concentration and the height of the wave.[2]

When a divalent instead of a trivalent cobalt (hexammine) salt was used Brdicka also noticed two "protein waves" which again reached a limiting value at a certain protein concentration. In general, the waves found with a trivalent cobalt solution were larger than with a divalent cobalt

[3] R. Brdicka, *Collection Czechoslov. Chem. Communs.*, **5**, 238 (1933).
[4] C. Tropp, L. Jühling, and F. Geiger, *Z. physiol. Chem.*, **262**, 225 (1939).

solution at the same total cobalt concentration. By making sets of calibration curves the catalytic waves can be employed for the quantitative determination of sulfur-containing proteins.

Brdicka[5] also found that cysteine and cystine produce a pronounced single catalytic wave in an ammonia buffer containing divalent cobalt or nickel, but not in a buffer containing trivalent cobalt. This catalytic wave occurs at a more negative potential than corresponds to the reduction of cystine to cysteine, and, therefore, must be caused by the reduced form, cysteine. The waves found with cystine were identical with those ob-

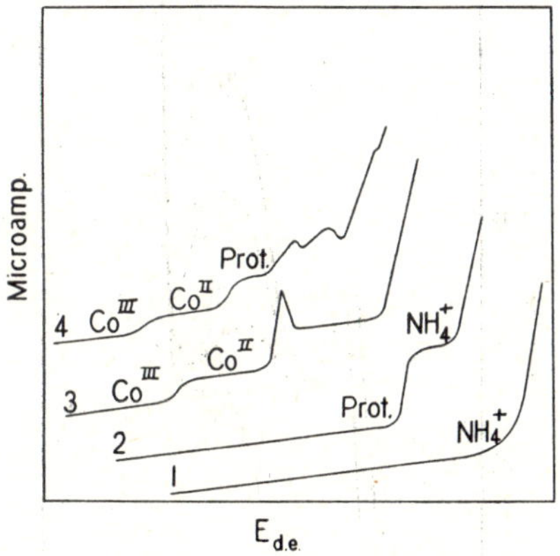

Fig. XLVI-1. Catalytic protein waves in 0.1 *N* ammonium chloride solution: (*1*) residual current of the solution above; (*2*) in the presence of a trace of protein; (*3*) wave of an ammoniacal cobalt solution; and (*4*) in the presence of a trace of protein.

tained with solutions of cysteine having double the concentration (1 molecule of cystine yields 2 molecules of cysteine on reduction at the electrode). The waves obtained with cystine were 500 times as great as those corresponding to the reduction of cystine to cysteine, indicating clearly the catalytic nature of the wave. Brdicka[5,6] attributes the catalytic cysteine wave to the fact that cysteine forms complex compounds with divalent cobalt. In these complexes there is a coordination bond between the cobalt (or nickel) and the sulfhydryl group. This bond weakens (activates) the bond between the sulfur and hydrogen in the sulfhydryl group and the deposition of hydrogen at the dropping mercury electrode is facilitated.

---

[5] R. Brdicka, *Collection Czechoslov. Chem. Communs.*, **4,** 148 (1933).
[6] R. Brdicka, *Biochem. Z.*, **272,** 104 (1934).

2. CATALYTIC WAVES OF CYSTEINE AND OTHER MERCAPTOACIDS OF SMALL MOLECULAR WEIGHT

Analytical Applications. Since the waves caused by cysteine (and cystine) in an ammoniacal cobalt solution are catalytic, minute traces of these compounds can be determined polarographically. The best conditions are obtained with a medium which is 0.001 to 0.002 *N* in cobaltous chloride, 0.1 *N* in ammonium chloride, and 0.1 *N* in ammonia. The cobalt solution should be added to the air-free ammonia buffer just before the experiments

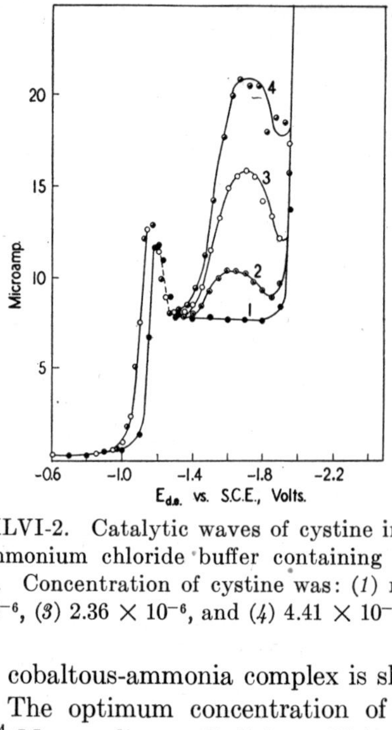

Fig. XLVI-2. Catalytic waves of cystine in an ammonia-ammonium chloride buffer containing a cobaltous salt. Concentration of cystine was: (*1*) none, (*2*) 1.00×10^{-6}, (*3*) 2.36×10^{-6}, and (*4*) 4.41×10^{-6} *M*.

are run, because the cobaltous-ammonia complex is slowly air-oxidized to the cobaltic state. The optimum concentration of cystine is between 5×10^{-6} *M* and 10^{-4} *M* according to Brdicka. This corresponds to 0.012 to 0.24 mg. of cystine in 10 ml. of solution. As the determination can easily be made with 1 ml. of solution it is possible to determine quantities of cysteine of the order of 1 to 2 micrograms. We found better limits of concentration between 10^{-6} and 2×10^{-5} *M* cystine. A set of catalytic cystine waves, determined by the authors with a manual apparatus, is shown in Fig. XLVI-2. Curve *1* was obtained with the cobaltous test solution without any cystine present. The other curves were obtained after addition of the amounts of cystine indicated. It will be noted that the deposition potential of the cobalt ion was shifted about 50 mv. to a more positive value when cystine was present. The catalytic waves were measured at a potential of -1.7 v.

The height of the catalytic waves was found to increase with increasing cystine concentration. The relation between the height of the catalytic wave and the concentration of cystine is shown in Fig. XLVI-3. With concentrations of cystine up to about 4×10^{-6} M the height of the wave was directly proportional to the concentration of cystine, but with higher concentrations the catalytic wave became proportionately less than at lower concentrations. From the curve in Fig. XLVI-3 it appears that the

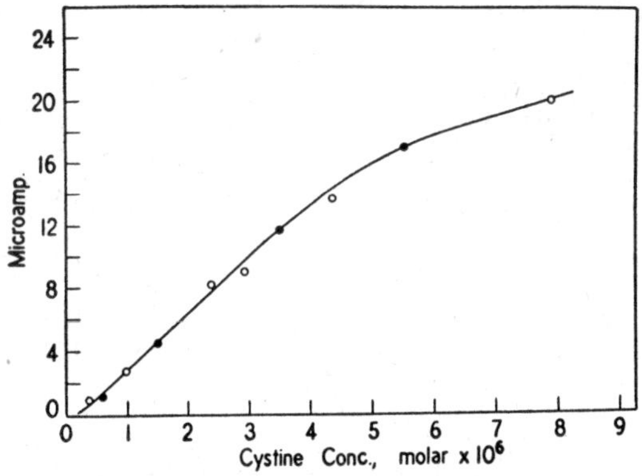

Fig. XLVI-3. Relation between the height of the catalytic cystine wave and the concentration of cystine in 0.001 M $CoCl_2$ + 0.1 N NH_4OH + 0.1 N NH_4Cl; ○ no gelatin present; ● 0.005 per cent gelatin present.

catalytic wave approaches a maximal value, independent of further additions of cystine, at a concentration of the latter somewhat greater than 10^{-5} M. From Fig. XLVI-2 it is apparent that a distinct maximum occurs on the reduction wave of cobalt. The authors found that this maximum is completely eliminated by the addition of a trace (0.005 per cent) of gelatin, which does not affect the height of the catalytic waves.

In Chapter XV, which deals with kinetic currents, it was noted that typical kinetic currents are not diffusion controlled. The height of such waves is independent of the height of the mercury in the reservoir. In agreement with this it is found[7] that the catalytic cysteine and protein waves are, within wide limits, independent of the height of the mercury

[7] O. H. Müller and J. S. Davis, Jr., *J. Biol. Chem.*, **159**, 667 (1945). W. Stricks, University of Minnesota, Unpublished results.

level. On the other hand, the cobalt wave is diffusion controlled and its height varies with the square root of the height of the mercury level. Therefore, for a given capillary and a given medium, the ratio of the catalytic wave to the cobalt wave varies with the height of the mercury in the reservoir, but the height of the catalytic wave is not affected (except for the $t^{1/6}$ effect on the diffusion current of cobalt with change of potential).

Since traces of cystine and cysteine can be determined polarographicallly, Brdicka[8] employed this method for the microdetermination of these compounds in the hydrolyzates of proteins and in the study of protein decomposition. After hydrolysis for 8 hours with boiling 5 *N* hydrochloric acid (reflux), he determined cystine in sheep wool, human hair, serum, blood albumin, beef, sheep brains, rye flour, egg albumin, and pituitary and thyroid hormones. Gelatin and pure silk, which do not contain sulfhydryl groups, did not show catalytic waves during or after the hydrolysis. Brdicka found it possible to determine cystine in as little as 0.5 mg. of human hair.

The catalytic waves are not specific for cystine and cysteine, but are also given by other compounds containing sulfhydryl or disulfide groups. For example, Brdicka[9] found that cysteylglycine and thioglycolic acid are also capable of giving the catalytic waves in the presence of divalent cobalt. Glutathione produced small waves, but Brdicka concluded that these were probably due to some cysteylglycine formed from the glutathione by hydrolysis, and not to the glutathione itself. The behavior of β,β'-dichlorosulfide (mustard gas) is interesting[10] although not completely understood. This compound does not yield a catalytic wave in the ammoniacal cobalt solution but, after it has been boiled in a concentrated ammonia solution for one-half hour, the solution produces catalytic waves with ammoniacal cobaltous and *cobaltic* salts. As stated before, cysteine does not yield a wave with an ammoniacal cobaltic solution, but proteins do. The decomposition products obtained by boiling the mustard gas with concentrated ammonia have not yet been isolated. The procedure can be used for the detection of mustard gas.

It was found by Stern and Beach[11] that homocystine and homocysteine give catalytic waves identical with those of cystine. Smith and Rodden[12]

[8] R. Brdicka, *Collection Czechoslov. Chem. Communs.*, **5**, 238 (1933).
[9] R. Brdicka, *Collection Czechoslov. Chem. Communs.*, **5**, 148 (1933).
[10] R. Brdicka, *Collection Czechoslov Chem. Communs.*, **9**, 76 (1937).
[11] A. Stern and E. F. Beach, *Proc. Soc. Exptl. Biol. Med.*, **43**, 104 (1940).
[12] E. R. Smith and C. J. Rodden, *J. Research Natl. Bur. Standards*, **22**, 669 (1939).

report that the following sulfur-containing compounds: methionine CH_3—S—CH_2—CH_2—CH(NH_2)—COOH, djenkolic acid CH_2(S—CH_2—CH-(NH_2)—COOH$)_2$, and benzylcysteine, $C_6H_6CH_2$—S—CH_2—CH(NH$)_2$—COOH, do not yield the catalytic cystine waves.

It is interesting that some sulfur-free amino acids give catalytic waves in an ammoniacal cobalt buffer.

In the application of the catalytic wave to the determination of cystine in biological materials it is of importance to know whether the height of the wave will be affected by other substances apt to be present. Sladek and Lipschütz[13] investigated, in this connection, the behavior of glucose, acetic acid, urea, cholesterol, lecithin, glycine, α- and β-alanine, tyrosine, lysine, arginine, β-phenyl-α-alanine, β-phenyl-β-alanine, tryptophane, and histidine. Brdicka[9] had already shown that glycine, asparagine, creatine, creatinine, arginine, leucine, and tyrosine did not produce catalytic waves. Sladek and Lipschütz found that tryptophane, histidine, arginine, β-phenyl-α-alanine, and β-phenyl-β-alanine exert a suppressive effect on the cysteine waves. The behavior of β-phenyl-β-alanine is striking; it gives a pronounced catalytic wave with a maximum at a potential 0.4 v. more negative than the cobalt maximum, and 0.2 v. more negative than the cysteine maximum. The same authors also found that the above amino acids, with the exception of β-phenyl-β-alanine, greatly affect the cobalt maximum in an unbuffered solution which was 0.1 *N* in potassium chloride and 0.002 *N* in cobalt chloride.

Roberts[14] reported that histidine produces a complex set of catalytic waves in an ammoniacal cobalt buffer. The first wave starts at −1.25 v.; its height was found to be proportional to the histidine concentration. A second wave at −1.55 v. is characterized by an acute maximum.

Smith and Rodden[12] noticed that the cysteine waves are not affected by methionine or benzylcysteine at concentrations up to more than twice that of the cysteine. Djenkolic acid, on the other hand, reduces the height of the cysteine maximum (and also that of the cobalt maximum) without changing the form or location of the wave. Stern, Beach, and Macy[15] also found that methionine does not affect the catalytic cysteine wave, unless its concentration is extremely high with regard to cystine (cysteine). Since various amino acids affect the cystine wave, Stern *et al.* made calibration curves for each protein hydrolyzate which they examined. These calibration curves were obtained by adding known small amounts of a standard

[13] J. Sladek and M. Lipschütz, *Collection Czechoslov. Chem. Communs.*, **6,** 487 (1934).
[14] E. R. Roberts, *Trans. Faraday Soc.*, **37,** 353 (1941).
[15] A. Stern, E. F. Beach, and I. G. Macy, *J. Biol. Chem.*, **130,** 733 (1939).

cystine solution to the hydrolyzate and determining the relative changes in the height of the polarographic waves. By extrapolation they determined the original amount of cystine present in the hydrolyzate. An accuracy of ±5 per cent was claimed for solutions containing only 12 to 24 micrograms of cystine per milliliter. Their polarographic results on the cystine content of hydrolyzates from casein, edestin, beef globin, sheep globin, horse globin, and human globin compared favorably with those obtained by a gravimetric technique. It should be added that homocysteine behaves in the same way as cysteine and cannot be differentiated from the latter polarographically.

A comparison of the colorimetric and polarographic methods for determining the cystine content of insulin has been reported by Sullivan, Hess, and Smith.[16] They used two crystalline samples of insulin which were subjected to hydrolysis. In the determination of the calibration curves they used synthetic mixtures of the amino acids present in insulin with different amounts of cystine and likewise subjected them to acid hydrolysis. They found that between the colorimetric and polarographic methods there was a relative difference of 10 per cent, or an absolute difference of 1 per cent in the cystine content of the insulin.

Salac[17] applied the polarographic method to the determination of cystine in beer.

In conclusion it may be mentioned that, although cysteine and cystine behave identically regarding the catalytic wave, it is possible to differentiate between the two substances. Monoiodoacetate ion in an alkaline medium reacts with the sulfhydryl group of simple sulfur compounds:

$$—SH + ICH_2COO^- \rightarrow —SCH_2COO^- + HI$$

$$(\text{or } —S^- + ICH_2COO^- \rightarrow —SCH_2COO^- + I^-)$$

Ten minutes after addition of monoiodoacetate ion to an alkaline cysteine solution Brdicka[18] found that the cysteine did not yield the catalytic wave any more, while cystine remained unaffected.

It is known that several drugs, such as quinone derivatives, penicillin, and other mold products, are inactivated by cysteine. Page and Waller[19] investigated the effect of some of these drugs on the catalytic cysteine wave

[16] M. X. Sullivan, W. C. Hess, and E. R. Smith, *J. Biol. Chem.*, **130,** 745 (1939).

[17] V. Salac, *Kvas*, **64,** 383 (1936); *Chem. Obzor*, **13,** 155 (1939); *Chem. Abstracts*, **33,** 6520 (1939).

[18] R. Brdicka, *Acta Intern. Vereinigung Krebsbekämpfung*, **3,** 13 (1938).

[19] J. E. Page and J. G. Waller, *Nature*, **157,** 838 (1946).

in ammoniacal cobaltous buffer. While equimolar neutral solutions of cysteine and certain quinone derivatives gave a fairly normal catalytic cysteine wave even after 24 hours, a strong suppressive effect on the cysteine wave was found when a 10-fold excess of the drugs was allowed to act on the cysteine before electrolysis. Of the mold products tested, patultin (clavacin) reacted more readily than crystalline sodium penicillin II (penicillin G); citrimin was least active.

From the above it is evident that, in complex media, several organic substances interfere with the catalytic cysteine wave. Stricks and Kolthoff[20] found that larger amounts of salts and traces of some heavy metals, especially silver and copper, reduced the catalytic wave. Nitrate in 0.1 *M* concentration eliminated the catalytic wave. Addition of 0.1 *M* ammonium sulfate to the standard medium reduced the wave 10 per cent, while the combination of 0.1 *M* ammonium sulfate and 0.1 *M* potassium chloride reduced it more than 20 per cent. The presence of 5×10^{-5} *M* silver in an equimolar cysteine solution reduced the wave by 13 per cent. Undoubtedly, the reason is that silver and copper form very stable complexes with cysteine.

3. ANALYTICAL APPLICATIONS OF CATALYTIC PROTEIN WAVES

For quantitative purposes use is made of the double catalytic waves obtained in an ammoniacal solution containing di- or trivalent cobalt. Taussig[21] and co-workers found the polarographic method very convenient in the determination of proteins in small volumes of cerebrospinal fluid. Insulin yields typical double waves just like other sulfur-containing proteins. Mulli and Werner[22] claimed that the biological activity of insulin samples can be estimated polarographically. However, Kanner and Reed[23] showed that, although the amount of protein present in an insulin sample can be determined polarographically, this amount of protein is not a measure of the physiological insulin activity. The activity, for example, may be destroyed by the application of heat, while the height of the protein waves is not affected thereby. Carruthers[24] and Carruthers and Suntzeff[25] found that purified cytochrome *c* gives a catalytic protein wave in a hexamminocobaltic chloride solution, 1 *N* in ammonium chloride and in

[20] W. Stricks and I. M. Kolthoff, Unpublished results.

[21] L. Taussig, J. Prokop, O. Skalickova, and R. Strubl, *Sbornik lakarsky*, **39**, 113 (1937); *Casopis Lekaru Ceskych*, **77**, 1314 (1938), quoted by Dr. G. Reed, Chicago.

[22] K. Mulli and H. A. Werner, *Deut. med. Wochschr.*, **63**, 1941 (1937).

[23] O. Kanner and G. Reed, *Proc. Soc. Exptl. Biol. Med.*, **42**, 387 (1939).

[24] C. Carruthers, *J. Biol. Chem.*, **171**, 641 (1947).

[25] C. Carruthers and V. Suntzeff, *Arch. Biochem.*, **17**, 261–267 (1948).

ammonium hydroxide. The wave is claimed to be proportional to the concentration of cytochrome *c* up to 30 mg. per liter. For greater concentrations a calibration curve was used in the quantitative determination of cytochrome *c* in different tissues, such as rat and mouse kidney, liver, and heart, mouse epidermis, etc. The results were found to be in agreement with those found spectrophotometrically. Catalytic waves of some other highly purified proteins, such as catalase, ascorbic acid, oxidase, prolactin, pepsin, etc., are also given by these authors. Using the polarographic method they investigated the role of cytochrome *c* in epidermal carcinogenesis in mice.

Interesting investigations on the polarographic determination of protein and cystine in urine have been carried out by Rosenthal.[26]

Using the polarographic method, Reed[27] demonstrated the presence of a cystine complex in urine from which cystine is liberated in an ammoniacal cobalt buffer. To 200 ml. of Brdicka's cobaltous buffer solution 0.5 ml. of urine was added and the mixture was polarographed at certain time intervals. A constant value of the catalytic cystine wave was obtained after 1 hour. It was found that, upon acid hydrolysis, the same amount of cystine was liberated as from the ammoniacal cobalt buffer.

Seuberling[28] determined the protein content of liquor cerebrospinalis by electrolyzing the diluted liquor in a lithium borate buffer of pH 10. Since the height of the catalytic wave is not strictly proportional to the amount of protein, a calibration curve was taken with a liquor whose protein content had been determined gravimetrically. However, a direct comparison between the wave height and the gravimetric determination of proteins is hardly possible for the wave height is affected not only by the amount, but also by the composition, of the protein. The polarographic method is considered to be useful for obtaining information on the amount of polarographically active cystine groups in the liquor.

4. EFFECT OF DENATURATION OF PROTEINS ON THE CATALYTIC WAVES

It is known that disulfide or sulfhydryl groups are liberated when the protein molecule is denatured. The polarographic examination of protein solutions during the denaturation process shows, in most cases, an increase in the catalytic protein waves. This indicates that polarographically active groups are liberated on denaturation. The change in the structure of the

[26] H. G. Rosenthal, *Mikrochemie*, **22,** 233 (1937).
[27] G. Reed, *J. Biol. Chem.*, **142,** 61–64 (1942).
[28] O. Seuberling, *Klin. Wochschr.*, **16,** 644 (1937).

denatured protein molecule becomes evident from the change in the shape of the catalytic double wave.

Wenig and Jirovec[29] investigated polarographically the effect of ultraviolet radiation upon proteins and found an increase in both of the double protein waves. The second wave increased more than the first and the ratio of the two reached a maximum after a certain time of radiation. After this maximum had been reached the protein started to precipitate; both waves then decreased, but their ratio remained the same. The nature of the two waves is still not known.

Schmidt[30] showed that the effect of ultraviolet light on protein solutions increases with decreasing concentration of the protein. No polarographic effect was found on irradiation of undiluted serum. Albumins are more affected by irradiation than globulins.

Fiala[31] carried out polarographic investigations of the ultraviolet irradiation of albumin solutions in the presence and absence of oxygen and did not find any difference in the character of the protein waves. He also studied the oxygen wave in 0.5 per cent gelatin solution during irradiation with light of wave lengths shorter and longer than 3000 A. Over both ranges of wave length a reduction in the height of the oxygen wave was found while the vessel was tightly closed. Fiala[32] also reported polarographic experiments on the photodynamic action of organic dyes on proteins.

Interesting irradiation experiments with serum protein solutions were carried out by Heeren and Seuberling.[33] While the effects obtained with ultraviolet radiation were found to be the same as those reported by Wenig and Jirovec, irradiation of protein solutions with x-rays up to an intensity of 100,000 r. did not show any appreciable effect on the height and shape of the catalytic waves. This result is remarkable since ultraviolet and x-rays are similar in their biological action.

Experiments carried out by Dr. W. Stricks at the University of Minnesota showed that more significant tests could be obtained when the irradiated serum was polarographed after deporteination with sulfosalicylic acid. In one case the maximum catalytic wave height of the whole serum increased by 60 per cent after ultraviolet irradiation, while the wave height obtained with the deproteinated serum was 11 times as great as that of the filtrate wave of the unradiated serum at the proper dilution.

---

[29] K. Wenig and O. Jirovec, *Biochem. Z.*, **295,** 405 (1938).

[30] H. W. Schmidt, *Biochem. Z.*, **306,** 167–176 (1940).

[31] S. Fiala, *Chem. Listy*, **40,** 82–86 (1946); *Biochem. Z.*, **318,** 67–73 (1947).

[32] S. Fiala, *Chem. Listy*, **39,** 14–19 (1945).

[33] J. P. Heeren and O. Seuberling, *Strahlentherapie*, **67,** 130 (1940).

On subjecting human serum to denaturation in 0.2 *N* potassium hydroxide, Brdicka[34] and Brdicka and Klumpar[35] found that the catalytic waves of the whole serum increased with time of denaturation up to a maximum value after about 20 minutes. Upon longer periods of denaturation a cloudiness appeared which Brdicka attributed to precipitation of the denatured protein; the height of the catalytic waves started to decrease at the same time.

Following Brdicka's procedure (*v.i.*) the authors[36] observed the maximum catalytic effect after about 15 to 30 minutes. Probably the "room temperature" during our experiments was higher than that in Brdicka's experiments, which would accelerate the denaturation. We also found that the catalytic waves obtained with whole blood were only about one-fourth as large as those obtained with the corresponding serum.

Crossley *et al.*[37] denatured dog sera with 0.7 *N* potassium hydroxide and polarographed the deproteinated sera using sulfosalicylic acid as the deproteinating agent. They found a decrease in wave height during the 5 to 10 minutes of alkali treatment, followed by a rise, until a fairly constant wave height was reached after 60 minutes.

Winzler and collaborators[38] obtained higher sulfosalicylic acid filtrate waves with alkali-denatured human serum but smaller filtrate waves with alkali-denatured rat or rabbit serum, regardless of how time or strength of alkali were varied.

Experiments with mouse serum carried out by W. Stricks at the University of Minnesota showed that alkali-treated (0.1 *N* KOH, 1 hr.) mouse serum gave smaller waves after deproteination with sulfosalicylic acid. The alkali-denatured whole serum gave 70 to 80 per cent higher waves as compared with the waves of untreated whole serum.

In present work carried out at the University of Minnesota, human serum is denatured in approximately 0.1 *N* potassium hydroxide at 25° C. After various lengths of time of alkali denaturation, the serum is deproteinated with sulfosalicylic acid. With normal sera the height of the waves obtained with the deproteinated sera decreased with the time of denaturation within the first 30 minutes.

An increase in the height and a change of the shape of the protein double

[34] R. Brdicka, *Biol. Listy*, **22,** 39 (1937); *J. chim. phys.*, **35,** 89 (1938).

[35] R. Brdicka and J. Klumpar, *Casopis Ceskoslov. Lekarnictva*, **17,** 234 (1937).

[36] I. M. Kolthoff and J. J. Lingane, Report submitted to Committee on Cancer Research of the University of Minnesota, January, 1939.

[37] M. L. Crossley, R. H. Kienle, B. Vassel, and G. L. Christopher, *J. Lab. Clin. Med.*, **27,** 213–222 (1941).

[38] R. J. Winzler, D. Burk, and M. Hesselbach, *J. Natl. Cancer Inst.*, **4,** 417 (1944).

wave is also observed when the proteins are acted upon by proteolytic enzymes. Brdicka[39] who carried out experiments with pepsin in 0.05 *N* hydrochloric acid at 40° C., found that the height of the catalytic waves depends upon the time of denaturation. He found a maximum after 30 minutes. The authors, using Brdicka's procedure, found a maximum height after 12 minutes, after which time the height remained constant. Hence, unlike the alkaline denaturation, the time of acid denaturation with pepsin is not at all critical, provided it is at least 15 minutes. The action of enzymes causes a much greater increase in the protein waves than was observed in other denaturation experiments. Crossley *et al.*[37] studied the rate of pepsin and trypsin digestion of dog sera. The deproteinated sera gave an increase in wave height similar to that given by the whole sera. Trypsin was found to act more slowly than pepsin.

Tropp[40] found that insulin, unlike other proteins, when treated with pepsin or trypsin shows a decrease in the catalytic waves. The action of these materials on insulin was also studied by Herrback.[41] Since insulin loses its physiological activity during the action of these enzymes, several authors, like Mulli *et al.*,[22] Kanner *et al.*,[23] and others,[42, 43] investigated the relationship between the physiological activity and the height of the polarographic double wave of insulin.

Tropp and his co-workers[44] studied the heat denaturation of different proteins in the presence of urea and found an increase in the height of the protein waves. Tropp *et al.* believe that the second of the double waves is caused by sulfhydryl groups in the proteins. More systematic investigations regarding the characteristics of the double waves as related to the structure of the proteins are desirable. In an attempt to find a characteristic polarographic quantity for different proteins Tropp[50] introduced the quotient h_2/h_1 of the heights of the maximum peaks of the second (h_2) and the first (h_1) maximum of the catalytic wave. This ratio was determined for solutions of various proteins at different concentrations. In a

---

[39] R. Brdicka, *Mikrochemie*, **15,** 167 (1934).

[40] C. Tropp, *Klin. Wochenschr.*, **17,** 465 (1938).

[41] X. X. Herrback, Inaugural Dissertation, University of Würzburg, 1944.

[42] P. Zadina, *Biol. Listy*, **22,** 122 (1937).

[43] K. Hallas-Moller, Dissertation, Nyt Nordisk Forlag, Copenhagen, 1945.

[44] C. Tropp, *Z. physiol. Chem.*, **262,** 199 (1939). L. Jühling, C. Tropp, and E. Woehlisch, *ibid.*, **262,** 210 (1939). C. Tropp, L. Jühling, and F. Geiger, *ibid.*, **262,** 225 (1939). C. Tropp, *Naturwissenschaften*, **26,** 548 (1938).

[50] C. Tropp, *Ber. Phys.-chem.-med. Ges. Würzburg*, **64,** 178 (1941). C. Tropp and F. Geiger, *Z. physiol. Chem.*, **272,** 121 (1942); *ibid.*, **272,** 134 (1942). C. Tropp and W. Stoye, *ibid.*, **275,** 80 (1942). C. Tropp, F. Geiger and W. Stoye, *ibid.*, **277,** 192 (1943). C. Tropp and W. Herrback, *ibid.*, **281,** 50–57 (1944).

few cases the quotient h_2/h_1 was also determined for protein solutions after treatment with pepsin or urea.

Waldschmidt-Leitz and Mayer[51] polarographed deproteinated human sera after heat treatment. Crossley *et al.*[37] used 8 *M* solutions of urea and guanidine hydrochloride as denaturing agents for dog sera and found that these compounds inhibited the formation of the polarographic wave to such an extent as to preclude their use for polarographic denaturation purposes.

Denaturation experiments with human sera in 10 *M* urea and 3.5 *M* guanidine hydrochloride solutions at 25° C. were performed by W. Stricks at the University of Minnesota. The catalytic double waves were not found to be inhibited in the presence of the above denaturing agents. With both denaturing agents a marked increase in the catalytic waves of the whole denatured serum was found, guanidine hydrochloride being more effective as a denaturing agent than urea. The appearance of the waves after denaturation with guanidine hydrochloride was found to be similar to that after pepsin treatment.

Extensive experiments on the polarographic behavior of cow and mouse milk were carried out by Stricks and Kolthoff.[52] Milk and milk sera which had been subjected to various denaturation processes were found to give protein double waves similar to those obtained with blood sera. Since pure casein dissolved in dilute alkali gives extremely small catalytic waves, proteins other than casein in the milk must cause its polarographic activity. Casein, after hydrolysis with hydrochloric acid, gives characteristic catalytic cysteine waves.

Fiala[53] investigated polarographically the effect of formaldehyde on horse pseudoglobulin. He found a reduction in the height of the double wave. From this finding it is concluded that the action of formaldehyde on proteins is a condensation rather than a hydrolysis reaction.

5. SEROLOGICAL APPLICATIONS OF CATALYTIC WAVES FOR DIAGNOSIS OF CANCER

(a) Methods

The stimulus to serological research with the dropping electrode was given by the work of Purr and Russell[54] who studied the activating action

[51] E. Waldschmidt-Leitz and K. Mayer, *Z. physiol. Chem.*, **261,** Heft 1, 2 (1939).
[52] W. Stricks and I. M. Kolthoff, *International Polarography Congress, Prague,* **1,** 386 (1951).
[53] S. Fiala, *Naturwissenschaften,* **31,** 370 (1943).
[54] A. Purr and M. Russell, *Z. physiol. Chem.,* **228,** 198 (1934).

of blood and serum on certain enzymes. They found that blood considerably increased the proteolytic properties of inactivated papain and cathepsin. Moreover, they discovered that blood from patients with cancer exerted a smaller activating action than did normal blood. Purr and Russell ascribed this activating action of blood to the sulfhydryl groups of the blood proteins on the basis of results of previous investigators[55] and concluded that carcinoma blood has a reduced sulfhydryl activity. Waldschmidt-Leitz and co-workers[56] established a similar reaction between carcinoma and normal sera and attempted to use this relation diagnostically. They found that the activating effect of carcinoma serum was smaller than and often only one-half of that of normal serum.

In 1936 Brdicka became acquainted with the work of Purr and Russell and, as the discoverer of the catalytic sulfhydryl waves, it was logical that he should be the originator of the use of the polarograph in serological cancer diagnosis.[57] He added equal amounts of fresh carcinoma serum or normal serum to an ammoniacal cobalt solution and found that the catalytic waves obtained with carcinoma sera in general were smaller than those observed with normal sera. However, in many cases the difference was too small to be of practical significance. On using denatured sera the differences became more pronounced. Alkali, pepsin, and heat denaturation were mainly used for this purpose.

The protein waves of denatured carcinoma sera were found to be much smaller than those of normal sera. In order to prove that the decrease in catalytic activity of carcinoma serum was due to the presence of fewer disulfide groups, or to inhibition in the release of these groups on denaturation, Brdicka treated several samples of sera with boiling 5 *N* hydrochloric acid. After complete hydrolysis he found that the hydrolyzates from carcinoma sera contained less cystine than those from normal sera.

When this method was tested on a larger number of samples, deviations from the general were sometimes observed which were attributed to a variation in the protein content of the serum For this reason Tropp[58] proposed to determine also the total protein content. In order to get

[55] W. Grassmann, H. Dyckerhoff, and O. V. Schoenebeck, *Z. physiol. Chem.*, **186,** 193 (1929). E. Waldschmidt-Leitz and A. Purr, *ibid.*, **198,** 260 (1931). A. Purr, *Biochem. J.*, **29,** 5, 13 (1935).

[56] E. Waldschmidt-Leitz, *Angew. Chem.*, **51,** 916 (1936). E. Waldschmidt-Leitz, O. Conrath, and J. Gloeditsch, *Naturwissenschaften*, **25,** 60 (1937).

[57] R. Brdicka, *Nature*, **139,** 330, 1020 (1937); *Acta Unio Intern. contra Cancrum*, **3,** 13 (1938); *Compt. rend. soc. biol.*, **128,** 54 (1938); *J. chim. phys.*, **35,** 89 (1938).

[58] C. Tropp, *Klin. Wochenschr.*, **17,** 1141 (1938).

comparable results he plotted the height of the polarographic waves against the total protein content.

In cases of malignant neoplasmas Sartori and Stolfi[59] found that the sera differed sufficiently from those of normal individuals if fresh sera were used for the polarographic tests. On aging the polarographic differences between normal and abnormal sera become indistinct.

A deeper insight into the pathological changes in the serum proteins was obtained by Brdicka,[60] in cooperation with K. Mayer, on using partially deproteinated serum.

After treating serum with a deproteinating agent (sulfosalicylic acid), catalytic polarographic activity was observed in the filtrate. The catalytic waves were found both with divalent and trivalent cobalt and, therefore, cannot be attributed to cystine (which does not yield a catalytic wave with trivalent cobalt), but must be due to high molecular weight compounds which are not precipitated by sulfosalicylic acid.

As indicated by Brdicka[60] these compounds, which are nondialyzable, contain cystine and tyrosine and are precipitated by tungstic and metaphosphoric acids, colloidal ferric oxide, and ethanol. As was found by Winzler *et al.*,[38] they are also precipitated by about 70 per cent saturated ammonium sulfate. They are not coagulated by heat and not precipitated by trichloroacetic acid. Winzler *et al.* also found that the isolated material contains a large amount of sulfur not accounted for by cystine or methionine, indicating that it has combined with sulfosalicylic acid.

There is experimental evidence that these compounds are degradation products of serum proteins. Thus, an arbitrarily high polarographic filtrate wave can be obtained by proteolyzing or hydrolyzing a normal serum to a certain extent and deproteinating with sulfosalicylic acid.

While the catalytic waves of whole carcinoma sera in general were lower than those of normal sera, *the opposite was found in sulfosalicylic acid filtrate*, even though the total number of disulfide groups in the carcinoma sera is less than in the normal sera. Brdicka[61] believes that the increased "filtrate wave" found in carcinoma sera may have its origin in some phenomenon similar to that brought about by Abderhalden's protective enzyme, or Abwehrferment, which is proteolytic to "foreign" proteins in blood. Brdicka believes that a malignant tumor may produce such "for-

[59] G. Sartori and G. Stolfi, *Ricerca sci.*, **11,** 660–661, Sept. 1940; *Chem. Zentr.*, *I*, **1941,** 784.

[60] R. Brdicka, *Klin. Wochenschr.*, **18,** 305 (1939); *Acta Radiol. Cancerol. Bohemiae Moraviae*, **2,** 7 (1939). R. Brdicka, J. Klumpar, and F. V. Novak, *ibid.*, **2,** 27 (1939).

[61] R. Brdicka, *Acta intern. Ver. Krebsbekämpf.*, **3,** 13 (1938).

eign" proteins which would then be split to yield polarographically active groups.

Experiments with alkali-denatured normal and cancerous sera are also being carried out at the University of Minnesota. The sera are deproteinated with sulfosalicylic acid after 30 seconds, 30, 60, 90, and 120 minutes of alkali action. While most of the normal deproteinated sera give a marked decrease in the height of the catalytic waves during the first 30 minutes of denaturation, the cancerous sera show an increase in wave height of about 50 per cent during the first 30 minutes, after which time the wave height decreases and attains a fairly constant value after 90 minutes of denaturation. Thus, the rate of alkali denaturation as observed from catalytic waves of the deproteinated sera seems to differentiate between sera of normal and cancerous individuals. These studies are being continued at present and will be published after a sufficient number of normal and pathological sera have been investigated.

The introduction of the polarographic protein test into investigations of the Abderhalden reaction might be of great interest.[62–64]

Besides sulfosalicylic acid filtrate tests, Merten and co-workers[65] also used alcohol filtrate tests of heat-denatured sera. Unlike sulfosalicylic acid filtrate waves, alcohol filtrate waves of cancerous sera were found to be lower than those of normal sera. A factor:

$$F = \frac{\text{per cent increase in SSF above normal}}{\text{per cent decrease in AF below normal}}$$

was considered to be useful for the quantitative comparison of different sera. SSF denotes the maximum height of the sulfosalicylic acid filtrate wave and AF the maximum height of the alcohol filtrate wave.

More recently Müller and Davis[66] proposed the introduction of the "protein index" as a method of characterizing and comparing polarographic results obtained with various blood sera. For this purpose two tests are suggested—the sulfosalicylic acid filtrate test and the "digest test" in which the polarogram of the whole alkali-denatured serum is taken. The ratio of the two maximum wave heights obtained in these two tests multi-

[62] V. Podrouzek, *Fermentforschung.*, **17,** 53–62 (1942).

[63] E. Abderhalden, R. Abderhalden, and J. Klumpar, *Fermentforschung*, **17,** 600–609 (1945).

[64] V. Podrouzek and O. Jonas, *Casopis Lekaru Ceskych*, **84,** 235–238 (1945).

[65] R. Merten and W. Spiegelhoff, *Z. klin. Med.*, **138,** 42 (1940). R. Merten, *ibid.*, **139,** 133–144 (1941). R. Merten and J. Bernot, *ibid.*, **139,** 145–156 (1941).

[66] O. H. Müller and J. S. Davis, *J. Biol. Chem.*, **159,** 667 (1945); *Arch. Biochem.*, **15,** 39 (1947).

plied by a suitable factor is denoted as the "protein index." This quotient was found to be practically independent of small variations in room temperature. It is claimed to be unaffected by considerable changes in the drop time and drop size of the mercury cathode. No significant differences between the indices of men and women were found. The index does not vary appreciably with age and is practically unaffected during menstruation or pregnancy. This constancy exists in spite of the fact that the filtrate and digest values alone may show considerable variations.

The protein index of a normal individual was found to be between 1 to 5 while it was found to be 20 for an abnormal human plasma. Portillo and Ortego[67] successfully applied Müller's method.

A different approach to the polarographic study of carcinoma sera was made by Koltljar and Podrouzek.[68] It is known that normal serum is able to decompose fibrin prepared from carcinoma blood and that carcinoma serum is able to decompose fibrin from normal blood. The above authors used the polarographic method for the detection of proteolysis. They obtained an asceptic ultrafiltrate of carcinoma serum which they then incubated at 37° C. with normal and with carcinoma fibrin. After incubation, the samples were filtered through hard filter paper and a certain volume of the filtrates was added to Brdicka's cobaltous test solution. A control curve run with the ultrafiltrate without the addition of any fibrin showed that the membrane filter used for the ultrafiltration does let some protein through, as indicated by small protein waves. The filtrate from the sample containing carcinoma ultrafiltrate and carcinoma fibrin showed a very slight increase in the protein wave over the control. However, the filtrate from the sample containing carcinoma ultrafiltrate and normal fibrin showed a pronounced increase in the protein wave.

(b) Procedures for Serological Tests according to Brdicka and Modifications Thereof

Standard Solutions. 0.008 N $CoCl_2$; 0.03 N $Co(NH_3)_6Cl_3$, 1 N NH_4Cl, 1 N NH_3; 0.02 M cysteine; 0.02 M cystine in 1 N KOH (not stable); and 20 per cent sulfosalicylic acid.

Cobalt Test Solutions. (*a*) *Cobaltous Solution.* 1.6×10^{-3} N $CoCl_2$, 0.1 N NH_3, 0.1 N NH_4Cl, *freshly* prepared by mixing 20 ml. of standard $CoCl_2$ solution, 10 ml. of NH_4Cl, 60 ml. of water, and 10 ml. of 1 N NH_3. Ammonia is added last to prevent precipitation of part of the cobalt.

---

[67] R. Portillo and M. Ortego, *Anales fís. y quím. Madrid*, **43,** 855 (1947); *Chem. Abstracts*, **42,** 3069 (1948).

[68] A. M. Koltljar and V. Podrouzek, *Nature*, **142,** 872 (1938).

(*b*) *Cobaltic Solution.* 3×10^{-3} *N* $Co(NH_3)_6Cl_3$, 0.1 *N* NH_4Cl, and 1 *N* NH_3. This is prepared by mixing 10 ml. of standard $Co(NH_3)_6Cl_3$, 10 ml. of NH_4Cl, 60 ml. of water, and 20 ml. of 5 *N* NH_3.

Crossley *et al.*[69] use the same cobalt concentration, but 0.05 *N* NH_4Cl and 0.15 *N* NH_3.

Winzler *et al.*[70] use a buffer with the following composition: 3×10^{-3} *N* $Co(NH_3)_6Cl_3$, 0.1 *N* NH_4Cl, 0.7 *N* NH_4OH, and 0.1 *N* KCl. 1 ml. of saturated caffeine solution is added per 100 ml. of the reagent to suppress the oxygen and cobalt maxima.

Müller[66] works with a solution which is 3×10^{-3} *N* in $Co(NH_3)_6Cl_3$, 0.1 *N* in NH_4Cl, and 0.8 *N* in NH_4OH.

Alkaline Denaturation. Mix 0.5 ml. of serum with 0.5 ml. of water and 0.25 ml. *N* KOH. Add 0.1 ml. of this solution to 20-ml. portions of the cobaltous test solution after standing for 5, 10, 15, and 20 minutes. Müller employes the same alkaline denaturation mixture but uses oxalated plasma instead of blood serum.

Acid–Pepsin Denaturation. Dissolve 0.1 g. of U. S. P. pepsin in 100 ml. of 0.05 *N* hydrochloric acid. Transfer 5-ml. portions of this solution into test tubes which are kept in a water bath at 40° C. Add 0.2 ml. of the serum to the test tube and allow to stand for one-half hour at 40° C. Then add 0.2 ml. of this mixture to 10 ml. of cobaltous test solution and determine the catalytic wave.

Deproteination of Serum with Sulfosalicylic Acid. Mix 0.4 ml. of serum with 1 ml. of 0.1 *N* KOH. After one-half hour add 1 ml. of 20 per cent sulfosalicylic acid solution and filter through hard paper. Add 0.5 ml. of the filtrate to 5 ml. of the cobaltic test solution.

The filtrate test can also be performed with blood instead of serum. The following procedure is recommended by Brdicka: 0.1 ml. of blood is added to 1 ml. of 0.1 *N* KOH. After 45 minutes add 1 ml. of 5 per cent sulfosalicylic acid solution and after 10 minutes filter through hard paper. Add 1 ml. of the clear filtrate to 2 ml. of the cobaltic solution.

Crossley uses Brdicka's procedure for sera while Winzler *et al.* apply blood without previous denaturation. They use the following procedure: Add 0.2 ml. of blood to 1.8 ml. of isotonic salt solution. To this mixture add 2 ml. of 10 per cent sulfosalicylic acid. Filter after 20 minutes through a Whatman No. 5 paper and add 1 ml. of the filtrate to 2 ml. of the cobaltic reagent.

---

[69] M. L. Crossley, R. H. Kienle, B. Vassel, and G. L. Christopher, *J. Lab. Clin. Med.*, **26**, 1500 (1941).

[70] R. J. Winzler, D. Burk, and M. Hesselbach, *J. Natl. Cancer Inst.*, **4**, 417 (1944).

According to Müller 0.5 ml. of oxalated plasma is mixed with 1 ml. of water and 0.1 ml. of 1 *N* KOH. The mixture is allowed to stand for 30 minutes at room temperature. Then 1 ml. of 20 per cent sulfosalicylic acid solution is added and the mixture is filtered after 10 minutes. 0.5 ml. of the filtrate is added to 5 ml. of the cobaltic solution.

Winzler *et al.*[70] pointed out that the procedures must always be followed exactly in order to obtain reproducible and comparable results. Thus, among other factors, they found that the filter paper used should always be of the same size and quality since the paper absorbs some proteose.*

Experiments carried out at the University of Minnesota showed that shaking the same volume of sulfosalicylic acid filtrate with various quantities of finely divided filter paper resulted in a marked decrease in the height of the filtrate wave—this effect increasing with the amount of paper used. A similar effect was found on shaking the filtrate with alumina and fine glass powder. Reproducible results with deproteinated sera could easily be obtained by centrifuging the precipitated proteins instead of applying filtration. The following procedure was found to give the best results: 0.5 ml. of serum is placed in a 5-ml. volumetric flask and 1.25 ml. of 0.1 *N* potassium hydroxide solution is added. After standing for 30 minutes 1.5 ml. of 20 per cent sulfosalicylic acid is added. After 5 minutes the mixture is filled up to the mark with distilled water, transferred into a centrifuging tube, and centrifuged; 10 minutes of centrifuging gives a perfectly clear centrifugate. The centrifugate was found to give higher waves than the filtrate of the same concentration.

Waldschmidt-Leitz and Mayer[71] denatured a mixture of 0.5 ml. of serum and 1 ml. of water by heating it on a boiling water bath for 10 minutes. The mixture was well shaken, deproteinated with sulfosalicylic acid, and part of the filtrate added to the cobaltic test solution.

Merten *et al.*[65] apply heat denaturation to sera according to the procedure of Waldschmidt-Leitz and deproteinate by the addition of 2.5 ml. of absolute alcohol. The mixture is filtered and 0.5 ml. of the filtrate is added to 4.5 ml. of the cobaltic buffer.

Test for Cleavage Products after Denaturation with Acid–Pepsin. Dissolve 0.1 ml. of U. S. P. pepsin in 100 ml. of 0.1 *N* HCl. To 2 ml. of serum add 2 ml. of the pepsin solution, and heat in a water bath at 40° C. After various time intervals remove 1-ml. portions and precipitate the proteins

* Recently, evidence has been obtained that the substances responsible for the filtrate wave are not proteoses but have the characteristics of mucopolysaccharides (see C. B. Huggins, *Proc. First Conference Cancer, Diagnostic Tests*, 7 (1950).

[71] E. Waldschmidt-Leitz and K. Mayer, *Z. physiol. Chem.*, **261,** Heft 1, 2 (1939).

with 1 ml. of 20 per cent sulfosalicylic acid. Filter and add 0.5 ml. of the filtrate to 5 ml. of the cobaltic test solution.[72]

Cystine Determination in Serum Hydrolyzates. Add 1 ml. of serum to 20 ml. of 5 *N* hydrochloric acid. Boil the solution under a reflux condenser for 8 hours, dilute to 100 ml., and filter. Add 0.5 ml. of the filtrate to 10 ml. of the cobaltous test solution and compare the waves obtained with those produced by known amounts of cystine added to the test solution.

(c) Evaluation of Polarographic Tests in Connection with Cancer and Other Diseases

The first systematic research using the polarographic diagnosis of cancer was carried out by Brdicka in collaboration with Dr. F. N. Novak, Director of the Radiotherapeutic Institute, Prague. From 187 cases listed the polarographic diagnosis agreed in 102 out of 107 histologically ascertained carcinoma cases and 11 of 13 sarcoma cases, while only one of the 38 normal sera showed a positive polarographic reaction. Of the 29 noncarcinomatic diseases, 14 were polarographically positive and 15 negative.

In a recent excellent review with 84 references on polarographic cystine and protein tests, Brdicka[73] reported the results of the polarographic examination of 15,000 samples of different sera, mostly of patients suffering from cancer. The tests which, since 1939, have been carried out at the Radiotherapeutic Institute, Bulovka Hospital, Prague, showed that 20 per cent of polarographic results fail. Brdicka tested the sulfosalicylic acid filtrates of alkali-denatured sera in most of the tests.

Similar results were obtained at the Finsen Institute and Radium Station, Copenhagen,[74] where among the positive reactions some cases of hepatic disorder were ascertained.

Successful results of Brdicka's method were reported by Gallego and Ruiz.[75] Sera from patients with malignant tumors in evolution gave a positive polarographic test in 95.8 per cent, and a doubtful one in 4.2 per cent, of the cases. With malignant tumors which had been treated and had relapsed, 100 per cent of the cases were positive. Those not wholly cured but having a favorable evolution gave 100 per cent doubtful results.

[72] Compare also P. Meyer-Heck, *Z. Krebsforsch.*, **49,** 142, 560 (1939), who tests with the cobaltic and cobaltous test solutions. The waves obtained with carcinoma sera are smaller than those with normal sera.

[73] R. Brdicka, *Research*, **1,** 1 (1947).

[74] F. Bergh, O. M. Henriques, and J. Schousbon, *Nature*, **141,** 751 (1938).

[75] A. S. Ruiz and J. L. Gallego, (Inst. Santiago Ramony Cajal, Madrid) *Rev. españ. fisiol.*, **1,** 203, 31 (1945); *Chem. Abstracts*, **41,** 3857 (1947); *Anales fís. y quím. Madrid*, **41,** 1211–1236 (1945); *Chem. Abstracts*, **41,** 4567 (1947).

With benign tumors the result was negative in 80 per cent, and doubtful in 20 per cent of the cases. With healthy individuals the test was 100 per cent negative. In tuberculosis, typhoid, and lymphogranulomatosis, the results were 100 per cent positive.

Several other investigators have studied the polarographic method for the detection of cancer. Some of their results are summarized below.

Acid–Pepsin Method. Meyer-Heck[76] investigated 35 cases of carcinoma and sarcoma: 71 per cent gave a positive polarographic result; controls, *i. e.*, no cancer or carcinoma, 28 cases of inflammation, 50 per cent positive, and 45 cases without inflammation, 11 per cent positive. Bernhard[77] found 87 per cent of tests positive in cases of carcinoma and sarcoma; 22.5 per cent of controls positive. Tropp,[78] who corrected for the protein content, found the following results: 59 cases of carcinoma and sarcoma, 90 per cent positive; controls, inflammations gave a positive polarographic test. Wedemeyer and Daur[79] using Tropp's method report the following: 97 cases of carcinoma and sarcoma, 99 per cent positive; controls, of 53 cases two gave a positive test (inflammation and fever). Kolthoff and Lingane[80] in a limited number of cases (8 carcinoma and sarcoma; 8 controls) did not find a difference between the averages of the two groups. Rusch, Klatt, Meloche, and Dirksen[81] report the results of 150 cases of normal, noncancerous, and cancerous human beings and rats. They found considerable overlapping in all groups, making it impossible to distinguish any individual case as being representative of normal, cancerous, or noncancerous conditions. They separated the albumin from the globulin in the serum and found that only the albumin fraction yielded a typical wave after acid hydrolysis; on the other hand, the protein wave was not observed with the globulin fraction, although the serum globulin contains cystine. In agreement with Wedemeyer and Daur[82] the above authors found a close parallelism between the amount of serum albumin and the height of the catalytic protein waves in the serum. They conclude that the polarographic method is of limited value in cancer diagnosis since changes in serum albumin are by no means specific for neoplastic diseases.

---

[76] P. Meyer-Heck, *Z. Krebsforsch.*, **49,** 142, 560 (1939).

[77] Bernhard, *Lecture Surgeon Congress*, Berlin, 1938; quoted by Meyer-Heck; *Deut. med. Wochenschr.*, **1939,** 596; *Arch. klin. Chir.*, **193,** 543 (1938).

[78] C. Tropp, *Klin. Wochenschr.*, **17,** 1141 (1938).

[79] H. E. Wedemeyer and T. Daur, *Z. Krebsforsch.*, **49,** 10 (1939).

[80] I. M. Kolthoff and J. J. Lingane, *Report to Committee on Cancer Research*, University of Minnesota, January, 1939.

[81] H. P. Rusch, T. Klatt, V. W. Meloche, and A. J. Dirksen, *Proc. Exptl. Biol. Med.*, **44,** 362 (1940); *Trans. Electrochem. Soc.*, **79,** 413 (1941).

[82] H. E. Wedemeyer and T. Daur, *Z. Krebsforsch.*, **49,** 10 (1939).

Contrary to the findings of Rusch *et al.*, Wenig[83] and Schmidt[84] obtained characteristic protein waves with globulin fractions of human and horse sera. These results were also confirmed by experiments with human and mouse sera carried out at the University of Minnesota.

Alkaline Denaturation. Brdicka[85]: 182 cases of carcinoma and sarcoma, 97 per cent positive; of 28 suspected cases of carcinoma, 24 cases positive; controls, 7 out of 14 cases positive. Meyer-Heck[76]: 21 cases of inflammation 100 per cent positive; no inflammation, 15 cases, 27 per cent positive. More favorable were the results obtained by Walker and Reimann[86] who investigated the sera from 386 patients. The sera from the majority of their cancer cases yielded catalytic waves which fell below the limit of the waves given by normal sera. Without exception they found that the larger the malignant tumor, the greater was the deviation from normal. In addition to finding that fever causes a deviation from the normal, they reported that there is a possibility that pneumonia and tuberculosis also may cause a deviation, though the number of cases tested was much too small to be conclusive. The results obtained in a small number of cases by the authors again were very inconclusive.

Heat Denaturation and Deproteination. Using their own procedure Waldschmidt-Leitz and Mayer[71] found, of 500 cases of carcinoma and sarcoma, 96 per cent positive. Favorable results were also obtained by Felkel[87]: 285 cases of carcinoma and sarcoma, 90 per cent positive; controls, 64 cases of inflammation, 83 per cent positive, and 45 cases without inflammation, 9 per cent positive. In another paper Felkel[88] reports on 176 cancer cases, with 92.6 per cent positive, and 109 recurrent, with 88.9 per cent positive. Schäfer[89] who applied the Waldschmidt-Leitz-Mayer method, found agreement between the clinical cancer diagnosis and the polarographic test in 92 per cent of the cases investigated.

Griessmann *et al.*[90] tested about 1000 cases of cancer by the Waldschmidt-Leitz method in combination with other tests and found 85 per cent to be positive. Abel[91] applied this method to 85 cancerous sera: 55 were positive, 18 negative, 2 uncertain, and 10 could not be tested because hemolysis occurred. Meyer-Heck[76] found 83 per cent positive in 77 cases of car-

[83] K. Wenig and O. Jirovac, *Biochem. Z.*, **295,** 405 (1938).
[84] H. W. Schmidt, *Biochem. Z.*, **306,** 167–176 (1940).
[85] R. Brdicka, *Nature*, **139,** 330, 1020 (1937).
[86] A. C. Walker and S. P. Reimann, *Am. J. Cancer*, **37,** 585 (1939).
[87] R. K. Felkel, *Med. Klin.*, **34,** 840 (1938).
[88] R. K. Felkel, *Zentr. Gynäkol.*, **63,** 647 (1939).
[89] Schäfer, *Diss. Chir. Univ.-Klin. Giessen*, (1938).
[90] H. Griessmann, K. Kohler, and W. Sohnel, *Chirurg.*, **10,** 609 (1938).
[91] W. Abel, *Chirurg.*, **12,** 585–88 (1940).

cinoma and sarcoma. In the controls, 66 cases of inflammation were 65 per cent positive, and 83 cases without inflammation were 11 per cent positive.

Mahlo[92] applied the polarographic filtrate test to the serum of individuals with chronic gastritis and found that this disease may produce normal polarographic curves or curves which are similar to those obtained from patients with stomach cancer. Chystrek[93] investigated 48 cancer cases, of which 92 per cent were positive.

In an attempt to make the polarographic test more specific for cancer, Waldschmidt-Leitz and Mayer[94] fractionated the dialyzed sulfosalicylic acid filtrate with alcohol of different concentrations. Schmidt[95] investigated 274 pathological sera, applying Brdicka's and Waldschmidt-Leitz's tests. Out of 106 Brdicka tests 64.4 per cent gave an essential increase in the height of the filtrate wave, 22.5 per cent gave a small increase, and 13.1 per cent were normal. The alcohol precipitation method was applied to 168 sera but did not indicate an improvement in the polarographic cancer diagnosis.

Similar results were found by Merten and co-workers[65] who investigated a great number of human sera. Meyer-Heck[96] also used the combination of sulfosalicylic acid filtrate and alcohol filtrate tests on heat-denatured sera. He reported 237 control cases, including all kinds of diseases except tumors: 66.6 per cent gave a positive cancer SSF test and 65.3 per cent a positive cancer AF test. Out of 37 cases of malignant neoplasms 80 per cent were found to be positive with both SSF and AF tests. The additional alcohol test did not improve the specificity of the polarographic cancer test. The combination of sulfosalicylic acid and alcohol filtrates was also used by Albers[97]: in 32 cases of cancer, 7 were negative or doubtful; in 101 controls (different diseases), 49 were found to be positive.

Mayer[98] deproteinated heat-denatured sera with sulfosalicylic acid and fractionated the dialyzed filtrates with 66 and 90 per cent alcohol. The two fractions were analyzed and considered to be of a mucoid-like nature.

Experiments with deproteinated normal and cancerous sera carried out at the University of Minnesota revealed that studies on the rate of alkali denaturation differentiated between the sera of healthy and cancerous

[92] A. Mahlo, *Deut. med. Wochenschr.*, **65,** 211 (1939).
[93] E. Chystrek, *Deut. med. Wochenschr.*, **66,** 1190 (1940).
[94] E. Waldschmidt-Leitz and R. Mayer, *Z. angew. Chem.*, **51,** 324 (1938).
[95] H. W. Schmidt, *Z. Krebsforsch.*, **50,** 390 (1940).
[96] P. Meyer-Heck, *Z. Krebsforsch.*, **49,** 560–572 (1939); **52,** 144–157 (1941).
[97] D. Albers, *Biochem. Z.*, **306,** 236 (1940).
[98] K. Mayer, *Z. physiol. Chem.*, **275,** 16–24 (1942).

persons. Denaturation was carried out with 0.1 *N* potassium hydroxide. The sera were deproteinated with sulfosalicylic acid after various periods of time of contact of the serum with the potassium hydroxide. The deproteinated solutions were polarographed in a cobaltic buffer. Normal sera "filtrates" were found to decrease in height during the first 30 minutes of alkali denaturation after which time an increase in wave height ensued. After about 2 hours of alkali action the waves remained unchanged on prolonged contact with alkali. The behavior of cancerous sera was markedly different. These sera, in most of the experiments, gave a characteristic increase in the height of the peptose ("filtrate") wave during the first 30 minutes and then a decrease. After about 2 hours of contact with alkali the waves did not change in height.

Zbarskii and El'piner[99] investigated 24 cancer cases, of which 20 cases were found to be positive. Patients with other diseases also gave positive results. Reboul and co-workers[100] used Brdicka's method but replaced the galvanometer by a cathode-ray oscillograph. 20 cases were investigated, of which the results were 100 per cent positive for cancer but also positive in 20–25 per cent of normal persons.

It is seen that the polarographic serum reaction has been the subject of numerous investigations. Since it is beyond the scope of this book to give a complete report of all the experimental work, references to papers not reviewed in this section are listed below:

E. Hamamoto, *Oriental J. Diseases Infants*, **19,** 18–19 (1936).
Nordmark-Werke, *Materia medica Nordmark*, **8,** 25–28, 73–75 (1938).
F. Bergh, O. M. Heurignes, and C. G. Wolffbrandt, *Nature*, **142,** 212 (1938).
E. Kilianowia, O. Maly, and V. Moravek, *Acta Rediol. Cancerol. Bohemico Slovenica*, **1,** 134 (1938).
E. Poli, reviewed in *Z. Krebsforschung*, **50,** 332 (1940).
L. Hauss and J. Kockelmayer, *Rev. Actual. Therapeut.*, **15,** No. 2 (1939).
H. Dietel, *Z. Geburthilfe und Frauenheilkunde*, **2,** 99–105 (1940).
V. Podrouzek and D. Krkoskova, *Casopis Ceskoslov. Lekarnictva*, **20,** 211–212 (1940).
R. Brdicka, *Ccasopis Ceskoslov. Lekarnictva*, **21,** 214–219 (1941).
E. R. Roberts, *Sci. J. Roy. Coll. Sci.*, **11,** 10–18 (1941).
H. W. Schmidt, *Klin. Wochschr.*, **275,** 198–201 (1942).
M. Fort, R. Brdicka, and K. M. Voriskova, *Casopis Lěkaru Ceskych*, **81,** 1181, 1211 (1942); **82,** 432, 474, 1391 (1943); **83,** 1222 (1944).
O. Jonas, *Casopis Lěkaru Ceskych*, **82,** 1199 (1943).

---

[99] B. I. Zbarskii and I. E. El'piner, (*Acad. Med. Moscow*) *Byull. Eksptl. Biol. Med.*, **24,** No. 1, 22–25 (1947); *Chem. Abstracts*, **42,** 3843c (1948).

[100] G. Reboul, F. Bon, and J. A. Reboul, *Compt. rend.* **225,** 89–91 (1947). J. Reboul, G. Reboul, and M. Dargent, *Bull. assoc. franç. étude cancer*, **35,** 193–200 (1948); *Chem. Abstracts*, **43,** 3514h (1949).

J. Melka, *Casopis Lěkaru Ceskych*, **82,** 1358 (1943).
J. Nosek, *Casopis Lěkaru Ceskych*, **82,** 724 (1943).
S. Österlind, *Svenska Lökar-Tidningen*, No. 40, 10 (1945); No. 8, 16 (1947).
M. Hrdlicka and J. Melka, *Casopis Lěkaru Ceskych*, **85,** 506 (1946).
I. E. El'piner, *Advances in Modern Biol. U. S. S. R.*, **23,** 161–180 (1947).
V. Podrouzek, *Rept. Proc. 4th Intern. Congr. Microbiol.*, **1947,** 336–337 (1949).

From all the literature it is clear that the most reliable results are obtained after deproteination with sulfosalicylic acid. The height of the filtrate wave increases with development of the malignant disease. The clinical findings are usually closely related to the polarographic test. When the tumor is surgically removed or cured by irradiation the test becomes negative unless the organism suffers from metastasis. The starting recidive is usually indicated by a gradual increase in the filtrate wave. Therefore, the test is important for the control of cured patients. Increase in the filtrate wave indicates a pathological change in the organism and requires clinical examination of the patient. It should be emphasized that:

(*a*) *Not in all cases of cancer or sarcoma is a positive test obtained.*

(*b*) A positive test is not specific for carcinoma or sarcoma as it is often obtained in cases of inflammation.

Considering the variation in the height of the filtrate wave with time of denaturation, conclusions can be made as to the character of the disease. Inflammation, infections, or other disease of noncancerous origin show, in general, a progressive decrease in the filtrate wave after suitable treatment, whereas under similar conditions an increase in the filtrate wave is observed for a malignant process. Thus, Mecl[101] makes use of the polarographic test and recommends surgical intervention for patients suffering from stomach ulcers if the polarographic test does not indicate an improvement during the conservative therapeutic treatment.

No doubt more systematic work is still necessary before the polarographic method can be used as a reliable test for the diagnosis of cancer. However, the results of the polarographic methods have proved valuable when used together with the clinical findings.

In addition to the use of polarographic tests in clinical practice they can also be widely applied in experimental pathology. Thus Winzler and Burk[102] investigated sulfosalicylic acid filtrates of whole blood from rats and rabbits with cancer and other diseases and found a remarkable increase in the filtrate wave in pathological cases.

---

[101] A. Mecl, *Casopis Lekaru Ceskych*, **83,** 1014 (1944); **85,** 715 (1946).

[102] R. J. Winzler and D. Burk, *Cancer Research*, **3,** 135 (1943); *J. Natl. Cancer Inst.*, **4,** 417–428 (1944).

Westfall *et al.*[103] studied the deproteinized sera of rabbits before and after implantation with the Brown-Pearce carcinoma. It was found that there is an increase in wave height with well-developed tumors. Only in extreme cases could a pronounced correlation between the polarographic activity and the size of the tumor or metastatic involvement be detected. No qualitative differences in the character of the polarographic filtrate waves of serum from normal and tumor-bearing animals were found. Zbarskii *et al.*[99] also found that the blood of rabbits with Brown-Pearce carcinoma gave significantly higher waves as compared with the controls.

Carcinoma in mice and rats was also studied by Hatschek[104] and Miner.[105] Sera of dead mice were investigated by Melka.[106] Rush and Miner[105] polarographed sera of rats fed with *p*-dimethylaminoazobenzene and found a reduction in the level of the sulfur-containing amino acids.

Crossley and co-workers[107] made use of the polarographic method in the study of normal and type I pneumococcus-infected dog sera. The height of the sum of the two catalytic waves obtained with whole sera in a cobaltic buffer was found to decrease during infection and return to normal with recovery. The sulfosalicylic acid filtrate waves increase in height, and change their shape during infection. With an increase in wave height during illness a corresponding increase in nitrogen values was found. The cystine content of the hydrolyzates of dog sera after infection was determined by the colorimetric and polarographic methods. Although the colorimetric procedure gave higher values than the polarographic method, the relative changes found by both methods were comparable. The cystine content of the serum hydrolyzates decreased during infection and returned to normal upon recovery. The percentage decrease in cystine concentration of the hydrolyzates was paralleled by the percentage fall in the total nitrogen content. The rate of denaturation of serum proteins from normal and infected dogs was studied polarographically. While no appreciable differences between normal and infected sera were found when detergents or pepsin and trypsin were used, the rate of alkali denaturation was found to be markedly different for normal and infected sera. After various

---

[103] B. B. Westfall, J. W. Thompson, D. Burk, R. Bishop, G. Jarrels, and R. Moore, *J. Natl. Cancer Inst.*, **5,** 407–414 (1945).

[104] R. Hatschek and F. Poupe, *Casopis Lekaru Ceskych*, **82,** 1544 (1942).

[105] L. D. Miner and H. P. Rush, *Science*, **94,** 468 (1941).

[106] J. Melka, *Casopis Lekaru Ceskych*, **83,** 1574–1578 (1944).

[107] M. L. Crossley, R. H. Kienle, B. Vassel, and G. L. Christopher, *J. Lab. Clin. Med.*, **26,** 1500–1505 (1941). M. L. Crossley, B. Vassel, and G. L. Christopher, *ibid.*, **26,** 1635–1642 (1941). M. L. Crossley, R. H. Kienle, B. Vassel, and G. L. Christopher, *ibid.*, **27,** 213–222 (1941). M. L. Crossley, *Science*, **93,** 409–412 (1941).

periods of denaturation in 0.7 *N* sodium hydroxide the proteins were precipitated with sulfosalicylic acid. The filtrate was electrolyzed in a cobaltic buffer.

6. ACTIVATION OF HYDROGEN PEROXIDE BY HEMOGLOBIN AND HEMATIN

The reduction wave of hydrogen peroxide is quite flat, and the reduction of this compound at the dropping electrode is irreversible. Brdicka and Tropp[108] found that the hydrogen peroxide wave starts at more positive potentials in the presence of traces of hemoglobin and hematin. On adding increasing amounts of hemolyzed blood particles to an air-saturated physiological sodium chloride solution the peroxide wave approached closer and closer to the first wave of oxygen and finally coincided with the latter. Concentrations of 2 micrograms of hemoglobin and 8 micrograms of hematin per milliliter could be detected by this effect. Kopropoporphyrin I, under similar conditions, had no effect on the peroxide wave. In order to show the effect of hemoglobin and hematin and to use it for quantitative purposes it is best to work with solutions saturated with air. Solutions of hydrogen peroxide give similar effects, but the peroxide in solution is rapidly decomposed by the activating compounds. On the other hand, when working with dissolved oxygen the hydrogen peroxide is produced at the electrode (first wave of oxygen corresponds to reduction to peroxide, see Chap. XXXIII), and the results in the presence of the activating compounds are perfectly reproducible.

Brdicka and Tropp found that traces of cyanide prevent the activating effect of hemoglobin and hematin. (No use of this fact has yet been made in the detection and determination of traces of cyanide.)

In the application of the activating effect to the quantitative determination of hemoglobin and hematin it is recommended that the height of the catalytic wave be measured and compared with that of solutions of these two substances at known concentrations in the same medium (same pH). As the catalytic effect depends upon the pH the use of buffer solutions is imperative. Brdicka and Tropp recommend borax (pH about 9). Strnak[109] found the effect most pronounced in a buffer solution which is 0.1 *N* in ammonium chloride and about 1.6 *N* in ammonia. In this way one part of blood in 150,000 parts of solution could be detected. In the quantitative determination of hemoglobin and hematin perhaps the best results are obtained when the area in the polarograms between the "nor-

[108] R. Brdicka and C. Tropp, *Biochem. Z.*, **289,** 301 (1936).

[109] F. Strnak, *Collection Czechoslov. Chem. Communs.*, **11,** 391 (1939).

mal" hydrogen peroxide waves and those in the presence of the activating substances are measured. When these areas are plotted against the concentrations of hemoglobin or hematin Brdicka and Tropp obtained curves which resemble the ordinary types of adsorption isotherms.

The catalytic effect is caused by an activation of hydrogen peroxide by the iron porphyrin derivatives; probably, the peroxide adds to the hemoglobin or hematin. Cyanide saturates the coordination spheres of iron and, therefore, prevents the catalysis.

Swedin[110] found that catalase peroxidase and cytochrome *c* also catalyze the cathodic reduction of hydrogen peroxide in a phosphate buffer of pH 7.38. The catalytic effect is found to be proportional to the hemin content of the solution. This is shown by a comparison of the results obtained for different hemin proteids. From the lowest concentration of the respective hemin proteid at which a catalytic effect can still be observed the hemin content of the solution is computed. The hemin molarity of all solutions is found to be of the same order of magnitude at the concentration at which the catalytic effect on the reduction of hydrogen peroxide just sets in. It is concluded that this effect is not specific to the respective ferment but may be ascribed entirely to the hemin component of the proteid. The results obtained by Swedin are given in Table XLVI-1.

TABLE XLVI-1

LOWEST CONCENTRATION OF HEMIN FOR CATALYTIC EFFECT

Substance	Mol. wt.	No. of hemin molecules per hemin proteid molecule	Lowest concn. for catalytic effect		
			γ/ml.	Moles/ml. × 10^{12}	Hemin molarity, × 10^7, moles/liter
Hemoglobin	67,000	4	2[a]	29.8	1.19
Hematin	592	1	0.08[a]	135	1.35
Cytochrome *c*	13,000	1	4	307	3.07
Peroxidase	44,100	1	10	226	2.26
Catalase	225,000	3	10	44.4	1.33

[a] According to Brdicka and Tropp.[108]

Bilirubin, $C_{33}H_{36}N_4O_6$. Bilirubin is the main constituent of the bile pigment; it is easily soluble in an alkaline medium and is then readily oxidized by air, giving rise to the formation of blue oxidation products.

According to Fischer and Postowsky[111] bilirubin yields mesobilirubin

[110] B. Swedin, *Acta Chem. Scand.*, **1,** 500–502 (1947).

[111] H. Fischer and J. J. Postowsky, *Z. physiol. Chem.*, **152,** 300 (1926).

upon catalytic reduction:

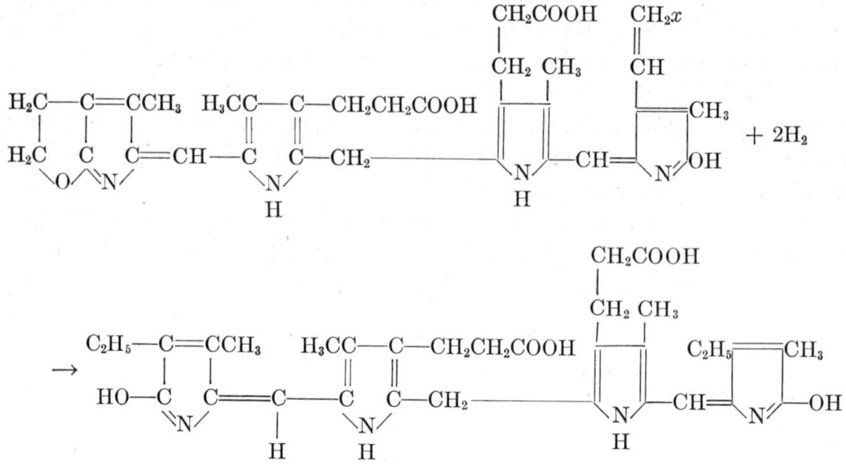

Tachi[112] studied the polarographic reduction in alkaline buffers. In 0.1 N sodium hydroxide (pH = 13) he found the diffusion current proportional to the concentration in a range between 2×10^{-4} and 2×10^{-5} M. It is peculiar and not explained that the diffusion current increases with decreasing pH. Thus, he found the following values of the diffusion current (in microamperes) at different pH values in 2×10^{-4} M solutions: 0.49 at pH = 13, 0.87 at pH = 12, 0.78 at pH = 11, 0.66 at pH = 10, 0.83 at pH = 9, 0.95 at pH = 8. The reduction potential (tangent method) was found at −1.46 v. (N.C.E.) and is independent of the pH, when the latter is greater than 10.

After air oxidation of bilirubin various waves were found in alkaline media which were attributed to at least two reducible compounds.

7. VITAMIN B FACTORS

Most of the vitamins, including some vitamin B factors, have been discussed in organic polarography. In this section some investigations of more specific biological importance are discussed. A preliminary study of the polarographic determination of riboflavin and other vitamin B factors was made by Lingane and Davis.[113] Their results indicate that the polarographic method promises to be particularly valuable for the determination of riboflavin in the presence of the other members of the vitamin B group.

[112] I. Tachi, *Mem. Coll. Agr. Kyoto Imp. Univ.*, **42,** 54 (1938).
[113] J. J. Lingane and O. L. Davis, *J. Biol. Chem.*, **137,** 567 (1941).

Table XLVI-2 lists the half-wave potentials of vitamin B factors in 0.1 *M* buffers at various pH values and it is evident that riboflavin yields a wave before any of the other members of the group are reduced. The fact that the half-wave potential of riboflavin in an unbuffered potassium chloride solution is more positive than that in the buffer of pH 5.8 is difficult to explain, especially since hydrogen ions are used up in the reduction.

The diffusion current of riboflavin was found to depend somewhat upon the pH and the type of buffer used. Therefore, calibration data for a given capillary must be obtained under the same conditions that prevail in an actual determination. Good waves were found over a pH range from

TABLE XLVI-2

HALF-WAVE POTENTIALS (*vs.* S.C.E.) OF THE VITAMIN B FACTORS AT VARIOUS pH VALUES AT 25° (LINGANE AND DAVIS)

Substance	pH 5.8	pH 7.2	pH 9.0	0.1 *N* KCl
Riboflavin	−0.41	−0.47	−0.54	−0.35
Thiamin	[a]	−1.3	[b]	−1.25
Nicotinic acid	[a]	−1.56	−1.6	−1.7
Pyridoxin	[a]	[a]	..	−1.8, −2.0[c, d]
Pantothenic acid	[a]	[a]	..	−2.0[c]

[a] Hydrogen wave interferes.
[b] Vitamin unstable.
[c] In 0.1 *M* tetramethylammonium bromide.
[d] Double wave.

4 to 9 (see Fig. XLVI-4). However, the vitamin decomposes in an alkaline medium and the optimum pH for its determination is in the neighborhood of 7. The diffusion current, after correction for i_r, was proportional to the concentration over the concentration range from 10^{-5} to 10^{-4} *M* (4 to 50 p.p.m.).

Von Müralt obtained a characteristic wave near −1.9 v. in the polarograms of extracts of stimulated cholinergic nerves,[114] solutions used for bathing such nerves,[115] and extracts of heart fluids[116] after vagal stimulation. He attributed these waves to the presence of thiamin or a related compound. Since the concentration of the material set free was much lower than the lowest concentration of thiamin detectable polarographically,

[114] A. von Müralt, *Pflügers Arch. ges. Physiol.*, **245,** 604 (1942); *Verhandl. Vereins Schweiz. Physiol. Pharmakol.*, **19,** Tagung, July 23 (1941).
[115] A. von Müralt, *Helv. Physiol. Acta*, **1,** C20 (1943).
[116] A. von Müralt, *Nature*, **154,** 767 (1944).

von Müralt ascribed this wave to a catalytic effect of the thiamin or its derivative at the dropping electrode. Wollenberger[117] electrolyzed highly diluted thiamin solutions in different buffers and obtained the best catalytic waves in phosphate buffers. The height of these catalytic waves was found to be nearly proportional to the logarithm of the hydrogen ion activity and similar to other catalytic waves; the relation of current to thiamin concentration could be expressed by a Langmuir adsorption isotherm. The catalytic current obtained in a buffer of pH 5.9 is about 4000 times as large as the diffusion current of thiamin of the same concentration.

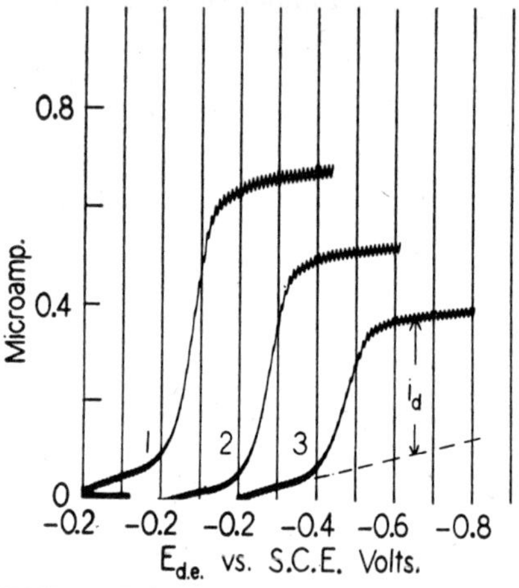

Fig. XLVI-4. Polarograms of riboflavin (vitamin B_2) in a 0.1 *M* phosphate buffer of pH = 7.2. Concentrations of riboflavin were: (*1*) 0.133, (*2*) 0.106, and (*3*) 0.0798 millimole per liter, corresponding to 50, 40, and 30 p.p.m.

Zambotti and Ferranto[118] used the polarographic method for the determination of the surface activity of vitamin B_1. Zambotti[119] also investigated the surface activity of the pyrimidine group of vitamin B_1. A solution of 2-methyl-4-amino-5-methylaminopyrimidine·2HCl (20 mg. per 100 ml.) was found to lower the maximum oxygen absorption.

Nicotinic Acid.* Lingane and Davis found a well-defined and easily

---

[117] A. Wollenberger, *Science*, **101**, 386 (1945).

[118] V. Zambotti and A. Ferrante, *Boll. soc. ital. biol. sper.*, **14**, 689–690 (1939); *Chem. Abstracts*, **34**, 6949[6] (1940).

[119] V. Zambotti, *Atti soc. med. chir. Padova*, **18**, 97–98 (1940); *Chem. Abstracts*, **38**, 1772[4] (1944).

* See also p. 814.

measurable wave in a tetramethylammonium borate buffer of pH = 9. In buffers of pH less than 7 the discharge of hydrogen masks the wave of nicotinic acid. A fairly well-defined wave is obtained in unbuffered solutions of potassium chloride, after the nicotinic acid is neutralized with sodium hydroxide.

From the above it is evident that an unbuffered solution of 0.1 *N* potassium chloride (pH = 7) is the best supporting medium for the simultaneous determination of riboflavin, thiamin, and nicotinic acid. This is

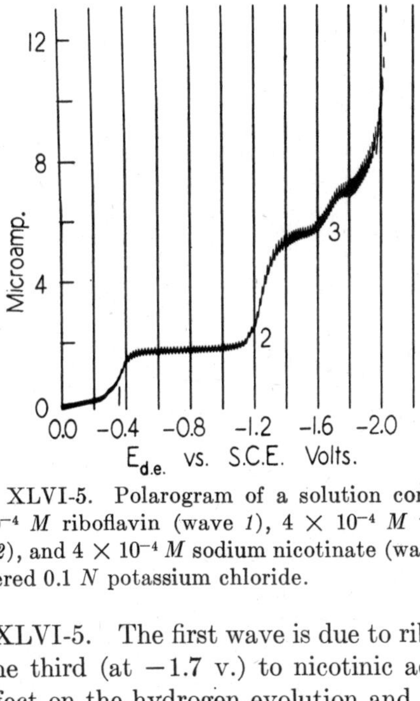

Fig. XLVI-5. Polarogram of a solution containing 3×10^{-4} *M* riboflavin (wave *1*), 4×10^{-4} *M* thiamin (wave *2*), and 4×10^{-4} *M* sodium nicotinate (wave *3*), in unbuffered 0.1 *N* potassium chloride.

illustrated in Fig. XLVI-5. The first wave is due to riboflavin, the second to thiamin, and the third (at −1.7 v.) to nicotinic acid.

The catalytic effect on the hydrogen evolution and the reduction wave of some *pyridine derivatives*, particularly of the amide of nicotinic acid, were investigated by Knobloch.[120] The catalyzed wave caused by the nicotinic acid amide appears at a pH of about 7.24, increases in height with acidity, and becomes indistinct at pH values lower than 4.78. The occurrence of the wave in neutral or acid solutions only indicates that the catalytic effect is due to the dissociated form of the amide (cation). The height of the catalytic wave increases proportionally to the capacity of the buffer and varies with the concentration of the catalyst according to Langmuir's

[120] E. Knobloch, *Collection Czechoslov. Chem. Communs.*, **12**, 407–422 (1947).

adsorption isotherm. Since the catalyzed wave occurs at a more negative potential (−1.8 v.) than that at which the amide is reduced, it is apparent that the reduced form of the amide is responsible for the catalytic effect. Assuming that the height of the catalyzed wave is directly proportional to the concentration of the cation of the amide, the dissociation constant of the latter can be found from a plot of the height of the catalytic wave *vs.* pH. Thus, the constant for the reduced amide was calculated to be $K_r = 5 \times 10^{-5}$ in the Brönsted sense. The potentiometric titration of the nicotinic acid amide gave the dissociation constant of the normal

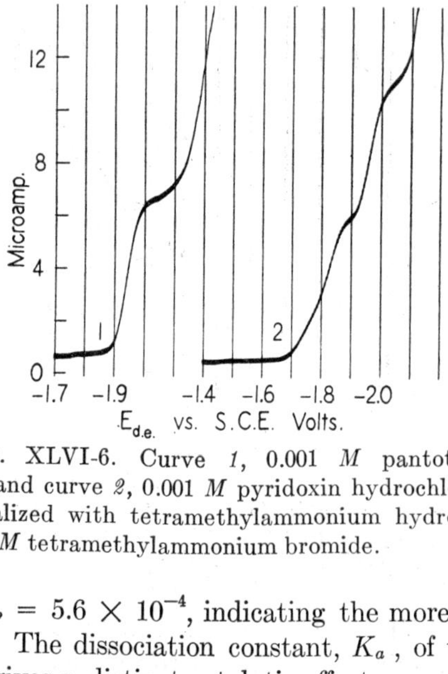

Fig. XLVI-6. Curve *1*, 0.001 *M* pantothenic acid, and curve *2*, 0.001 *M* pyridoxin hydrochloride; neutralized with tetramethylammonium hydroxide, in 0.1 *M* tetramethylammonium bromide.

(oxidized) form, $K_o = 5.6 \times 10^{-4}$, indicating the more basic character of the reduced form. The dissociation constant, K_a, of the pyridinium ion is 3.3×10^{-6}. It gives a distinct catalytic effect even at a pH of 9 where the amide gives no catalytic wave at all. This is additional evidence that this effect is due to the cation of the base.

The amide of nicotinic acid, as well as other reducible pyridine derivatives, can be determined polarographically with great accuracy in alkaline (0.1 *N* NaOH) or neutral (0.1 *N* KCl) solutions (see further, p. 814).

Pantothenic Acid and Pyridoxine. Lingane and Davis found these substances quite difficult to reduce, although they obtained fairly satisfactory waves by using tetramethylammonium bromide as supporting electrolyte, as shown in Fig. XLVI-6. It is necessary to neutralize the pantothenic acid and the pyridoxine hydrochloride with tetramethylammonium hydroxide in order to prevent the appearance of an interfering hydrogen wave.

8. LIPIDES

Carruthers and Suntzeff[121] found that lipides extracted from mouse tissues give cathodic waves at the mercury dropping electrode. Mixtures of dioxane and water were found to be the most suitable solvent for electrolysis and tetra-*n*-butylammonium iodide was used as the supporting electrolyte. Lipides were extracted with alcohol–ether mixtures from normal or hyperplastic epidermis and from carcinomas. Lipides from normal or hyperplastic epidermis extracted with alcohol–ether mixtures containing 40, 50, 60, or 75 per cent ether, gave a double wave. The half-wave potentials of the two waves were found to be independent of the concentration of the lipide and of the dioxane concentration between 62.5 and 37.5 per cent. On the other hand, a single wave was obtained with lipides obtained from carcinoma by extraction with alcohol–ether mixtures containing 60 and 75 per cent ether. The wave could be found occasionally only if the extractant contained 50 per cent or less ether. The half-wave potential of this single wave was nearly the same as that of the second part of the double wave of epidermis. Other tests with lipides which were performed after prolonged treatment of the tissues with 95 per cent alcohol gave similar differences between the epidermis and carcinoma. The differences found in the polarographic behavior of the material from mouse epidermis and from the tumors are considered to be an indication that an alteration in the structure of a lipide occurs during the process of epidermal carcinogenesis in mice.

[121] C. Carruthers and V. Suntzeff, *Science*, **108**, 450–454 (1948); *Cancer Research*, **9**, 210 (1949).

Part Six

AMPEROMETRIC TITRATIONS

CHAPTER XLVII

Amperometric Titrations

1. INTRODUCTION

The classical types of electrometric titration are the potentiometric and conductometric methods. In potentiometric titrations the potential of a suitable indicator electrode is measured during the titration, and the end point is characterized by a more or less pronounced change in the potential. In conductometric titrations the electrical conductance of the solution is measured during the titration, and the end point is found graphically as the point of intersection of straight lines giving the changes in conductance before and after the equivalence point. To these two methods may now be added another which we call "amperometric titration." In amperometric titrations the *current* which passes through the titration cell between an indicator electrode and an appropriate depolarized reference electrode at a suitable applied e.m.f. is measured as a function of the volume of a suitable titrating solution. Quite generally, the end point of an amperometric titration is found as the point of intersection of two lines giving the change in the current before and after the equivalence point.

It is of historical interest to mention that the principle of amperometric titrations was proposed as early as 1897 by Salomon[1] and it was later used by Nernst and Merriam.[2] It is peculiar that these papers have so long escaped the attention of analytical chemists. It was shown by these early workers that the "diffusion current" is proportional to the concentration of the electroreducible substance. In a paper entitled, "On a Galvanometric Titration Method," Salomon described the amperometric titration of chloride with silver nitrate. He placed two silver electrodes in a dilute potassium chloride solution, applied an e.m.f. of 0.1 v. and titrated with silver nitrate solution. Up to the equivalence point the "residual current" was very small, but at the equivalence point both electrodes became depolarized, and a sudden large deflection of the galvanometer was observed. Nernst and Merriam[2] placed two palladium electrodes,

[1] E. Salomon, *Z. physik. Chem.*, **24,** 55 (1897); **25,** 366 (1898); *Z. Elektrochem.*, **4** 71 (1897).

[2] W. Nernst and E. S. Merriam, *Z. physik. Chem.*, **53,** 235 (1905).

saturated with hydrogen, in solutions of various acidities and alkalinities and determined the current which passed on application of a small e.m.f. In the titration of an acid with a base they obtained a V-shaped titration curve with a minimum current at the end point. Fresenius[3] later investigated the possibility of determining small hydrogen ion concentrations by use of the above principle, but in this particular instance the method is of little importance since simpler and better methods are available.

The "galvanometric titration" method proposed by Salomon in 1897 resembles very closely the so-called "dead-stop end point method" of of Foulk and Bawden.[4] In their application of this method to various oxidation-reduction titrations, Guzman and Rancano[5] proposed for it the name "depolarimetry," since the end point is usually characterized by a sudden depolarization or polarization of one or of both of the electrodes. The above titration techniques are not amperometric in the strict sense of the word because the current is not measured throughout the entire titration, but the end point is simply noted by the sudden increase or sudden cessation of current. Heyrovsky and Berezicky[6] used a depolarized anode and the dropping mercury electrode as indicator electrode in the titration of barium with sulfate and suggested the name *polarographic titration.* Majer,[7] discussing the titration of lead with sulfate and some general characteristics of the method, introduced the term *polarometric titrations.* As it is the current which is measured during the titration, we prefer to call this type of titration *amperometric titration*, in analogy to the terms potentiometric and conductometric titrations.

2. AMPEROMETRIC TITRATIONS WITH THE DROPPING MERCURY ELECTRODE

In order to explain the principle of amperometric titrations we will first consider the current–voltage curve obtained with a solution of an electroreducible substance, such as lead ion, at the dropping mercury electrode. As a reference electrode we may use an outside saturated calomel electrode connected to the titration cell by a low resistance salt bridge (see Chap. XVII). When the lead solution contains an excess of indifferent electrolyte we obtain a current–voltage curve like that shown in Fig. XLVII-1. The

[3] L. R. Fresenius, *Z. physik. Chem.*, **80**, 481 (1912).

[4] C. W. Foulk and A. T. Bawden, *J. Am. Chem. Soc.*, **48**, 2045 (1926).

[5] J. Guzman and A. Rancano, *Anales soc. españ. fís quím.*, **32**, 590 (1934). J. Guzman, *ibid.*, **33**, 109 (1935); see also *Z. anal. Chem.*, **103**, 445 (1935).

[6] J. Heyrovsky and S. Berezicky, *Collection Czechoslov. Chem. Communs.*, **1**, 19 (1929).

[7] V. Majer, *Z. Elektrochem.*, **42**, 120, 122 (1936).

current, i, is plotted on the ordinate and the potential of the dropping electrode, $E_{d.e.}$, is plotted on the abscissa. When $E_{d.e.}$ is greater than A the current remains practically constant until $E_{d.e.}$ becomes so large that the reduction of some other substance present in the solution is initiated. In Fig. XLVII-1 this happens at point B. The constant current is the diffusion current, i_d, which is proportional to the concentration of the electroreducible substance.

Suppose that we titrate a solution of lead ions with a reagent such as oxalate, which forms an insoluble lead salt and which is not reduced at an applied e.m.f. between A and B. An e.m.f. which has a value between A and B is applied to the cell. Upon addition of the oxalate solution the lead is precipitated and the diffusion current of lead decreases in proportion

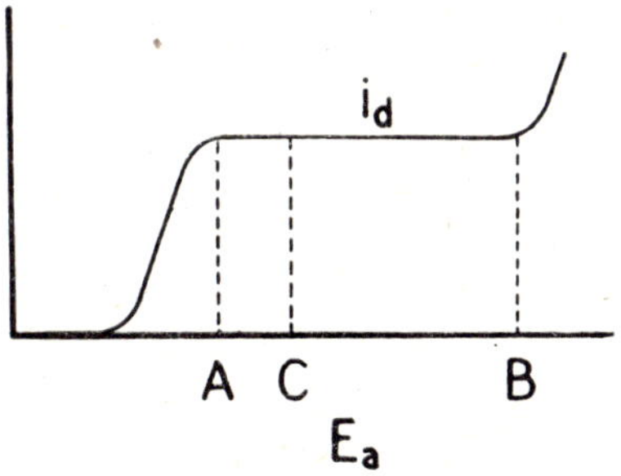

Fig. XLVII-1. Schematic current-voltage curve with the dropping electrode.

to the decrease in the lead concentration. If we neglect for the moment the solubility of the lead oxalate, the current decreases during the titration until it becomes equal to the very small value of the residual current of the medium at the end point. Upon further addition of oxalate the current will remain unchanged. Plotting the current against the volume of reagent added does not usually yield a perfectly straight line, but the line has a slight curvature due to the fact that the solution is diluted by the addition of the reagent. This dilution effect can be easily corrected for by multiplying the measured values of the current by a term $(V + v)/V$, where V denotes the original volume of the solution titrated and v the volume of reagent added at any point. In Fig. XLVII-2 are shown the theoretical titration lines for the titration of a 0.01 M lead solution with 0.01 M oxalate (curve *1*), and with 0.1 M oxalate (curve *2*). For the sake of convenience the residual current has been assumed to be equal to zero, but the same considerations hold when the residual current is not negligibly small. Curve *3* is a straight line and was obtained from *1* and *2* by applying the volume correction:

$$i_{\text{corr.}} = i_{\text{meas.}} \left(\frac{V + v}{V} \right)$$

It is seen that curve *1* corresponding to the more dilute oxalate solution deviates much more from a straight line than curve *2*. In order to keep the dilution effect very small it is necessary to use a reagent which is 10 to 20 times more concentrated than the solution of the substance to be titrated.

Quite generally, when the solubility of the reaction product is negligibly

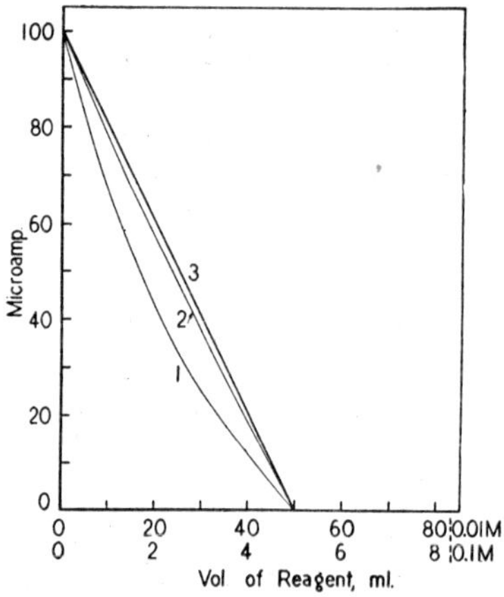

Fig. XLVII-2. Schematic amperometric titration curves of 50 ml. of 0.01 *M* lead ion with 0.01 *M* (curve *1*) and 0.1 *M* (curve *2*) oxalate ion. Curve *3* is corrected for volume change.

small it suffices to measure some four or five values of the current about 10 to 20 per cent before and after the end point. Under such conditions it is hardly necessary to apply the volume corrections if the reagent is 10 to 20 times more concentrated than the substance titrated. On the other hand, when the solubility is not negligible (*vide infra*) measurements close to the equivalence point have no practical value, and measurements are made only at points considerably removed from the end point where the solubility is negligibly small as a result of the common ion effect. Under such conditions the volume correction must always be applied. In the following discussion the titration lines shown have all been corrected for the dilution effect, unless stated otherwise.

From the above it is evident that this method of finding the end point is quite different from that used in potentiometric titrations and resembles much more the graphical method used in conductometric titrations. However, the theory underlying the latter type of titration is quite different from that of amperometric titrations. At this point it may also be mentioned that usually enough of an indifferent salt is added at the beginning of the titration to eliminate the migration current (see Chap. VII). (An indifferent salt is a salt which is not reduced at the particular applied e.m.f. employed for the titration.)

We have discussed the case in which an electroreducible substance is titrated with a reagent which is not electroreducible at the particular e.m.f. employed. In the converse case in which a nonreducible substance is titrated with a reducible reagent, the current remains equal to the residual current until the end point is reached, and then increases from

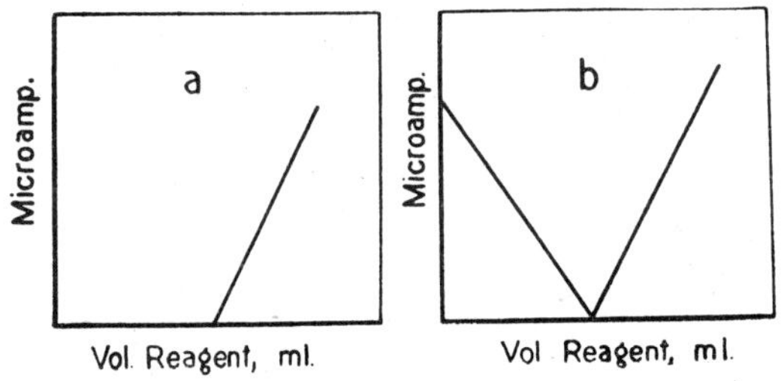

Fig. XLVII-3. (*a*) Amperometric titration curve of a nonreducible substance with a reducible reagent; (*b*) titration curve when the substance titrated and the reagent are both reducible.

there on as shown in Fig. XLVII-3*a*. When the substance to be titrated and the reagent are both electroreducible at the particular applied e.m.f. employed the current decreases until the end point is reached, and then increases again when an excess of the reagent has been added as shown in Fig. XLVII-3*b*.

It is well known that several anions, such as halides, thiosulfate, sulfite, thiocyanate, and sulfide, yield "anodic" waves at the dropping mercury electrode. This fact can be used for the amperometric titration of some of these ions with a suitable reagent. For example, the anodic wave of iodide (*2*) and the cathodic wave of mercuric nitrate (*1*) (maximum suppressed with gelatin) are shown schematically in Figure XLVII-4. It is seen that at a potential of the dropping electrode more positive than *B* the anodic diffusion current of iodide ion is obtained, while at values of $E_{d.e.}$ more negative than *A* the cathodic diffusion current of mercuric ion is measured. If an iodide solution were titrated with a solution of mercuric nitrate, with the potential of the dropping electrode adjusted to a value between *A* and *B*, the originally negative diffusion current of iodide should decrease during the titration, become zero at the end point, and change direction (become positive) upon addition of an

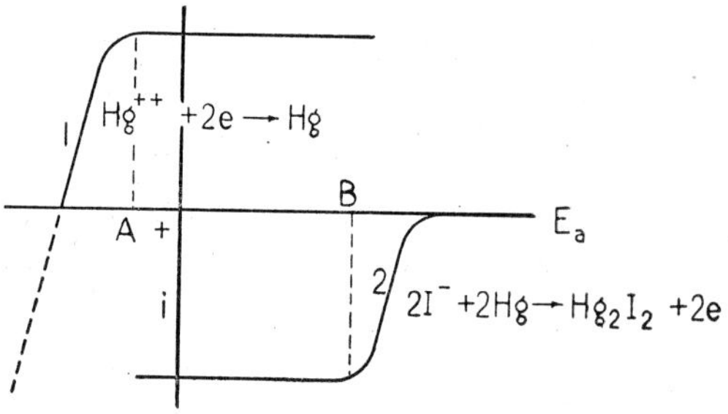

Fig. XLVII-4. Schematic diagram of the anodic wave of iodide ion (curve *2*), and the cathodic wave of mercuric ion (curve *1*).

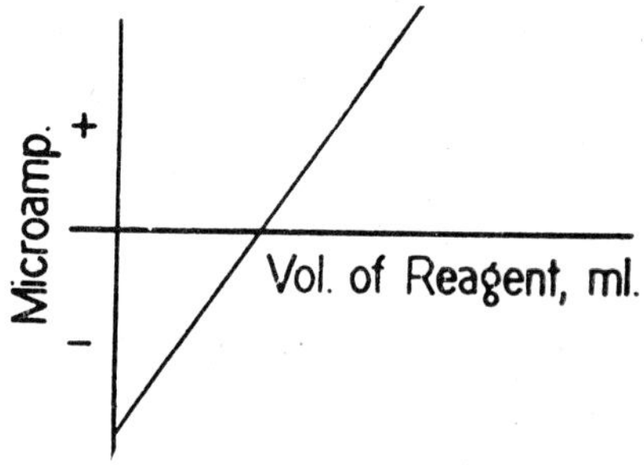

Fig. XLVII-5. Theoretical titration line of iodide ion titrated with mercuric ion. (A similar titration curve is obtained in the titration of chloride ion with silver ion, and in the titration of titanous salts in tartrate solution with various oxidizing agents.)

excess of mercuric nitrate. The titration line obtained would be as shown in Fig. XLVII-5. Actually, it is found that the diffusion current of iodide is not strictly proportional to the iodide ion concentration and for this reason the experimental titration line is not quite straight during the early stages of the titration (see Figure XLVII-22); although close to the end point the points do lie on a straight line. In general, the slope of the line before and after the end point is not the same because the diffusion coefficient of the substance being titrated is usually different from that of the reagent. For example, the diffusion current of mercuric mercury is smaller than that of iodide at the same normality, and this is the reason that the slope of the titration line in Fig. XLVII-22 becomes smaller after the end point has been passed. From a practical viewpoint, it is very simple to add mercuric nitrate solution until the current has just become equal to zero. This is analogous to a potentiometric titration carried out to the "equivalence point potential."

When certain conditions are fulfilled it is also possible to perform amperometric *oxidation-reduction* titrations with the dropping electrode. In order to make the principle clear we will first consider the waves obtained with a mixture of titanous and titanic titanium, which are represented in

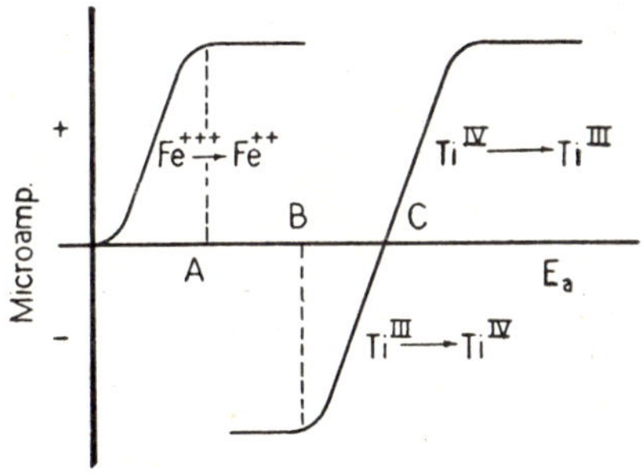

Fig. XLVII-6. Schematic current–voltage curves of a mixture of titanic and titanous titanium and ferric iron, in a tartrate medium.

Fig. XLVII-6. In order to get a normal reduction wave of the Ti^{++++} it is necessary to add some salt which will form a complex with the titanic ion, such as citrate or tartrate.[8] At a potential of the dropping electrode more negative than *C* the titanic ion complex is reduced, while the titanous titanium is oxidized when $E_{d.e.}$ is more positive than *C*. When the potential is more positive than *B* the diffusion current due to the oxidation of titanous titanium is obtained. In Fig. XLVII-6 the reduction wave of ferric iron is also shown schematically. It is seen that when $E_{d.e.}$ is more negative than A the diffusion current of ferric iron is obtained. If we titrate a titanous solution with ferric iron, with the potential of the dropping electrode set to a value between *A* and *B*, it is readily seen that we should obtain a titration line of the same type as that in Fig. XLVII-5. Actually, Spalenka[9] has carried out the titration of titanous solutions with several

[8] R. Strubl, *Collection Czechoslov. Chem. Communs.*, **10**, 475 (1938).
[9] M. Spalenka, *Collection Czechoslov. Chem. Communs.*, **11**, 146 (1939).

oxidizing agents, such as ferric iron, chromate, bromate, ferricyanide, and iodate. He obtained titration lines of the same general form as that shown in Fig. XLVII-5.

It should be emphasized again that the slope of the line before the end point will be somewhat different from that after the end point. If the diffusion coefficient of the oxidizing agent is greater than that of the titanous titanium, the slope of the line after the end point will be steeper than before. The converse will be true if the oxidizing agent has a smaller diffusion coefficient than the titanous titanium.

The type of amperometric titration described so far is based on the fact that either the substance titrated or the reagent or both produces a wave (either cathodic or anodic) with a well-defined diffusion current. In case neither the substance titrated nor the reagent yields a wave it is, under suitable conditions, still possible to carry out an amperometric titration by adding a "polarographic indicator" for the detection of the end point. This principle was introduced by Ringbom and Wilkman.[10] The indicator substance must yield a diffusion current at the applied voltage. Under ideal conditions it should react with the reagent after all of the titrated substance has reacted; this condition is hardly ever fulfilled. Still the end point can be found when only part of the indicator reacts during the main reaction. As an illustration consider the titration of aluminum with fluoride[10] with ferric iron as indicator (see p. 475). With the S.C.E. as reference electrode ferric iron yields a diffusion current at zero applied voltage. Both aluminum and iron form a complex with fluoride, but the aluminum complex is considerably more stable than that of the iron. In a medium of 50 per cent ethanol nearly all the aluminum reacts with the fluoride before the iron does. During the titration the diffusion current of iron decreases slowly (by reaction with fluoride), and very rapidly near the end point. The location of the end point corresponds to a zero current. (In this example a negative current is found after the end point because the residual current is negative under the specified conditions.) It is seen that the end point corresponds to the sum of aluminum and the added amount of the indicator ferric iron. In order to find the true end point in the aluminum titration a correction must be applied for the amount of reagent bound by the indicator. From a practical viewpoint the application of this correction is a disadvantage.

Quite generally, the application of the indicator method will remain limited because, in general, it will be hard to satisfy the condition that the bulk of the indicator must not react with the reagent until practically all of the substance to be titrated has reacted.

---

[10] A. Ringbom and B. Wilkman, *Acta Chem. Scand.*, **3**, 22 (1949).

Other examples of titration with indicators are those of calcium or magnesium with fluoride (ferric iron as indicator) or calcium with oxalate. In the latter titration a relatively large amount of cadmium is added as the indicator. The solubility product of cadmium oxalate is about 8 times larger than that of calcium oxalate. Hence both precipitate together upon addition of oxalate to a mixture of calcium and cadmium. The titration is carried out in 50 per cent ethanol. With 0.01 *M* calcium solution (which was made 0.008 *M* in cadmium) an accuracy of the order of 1 per cent was found.

Another type of amperometric titration in which no chemical reaction occurs in the solution between the substance titrated and the titrant added, but in which the cathodic (or anodic) current of the substance "titrated" is *compensated* by the anodic (or cathodic) current of the titrant, is mentioned briefly in Chap. VI. Under these conditions the following relation holds at the end point, where the current is zero:

$$\frac{C(\text{substance titrated})}{C(\text{titrant})} = \left(\frac{D(\text{titrant})}{D(\text{substance titrated})}\right)^{1/2} = R$$

Kolthoff and Miller[11] "titrated" an oxygen solution in 0.1 *N* sodium hydroxide in this way at a potential of −0.45 v. with a sodium sulfide solution until the current became zero.

Lingane[12] utilized the same principle in the "titration" of stannous tin with cupric copper. Stannous tin in an acidic tartrate solution produces an anodic wave with a half-wave potential of about −0.3 v. Cupric copper in the same medium produces a reduction wave to metallic copper with a half-wave potential of −0.1 v. One would expect that the stannous solution would reduce the cupric solution. However, Lingane found that this reaction does not occur at a measurable speed at room temperature. Still, it is possible to titrate Sn^{++} in a tartrate medium with a cupric salt solution by taking advantage of the fact that in the potential range from −0.1 to −0.3 v. the cathodic diffusion current of copper *compensates* the anodic diffusion current of tin.

In the titration of stannous solutions with copper, Lingane found that the value of $(D_{Sn}/D_{Cu})^{1/2}$ varied with the concentration of tin titrated. With tin concentrations smaller than 0.001 *M*, *R* was found equal to 1.02. When C_{Sn} was 0.002 *M*, *R* was found to be 1.00, when C_{Sn} was 0.004 *M*, *R* was 0.98. The change in the value of *R* is not explained. Lingane carried out the titration of stannous tin in a medium which was 0.38 *M* in disodium tartrate, 0.12 *M* in sodium hydrogen tartrate, 0.12 *M* in sodium

[11] I. M. Kolthoff and C. S. Miller, *J. Am. Chem. Soc.*, **62,** 2171 (1940).

[12] J. J. Lingane, *J. Am. Chem. Soc.*, **65,** 866 (1943).

chloride, and 0.01 per cent in gelatin (pH 4.3). Cupric sulfate solution (about 0.01 *M*) was added from a microburet. For the titration of quantities of tin less than 10 mg. per 50 ml., the procedure seems to be of special practical advantage.

Limits of Solubility in Amperometric Titrations. One of the advantages of amperometric titrations is that they can be carried out in cases in which the solubility relations are such that the potentiometric or the indicator methods no longer yield good results. The reason for this is that in amperometric titrations readings near the equivalence point have no special significance. When the readings are recorded in regions where the solubility is repressed more or less quantitatively by the common ion effect, two straight lines will be found which intersect at the equivalence point. Let us first consider a precipitate AB of which only the cation A^{n+} or the anion B^{n-} is electoreducible at a particular applied e.m.f. We will assume that good results can be obtained when the solubility at 25 per cent before or after the equivalence point is equal to or less than 1 per cent of the original concentration of A or B. For the sake of simplicity we will neglect the volume changes during the titration. When the original concentration of A^{n+} is equal to c, the following condition should be fulfilled. At 25 per cent before the equivalence point, when A^{n+} is being titrated with B^{n-}:

$$[B^{n-}] \lesseqgtr \frac{c}{100}$$

$$[A^{n+}] = \frac{c}{4} + \frac{c}{100} \cong \frac{c}{4}$$

In order to have the above condition satisfied:

$$[A^{n+}][B^{n-}] = S_{AB} \lesseqgtr \frac{c^2}{400} \qquad (1)$$

Employing the above assumption we have calculated in Table XLVII-1 the upper values of the solubility product, and the solubility of the salt AB at which the amperometric titration still yields good results.

The figures in Table XLVII-1 are based on rather arbitrary assumptions and hence they have no exact significance, but they do give the *order* of the limits of the solubility product and of the solubility at the indicated concentration c at which the amperometric titration should yield good results, *provided solubility equilibrium is attained rapidly during the titration.* If we had assumed that accurate results can be obtained when the solubility is less than 1 per cent of c at 10 per cent before or after the equivalence point the limit of the solubility product would have been found 2.5 times smaller than given in Table XLVII-1. The limiting solubility product would be two times greater than the figures in the table if the above condition were only fulfilled 50 per cent before or after the equivalence point.

The above calculations are given for the case in which only either A or B is electroreducible at the applied e.m.f. If both are reducible, eq. 1 has to be changed. Assuming again that the same conditions have to be fulfilled as given in Table XLVII-1, we find that:

$$S_{AB} = \frac{c}{(b+1)100} \times \frac{c}{4} = \frac{c^2}{400(b+1)} \tag{2}$$

The constant b is the ratio of the diffusion currents of B^{n-} and A^{n+} at the same molar concentration. For example, in the case of silver iodate, the

TABLE XLVII-1

UPPER LIMITS OF THE SOLUBILITY PRODUCT AND OF THE SOLUBILITY OF AB AT WHICH AMPEROMETRIC TITRATION IS JUST POSSIBLE. SOLUBILITY AT 25 PER CENT BEFORE OR AFTER THE EQUIVALENCE POINT IS EQUAL TO 0.01 c, c BEING THE ORIGINAL CONCENTRATION OF A OR OF B

Original concentration of A or B, molar	Upper limit of solubility product, S_{AB}	Upper limit of solubility, S
0.01	$< 2.5 \times 10^{-7}$	$< 5 \times 10^{-4}$
0.001	$< 2.5 \times 10^{-9}$	$< 5 \times 10^{-5}$
0.0002	$< 10^{-11}$	$< 10^{-5}$
0.0001	$< 2.5 \times 10^{-11}$	$< 5 \times 10^{-6}$

iodate ion in its reduction to iodide (6 electrons) yields a diffusion current which is 5 times greater than that of silver (one electron).

If the precipitate has the composition A_2B or AB_2 the relation between the solubility product and the concentration is no longer given by eqs. 1 or 2. If either A or B is electroreducible we find that the limiting condition is:

$$S_{A_2B} \lesseqgtr \frac{c^2}{80{,}000} \tag{3}$$

In eq. 3 the concentration c is expressed in *normality* and not in molarity. If both A and B are electroreducible, and at the same normality the diffusion current of B^- is b times greater than that of A^{++} (precipitate AB_2), it is calculated that:

$$S \lesseqgtr \frac{c^2}{(b+1)^2 80{,}000} \tag{4}$$

In order that the titration be practicable, solubility equilibrium must be attained relatively quickly after each addition of reagent. When this is not the case addition of an organic solvent, such as ethanol or acetone, as a rule has a favorable effect. The solubility of most inorganic precipitates is decreased by the addition of organic solvents, and solubility equilibrium is attained more rapidly than in a purely aqueous medium.

Location of Minimum Current with Regard to Equivalence Point in a Precipitation Reaction. If only one of the two constituents of the precipitate A_nB_m is reducible the current does not pass through a minimum during the titration. If A is reduced the current approaches asymptotically the residual current of the medium with an excess of reagent. On the other hand, if both A and B are reducible a minimum value of the current occurs at or near the equivalence point. Theoretically, the minimum only coincides with the equivalence point when the precipitate has the composition AB and the diffusion currents of A and B are the same at equal concentrations. If the latter condition is not fulfilled, or when the precipitate is of unsymmetrical valence type, *e. g.*, A_2B or AB_2, etc., the titration curve at both sides of the equivalence point is not symmetrical.

Let us consider a precipitate of the type AB the cation of which produces a much smaller diffusion current than the anion at the same molarity, for example, silver iodate. At the same molarity the diffusion current of iodate is about 5 times greater than that of silver. The solubility product of silver iodate is equal to 4.9×10^{-8} in a medium which is 0.1 M in potassium nitrate and 0.1 M in acetic acid. The diffusion current, i_{Ag}, produced by the silver ions is:

$$i_{Ag} = k_{Ag}\,[Ag^+] \tag{5}$$

in which k_{Ag} is a constant and $[Ag^+]$ is the silver ion concentration. Similarly we find for the diffusion current i_{IO_3} of iodate:

$$i_{IO_3^-} = k_{IO_3}\,[IO_3^-] = 5k_{Ag}\,[IO_3^-] \tag{6}$$

The total diffusion current due to a saturated solution of silver iodate then is:

$$i_{AgIO_3} = 6k_{Ag}\,s \tag{7}$$

in which s is the molar solubility of the silver iodate.

When we add to a saturated solution of silver iodate enough silver nitrate to produce a silver ion concentration of 10^{-4} M, the total silver ion concentration is $[Ag^+] = 10^{-4} + s$, where s is the molar solubility in the resulting solution. Then we find:

$$[Ag^+] [IO_3^-] = (10^{-4} + s)s = S_{AgIO_3} = 4.9 \times 10^{-8} \quad (8)$$

and:

$$s = 1.77 \times 10^{-4} \quad (9)$$

At the equivalence point the solubility of silver iodate is equal to $\sqrt{S_{AgIO_3}}$, or 2.21×10^{-4} M, and the current at the equivalence point is:

$$i_{AgIO_3} = 6k_{Ag} \times s = 6k_{Ag} \times 2.21 \times 10^{-4} = 13.26 \times k_{Ag} \times 10^{-4} \quad (10)$$

When the excess of silver added is equal to 10^{-4}, we find:

$$i = k_{Ag} \times 10^{-4} + 6k_{Ag} \times 1.77 \times 10^{-4} = 11.62\ k_{Ag} \times 10^{-4} \quad (11)$$

Hence with this slight excess of silver the current is smaller than at the equivalence point. Carrying out the above calculations for different excesses of silver we find that a minimum current occurs when the excess of silver is about 4×10^{-4} M.

It is simpler and more exact to derive the location of the minimum in the following way.[13] Consider a precipitate of the composition AB, whose solubility product is S. Let V be the volume of solution containing one mole of A, and assume that the correction due to dilution during the titration is negligibly small. Consider a point during the titration when x moles of B have been added, y moles of AB have been formed, and the concentrations of A and B are C_A and C_B, respectively. It is evident that the following relations hold:

$$C_A \cdot C_B = S \quad (12)$$

$$C_A \cdot V + y = 1 \quad (13)$$

$$C_B \cdot V + y = x \quad (14)$$

From these equations it readily follows that:

$$C_B^2 + C_B \frac{1 - x}{V} - S = 0 \quad (15)$$

$$C_A^2 - C_A \frac{1 - x}{V} - S = 0 \quad (16)$$

The observed current at this point will be the sum of the diffusion currents

[13] I. M. Kolthoff and H. A. Laitinen, *Rec. trav. chim.*, **59**, 922 (1940).

of A and B, and is given by:

$$i = k_A C_A + k_B C_B \tag{17}$$

where k_A and k_B are the diffusion current constants of A and B, respectively. Solving for C_A and C_B from eqs. 15 and 16 and substituting in eq. 17 we obtain:

$$i = (k_A - k_B)\left(\frac{1 - x}{2V} + \frac{k_A + k_B}{2}\right)\sqrt{\frac{1 - x^2}{V} + 4S} \tag{18}$$

Eq. 18 is perfectly general for a precipitate of composition AB. If both A and B are reducible at the particular applied e.m.f., k_A and k_B are positive numbers. If either A or B is not reducible, k_A or k_B becomes equal to zero, and there is no minimum in the titration curve.

To solve for the point of minimum current, let the slope $di/dx = 0$. From eq. 18:

$$\frac{di}{dx} = -\left(\frac{k_A - k_B}{2V}\right) - \frac{(k_A + k_B)V^2}{2(1 - x)} \cdot \frac{1}{\sqrt{\frac{(1 - x)^2}{V} + 4S}} \tag{19}$$

and at the point of minimum current:

$$(k_A + k_B)\left(\frac{1 - x}{V}\right)_{\text{min.}} = -(k_A - k_B)\sqrt{\frac{(1 - x^2)}{V} + 4S} \tag{20}$$

Solving for $\left(\frac{1 - x}{V}\right)$, we have

$$\left(\frac{1 - x}{V}\right)_{\text{min.}} = -(k_A - k_B)\sqrt{\frac{S}{k_A k_B}} \tag{21}$$

According to eq. 21, the deviation of the minimum current point from the equivalence point ($x = 1$) depends on the solubility of the precipitate and the difference between the diffusion current constants of A and B.

We will now consider a precipitate of composition AB_2 whose solubility product is S. Again, let V be the volume of solution containing one mole of A, and consider a point during the titration where x moles of B have been added, and y moles of AB_2 have been formed. The equations for this case are:

$$C_A \cdot C_B^2 = S \tag{22}$$

$$C_A \cdot V + y = 1 \tag{23}$$

$$C_B \cdot V + 2y = x \tag{24}$$

Eliminating y, we have:

$$C_B^3 + \frac{2 - x}{V} \cdot C_B^2 - 2S = 0 \tag{25}$$

$$4C_A - 4\left(\frac{2 - x}{V}\right) \cdot C_A^2 + \left(\frac{2 - x}{V}\right)^2 C_A - S = 0 \tag{26}$$

Since exact solutions of eqs. 25 and 26 in general are very involved, the theoretical titration curve is best obtained for each special case by solving eqs. 25 and 26 by approximate methods and substituting in eq. 17 to find the current for each value of x.

These relations have been checked experimentally by Kolthoff and Laitinen in the case of silver iodate. They determined the diffusion current of saturated solutions of pure silver iodate in a medium which was 0.1 N in potassium nitrate and 0.1 N in acetic acid in the presence of varying concentrations of silver nitrate and potassium iodate at 25° C. The acetic acid served to make the solution sufficiently acid to shift the reduction wave of iodate ion to a sufficiently positive potential so that the diffusion current of iodate could be obtained in a convenient range of potentials (−0.8 to −1.0 v. *vs.* S.C.E.). The direct interaction of silver ions with mercury from the dropping electrode was prevented by collecting the mercury drops in a small cup connected to a capillary tube through which the mercury could be drawn from the cell. The diffusion current constants of silver and iodate (k_{Ag} and k_{IO_3}) were determined by measuring the diffusion currents of various concentrations of silver nitrate and potassium nitrate and potassium iodate separately in the same potassium nitrate–acetic acid solution. In this way the solubility of silver iodate in the above medium was found to be 2.22×10^{-4} M, corresponding to a solubility product $S = 4.93 \times 10^{-8}$. The theoretical titration curve was calculated from eq. 18 and is plotted in Fig. XLVII-7.

The experimental points are in agreement with the calculated curve over a range of excess concentrations from 1.5×10[illegible] M silver to 3×10^{-4} M iodate. According to eq. 21, the theoretical minimum in the titration curve should occur at an excess concentration of 3.8×10^{-4} M silver. The minimum is extremely flat and was found at an excess concentration of 4.0×10^{-4} M silver, in good agreement with the theoretical value.

3. TECHNIQUE OF AN AMPEROMETRIC TITRATION

For the sake of economy and convenience it is recommended always to use an outside reference electrode instead of a large layer of mercury on the bottom of the titration cell (Chap. XVII). A saturated calomel electrode is the most convenient reference electrode, although any other depolarized electrode (silver–silver chloride, etc.) may be used. In view of the fact that the currents which pass through the cell during the titration are relatively small such reference electrodes remain practically depolarized during the titration. A titration cell of the type shown in Fig. XLVII-8 has proved to be very useful.

The connection between the solution in the cell and the reference electrode is made with an agar salt bridge, the gel being 3 per cent in agar and saturated with potassium chloride. In titrations in which chloride inter-

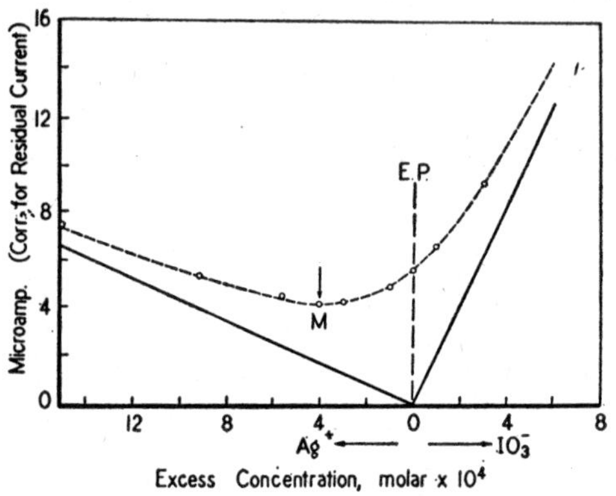

Fig. XLVII-7. Titration of silver ion with iodate ion. The dotted curve is the theoretical titration curve allowing for the solubility of silver iodate, and the circles are experimental points. The straight lines are the hypothetical titration lines that would be obtained if the solubility of silver iodate were negligibly small. The current values have been corrected for residual current.

feres the connection is made with an agar potassium nitrate bridge. The solution to be titrated is placed in the left-hand compartment. The wide horizontal tube connecting the two compartments is filled with an agar-salt gel. The liquid gel is placed in the connecting arm as shown in Fig. XLVII-9; the cork plunger prevents the outflow of the gel into the second compartment. The lower part of this compartment up to the top of the connecting tube is also filled with the gel. The reference electrode can be placed in the right-hand compartment, or the latter can be filled with saturated potassium chloride solution and connection with the reference electrode made with another salt bridge in the usual way. The resistance between the left-hand compartment and the reference electrode is usually negligibly small.

The titration cell is closed with a rubber stopper (not shown in Fig. XLVII-8) which has a number of holes, allowing the introduction of the tip of the buret, the capillary of the dropping electrode (drop time about 3 seconds), and an inlet and outlet tube for nitrogen. In Fig. XLVII-8

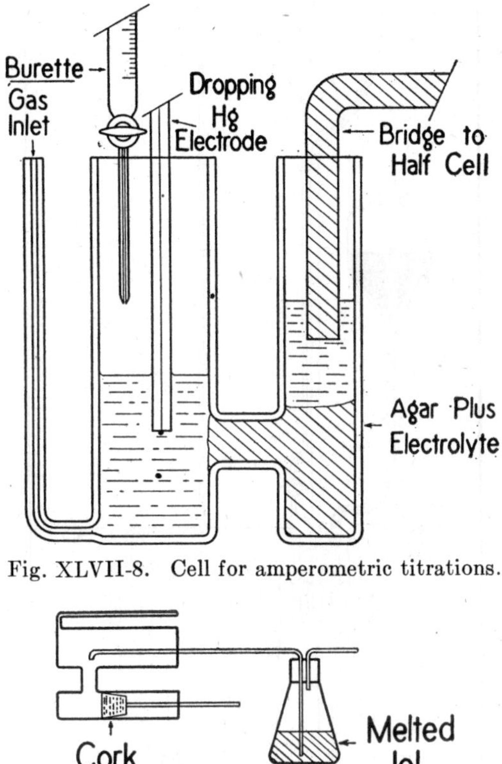

Fig. XLVII-8. Cell for amperometric titrations.

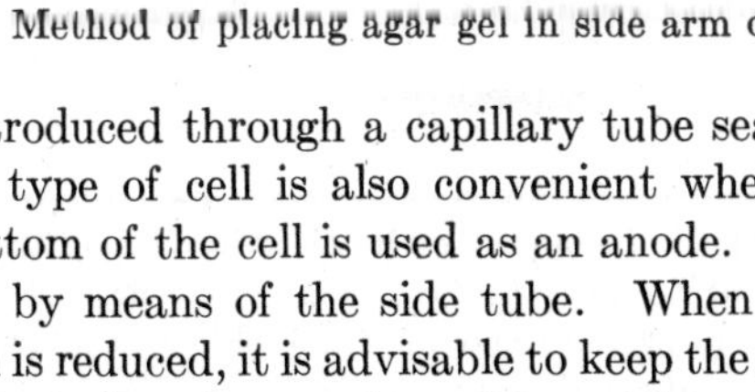

Fig. XLVII-9. Method of placing agar gel in side arm of titration cell.

the inert gas is introduced through a capillary tube sealed to the bottom of the cell. This type of cell is also convenient when a large layer of mercury in the bottom of the cell is used as an anode. Electrical connection is then made by means of the side tube. When the applied e.m.f. is such that oxygen is reduced, it is advisable to keep the air removed during the titration. This is done by passing nitrogen through the solution for 10 minutes, before starting the titration, and passing it through for 2 minutes after each addition of reagent. In routine work it is simpler to store, and introduce the reagent into the titration cell in the absence of air, as shown in Fig. XLVII-10.

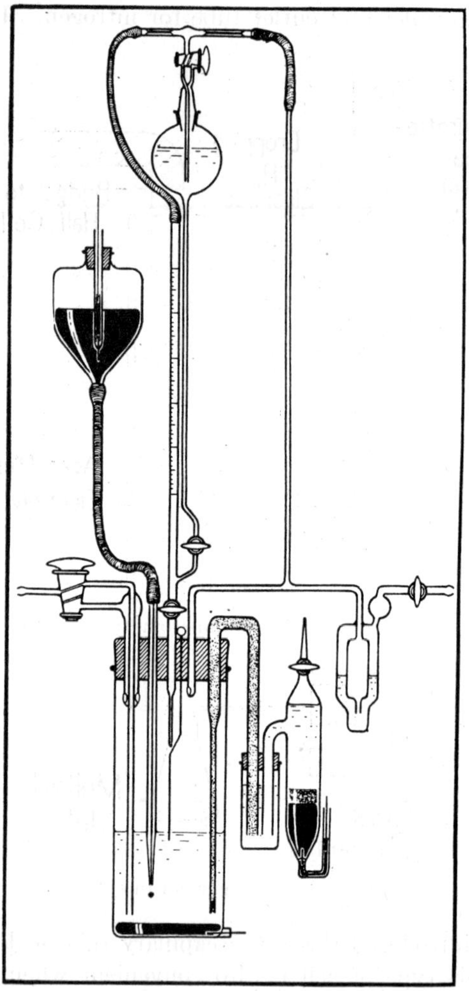

Fig. XLVII-10. Amperometric titration apparatus (according to Kolthoff and Langer, *J. Am. Chem. Soc.*, **62,** 211 (1940)). The reagent solution is contained in a microburet and freed from dissolved air with nitrogen before a titration. The two-way gas inlet in the left side of the cell permits the nitrogen to be bubbled either through the solution or over its surface.

Stock[14] described a simple polarizing unit for amperometric titrations. When using dropping mercury as indicator electrode, the growth and fall of the mercury drops causes the galvanometer spot to oscillate. However, a condenser of large capacity connected across the galvanometer shunts damps the oscillations (see Chap. XVI). A galvanometer of comparatively short period (3 sec. or even less) may then be easily read. For adequate damping, the parallel resistance of the galvanometer and its shunt, which depends upon the setting of the shunt, must not be too small.

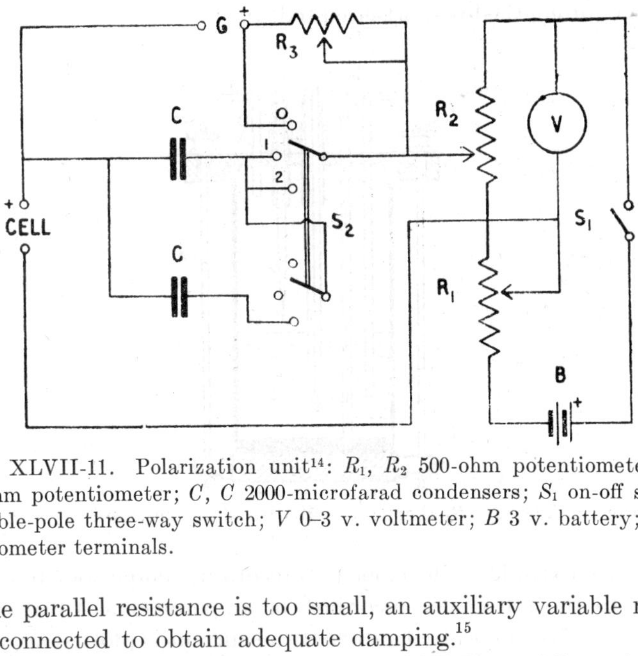

Fig. XLVII-11. Polarization unit[14]: R_1, R_2 500-ohm potentiometers; R_3 5000-ohm potentiometer; *C*, *C* 2000-microfarad condensers; S_1 on-off switch; S_2 double-pole three-way switch; *V* 0–3 v. voltmeter; *B* 3 v. battery; and *G* galvanometer terminals.

When the parallel resistance is too small, an auxiliary variable resistance may be connected to obtain adequate damping.[15]

Circuit arrangements are shown in Fig. XLVII-11. Most of the components are standard radio parts. A constant voltage drop, adjusted by rheostat R_1 and indicated by voltmeter *V*, is maintained across potentiometer R_2. The latter is provided with a graduated scale calibrated in the usual way and controls the voltage applied to the electrode system in the titration cell. Since the voltmeter is always adjusted to the same reading, it must be precise only at this reading. Accordingly, a compara-

[14] J. T. Stock, *Analyst,* **71,** 585 (1946).

[15] M. A. Fill and J. T. Stock, *Trans. Faraday Soc.*, **40,** 502 (1944).

tively cheap instrument may be used,[16] or a meter designed for other purposes may be adapted. Thus the present unit employs an available 0–500 microammeter, suitably equipped with a series resistance, R_1 being adjusted to give exactly full-scale deflection. A three-way switch S_2 enables one or both of the condensers to be connected. With the switch in position O, no damping occurs. The condensers should be of good quality and should be connected with due regard for their polarity markings. Rheostat R_3 enables adequate damping to be obtained if the parallel resistance of the galvanometer and shunt is small; when switch S_2 is in position O, this rheostat is automatically short-circuited.

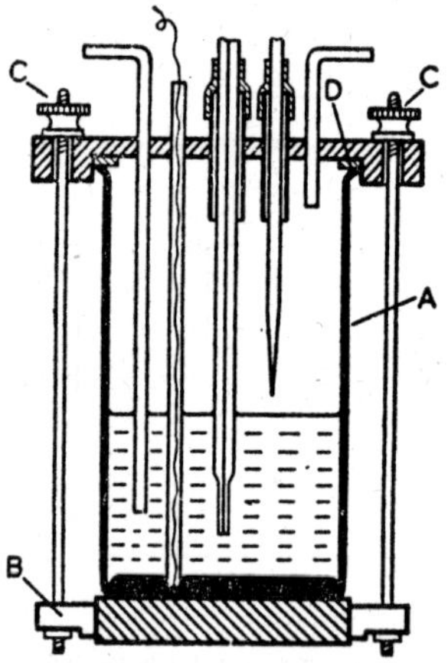

Fig. XLVII-12. Beaker-type cell with clamp-on cap.[18]

Barredo and Taylor[17] developed a recording potentiometer which is adaptable to the automatic recording of amperometric titrations. The reagent is added at a constant rate and the current is plotted against time. A break occurs at the end point.

Owing to the danger of breaking fragile electrode capillaries, buret jets, and the like, on attempting removal from rubber stoppers, Stock[18] prefers to use for beaker-type cells the Bakelite clamp-on cap shown in Fig. XLVII-12. The electrode capillary or buret jet passes through a short glass tube cemented into the cap, a short sleeve of rubber tubing preventing ingress of air. A tall-form 100-ml. Pyrex beaker A is used as the cell, and is retained by stirrup B. The thumbscrews CC enable the lip of the

[16] O. H. Müller, *J. Chem. Ed.*, **18**, 111 (1941).
[17] J. M. Gonzalez Barredo and J. K. Taylor, *Trans. Electrochem. Soc.*, **92**, 437 (1947).
[18] J. T. Stock, *Analyst*, **72**, 291 (1947).

beaker to be lightly pressed into rubber joint *D*, thus forming a gas-tight seal.

Langer[19] has described both macro- and micro-scale cells. The latter type is depicted in Fig. XLVII-13. It incorporates a saturated calomel reference electrode which surrounds the titration chamber, junction with the test solution being made through ground joint *A*.

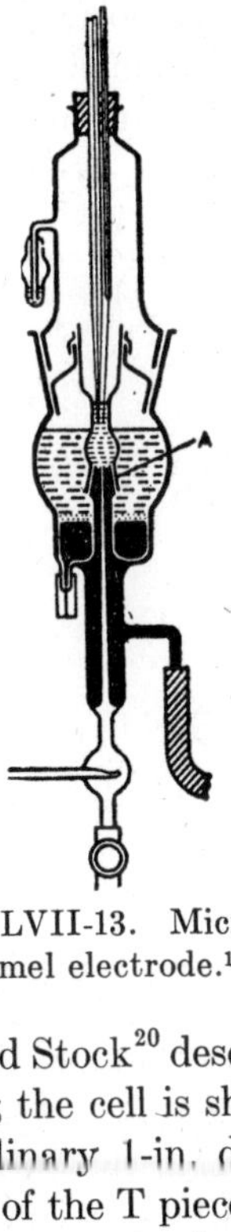

Fig. XLVII-13. Microcell incorporating calomel electrode.[19]

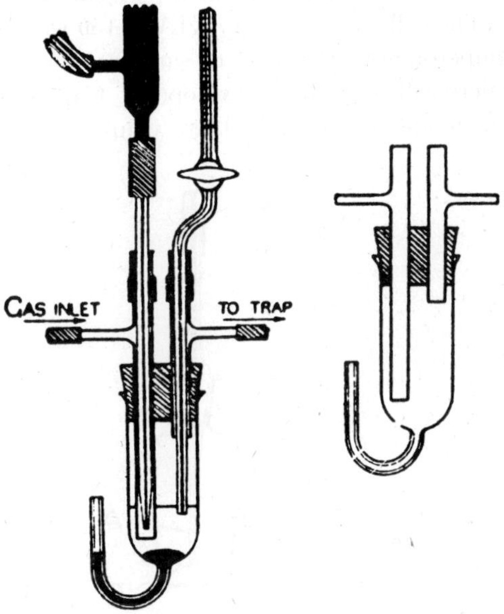

Fig. XLVII-14. Titration cell.[20]

Fill and Stock[20] describe a simple cell which permits the titration of 5-ml. portions; the cell is shown in Fig. XLVII-14. The outer portion is made from ordinary 1-in. diameter Pyrex boiling tube. The straight-through portions of the T pieces are so constructed that the dropping electrode and the jet of the microburet have a clearance of about 0.5 mm. within them. The annular spaces thus formed are ample for the entry and egress of the stream of nitrogen or hydrogen used initially to remove dissolved oxygen and subsequently to stir the solution after each addition of reagent. The longer T piece effectively protects the tip of the dropping electrode from accidental damage. Short jointing sleeves of rubber tubing permit ready removal of the dropping electrode and buret when desired.

[19] A. Langer, *Ind. Eng. Chem., Anal. Ed.*, **17**, 454 (1945).
[20] M. A. Fill and J. T. Stock, *Analyst*, **69**, 178 (1944).

A mercury pool is used as a quiet internal electrode. It was shown by Majer[21] that, in the presence of halide ions, the pool remains at constant potential, provided that its area is at least 1 $cm.^2$. In the titration cell a small amount of mercury (1–1.5 ml.) gives a pool of ample area. The requisite quantity of mercury is measured into the cell by means of the suctionless pipet shown in Fig. XLVII-15*a*. In a later paper Stock[22] described a titration cell which permits the titration of 0.5–1 ml. volumes.

The cell shown in Fig. XLVII-15*b* was devised for the polarography and amperometric titration of samples too small to be examined in the semimicro cell previously developed. Cap *A*, body *B*, and solution container *C* are made of Pyrex and are connected by ground joints. Several inter-

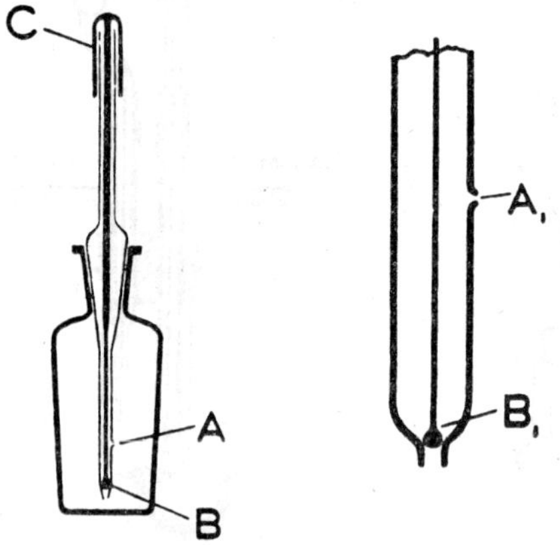

Fig. XLVII-15*a*. Suctionless pipet for titration cell.[20]

changeable solution containers are provided, so that a series of samples may be examined without dismantling the apparatus. Dropping mercury electrode *D* is held in position by a sleeve of rubber tubing and fits snugly in the tube surmounting the cap. Similar sleeves retain the outlet of the microburet *E*[23] and the gas inlet tube *F*. The extremities of the microburet and gas inlet tube are arranged tangentially as shown in (b) of the figure. Entry of gas produces a spinning motion in the liquid,[24] promoting mixing and elimination of dissolved oxygen. Since the immersion of the tip of the gas inlet tube need not exceed 1–2 mm., spurting is greatly re-

[21] V. Majer, *Collection Czechoslov. Chem. Communs.*, **7**, 146, 215 (1935); **9**, 360 (1937).

[22] J. T. Stock, *Analyst*, **71**, 583 (1946). See also J. T. Stock and M. A. Fill, *Metallurgia*, **30**, 277 (1946).

[23] J. T. Stock and M. A. Fill, *Metallurgia*, **31**, 103 (1944).

[24] M. Spalenka, *Collection Czechoslva. Chem. Communs.*, **11**, 146 (1939).

duced. Gas escapes through a trap attached to the side tube of *A*, the the narrow annulus around the dropping mercury electrode being sufficient to allow this.

Mercury is introduced into container *C* up to a mark etched on the walls. The solution to be examined (0.5–1 ml.) is then introduced and *C* is attached to the body of the cell. Electrical connection to the mercury pool is made by platinum wire *G*. After eliminating dissolved oxygen, the amperometric titration is carried out. The container is then detached, the tips of the dropping mercury electrode, gas inlet tube, and microburet are wiped with strips of filter paper, and another previously filled container is placed in position. For polarographic measurements the microburet is replaced by a plug of glass rod, or by a pencil-type saturated calomel electrode,[25] the tip of which is immersed in the liquid in the cell. The potential of the mercury pool may thus be examined.

The flow of mercury from the dropping electrode is normally insufficient to cause an excessive rise in the level in *C*. When the examination is prolonged, *e. g.*, when using sluggish precipitants, the solution container shown in (*c*) of Fig. XLVII-15*b*, designed to minimize the rise in level, is used.

Laitinen and Burdett[26] developed the cell shown in Fig. XLVII-16 for amperometric titrations with a continuous stream of inert gas bubbling through the solution for continuous stirring. The dropping electrode is shielded from the deleterious effect of stirring by a double glass mantle which is perforated in such a way that quick mixing of the inner and outer solutions is obtained. The inert gas is introduced through the large sintered glass disc (coarse porosity) in the bottom of the cell, which provides wide dispersion of the bubbles and, consequently, very rapid removal of dissolved oxygen. Connection to an external calomel reference electrode is made through a salt bridge whose end is closed by a sintered glass disc. This cell eliminates the inconvenience of starting and stopping the gas flow after each addition of titrant, and thus permits more rapid titrations than with the usual type of cell.

For reasons already mentioned the reagent solution should be 10 to 20 times more concentrated than the solution to be titrated, and it is best added from a microburet. If the substance to be titrated is electroreducible enough electrolyte is added to eliminate the migration current. If the substance titrated is not and the reagent is reducible the addition of indifferent electrolyte is, in general, not necessary, as enough electrolyte is

---

[25] P. W. West and E. S. Amis, *Science*, **101,** 71 (1945).

[26] H. A. Laitinen and L. W. Burdett, *Anal. Chem.*, **22,** 833 (1950).

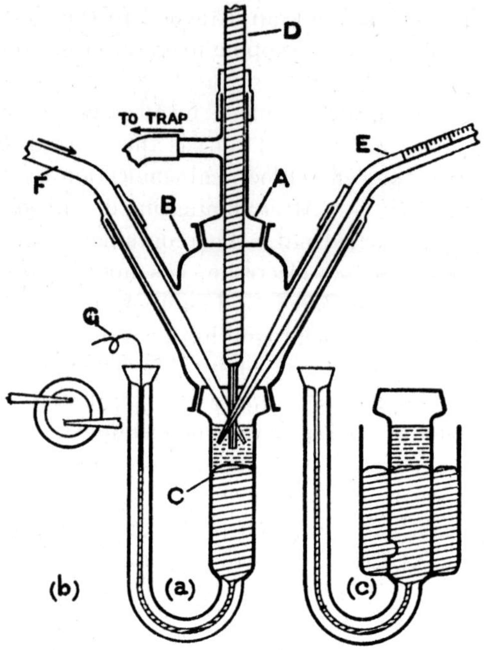

Fig. XLVII-15*b*. Semimicro titration cell.[23]

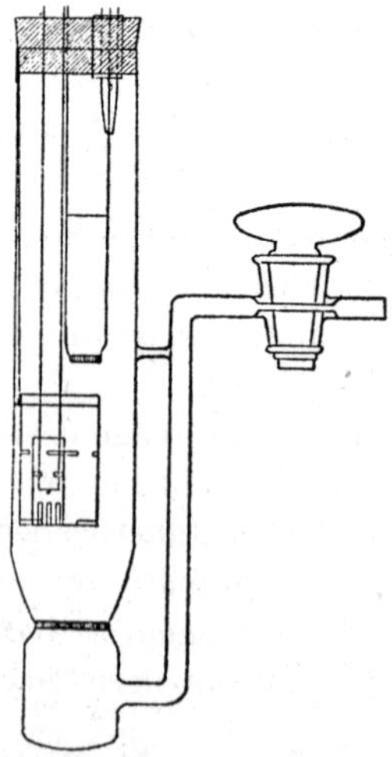

Fig. XLVII-16. Laitinen-Burdett cell for amperometric titrations.[26]

formed during the titration to eliminate the migration current beyond the end point. The addition of indifferent electrolyte is also unnecessary if the substance titrated is electroreducible but only measurements in the neighborhood of the equivalence point are recorded. If lead nitrate, for example, is titrated with potassium chromate the equivalent concentration of the reaction product, potassium nitrate, say at ten per cent before the end point is about ten times greater than that of lead. This excess is large enough practically to eliminate the migration current of lead.

Under certain circumstances it may be desirable to add a suitable maximum suppressor. As such, a small amount of gelatin or a little thymol is usually very effective. In general dyestuffs are also very effective in suppressing maxima, but in precipitation reactions, the presence of a dye is sometimes undesirable, as it may be coprecipitated during the titration. It is not necessary to perform the titration with the cell in a thermostat, but care should be observed not to heat the solution by taking the titration cell in the hand during the titration. Straight titration lines will not be obtained if the temperature fluctuates in an irregular way during the titration, because the diffusion current increases by about 2 per cent per degree increase in temperatures.

When the characteristics of the waves of the substance to be titrated and the reagent are not known it is always necessary first to determine the current–voltage curves of both in the medium in which the titration is being carried out. The applied e.m.f. is then adjusted at the beginning of the titration to such a value that the diffusion current of the substance to be titrated, or of the reagent, or of both, is obtained. It is of practical convenience when the diffusion current region extends over a considerable range of potentials, because then the applied e.m.f. need only be adjusted approximately. It is not necessary in such cases to adjust E_a with the aid of a potentiometer but its value can be read on a voltmeter.

Instead of using a battery and some kind of potential divider it is still simpler to use as a source of the applied e.m.f. a suitable cell of known e.m.f. This principle has been applied by Petering and Daniels[27] in the simple polarographic determination of oxygen. With a few of these standard cells and a few different reference electrodes it is possible to cover a wide range of applied e.m.f. values and to select the proper combination which yields the desired applied e.m.f.

The experimental arrangement becomes still simpler when the titration can be made at an applied e.m.f. of zero. As an example consider the

[27] M. G. Petering and F. Daniels, *J. Am. Chem. Soc.*, **60**, 2796 (1938). See also the increment principle of R. H. Muller and J. F. Petras, *ibid.*, **60**, 2990 (1938).

cell consisting of the dropping electrode in a dilute mercurous nitrate solution, containing an excess of indifferent electrolyte, and the saturated calomel electrode as a reference electrode. The potential of the dropping electrode in this solution is positive with regard to the calomel electrode and if the cell is simply short-circuited a diffusion current corresponding to the concentration of the mercurous mercury is obtained. Upon titration of the mercurous mercury, say with chloride, the current decreases during the titration and becomes practically zero at and beyond the end point. In the titration of chloride with mercurous nitrate the current remains practically zero until the end point and then increases. Some examples of such titrations in which E_a is zero are given later on.

In the above example the potential of the reference electrode was more negative than that of the dropping electrode at the point at which the diffusion current of mercurous mercury is obtained. If we consider the cell consisting of the dropping electrode in a dilute solution of thallous or lead ions connected to a saturated calomel electrode, no galvanic current flows on short-circuit because the potential of the dropping electrode required for the discharge of thallium or lead is more negative than that of the calomel electrode. If we replace the latter by an electrode which has a much more negative potential, *e. g.*, a mercury-potassium cyanide half cell, it should be possible to carry out the titration of thallium or lead at an applied e.m.f. of zero. It is to be expected that half cells containing zinc or cadmium amalgams in solutions of zinc or cadmium salts would also be useful for the above purpose.

Amperometric titrations of substances, or with reagents, which oxidize mercury can be carried out. In such cases it is essential to use an outside reference electrode, and not an internal quiet pool of mercury, as the second electrode. As the speed of oxidation of mercury as a rule is very small, and the rate is proportional to the magnitude of surface exposed to the solution, an error caused by interaction of the drops of mercury collecting on the bottom of the cell and the oxidizing agent is usually negligibly small. If necessary, the error can be minimized by collecting the falling drops of mercury in a narrow glass container. Successive amounts of reagent are added and four to five measurements of the current are recorded, say in the range between 10 to 20 per cent before and after the end point if the solubility of the reaction product is negligibly small. If the solubility is pronounced, measurements are recorded during the earlier stages of the titration, and after the addition of relatively large excesses of reagent, when the solubility as a result of the common ion effect has become negligibly small. The readings of the current after correction for the dilution effect $(V + v)/V$ are plotted against the volume of reagent added. The point of intersection of the two straight lines is the end point.

In cases in which only the reagent or the substance titrated yields a diffusion current it has been suggested[28] to titrate to the point where the galvanometer just gives a distinct deflection or the deflection becomes equal to zero, or a minimum (dead-stop!). In general, this practice is not recommended, as difficulties will be encountered when the solubility of the precipitate is not negligibly small.

4. APPLICATIONS OF AMPEROMETRIC TITRATIONS WITH THE DROPPING ELECTRODE

The first amperometric titrations were performed by Heyrovsky and Berezicky.[29] They titrated solutions of barium and strontium with sulfate, and of calcium with oxalate. They were especially interested in the decrease in the "solubility" at the end point when the suspension was allowed to age. Actually, the decrease which they measured does not correspond to the decrease in the solubility, as the phenomena are more involved due to a purification of the precipitates on aging. Neuberger[30] titrated phosphate with bismuth oxyperchlorate ($BiOClO_4$) at an applied e.m.f. at which bismuth yields a diffusion current. More systematic investigations which are still in the beginning stage have been carried out by Kolthoff and coworkers. Some of the results are summarized below.

Titration of Lead with Dichromate or Chromate. The titration of lead with dichromate yields accurate results, even in very dilute solutions.[31] As an example, we give in Fig. XLVII-17 the results obtained by Kolthoff and Pan in the titration of 50 ml. 0.01 *M* lead nitrate, which was 0.01 *N* in potassium nitrate, with 0.05 *M* potassium dichromate at a potential of −1.0 v. (*vs.* S.C.E.). At this potential both lead and dichromate ($Cr^{++++++} \rightarrow Cr^{+++}$) yield diffusion currents. The dotted lines give the values of the current as actually determined, and the solid lines are the values after correction for the dilution effect. It is seen that the two corrected straight lines intersect exactly at the equivalence point. The reproducibility and accuracy of this titration were found to be excellent; the end point in four different titrations was at 5.00 ± 0.01 ml. of dichromate.

---

28 A. Neuberger, *Z. anal. Chem.*, **116,** 1 (1939); *Arch. Eisenhüttenw.*, **13,** 171 (1939). G. Thanheiser and J. Willems, *ibid.*, **13,** 73 (1939).

29 J. Heyrovsky and S. Berezicky, *Collection Czechoslov. Chem. Communs.*, **1,** 19 (1929).

30 A. Neuberger, *Z. anal. Chem.*, **116,** 1 (1939).

31 I. M. Kolthoff and Y. D. Pan, *J. Am. Chem. Soc.*, **61,** 3402 (1939).

From Fig. XLVII-17 it is seen that the residual current at the equivalence point is quite appreciable, and that both the precipitation and the reagent lines are curved in the immediate neighborhood of the end point. The residual current at the end point is to be attributed to dissolved lead chromate, but the straight lines intersect at a point where the residual current is smaller than 1 microamp. The relatively large solubility of

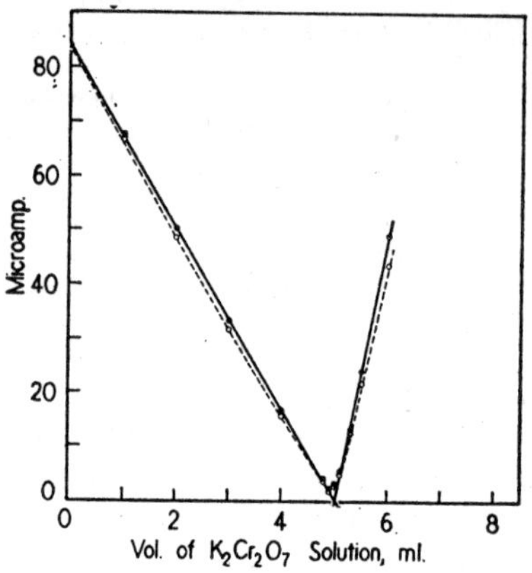

Fig. XLVII-17. Amperometric titration curve of 50 ml. of 0.01 *M* lead nitrate with 0.05 *M* potassium dichromate. Titrated at $E_{d.e.} = -1.0$ v. *vs.* S.C.E. The dotted curve shows the experimental points, and the solid straight lines were obtained after correction for the dilution effect.

lead chromate in the above experiment is due to the free acid which is formed during the titration:

$$Cr_2O_7^{--} + 2Pb^{++} + H_2O \rightleftharpoons 2PbCrO_4 + 2H^+$$

When neutral lead nitrate solutions are titrated with potassium chromate instead of dichromate the solubility of the lead chromate is negligibly small; the residual current was found to be less than 1 microamp. However, in titrations with chromate in neutral solutions the end point was always found to be 0.8 to 1 per cent before the true equivalence point, the deviation undoubtedly being due to coprecipitation of basic lead chromate with the lead chromate. Therefore, it is better to perform the titration in a slightly acid medium.

Even in solutions which are 0.1 *M* in perchloric acid excellent results were obtained, although the solubility near the end point is very marked. For example, in 0.1 *N* perchloric acid a residual current of 10.1 microamp. was observed at the end point. Good results were also obtained in acetate buffers. In buffer solutions with an acid reaction, and in stronger acid solutions, the titration can be performed at an applied e.m.f. of zero. Under these conditions the lead is not reduced, but the dichromate yields a diffusion current. As an example the results of a titration of 0.01 *M* lead with 0.05 *M* dichromate in an acetate buffer with a pH of 4.2 at $E_a = 0$ are given in Fig. XLVII-18. The performance of the lead titration at zero applied e.m.f. is of particular advantage in cases in which

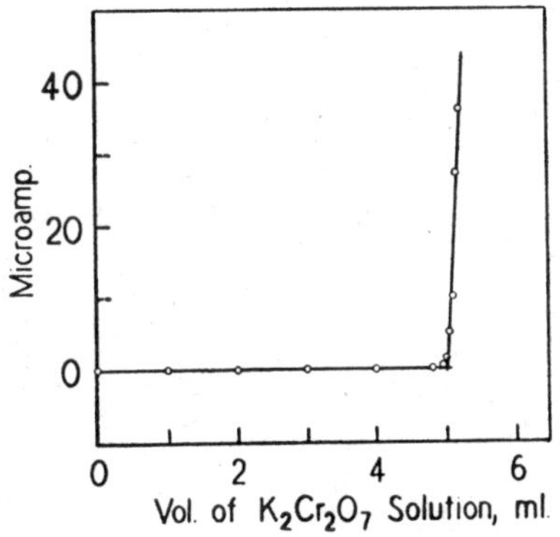

Fig. XLVII-18. Titration of 50 ml. of 0.01 *M* lead nitrate with 0.05 *M* potassium dichromate, at an applied e.m.f. of zero in an acetate buffer of pH = 4.2.

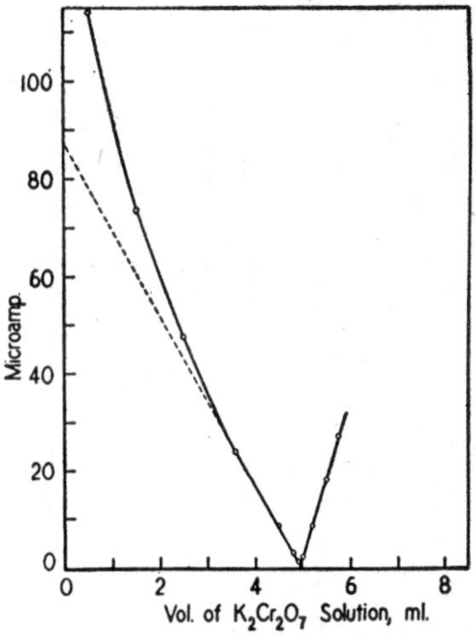

Fig. XLVII-19. Influence of the migration current in the titration of 50 ml. of 0.01 *M* lead nitrate with 0.05 *M* of dichromate without any supporting electrolyte. Titration performed at $E_{d.e.} = -1.0$ v. *vs.* S.C.E.

the solution contains substances which are reduced at potentials at which lead yields a diffusion current. For example, at $E_a = 0$ it is not necessary to remove dissolved oxygen from the solution, as the oxygen is not reduced at this potential. Even the titration of 0.001 *M* lead solutions can be carried out quickly and accurately; the results obtained agreed within 0.3 per cent with the theoretical.

In the theoretical discussion it was mentioned that the precipitation line (after correction for the dilution effect) is straight from the very beginning of the titration if enough indifferent salt is added to eliminate the migration current. In order to illustrate the effect of the migration current we give in Fig. XLVII-19 the results of a titration of 0.01 *M* lead with 0.05 *M* potassium dichromate at $E_a = -1.0$ v. without any supporting electrolyte present. The current at the beginning of the titration was much greater than the diffusion current of lead, since it also included the migration current. During the titration the lead concentration decreases, whereas the concentrations of the reaction products, potassium nitrate, and nitric acid increase. Consequently, the contribution of the migration current becomes less and less upon successive additions of reagent. During the precipitation of the last 20 per cent of the lead the currents practically correspond to the diffusion currents of the dissolved lead.

From an analytical viewpoint it is of interest to say a few words regarding the titration of lead in the presence of barium. Although the chromates of both of these cations are slightly soluble, barium chromate differs from the lead salt in that it is freely soluble in dilute mineral acids. In a medium which was 0.01 *N* in perchloric acid lead could be titrated accurately with chromate or dichromate, even when the concentration of barium was ten times greater than that of lead. The titration can be carried out either at an applied e.m.f. of zero or at a potential of -1.0 v.; in the latter case both the lead and the reagent are reduced.

In a neutral medium both lead and barium chromate are insoluble. As the solubility product of lead chromate is much less than that of the barium salt, one might expect that upon addition of chromate to a mixture of a lead and a barium salt the former will precipitate first quantitatively, and then the barium chromate will precipitate. Such titrations have been carried out at $E_a = 1.0$ v., at which applied e.m.f. both lead and chromate yield a diffusion current, while barium is not reduced. The mixture used was 0.01 *M* in lead and 0.01 *M* in barium. It is seen from Fig. XLVII-20 that a marked coprecipitation of barium chromate occurs during the precipitation of the lead chromate. The first end point (complete precipitation of lead) is at 3.1 ml. which is 20 per cent after the true equivalence point (2.5 ml.). The second end point, corresponding to complete precipitation of lead and barium, was found at the correct location (5.00 ml.). Incidentally, it may be mentioned that the amperometric titration of barium with chromate in neutral solution yields excellent results.

Titration of Barium with Chromate. Kolthoff and Gregor[32] carried out the titration of 0.001 *M* barium in the presence of ethanol which lowers the solubility of barium chromate. In 20, 30, and 50 per cent ethanol the results were low by 2.4, 5.0, and 3.4 per cent, respectively. It is desirable to study this titration further in the presence of ethanol and a buffer with a pH of the order of 5 to 6.

---

[32] I. M. Kolthoff and H. P. Gregor, *Anal. Chem.*, **20**, 541 (1948).

Lead Nitrate as Reagent in Titration of Anions. Several anions yield slightly soluble salts with lead. If coprecipitation phenomena do not interfere it may be expected that these anions can be titrated amperometrically with lead.

Sulfate. From a practical point of view the most important titration of this type is that of sulfate with lead nitrate. This amperometric titration was first studied by Majer.[33] From his results it is hard to estimate the end point. Moreover, Majer's graphical method of locating the

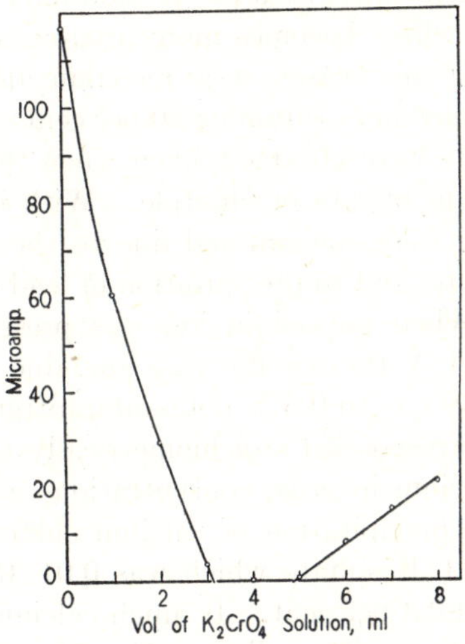

Fig. XLVII-20. Titration of a mixture of 0.01 *M* lead ion and 0.01 barium ion with potassium chromate in neutral medium. Titrated at $E_{d.e.}$ = −1.0 v. *vs.* S.C.E.

end point is not of much practical use, as the solubility of the lead sulfate in the medium should be known. Since the solubility varies greatly with the ionic strength of the medium the solubility of the precipitate under the working conditions is rarely known exactly. Spalenka[34] recommends that the titration be performed in a medium containing 30 to 40 per cent of ethanol in order to suppress the solubility of the lead sulfate. A more systematic analytical study has been made by Kolthoff and Pan.[35] In general, the titration is performed in the absence of oxygen at a potential of −1.2 v. (*vs.* S.C.E.), at which potential lead yields a diffusion current.

33 V. Majer, *Z. Elektrochem.*, **42,** 120, 123 (1936).
34 M. Spalenka, *Collection Czechoslov. Chem. Communs.*, **11,** 146 (1939).
35 I. M. Kolthoff and Y. D. Pan, *J. Am. Chem. Soc.*, **62,** 3332 (1940).

The titration lines have the general appearance shown in Figure XLVII-3*a*. It is not necessary to add an indifferent electrolyte, as the current does not increase markedly until an excess of lead is added, and the amount of salt formed during the titration is sufficient to suppress the migration current of lead completely.

Aqueous solutions of sulfates at concentrations of 0.01 *M* or greater can be titrated with 0.1 *M* lead nitrate with an accuracy of ±0.3 per cent. Points very close to the end point are rejected in the construction of the precipitation and reagent lines, because the solubility effect is very marked here. This solubility effect becomes more pronounced in the presence of indifferent salts. For this reason it is recommended to titrate 0.01 *M* sulfate solutions in a medium containing 20 per cent ethanol. Under these conditions good results were obtained, even when the solution was 0.1 *N* in sodium or potassium nitrate or chloride. With higher concentrations of chloride difficulties were encountered due to the large increase in the solubility of lead sulfate, and to precipitation of lead chloride with a slight excess of reagent. When potassium ion was present in concentrations much greater than 0.1 *N* the results were low, due to precipitation of a double salt $PbSO_4 \cdot K_2SO_4$. In 0.5 *N* potassium nitrate the location of the end point actually corresponded stoichiometrically to the composition of the double salt. Calcium in large concentrations in 20 per cent ethanol interfered due to the precipitation of calcium sulfate. However, it was possible to titrate 0.01 *M* sulfate which was 0.01 *M* in calcium chloride with an accuracy of ±0.3 per cent. If much calcium is present it can be removed by precipitation with sodium carbonate. The filtrate is made acid with nitric acid with methyl orange as indicator, and the titration is performed after addition of the ethanol.

Solutions which were 0.001 *M* in sulfate were titrated with 0.01 *M* lead nitrate in a medium containing 30 per cent ethanol with an accuracy of ±0.3 per cent. When the solution was 0.01 *N* in potassium nitrate the results were 1 per cent high, and in 0.05 *N* potassium nitrate, 2 per cent high. This is due to coprecipitation of lead nitrate with lead sulfate. In 0.1 *N* potassium nitrate the results were very low due to precipitation of the double salt.

In the reverse titration of 0.01 *M* lead nitrate in 20 per cent alcoholic medium with 0.1 *M* alkali sulfate the results were about 1 per cent low due to coprecipitation of lead nitrate. Since coprecipitation phenomena are not found in the titration of lead with potassium dichromate and the solubility of lead chromate is much smaller than that of lead sulfate it is preferable to titrate lead with dichromate instead of with sulfate.

An application of the above titration is described by Axford and Sugden[36] in the determination of sulfur trioxide (of the order of 2 per cent) in sulfur dioxide gas. The gas is absorbed in 1 *N* sodium hydroxide containing a few drops of glycerol to inhibit air oxidation of sulfite, and the solution is acidified with 1 *N* nitric acid. The sulfur dioxide is then blown out by passing through a stream of nitrogen for 30 minutes and the solution is rendered just alkaline to methyl orange. After addition of one-third volume of ethanol the mixture is titrated with lead nitrate.

Sulfur in coke can be determined by amperometric titration after oxidation to sulfate.[37]

Molybdate. Thanheiser and Willems[38] titrated molybdate with lead nitrate or perchlorate in the presence of formate or acetate. These authors used a layer of mercury as internal anode and worked at an applied e.m.f. of 0.6 v. At this voltage molybdate is not reduced whereas lead yields a diffusion current. Thanheiser and Willems did not find the end point, graphically, but determined it from the point where the current suddenly increased due to the presence of an excess of lead. For reasons already mentioned, we prefer to determine the end point graphically. Thanheiser and Willems performed the titration at 80° C. in the absence of oxygen. The results were improved by the addition of about 15 per cent alcohol.

Procedure. Introduce 20 ml. of the solution, containing 0.15 to 9.0 mg. of molybdate and no more than 2 g. of potassium hydroxide (*vide infra*) into the titration cell and neutralize with 6 *N* nitric acid to a pH of 5, using chlorphenol red or methyl red as indicator. Add 5 ml. of a 10 per cent sodium formate solution and 5 ml. of ethanol, heat to 80° C., and pass nitrogen through for 2 minutes. Apply an e.m.f. of 0.6 v. and titrate with lead perchlorate (or lead nitrate) of such a concentration that 1 ml. corresponds to 0.5 mg. of molybdenum.

Thanheiser and Willems report good results, but judging from the values given it seems that the accuracy is only of the order of a few per cent. They applied the method to the determination of molybdenum in steel. It is necessary to remove the iron before the titration is made.

Procedure for the Determination of Molybdenum in Steel (Thanheiser and Willems). Weight 0.1 to 0.3 g. of steel into a 50-ml. beaker and add 2 to 4 ml. of 60 per cent perchloric acid. After all the steel is dissolved, dilute with water to 10 ml., heat, and neutralize with 2 *N* potassium hydroxide, until the solution is reddish brown. Add 2 ml. of 15 per cent ferrous chloride solution to reduce chromate and vanadate. Add the mixture with stirring to 20 ml. of a hot 10 per cent solution of potassium hydroxide, and add some graphite and boil for 5 minutes. The volume after the heating should not be greater than 25 ml. Cool, transfer to a 50-ml. volumetric flask,

[36] D. W. E. Axford and T. M. Sugden, *J. Chem. Soc.*, **1946,** 901.
[37] G. A. Butenko and V. M. Pindas, *Zavodskaya Lab.*, **9,** 634 (1940).
[38] J. Thanheiser and J. Willems, *Arch. Eisenhüttenw.*, **13,** 73 (1939).

and fill up to the mark. Introduce 20 to 30 ml. of the filtrate into the titration cell, neutralize, and proceed as described above.

If the steel contains tungsten the latter must first be removed. In this case the sample of the steel (0.2 to 0.4 g.) is dissolved in concentrated hydrochloric acid, 2 ml. of concentrated nitric acid is added, and the mixture heated for 10 minutes at 80° to 90° C. The solution is transferred to a 50-ml. volumetric flask, and, after cooling, made up to volume. An aliquot part of the filtrate is evaporated nearly to dryness, 2 ml. of 60 per cent perchloric acid is added, and the heating continued for 10 minutes until most of the acid is fumed off. The residue is treated as described in the general procedure.

Fluoride as Lead Chlorofluoride. Haul and Griess[39] investigated the precipitation of fluoride as lead chlorofluoride polarographically and gave a procedure for the amperometric titration of fluoride. Fluoride can also be precipitated as lead fluoride but, since the solubility of the chlorofluoride is less than that of the fluoride, precipitation as PbFCl is preferable. The solubility product of this compound at 18.5° C. is of the order of 2.2×10^{-8}. In order to decrease the solubility during the titration Haul and Griess used 50 per cent ethanol as a solvent.

Petrow and Nash[40] studied this titration systematically, defined the optimum conditions, and achieved an accuracy of the order of ±0.3 per cent with quantities of fluoride between about 5 and 50 mg. per 100 ml. Petrow and Nash found that addition of ethanol was unnecessary but that the initial pH of the fluoride solution must be in the range between 5.5 and 6.5 for optimum results. At lower pH values the results are low (5 to 8 per cent at pH 5.2) and at higher pH values the current readings are poorly defined due to local precipitation of lead hydroxide and its slow conversion to the chlorofluoride. The pH adjustment is made with dilute hydrochloric acid or sodium hydroxide using either the glass electrode or bromthymol blue as indicator. The titration is performed in a solution containing 0.1 *M* potassium or sodium chloride with the potential of the dropping electrode at −1.0 v. *vs.* S.C.E. It is not necessary to remove dissolved air. The current remains constant before the end point and then increases when an excess of lead ion had been added. Because of the appreciable solubility of lead chlorofluoride the titration curve shows considerable curvature in the region of the end point, but readings taken at a considerable distance on either side of the end point yield straight lines which intersect at the true end point. Anions which precipitate with lead ion interfere, but those, such as phosphate, which are precipitable by silver ion may be preliminarily removed with this reagent. Concentrations of

[39] R. Haul and W. Griess, *Z. anorg. Chem.*, **259**, 42 (1949).
[40] H. G. Petrow and L. K. Nash, *Anal. Chem.*, **22**, 1274 (1950).

nitrate ion greater than about 0.1 *M* interfere, apparently because of complexation with lead ion.

Chloride. Korshunov and Gurevich[41] titrated chloride with lead nitrate in the presence of enough alcohol to make the ethanol content at the end point 55 to 60 per cent. The method is of very limited practical use because many anions form slightly soluble lead salts in this medium.

Undoubtedly, many other anions, like ferrocyanide, sulfide, sulfite, etc., which yield insoluble lead salts can be titrated amperometrically with standard lead solution.

Titration of Nickel with Dimethylglyoxime. It is well known that nickel in an ammoniacal medium yields a red precipitate with dimethylglyoxime having the composition:

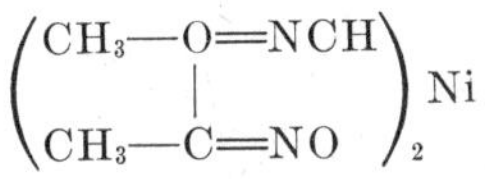

This reaction has been used for the amperometric titration of nickel.[42] The titration is best carried out in an ammoniacal medium at a potential of −1.85 v. Under these conditions the diffusion current of nickel, which is proportional to the concentration of the metal ion, is obtained. The behavior of the dimethylglyoxime reagent is peculiar. At $E_{d.e.}$ = 1.85 v. and at concentrations smaller than 0.002 *M* it yields a diffusion current proportional to the concentration. But at concentrations larger than 0.002 *M* the diffusion current of the dimethylglyoxime soon reaches a value above which it does not increase upon further increase in the concentration. Nevertheless, excellent results, precise and accurate within 0.4 per cent, were obtained in the titration of 0.01 to 0.001 *M* solutions of nickel with dimethylglyoxime when readings were made near the end point. As an illustration some data obtained in the titration of 50 ml. of 0.0001 *M*, 0.0002 *M*, and 0.0004 *M* nickel solutions, respectively, are given in Fig. XLVII 21. At these high dilutions the solubility of the precipitate is distinctly noticeable. However, by plotting the measurements more than 20 to 40 per cent removed from the equivalence point it is possible to get sufficient data to construct straight titration lines. At these high dilutions an accuracy of 1 to 2 per cent was obtained. Evidently, the method is also useful in micro work. Titrations have been performed with 10 ml. of solution containing 0.06 to 0.30 mg. of nickel with an accuracy of two to three per cent.

[41] I. A. Korshunov and A. G. Gurevich, *Zavodskaya Lab.*, **11,** 648 (1945).

[42] I. M. Kolthoff and A. Langer, *J. Am. Chem. Soc.*, **62,** 211 (1940).

Metal ions which are reduced at the applied e.m.f. used in the titration of nickel interfere when present in larger quantities, even if they do not form complexes with the reagent. Unfortunately, even relatively small amounts of cobalt exert an interfering action in the titration of nickel, because the cobalt forms a soluble and reducible complex with the dimethylglyoxime. When the amount of cobalt is more than 5 per cent of the nickel content a preliminary chemical separation has to be applied. The

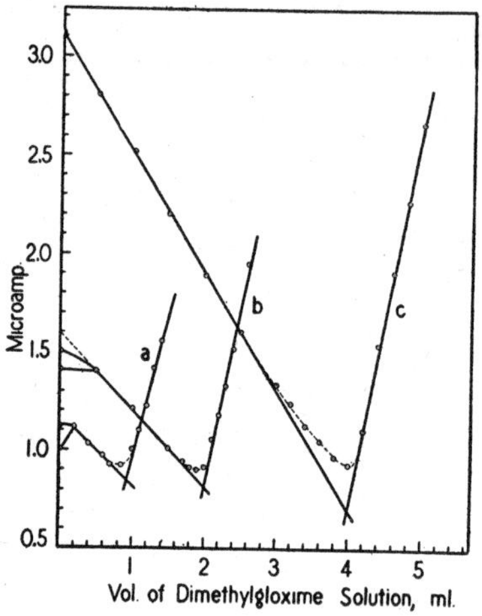

Fig. XLVII-21. Amperometric titration curves of (*a*) 0.0001, (*b*) 0.0002, and (*c*) 0.0004 *M* nickel solutions with an alcoholic dimethylglyoxime solution in an ammoniacal medium. Titration performed at $E_{d.e.}$ = −1.85 v. *vs.* S.C.E.

cobalt is precipitated as potassium cobaltinitrite. After destruction of the nitrite in the filtrate the nickel can be titrated. The method has also been applied to the determination of nickel in nickel steel. After removal of iron by the benzoate method the nickel content of a steel sample was found to be equal to 3.48 ± 0.04 per cent, in good agreement with the value of 3.47 per cent given by the Bureau of Standards.

Fill and Stock[43] titrated amounts of nickel between 0.018 and 0.58 mg. in a 5-ml. volume at an applied voltage of 1.6 v. with satisfactory accuracy.

Titration of Cobalt, Copper, and Palladium with α-Nitroso-β-naphthol. Several metals, such cobalt, copper, zirconium chromium, titanium, vanadium, tin, tungsten, palladium, uranium, and ferric iron, yield pre-

[43] M. A. Fill and J. T. Stock, *Analyst*, **69**, 178 (1944).

cipitates with α-nitroso-β-naphthol in an acetic acid-acetate buffer. Kolthoff and Langer[44] investigated the reaction between cobalt and this reagent. From a few experiments, they concluded that copper and palladium also can be determined by amperometric titration with α-nitroso-β-naphthol. Solutions of the latter yield a well-defined wave at the dropping electrode, the diffusion current being proportional to the concentration of the reagent. In acetate buffers the half-wave potential was found at about −0.08 v. (*vs.* S.C.E.), and in ammonia-ammonium chloride buffers at about −0.3 v. In acetate buffers the diffusion current was obtained at potentials more negative than −0.3 v. and in an ammoniacal medium at potentials more negative than −0.5 v. Therefore, when the amperometric titrations are performed at potentials more negative than −0.6 v., the reagent will yield its diffusion current in both acid and ammoniacal media. At such potentials divalent cobalt is not reduced and the current will remain equal to the residual current of the medium until all the cobalt is precipitated and then it increases when an excess of the α-nitroso-β-naphthol reagent is added (comp. Fig. XLVII-3*a*). On the other hand, copper and palladium yield diffusion currents at −0.6 v. and the lines obtained in the titration of these two metals are similar to those given in Fig. XLVII-3*b*. When the titration of cobalt is carried out at a potential of −1.6 v. (in the absence of a maximum suppressor) this metal also yields a diffusion current and the titration lines become similar to those of copper and palladium (Fig. XLVII-3*b*). However, it is advantageous to perform the titration at a potential of −0.6 v., because several other metals, such as zinc, aluminum, and manganese, are not reduced at this potential and do not affect the accuracy of the titration.

Titrations of cobalt have been carried out in acetate buffers and in an ammoniacal medium. In an acid medium the precipitate has a purple-red color, one cobalt ion combining with four molecules of the reagent. Apparently, the cobalt in the precipitate is present in the divalent form, and the precipitate, after drying, has the composition:

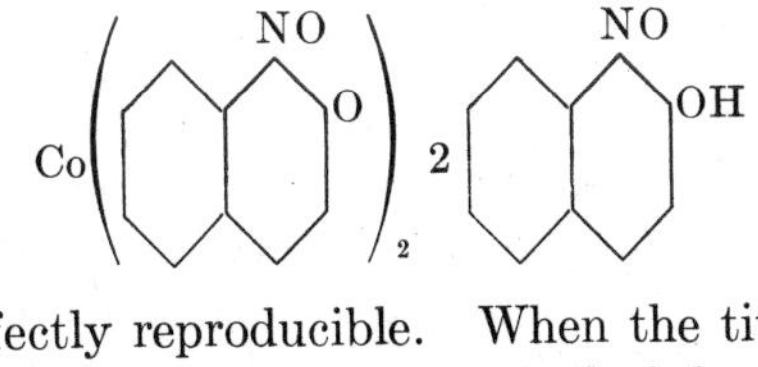

The results are perfectly reproducible. When the titration is carried out in an ammoniacal medium the precipitate is dark brown and contains one cobalt molecule to about three molecules of α-nitroso-β-naphthol. The

[44] I. M. Kolthoff and A. Langer, *J. Am. Chem. Soc.*, **62,** 3172 (1940).

results in an ammoniacal medium are not reproducible, probably because the cobalt is easily air oxidized to the trivalent state. For this reason it is best to perform the titration in an acetate buffer.

Copper behaves differently from cobalt; in both acid and ammoniacal media it yields a dark brown precipitate of the composition:

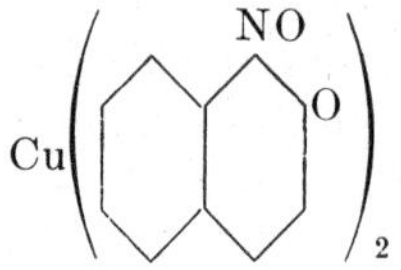

The reaction can be employed for the standardization of the reagent. However, in titrations of cobalt it is preferable to standardize the reagent against cobalt solutions of known concentration.

Reagent. A commercial preparation of β-nitroso-β-naphthol is recrystallized a few times from alcohol. An approximately 0.1 M solution of the purified reagent is prepared in 60 per cent acetic acid. If the solution is not clear it is filtered before use. It is standardized according to the procedure given below. The standard solution is stable for at least three to four weeks. In the titration of extremely dilute (about 0.0005 M) cobalt solutions a 0.01 M solution of the reagent in 50 per cent acetic acid is used.

Procedure. The cobalt solution (0.01 to 0.0005 M) which is about 0.1 to 0.2 M in sodium acetate and about 0.1 M in acetic acid is placed in the titration cell. Air is removed with nitrogen or hydrogen and the titration carried out at a potential of -0.6 v. (*vs.* S.C.E.). The reagent is added from a microburet.

Notes: 1. The titration is accurate and precise to within ± 0.5 per cent when dealing with 0.001 M or more concentrated cobalt solutions. A 0.0002 M cobalt solution can be titrated with an accuracy of 1 to 2 per cent.

2. Copper and palladium can be titrated by the same procedure. It should be emphasized that one cobalt ion combines with 4 molecules of reagent whereas copper and palladium combine with two. The possibility of titrating other metal ions which yield precipitates with the reagent has not yet been investigated. Such ions interfere with the titration of cobalt.

3. Zinc, aluminum, and manganese were found not to interfere. On the other hand, high results were found in the presence of nickel although the latter does not precipitate with the reagent.

Copper with Salicylaldoxime. Neuberger[45] added to the copper solution an excess of a 1 per cent alcoholic salicylaldoxime solution and titrated the excess with a standard copper solution at a potential at which copper yields a diffusion current. The titration is made in dilute acetic acid in the presence of ammonium chloride. No current is observed when there is still an excess of salicylaldoxime in the solution. The end point is marked by the sudden appearance of a current. In general, and especially when

[45] A. Neuberger, *Arch. Eisenhüttenw.*, **13**, 171 (1939).

dealing with dilute solutions, we prefer the graphical method for the end point determination. Analytical details are not given in Neuberger's paper.

Copper with α-Benzoinoxime (Cupron), $C_6H_5CHOHC{=}NOH{-}C_6H_5$. This titration was investigated by Langer.[46] At pH values smaller than 6 Langer found an "acid" reduction wave of benzoinoxime, the half-wave potential (*vs.* S.C.E.) of which is given by $E_{1/2} = -(0.71 + 0.06 \text{ pH})$ v. At pH values between 6 and 8 two waves, an "acid" and an "alkaline" wave, were found, the total wave height being approximately independent of the pH. At pH values greater than 8 only a single wave with a half-wave potential of approximately −1.63 v. appeared. Langer attributes the appearance of two waves to reduction of the un-ionized (acid wave) and ionized (alkaline wave) fractions of the benzoinoxime. Undoubtedly, the rate of transformation of the alkaline into the acid form also plays a part so that the height of each wave is not expected to correspond to the concentrations of the acid and alkaline forms in the bulk of the solution.

The titration of copper is carried out in 0.1 *N* ammonium chloride with an excess of ammonia. The solubility of the copper benzoinoxime increases appreciably with increasing ammonia concentrations. In order to get good results in the titration of 0.001 *M* copper solutions the concentration of ammonia should not be greater than 0.02 *M*. To eliminate maxima 5 drops of 10 per cent gelatin solution are added to 40 ml. of the solution to be titrated. A solution of the reagent (0.01 to 0.1 *M*) is prepared in 50 per cent ethanol and standardized amperometrically against copper solutions of known concentration.

When the titration is carried out at a potential of −1.0 v. the excess of reagent is not reduced (titration line as in Fig. XLVII-2). When it is performed at −1.7 v. the benzoinoxime is reduced (titration line as in Fig. XLVII-3*b*).

Interferences. Ferric iron is precipitated by ammonia and causes coprecipitation of copper; hence low results are found. The same is true in the presence of lead. Zinc present in amounts equal to five times the concentration of copper gave values 6 per cent too high in 0.1 *M* ammonia, but if present in smaller amounts than copper the results were 0.5 to 2 per cent high. Nickel was found to interfere strongly.

Fill and Stock[47] titrated quantities of copper between 0.022 and 0.63 mg. in a 5-ml. volume at a potential of −1.7 v. with satisfactory accuracy.

Copper and Zinc with Quinaldinic Acid. Stock[48] described procedures for titration of copper and zinc in concentrations of $1\text{–}3 \times 10^{-3}$ *M* with

[46] A. Langer, *Ind. Eng. Chem., Anal. Ed.*, **14**, 283 (1942).
[47] M. A. Fill and J. T. Stock, *Analyst*, **69**, 178 (1944).
[48] J. T. Stock, *J. Chem. Soc.*, **1949**, 1793. See also *Analyst*, **72**, 856 (1947).

with 2.5×10^{-2} *M* quinaldinic acid. The solution is buffered to a pH of 5.0, made 0.03 per cent in gelatin, and, after removal of oxygen, is titrated at 60° C. at an applied e.m.f. of −1.2 v. At room temperature the rate of precipitation is too small to make the titration practicable, although an excess of quinaldinic acid can be back titrated with copper (at room temperature). At an applied e.m.f. of −1.1 v., copper and the reagent yield a diffusion current, but zinc is not reduced. Thus, the titration line of copper is V-shaped and is like that of zinc. The accuracy of the titration is of the order of 1 per cent.

The procedure can be used for the stepwise titration of a mixture of copper and zinc at an applied e.m.f. of −1.06 v.; first the copper is precipitated and then the zinc. Copper can be determined in the presence of manganese, nickel, cobalt, and cadmium in a buffer of pH 2.1 at 60° C. (solution 0.03 per cent in gelatin) at an applied e.m.f. of 0.04 v. Under these conditions, copper yields a diffusion current, but the reagent does not produce a reduction current.

Note: Stock[49] showed that copper can also be titrated with quinoline-8-carboxylic acid. The reagent solution is prepared in 50 per cent aldehyde-free ethanol. The titration is carried out at room temperature at pH 5 (acetate buffer) in the presence of 0.03 per cent gelatin. At an applied e.m.f. of 1.35 v., V-shaped titration curves are obtained. The titration can also be carried out at 0.04 v. (V-shaped curves). Zinc, cadmium, ferrous iron, cobalt, nickel, and lead do not interfere.

Oxine as Reagent

Bismuth with 8-Hydroxyquinoline. Under specified conditions 8-hydroxyquinoline (oxine) forms precipitates with quite a number of metals. As mentioned on page 819 oxine can be reduced polarographically; however, the diffusion current appears at such negative potentials that it is preferable to determine the diffusion current of the metal ion which is being precipitated with oxine.

A preliminary study of the titration of bismuth with oxine has been made by Gillis *et al.*[50] They use a 2 per cent solution of oxine which is prepared by dissolving 2.0 g. of the reagent in 5 ml. of glacial acetic acid and diluting to 100 ml. with water. The pH of the bismuth solution must be carefully adjusted. Gillis *et al.* add to 10 ml. of approximately 0.05 *M* bismuth solution in 20 per cent nitric acid 2 g. of sodium tartrate. The

[49] J. T. Stock, *J. Chem. Soc.*, **1949,** 2470.

[50] J. Gillis, J. Eeckhout, and G. Standaert, *Mededeel. Koninkl. Vlaam. Acad. Wetenschap.*, **1940,** No. 7.

solution is neutralized with 1 *N* sodium hydroxide using methyl red as indicator. Then 7.33 g. of sodium acetate trihydrate and 0.16 ml. of 6 *N* acetic acid is added (pH is now 6.5) and the solution is titrated at a potential of 0.85 to 0.90 v. After each addition of reagent the mixture is stirred for 30 seconds and the current is measured after another 30 seconds. Gillis *et al.* do not remove dissolved oxygen, because they work with relatively concentrated solutions. It is not stated to what dilution the titration can be carried out. It also would be of interest to investigate the interference by other metal ions which may precipitate with oxine.

Zan'ko[51] titrates *zinc* and *copper* with oxine in an acetic acid-acetate buffer. *Magnesium* is best titrated in an ammonia-ammonium chloride buffer. Aluminum can be determined in both the acid and alkaline buffers. More systematic investigations have been carried out by Fujinaga.[52]

Magnesium is titrated in a medium which is 0.5 *M* in ammonium chloride and ammonia at a potential of −1.6 v. (S.C.E.). Results accurate to 0.7 per cent were obtained when the magnesium concentration was 0.01 *M* or greater. Small amounts of calcium do not interfere; larger amounts are made harmless by the addition of ammonium oxalate. The precipitate need not be removed. The method is applied to the determination of magnesium in brine, sea water, and aluminum alloys. The alloy is treated with sodium hydroxide and the residue dissolved in hydrochloric acid. An excess of ammonia is added and the mixture treated with hydrogen sulfide. The titration is carried out without removing the precipitate.

Copper, Cadmium, and Zinc. Stock[53] gives directions for the amperometric titration of *copper*, *cadmium*, and *zinc*. The titrations are carried out in acetate-hydrochloric acid buffers (over-all acetate concentrations, 0.2 *N*) in the presence of 0.02 per cent gelatin. Copper can be titrated at room temperature with an accuracy of better than per cent at an applied e.m.f. of 0.4 v. (L-type curves) in a concentration range between 0.25 and 2.5×10^{-3} *M* in a medium of pH 6.0. Because of the small solubility of the precipitate the titration of even 10^{-4} *M* copper is possible (accuracy about 5 per cent).

Although cadmium can be precipitated quantitatively at pH 6.0, the solubility of its oxinate and the rate of its precipitation are so much greater than that of copper that copper can be titrated in the presence of cadmium by the above procedure. After the precipitation of the copper the cadmium

---

[51] A. M. Zan'ko, *Dopovidi Akad. Nauk, U. R. S. R.*, **1940**, 27, 32. *Chem. Abstracts*, **37** 5921 (1943).

[52] F. Fujinaga, Kyoto University, Japan, Private communication.

[53] J. T. Stock, *Metallurgia*, **40**, 179, 229 (1949).

starts to precipitate, but this does not affect the shape of the titration line. Cadmium can be titrated in an acetate buffer of pH 6.0 at an applied e.m.f. of 0.9 v., at which cadmium yields a diffusion current, while the reagent does not produce a reduction current.

Zinc in concentrations of 1–2.5 $\times$ 10^{-3} M can be titrated at pH 5.0 at an applied e.m.f. of 1.4 v., at which V-type titration curves are obtained. The zinc oxinate is more soluble than the copper compound, and copper can be titrated in the presence of zinc by the above procedure at pH 6.0 if enough sodium potassium tartrate is added to make its concentration 0.07 M. In the acetate buffer in the absence of tartrate there is marked coprecipitation of zinc. The separate determination of cadmium and zinc in a mixture of the two was not found feasible, as the two ions precipitated together.

Fujinaga[52] titrates *copper* and *zinc* in an ammonia-ammonium chloride buffer at an applied e.m.f. of 1.6 v. Zinc can also be titrated in 0.1 N sodium hydroxide at an applied e.m.f. of -1.8 v.

Ferric Iron. Fujinaga dissolves the ferric iron in a large excess of ammonium carbonate (final concentration 1.5 to 2 N). Ammonia is added to this solution and the titration is carried out at a potential of -1.6 v.

Mixture of Ferric Iron, Copper, and Zinc. The sum of these three constiuents is titrated in the presence of an excess of ammonium carbonate (see section on ferric iron) and ammonia. To another sample of the solution ammonium chloride and ammonia are added, and the mixture is titrated without removing the hydrous ferric oxide (sum of copper and zinc). A third sample is heated to boiling, sodium hydroxide is added, and the mixture is cooled. Without removing the precipitate the dissolved zinc is titrated at a potential of -1.8 v.

Ferric Iron with Bromoxine. Sandberg[54] titrates ferric iron with bromoxine (5,7-dibromo-8-hydroxyquinoline) at 50° C.

Reagent. A saturated solution of bromoxine in air-free acetone at room temperature (3.8 g. per liter). The solution is stable for a month if protected from the light. It can be standardized against a ferric solution of known concentration.

Procedure. Remove air from the solution (0.05–1 mg. of ferric iron in 10 ml. of 0.010–0.025 N hydrochloric acid) with nitrogen and add 0.15 ml. of 0.1 per cent tylose. Titrate at 50° C. against the saturated calomel electrode and without any external applied potential. Readings can be made a few seconds after the addition of reagent.

---

[54] B. Sandberg, *Svensk Kem. Tid.* **58,** 197 (1946).

Notes: 1. The titration error is 2 per cent for 0.05 mg. of Fe, and not more than 0.7 per cent for more than 0.2 mg. of Fe.

2. Higher acid concentrations than specified may result in incomplete precipitation of iron, and lower concentrations (as well as titration at room temperature) may lead to precipitation of the reagent before it can react with ferric ion. The higher the concentration of iron, the greater the permissible concentration of acid can be.

3. The alkali metals, NH_4^+, alkaline earth metals, Zn^{++}, Mn^{++}, Co^{++}, Ni^{++}, Al^{+++}, Cr^{+++}, Ce^{+++}, and La^{+++} do not affect the titration. Titanium, tungsten, and molybdenum can be complexed with tartaric acid. Vanadate in appreciable amounts interferes and can not be complexed under the titration conditions. With the exception of copper, the metals of the hydrogen sulfide group have no effect except that some of them give diffusion currents. Sulfate is the only anion which interferes when present in large amounts.

4. Titration with bromoxine should also be applicable in the determination of small amounts of copper. Because of its specific character, it should be superior to benzoinoxine, since zinc, nickel, and other metals interfere when the latter is used

Cadmium with Naphthoquinoline. Sandberg[54] titrates cadmium in the presence of sulfuric acid and much iodide at a potential of −0.9 to −1.0 v. (S.C.E.). Cadmium yields a diffusion current and the reagent is not reduced.

Reagent. Solution in 0.1–0.5 *N* sulfuric solution. The solution is stable for at least 3 months in the dark (concentration not specified).

Procedure. Remove air from the cadmium solution which is 0.5 *N* in sulfuric acid. Then add sufficient potassium iodide to make its concentration about 1 *M*; the total volume is about 10 ml. Apply a potential of 0.9 v. and titrate at room temperature with the naphthoquinoline solution, while stirring by passing nitrogen into the solution. The current becomes constant about 3 minutes after the addition of reagent.

Notes: 1. 1–25 mg. of Cd in 10 ml. of solution can be titrated with an accuracy of 0.5 per cent. Solutions more dilute than 0.001 *M* in Cd can be titrated only approximately.

2. The concentration of sulfuric acid and iodide can be varied over a wide range.

3. Perchloric acid and nitric acid can be substituted for sulfuric acid. Even hydrochloric acid can be used if the iodide concentration is sufficiently high.

4. NH_4^+, Mg^{++}, Ca^{++}, Fe^{++}, Cr^{+++}, Al^{+++}, Ti^{++++}, Co^{++}, Ni^{++}, and Mn^{++} do not interfere. Fe^{+++} must be absent, since it oxidizes iodide to iodine, which gives a slightly soluble complex with naphthoquinoline. (Iron can be reduced with hydroxylamine.) Zinc is not coprecipitated, provided that the acid and iodide concentrations are kept low; for example, a 0.001 *M* solution of cadmium which is 2 *M* in zinc sulfate can be titrated with a relative error of 0.5 per cent.

5. Metals of the hydrogen sulfide group in general depolarize the mercury electrode (in the absence of complex-forming agents) and they must be removed. Ag, Hg, Pb, Cu, Bi, and Sb can be precipitated as the metals on iron wire in hot solution. Cadmium is not adsorbed. Any ferric salt formed can be reduced with hydroxyl-

amine sulfate, but generally this is not necessary. Arsenic in large amounts forms a passive layer on the iron wire which hinders the reduction. Small amounts of arsenic do not interfere in the titration; large quantities must be removed.

Copper and Ferric Iron with Cupferron. The ammonium salt of nitrosophenylhydroxylamine (cupferron) was introduced as an analytical reagent by Baudisch.[55] Introduced as a reagent for copper and ferric iron it was found later that quite a number of metal ions precipitate with the reagent. Koltoff and Liberti[56] investigated the use of this reagent in the amperometric titration of copper and ferric iron and devised a procedure for the titration of copper in the presence of iron. Although cupferron and nitrosophenylhydroxylamine are polarographically reducible it was preferred to carry out the titrations at potentials at which only the diffusion currents of copper and ferric iron, respectively, are measured.

Standard Solution of Cupferron. A commercial product of cupferron is recrystallized and the crystals are kept in the dark over ammonium carbonate. Under these conditions the solid does not decompose. A 0.02 *M* solution in water is prepared and stored in a dark bottle. This solution is stable for at least 2 weeks. The solution is standardized against a standard copper or ferric iron solution by one of the procedures described below.

Titration of Copper. The copper solution is placed in the titration cell and a phthalate, citrate, acetate, or tartrate buffer with a pH between 5 and 6 is added. The final concentration of the buffer is 0.05 *M* or less. Gelatin is added to a concentration of 0.01 per cent. Air is removed with nitrogen and the titration is carried out at a potential of −0.5 V. During the titration the precipitate separates in the form of very fine particles. Coagulation is promoted by vigorous stirring with a stream of nitrogen. The titration lines are of the type shown in Fig. XLVII-3*a*, and two moles of reagent are used per mole of cupric copper. An accuracy and a precision better than 1 per cent were obtained in the titration of 25 ml. of copper at concentrations between 5×10^{-4} and 2×10^{-3} *M*.

Notes: 1. No satisfactory results are obtained in unbuffered solutions which are slightly acid with hydrochloric acid. At a pH of 1 (0.1 *N* acid) scarcely any precipitation occurs. At a pH of 2 (0.01 *N* acid) a fine precipitate is formed, but the precipitation is not instantaneous. Since cupferron is decomposed in an acid medium high results (about 7 per cent in 0.01 *N* HCl) are found. If the buffer used can form a complex with copper its concentration should not be greater than 0.05 *M*. For example, in a 0.3 *M* citrate buffer the precipitation of copper from a 0.001 *M* solution was less complete than from 0.05 *M* citrate; no precipitation was obtained from a 0.05 *M* oxalate buffer (copper was present as a complex) or from an ammoniacal medium.

2. When the titration is carried out at a potential of −1.45 v. the diffusion current of cupferron is measured after the end point. In this case the titration diagram is of the type shown in Fig. XLVII-3*b*.

[55] O. Baudisch, *Chem. Ztg.*, **35**, 122 (1911).
[56] I. M. Kolthoff and A. Liberti, *Analyst*, **74**, 635 (1949).

Titration of Ferric Iron. Ferric cupferronate is less soluble than the corresponding copper compound. For this reason iron can be titrated at a lower pH than copper. In the procedure described below the titration cell is covered with a coat of black paint or wrapped in black paper to protect the solution from light. Ferric iron in tartrate or citrate is readily reduced by light.[57]

Procedure. A measured volume of the iron solutions is placed in the titration cell and a citrate or tartrate buffer with a pH between 3 and 4 is added. The final concentration of the buffer is 0.05 *M* or less. Gelatin is added to a concentration of 0.01 per cent. The mixture is made air-free and titrated at a potential of −0.5 v. The titration diagram is of the type shown in Fig. XLVII-3*a* and three moles of reagent are used per mole of ferric iron.

Notes: With 0.001 *M* or more concentrated ferric iron solutions the titration is accurate to 1 per cent. With 5×10^{-4} *M* solutions an accuracy of the order of 2 to 3 per cent was found. This low accuracy is due to the fact that the diffusion current of ferric iron in the buffers used is so small. The titration can also be carried out in 0.01 to 0.1 *N* hydrochloric acid (at an applied potential of −0.3 v.), but the bulky nature of the precipitate makes the measurement of the current difficult. The composition of the precipitate changes when the pH is raised from 5 to 7. No precipitate is formed in a medium containing ammonium tartrate and ammonia. In the recommended procedure the buffer concentration must be less than 0.1 *M*, as the complex which is formed increases the solubility of the precipitate.

1. Oxalate is unsuitable as a buffer because of the relatively large solubility of the precipitate in the presence of the complex which is formed.
2. When the titration is carried out at a potential of −1.4 v. the diffusion current of cupferron is measured after the end point.
3. Zinc, aluminum, chromium, manganese, and cadmium do not interfere when the iron is titrated in 0.1 *N* hydrochloric acid.
4. The titration of ferrous iron in a citrate buffer (pH 4 to 6) was investigated by measuring the anodic ferrous iron wave. Unsatisfactory results were obtained because of incomplete precipitation of the yellow ferrous cupferronate.

Copper in the Presence of Ferric Iron. When a mixture of ferric iron and copper is titrated in a citrate buffer of pH 5 the iron precipitates before the copper. However, the break in the titration line at the point of complete precipitation of iron is found with an error of the order of +8 per cent. The sum of iron and copper is found with an accuracy of 2–3 per cent. When the titration is carried out in 0.1 *N* hydrochloric acid the iron precipitates first. After its precipitation the copper current does not change until an appreciable excess of reagent is added. When the ratio of iron to copper is equal to or greater than 1 the iron can be determined with an accuracy of 1 per cent. Successful titrations of copper in

[57] See also J. J. Lingane, *J. Am. Chem. Soc.*, **68,** 2448 (1946).

the presence of large amounts of iron were carried out in the presence of potassium fluoride. The fluoride transforms the iron into the slightly soluble complex K_3FeF_6.

Procedure: To the mixture of iron and copper in a citrate buffer (0.05 *M*, pH 5) $KHF_2 \cdot 2H_2O$ is added to a concentration of 0.5 *M*. The copper is titrated with cupferron. Results accurate to 1 per cent were obtained even at a ratio of ferric iron to copper of 25.

Potassium with Dipicrylamine. Sandberg[58] gives directions for the titration of potassium with dipicrylamine.

Reagent. Air-free, approximately 0.07 *M* solution of sodium dipicrylaminate, standardized against potassium chloride solution of known concentration. The solution (even though supersaturated) is stable for as long as 2 months if no crystallization nuclei are introduced.

Procedure. Cool a monosodium phosphate-sodium hydroxide buffer solution (pH 12) of the sample to 0° and remove air. The solution should have a volume of 10 ml. and should be 0.01–0.15 *M* in potassium chloride. Apply a potential of 1.6 v. and titrate with the reagent solution, bubbling nitrogen through the solution for 5 minutes after the first addition of reagent, and for 2 minutes after succeeding additions. After precipitation of the potassium, the diffusion current of the reagent is measured.

Notes: 1. 4–120 mg. of K can be determined with an error not exceeding 0.4 per cent.

2. Amounts of sodium up to 15 times that of potassium do not interfere. In general, the permissible ratio Na/K increases with decreasing potassium concentration. More sodium can be present when the amperometric titration technique is used than when the gravimetric method is applied.

3. Lithium interferes less than sodium. Under certain conditions it may be advantageous to use lithium dipicrylaminate as the precipitant in place of the sodium salt.

4. Ammonium ion interferes, even in small quantities, but is readily removed by boiling with sodium hydroxide or, better, calcium hydroxide. Calcium scarcely interferes. In unfavorable cases, appreciable amounts of potassium can be coprecipitated with magnesium hydroxide. Zinc and manganese must be absent. Iron and aluminum can be precipitated with calcium hydroxide; the precipitate need not be filtered off. Mg, Mn, Zn, and Fe can be precipitated by oxine, first at pH 4.5, and then at pH 9.5; the oxine and ammonium salts are removed by volatilization

Calcium with Picrolonic Acid. Titration of Picrolonic Acid with Methylene Blue. Picrolonic acid, $C_{10}H_8O_5N_3$, yields a slightly soluble, crystalline precipitate with calcium. The solubility product of calcium picrolonate $[Ca^{++}] \cdot [\text{picrolonate}]^2$ was found to be of the order of 5×10^{-11} at 20° C.[59] Unfortunately, the precipitation of calcium with picrolonic acid

[58] B. Sandberg, *Svensk Kem. Tid.*, **58,** 197 (1946).

[59] G. Cohn and I. M. Kolthoff, *J. Biol. Chem.*, **147,** 705 (1943).

is too slow to be suitable for the amperometric titration of calcium. For this reason Cohn and Kolthoff[60] added an excess of picrolonic acid to the calcium solution and back-titrated the excess of reagent amperometrically with a standard solution of methylene blue. The reaction between methylene blue and picrolonic acid had been investigated earlier and made analytical use of by Bolliger.[61] He developed a method for the titration of picrolonic acid with methylene blue and applied this method to the indirect determination of calcium, magnesium, and organic bases by precipitation as picrolonates. Cohn and Kolthoff[60] titrated picrolonic acid amperometrically in a buffer solution which was 0.1 *M* in acetic acid, 0.0125 *M* in lithium acetate, and 0.1 *M* in lithium chloride with 0.01 *M* methylene blue solution.

At an applied voltage of 0.3 v. between the dropping electrode and a *mercury pool anode* methylene blue yields a diffusion current, but picrolonic acid is not yet reduced. As usual, nitrogen (or hydrogen) is passed through the solution to remove oxygen and to effect mixing before the measurement. The current remains equal to the residual one until the end point; after the end point the diffusion current of the excess of methylene blue is measured. The methylene blue solution is standardized against a standard solution of picrolonic acid.

In applying this method to the determination of calcium this cation is precipitated with an excess of picrolonic acid and the precipitate is removed by filtration. *Filter paper should not be used for this purpose* as picrolonic acid is strongly adsorbed on filter paper. Sintered glass crucibles are suitable for this filtration and no adsorption of the picrolonic acid occurs.

Procedure. Adjust the calcium content of the unknown to between 0.001 and 0.01 *M*. Add a quantity of buffer solution (1 *M* acetic acid, 0.125 *M* lithium acetate, and 1 *M* lithium chloride) such that it becomes diluted about 10 times in the final mixture. Add such a volume of standard 0.01 *M* picrolonic acid solution to 5 to 25 ml. of the unknown that the concentration of picrolonic acid after completed precipitation becomes at least 0.001 *M*. The amount of picrolonic acid shall not exceed the 4-fold molar concentration of the calcium. Keep the mixture overnight in an ice box when the calcium concentration is equal to or smaller than 5×10^{-3} *M*; when the calcium concentration is greater, keep it at room temperature (not above 20°). Filter through a sintered glass crucible and titrate an aliquot part with 0.01 *M* methylene blue according to the directions given above.

Instead of determining the excess of picrolonic acid by amperometric titration this can be done polarographically at a potential of −1.1 v. (*vs.* S.C.E.). Under these conditions the calcium picrolonate need not be removed by filtration.

---

[60] G. Cohn and I. M. Kolthoff, *J. Biol. Chem.*, **148,** 711 (1943).

[61] A. Bolliger, *Proc. Roy. Soc. New South Wales*, **67,** 240 (1933); **68,** 51, 197 (1934); *J. Biol. Chem.*, **107,** 229 (1934); *Analyst*, **64,** 416 (1939).

Notes: 1. The titration procedure yields results accurate within 2 per cent with solutions varying between 25 ml. 0.001 *M* and 10 ml. 0.01 *M* in calcium.

2. In general, alkali ions and magnesium do not interfere when their concentration is less than 0.05 to 0.06 *M* after dilution of the sample with the buffer and addition of the picrolonic acid solution. Phosphate in a concentration of 0.25 *M* or less and sulfate in a concentration less than 0.15 *M* do not interfere. Quite generally the recommended procedure yields good results in the determination of 0.001 to 0.01 *M* calcium solutions even when the concentrations of alkali, magnesium, sulfate, and phosphate ions are from 10 to 100 times larger than that of the calcium.

Phosphate with Uranyl Acetate. In buffered solutions uranyl ion reacts with phosphate ion with the formation of insoluble UO_2MPO_4 (M = Na, K, NH_4, $\frac{1}{2}$ Ca, etc.). In the proper medium uranyl yields a well-defined first diffusion current which is proportional to the uranyl concentration. Kolthoff and Cohn[62] made the reaction between uranyl and phosphate ions the basis of an amperometric titration of phosphate.

Reagents. *Uranyl Acetate.* A 0.1 *M* solution is prepared by dissolving 42.4 g. of C. P. uranyl acetate $UO_2(C_2H_3O_2)_2 \cdot 2H_2O$ in a 1-liter volumetric flask by shaking with about 300 ml. of water and 6–10 ml. of glacial acetic acid. After complete dissolution the flask is filled up to volume. The solution can be standardized by amperometric titration against monopotassium phosphate. For the titration of very dilute phosphate solutions the standard uranyl solution is diluted with water.

Monopotassium Phosphate. A 0.01 *M* solution of pure KH_2PO_4 is prepared by dissolving 1.3614 g. of the salt in water and diluting to 1 liter.

Potassium Chloride. A 1 to 2 *M* solution.

Procedure. The dissolved sample is diluted so that the concentration of phosphate in the final volume (all additions included) is not greater than 0.01 *M* or, if calcium (and sulfate) are present, not greater than 0.005 *M*. The concentration of calcium and sulfate must not be greater than 0.02 *M* and 0.01 *M*, respectively, or calcium sulfate may precipitate. If the solution is not neutral to bromcresol green, hydrochloric acid or sodium hydroxide is added until the indicator has a green color. About 0.5 ml. of 0.1 *M* acetic acid and enough potassium chloride are added to make the chloride concentration approximately 0.1 *M*. The solution is transferred to the polarographic titration cell, using a pool of mercury as an anode. Enough ethanol is added to make its concentration 20 per cent. Air is removed by passing nitrogen through, and the titration is carried out at an applied voltage of 0.7–0.8 v. The current remains very small (residual current) until all the phosphate is precipitated and then increases linearly with the excess of uranyl added.

Notes: 1. Concentrations of phosphate greater than 0.0003 *M* can be determined with an accuracy of 1 per cent.

2. Organic anions, like citrate, oxalate, acetate, tartrate, etc., which form complexes with uranyl ion, interfere. In general organic matter must be removed by ashing.

3. Magnesium and barium in relatively large concentrations have no effect upon

---

[62] I. M. Kolthoff and G. Cohn, *Ind. Eng. Chem., Anal. Ed.*, **14**, 412 (1942).

the results. Calcium in larger concentrations causes low results due to coprecipitation of calcium phosphate. However, when the concentration of calcium is smaller than 0.02 *M* and that of phosphate is 0.005 *M* or less, the error becomes negligibly small. In the presence of much calcium and sulfate a precipitate of calcium sulfate is formed which contains coprecipitated phosphate. In most practical cases the conditions specified in the procedure are satisfied. However, if appreciable amount of calcium sulfate precipitate, a separation must be made. This is most easily done by precipitation of calcium phosphate in an ammoniacal medium. The precipitate need not be washed.

4. Iron, both in the ferrous and ferric states, interferes. The simplest way to remove iron is by precipitation in hydrochloric acid medium with cupferron, and removal of the precipitate and the excess of reagent by shaking out with ether.

5. All metals, like lead, aluminum, and trivalent chromium, which precipitate with phosphate at a pH of about 3.5, interfere.

6. Pyrophosphate, arsenate, and vanadate ions precipitate with uranyl ion and interfere. *Arsenate* can be titrated in the same way as phosphate with uranyl ion.

Phosphate with Bismuth Oxyperchlorate Solution. Neuberger[63] performed the titration at a potential at which bismuth yields a diffusion current. Useful results were only obtained with relatively large amounts of phosphate (50 mg. of P_2O_5 or more), and even then the error was of the order of 1 per cent. Ferric iron interferes, even after reduction to the ferrous state, as the latter is easily air oxidized under the experimental conditions. It is not clear why Neuberger did not work in the absence of oxygen.

Fluoride with Thorium or Lanthanum. Langer[64] titrated neutral solutions of fluoride with thorium or lanthanum nitrates in a medium of 0.1 *N* alkali chloride or nitrate at a potential of −1.7 v. (between −1.6 and −1.8 v. *vs.* S.C.E.). Fluoride is not reducible and the current remains equal to the residual current until the end point. With an excess of thorium or lanthanum a current passes through. This, according to Langer, is caused by reduction of nitrate in the presence of thorium or lanthanum, but this seems very doubtful. Whatever the cause of the current, Langer found two straight lines which intersect at the end point. The reagent is standardized amperometrically with a standard fluoride solution. Fluoride solutions with a concentration of 0.01 to 0.001 *N* can be titrated with 0.1 *N* reagent (accuracy about 0.5 per cent); more dilute solutions are titrated with 0.01 *N* reagent. Large amounts of halides, nitrates, and perchlorates do not affect the accuracy. Sulfate, phosphate, arsenate, and substances with an alkaline reaction or which precipitate with the reagent, interfere. When thorium nitrate is used as a reagent, acetate ions should be absent

63 A. Neuberger, *Z. anal. Chem.*, **116,** 1 (1939).
64 A. Langer, *Ind. Eng. Chem., Anal. Ed.*, **12,** 411 (1940).

but they may be present with lanthanum nitrate as reagent. The titration needs further investigation.

Aluminum, Calcium, and Magnesium with Fluoride. In the first section of this chapter, the principle of amperometric titration methods with indicator substances, as proposed by Ringbom and Wilkman,[65] was discussed. These authors use the method for the titration of aluminum, calcium, and magnesium with fluoride, using ferric iron as indicator.

Aluminum. In neutral solution aluminum reacts with fluoride to form the stable complex AlF_6^{---} ion.

Procedure. To a 20–25 ml. sample containing 10–40 mg. of aluminum, add 0.5 ml. of 0.1 *M* ferric chloride, 25 ml. of alcohol, and an excess of solid sodium chloride. Remove dissolved oxygen, make connection with a saturated calomel electrode (an internal silver chloride electrode probably would be simpler), and titrate with 0.6 *M* sodium fluoride solution. The end point is found when the current is zero. A volume of fluoride equivalent to the amount of iron added is subtracted from the amount of fluoride consumed in the titration. The accuracy is stated to be of the order of 0.5 per cent. Interference caused by the presence of other substances are not discussed.

Calcium. The procedure is the same as that described for aluminum except that no sodium chloride is added. With 40–100 mg. of calcium in a volume of 50 ml. (50 per cent in ethanol) the accuracy is stated to be about 1 per cent; with smaller or larger amounts the error increases.

Magnesium. The dircct titration did not yield satisfactory results. Ringbom and Wilkman add the magnesium-containing sample to a measured excess of standard sodium fluoride solution, heat to boiling, cool, and add a measured excess of standard aluminum sulfate solution, 0.5 ml. of 0.1 *M* ferric chloride, and alcohol to a concentration of 50 per cent. After adding an excess of solid sodium chloride the excess of aluminum is titrated as described under aluminum. The accuracy is stated to be of the order of 1 per cent. Interference by other substances is not mentioned.

Ferrocyanide with Zinc. Spalenka[66] titrated in 0.2 *N* hydrochloric acid at a potential of −1.2 v., at which zinc yields a diffusion current. The zinc solution must be added to the ferrocyanide, and the end point is found after the precipitation of the ferrocyanide as $K_2Zn_3[Fe(CN)_6]_2$. This titration needs further investigation. Spalenka mentioned that the titration can also be performed in an ammoniacal medium, but the results were not very reproducible.

Titrations with Titanous Chloride. The principle on which ampero-

[65] A. Ringbom and B. Wilkman, *Acta Chem. Scand.*, **3**, 22 (1948).
[66] M. Spalenka, *Collection Czechoslov. Chem. Communs.*, **11**, 146 (1939).

metric titrations with titanous chloride are based has already been discussed. Strubl[67] performed the titrations in the presence of tartrate or citrate ions. Titanous titanium then yields an excellent anodic wave and titanic titanium a cathodic wave, at −0.44 v. *vs.* the S.C.E. Spalenka[66] did not use a separate anode, but employed a large pool of mercury in the cell as such. (In titrating with oxidizing agents which react with mercury it is certainly preferable to use an outside reference anode.) Spalenka added the oxidizing agent to the titanous chloride solution and titrated at an applied e.m.f. of about 0.3 v. Under this condition the titanous titanium yields an anodic diffusion current. The negative current decreases to zero at the end point and then becomes positive upon addition of an excess of reagent. Spalenka claims that, after correction for the dilution effect, the points before and after the end point all lie on one straight line. This cannot be true in general, as it would require that the diffusion current constants of titanous chloride and the various oxidizing agents all be equal. Spalenka titrated titanous chloride with ferric chloride, dichromate, iodate (rather complicated, first reduction to iodide, and at a later stage iodine is formed), chlorate (excess of chlorate does not yield a reduction wave), bromate, and ferricyanide, respectively, all with good results.

Chloride with Mercurous Nitrate. This titration[68] can be performed at an applied potential of zero (*vs.* S.C.E.). At this potential mercurous nitrate yields a diffusion current. It is necessary to add a relatively large amount of gelatin to the chloride solution in order to prevent a depolarizing effect of the colloidal calomel at the dropping electrode. The presence of gelatin is also desirable in order to eliminate the pronounced maximum in the reduction wave of mercurous ion. Fig. XLVII-22 shows the results obtained in the titration of a mixture of 40 ml. of 0.002 N hydrochloric in 0.1 N nitric acid and 2.5 ml. of 1 per cent gelatin with 0.02 M mercurous nitrate. The results were accurate within 0.3 per cent. Such a titration can be completed in 5 minutes. Solutions which were 0.001 N in chloride could be titrated with an accuracy of 1 per cent. Irregular results were found in the titration of chloride solutions more dilute than 0.001 N.

Iodide with Mercuric Nitrate. In Fig. XLVII-23 results are shown obtained in the titration of 50 ml. of 0.001 N potassium iodide in a solution which was 0.1 N in nitric acid and 0.05 per cent in gelatin with 0.01 M mercuric nitrate.[68] In the presence of an excess of iodide the negative diffusion current (anodic current) of iodide is measured, and with an excess

[67] R. Strubl, *Collection Czechoslov. Chem. Communs.*, **10,** 475 (1938). M. Spalenka, *ibid.*, **11,** 146 (1939).

[68] I. M. Kolthoff and C. S. Miller, *J. Am. Chem. Soc.*, **63,** 1405 (1941).

of reagent the cthodic diffusion current of mercuric nitrate is obtained. The point at which the current is zero corresponds to the end point. In Fig. XLVII-23 is also given the result of a potentiometric titration of the same solution, the values of the potential being plotted on the right-hand ordinate. The potentiometric end point coincided with 0.3 per cent with the amperometric end point.

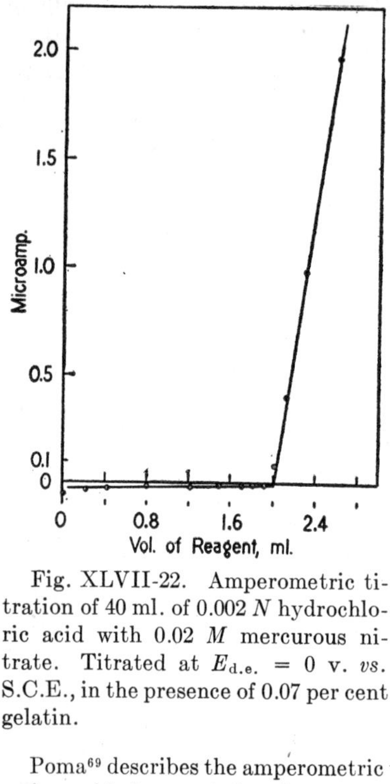

Fig. XLVII-22. Amperometric titration of 40 ml. of 0.002 *N* hydrochloric acid with 0.02 *M* mercurous nitrate. Titrated at $E_{d.e.} = 0$ v. *vs.* S.C.E., in the presence of 0.07 per cent gelatin.

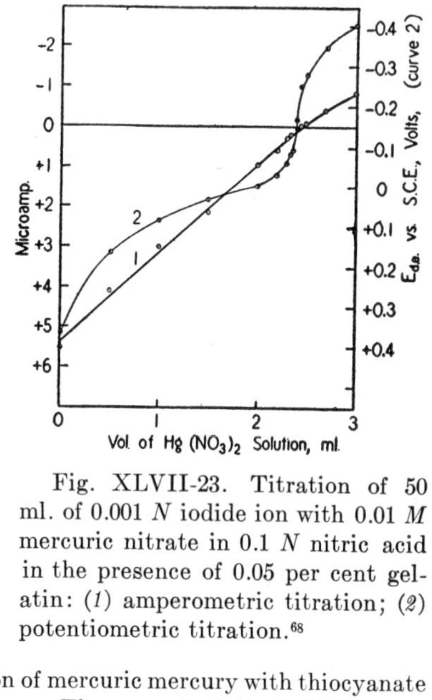

Fig. XLVII-23. Titration of 50 ml. of 0.001 *N* iodide ion with 0.01 *M* mercuric nitrate in 0.1 *N* nitric acid in the presence of 0.05 per cent gelatin: (*1*) amperometric titration; (*2*) potentiometric titration.[68]

Poma[69] describes the amperometric titration of mercuric mercury with thiocyanate and cyanide ions and also the reverse titrations. They are of relatively little practical significance since the applied potential during the titration depends on the kind and concentration of the reaction product. The titration with cyanide yields highly inaccurate results.

Aromatic Diamidines with Alizarin Sulfonate. Conn[70] found that diamidines of the type:

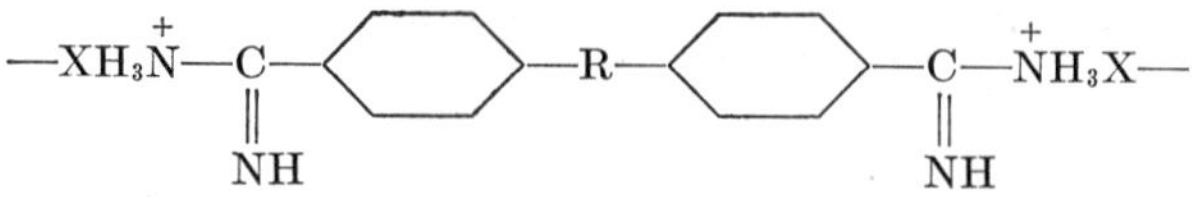

---

[69] K. Poma, *Bull. soc. chim. Belges*, **54,** 277 (1945).
[70] J. B. Conn, *Anal. Chem.*, **20,** 585 (1948).

(R may be —CH═CH—, —O—, or $(CH_2)_n$—O—) precipitate quantitatively with sodium alizarin sulfonate. The precipitates are extremely slightly soluble. Alizarin sulfonate at pH 7 gives a well-defined wave with $E_{1/2} = -0.67$ v. (S.C.E.) and the diffusion current is well developed at -0.9 v. The amperometric titration is carried out in a phosphate buffer of pH 7 at a potential of -0.9 v. The accuracy and reproducibility are excellent.

α-Tocopherol with Auric Chloride. Karrer[71] and his co-workers found that α-tocopherol (I) is quantitatively oxidized to tocopheryl quinone (II) by auric chloride:

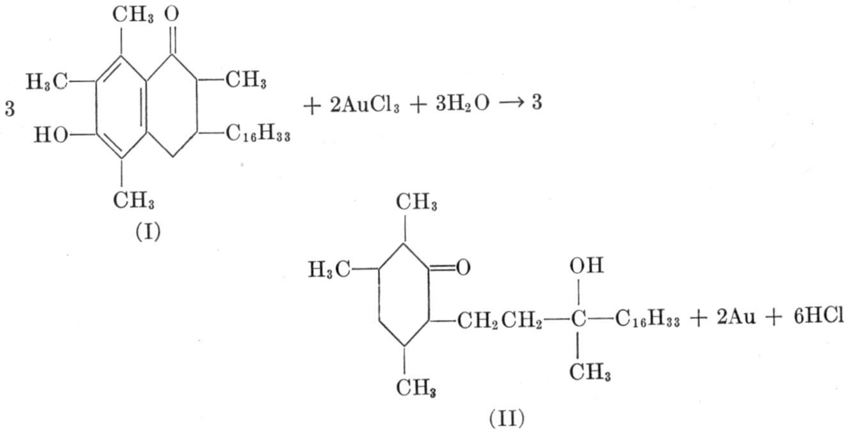

Smith, Kolthoff, and Spillane[72] made use of this reaction in the amperometric titration of α-tocopherol. The titration is best carried out in a 0.1 *M* benzoic acid-0.1 *M* sodium benzoate buffer in 75 per cent ethanol, which is 0.1 *M* in sodium chloride. At a potential of -0.075 v. (S.C.E.) the tocoquinone is not reduced while the auric gold at this potential yields a diffusion current which is proportional to the concentration of gold in the solution. When the titration is carried out at this potential the current remains equal to the residual current until the end point. Tocopherol in a concentration range between 1×10^{-3} and 3×10^{-4} *M* was determined with an accuracy of 0.5 per cent.

5. AMPEROMETRIC TITRATIONS WITH A PLATINUM MICROELECTRODE

From a practical viewpoint it is advantageous to carry out the titrations with a rotated electrode. This has the advantage that the diffusion

[71] P. Karrer, R. Escher, H. Fritzsche, H. Keller, B. H. Ringier, and H. Salomon, *Helv. Chim. Acta*, **21**, 939 (1938).

[72] L. I. Smith, I. M. Kolthoff, and L. J. Spillane, *J. Am. Chem. Soc.*, **64**, 646 (1942).

currents are much greater than in unstirred solutions at the same concentration and the current attains a constant value immediately. In order to obtain a linear relation between the current and the amount of reagent added the speed of stirring has to be kept constant during the titration. This was done by Laitinen and Kolthoff[73] by rotating the electrode with an electric motor whose speed could be regulated. The application of the wire electrode to amperometric titrations has not yet been studied systematically. It may be expected that such titrations will be of great practical importance in titrations with or of those oxidizing agents that yield a diffusion current at potentials at which oxygen is not yet reduced. Under these conditions the titration can be carried out quickly without the removal of oxygen.

Silver Nitrate as Reagent

Halides. Laitinen and Kolthoff[74] obtained good results in the titration of silver with chloride or *vice versa*, provided that gelatin was present to prevent depolarization of the colloidal silver chloride particles and that the concentration of silver or chloride titrated was equal to or greater than 0.01 *N*. Further studies by Laitinen *et al.*[75] revealed that results accurate to within 0.5 per cent were obtained in the presence of nitric or sulfuric acids and 0.1 per cent gelatin, when the chloride concentration was greater than 0.005 *N*. Using 50 per cent acetone as the titrating medium, the solubility of silver chloride is reduced to such a point that a 0.001 to 0.0005 *N* chloride solution can be titrated with an accuracy of about −3 per cent (presence of 0.1 per cent gelatin) and a 0.0001 *N* chloride solution with an error of 10 per cent. In 75 per cent acetone this error is reduced to 3 per cent.

Bromide and iodide can be titrated in the absence of gelatin. When the concentration of bromide is 0.01 *N* or greater the results are of the order of 1 per cent low. In aqueous solution 0.001 *N* bromide is titrated with an error of −3 per cent, and in 50 per cent acetone a 0.0001 *N* bromide solution gives a similar error. The results of the bromide titration remain unaffected by the presence of 0.01 *N* ammonia, but more ammonia interferes.

The results with iodide are comparable to those with bromide. However, iodide can also be titrated in a more strongly ammoniacal medium. With 0.01 *N* iodide, even 1 *N* ammonia has no effect, while with 0.001 *N*

[73] H. A. Laitinen and I. M. Kolthoff, *J. Phys. Chem.*, **45**, 1079 (1949).

[74] H. A. Laitinen and I. M. Kolthoff, *J. Phys. Chem.*, **45**, 1079 (1943).

[75] H. A. Laitinen, W. P. Jennings, and T. D. Parks, *Ind. Eng. Chem., Anal. Ed.*, **18**, 355 (1946).

iodide a 10 per cent error was found using 0.5 *N* ammonia, although 0.1 *N* ammonia hd no effect. In an ammoniacal medium the mercury–mercuric iodide–potassium iodide electrode[76] (*vide infra*) was used as the reference electrode.

Mixture of Halides. Silver bromide and silver iodide do not depolarize the rotating platinum electrode cathodically, while silver chloride particles cause a cathodic current even in the presence of a large excess of chloride but in the absence of gelatin. Use of this fact is made by Laitinen *et al.*[77] in the titration of bromide (or iodide, *vide infra*) in the presence of chloride. It was found that 0.01 *N* bromide could be titrated amperometrically in the presence of 0.1 *N* chloride in the presence of such electrolytes as nitric acid (0.8 *N*), or barium nitrate (0.4 *N*), or aluminum, etc.

Iodide is best titrated in the presence of bromide and/or chloride in 0.1 *N* ammonia using the mercury–mercuric iodide–potassium iodide half-cell (potential −0.23 v. *vs.* S.C.E.) as reference electrode. After the end point is passed, the solution is acidified to make it 0.8 *N* in nitric acid. The bromide is now titrated, using the S.C.E. as the reference electrode. After the end point 0.1 per cent of gelatin is added and the choride is titrated. The results are not as accurate as those obtained by some other method, but the advantage is that the results are found within a very short time and that dilute solutions can be analyzed without being concentrated.

Cyanide. Laitinen *et al.*[78] titrate cyanide amperometrically with silver nitrate in a solution which is 0.1 *N* in sodium hydroxide. The mercury–mercuric iodide–potassium iodide electrode (see section on mercaptans) is used as a reference electrode and no potential is applied to the system. The end point is found when all the cyanide has been transformed into the complex argentate $[Ag(CN)_2]$.

The method is accurate to within 0.2 per cent at concentrations between 0.1 and 0.001 *M* cyanide. At lower concentrations than 0.001 *M* the accuracy decreases, but a solution as dilute as 4×10^{-5} *M* cyanide could be titrated with an accuracy of 2.3 per cent and an end point was found at a concentration as small as 8×10^{-6} *M*. In general, results are accurate to within 0.5 per cent in the presence of a 50-fold excess of chloride or bromide and 1 *N* solutions of hydroxide, nitrate, or sulfate.

Mercaptans. Mercaptans can be titrated amperometrically with silver

---

[76] I. M. Kolthoff and W. E. Harris, *Ind. Eng. Chem., Anal. Ed.*, **18,** 161 (1946).

[77] H. A. Laitinen, W. P. Jennings and T. D. Parks, *Ind. Eng. Chem., Anal. Ed.*, **18,** 358 (1946).

[78] H. A. Laitinen, W. P. Jennings, and T. D. Parks, *Ind. Eng. Chem., Anal. Ed.*, **18,** 574 (1946).

nitrate in acid or ammoniacal media.[79] The ammoniacal medium is advantageous, since chloride does not interfere. Mercaptans (RSH) react with silver nitrate with the formation of insoluble silver mercaptide:

$$Ag^+ + RSH \rightarrow RSAg + H^+ \qquad \text{(acid)}$$

$$Ag(NH_3)_2 + RSH \rightarrow RSAg + NH_4^+ + NH_3 \qquad \text{(ammoniacal)}$$

For titrations of ammoniacal solutions a mercury–mercuric iodide half-cell is used as a reference electrode and no voltage need be applied. No current flows through the cell until the end point; after the end point has been reached the diffusion current of the excess silver ions is measured.

Mercury–Mercuric Iodide Electrode. The electrolyte solution for the reference half-cell is prepared by dissolving 4.2 g. of potassium iodide and 1.3 g. of mercuric iodide in 100 ml. of saturated potassium chloride solutions. A layer of mercury serves as the electrode. The potential is −0.23 v. (*vs.* S.C.E.).

The simple titration ensemble shown in Fig. XLVII-24 is recommended. The salt bridge, *E*, consists of some 60 cm. of about 6-mm. (inside diameter) soft rubber tubing filled with saturated potassium chloride solution. The rubber tubing is connected with a short length of glass tubing, *D*, filled with a gel of 3 per cent agar and 30 per cent potassium chloride. At the end of the glass tube, a coarsely sintered glass disc may be inserted. For further protection of the solution from contamination with iodide, the glass tube, *B*, having an agar or a finely sintered glass plug at its end, may be interposed. The electrolyte solution inside *B* can be easily rinsed out and replaced with fresh electrolyte whenever it has been in use long enough to become contaminated with iodide. It is essential that all sources of high resistance, such as air bubbles, be eliminated from the salt bridge.

The two half cells are short-circuited through a microammeter, G, *e. g.*, a Weston Electrical Instrument Corporation (Newark, N. J.) microammeter, Model 430, is very useful. The ensemble is generally convenient for rapid amperometric titrations with the rotating platinum electrode.

Procedure. In a 250-ml. beaker dilute a sample of mercaptan containing about 5 mg. of mercaptan sulfur to 100 ml. with 95 per cent ethanol. Make the solution about 0.25 *M* in ammonia and 0.01 to 0.1 *M* in some noninterfering electrolyte such as ammonium nitrate. Immerse the end of the salt bridge and the rotating platinum electrode in this solution and titrate with aqueous 0.005 *M* silver nitrate.

Make 2 or 3 readings of the microammeter before the end point. As long as silver is not in excess the current is zero or very small. After the end point the dif-

[79] I. M. Kolthoff and W. E. Harris, *Ind. Eng. Chem., Anal. Ed.*, **18,** 161 (1946).

fusion current corresponds to the diffusion current of the excess of silver. When the ammeter indicates that the end point has been passed, add 2 or 3 small increments of silver nitrate; read the current after each addition. If the electrode becomes sluggish, clean by wiping with a piece of cloth or between two fingers. Plot the

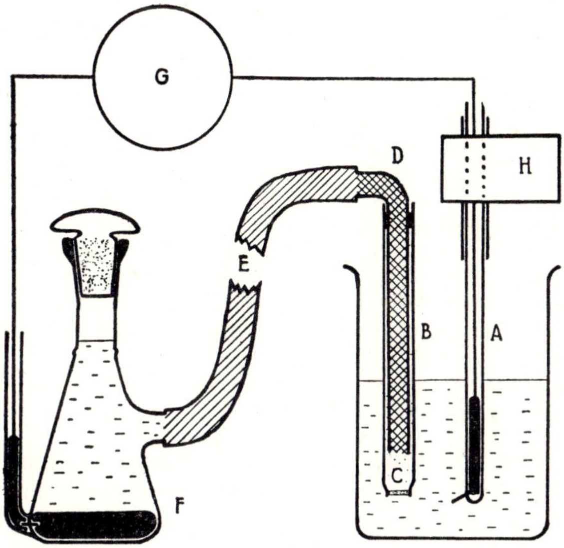

Fig. XLVII-24. Apparatus for amperometric titration.[79]

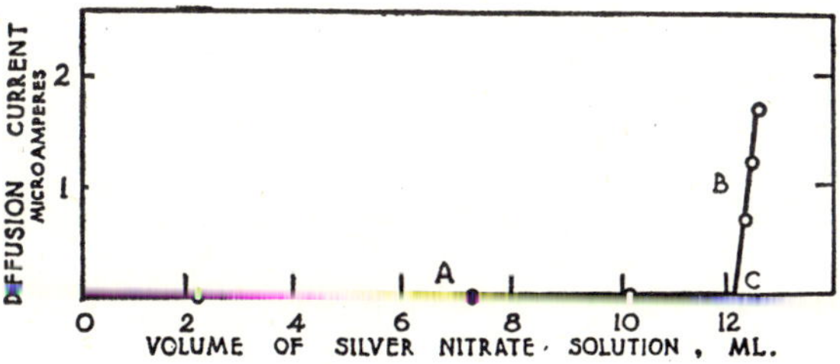

Fig. XLVII-25. Amperometric titration of 1.942 mg. of mercaptan sulfur with 0.00495 *N* silver nitrate.[79]

readings of the microammeter against the volume of silver nitrate added; the point of intersection *C* of the two straight lines corresponds with the end point. An example is shown in Fig. XLVII-25. A titration carried out by an experienced operator requires no more than 2 minutes.

Notes: 1. 5 mg. of mercaptan sulfur or more can be titrated with an accuracy of 0.2 per cent. Even 1 mg. of mercaptan sulfur can be determined with an accuracy of 1 per cent. A large number of primary and tertiary mercaptans (between C_2 and C_{18}) has been titrated with satisfactory accuracy.

2. Cyanide forms a stable complex with silver ions and interferes (see section on cyanide). Other ions, like iodide and sulfide, which form insoluble silver salts in an ammoniacal medium, interfere. Large amounts of chloride and small amounts of bromide do not interfere.

3. An indicator electrode may become insensitive or erratic after long use. As a rule, full sensitivity can be restored by wiping the electrode with a piece of cloth or even between two fingers. Thorough cleaning of the electrode with concentrated nitric acid is usually necessary only after it has been used for several hundred titrations.

4. Frank *et al.*[80] analyzed mercaptans by amperometric titrations and obtained results which were several per cent low. Apparently, part of the mercaptan was air-oxidized to disulfide before the titration. Strafford *et al.*[81] who tested mercaptonaphthalene and mercaptobenzthiazole amperometrically and by other methods showed that the low amperometric results were due to oxidation of the mercaptans in the course of the titration. Good results were obtained if the titration was carried out under nitrogen, using silver nitrate, freshly prepared with boiled water. The mercaptans are less susceptible to oxidation in neutral than in ammoniacal solutions.

5. The determination of mercaptans in lattices obtained by emulsion polymerizations is described by Kolthoff and Harris.[82]

6. Alkyl disulfides are reduced by amalgamated zinc to mercaptans which are titrated in the above manner.[83]

Sulfhydryl Groups in Amino Acids and Proteins. Cysteine and Cystine. Benesch and Benesch[84] applied the amperometric procedure of Kolthoff and Harris for the determination of mercaptans without essential change to the determination of —SH groups in thioglycolic acid, cysteine, glutathione, and ergothionine, as well as in certain proteins. Kolthoff and Stricks[85] titrated cysteine in aqueous solutions and obtained accurate results under proper conditions. Cystine does not react with silver under the conditions of the tiration; it can be determined after reduction to cysteine. This is best done with sodium amalgam in acid solution or by making use of the reaction between cystine and sodium sulfite in an ammoniacal medium. The sodium sulfite converts one-half of the cystine into cysteine and the other half into cysteine sulfonate. It also inhibits air oxidation of the cysteine in an alkaline medium:

$$RSSR + SO_3^{--} \rightleftharpoons RS^- + RSSO_3^-$$

[80] R. L. Frank, P. V. Smith, and F. E. Woodward, *J. Polymer Sci.*, **3,** 39 (1948).

[81] N. Strafford, F. R. Cropper, and A. Hamer, *Analyst*, **75,** 55 (1950).

[82] I. M. Kolthoff and W. E. Harris, *J. Polymer Sci.*, **2,** 41 (1947).

[83] I. M. Kolthoff, D. R. May, P. Morgan, H. A. Laitinen, and A. S. O'Brien, *Ind. Eng. Chem., Anal. Ed.*, **18,** 442 (1946).

[84] R. Benesch and R. E. Benesch, *Arch. Biochem.*, **19,** 35–45 (1948). See also E. B. Schoenbach, E. B. Armistead, and N. Weissman, *Proc. Soc. Exptl. Biol. Med.*, **73,** 44 (1950).

[85] I. M. Kolthoff and W. Stricks, *J. Am. Chem. Soc.*, **72,** 1952 (1950).

Procedures.

Determination of Cysteine. In a 150-ml. beaker, place 25–100 ml. of a solution which is 0.1 *M* in ammonia and 0.2–0.3 *M* in ammonium nitrate. Add so much of the cysteine solution that its concentration in the mixture is between 1 and 5×10^{-4} *M*. If the concentration of the solution to be titrated is less than 10^{-4} *M* in cysteine the electrolyte solution should be made air-free before adding the sample of cysteine. This is done either by bubbling nitrogen through the solution or by making the solution 0.025–0.05 *M* in sodium sulfite. Sodium sulfite may be used only if the sample does not contain cystine which reacts with the salt. Immerse the salt bridge and the rotating platinum electrode in the solution and titrate with silver nitrate of suitable concentration (0.005–0.001 *M*) at an applied potential of −0.3 v. *vs.* S.C.E.

Notes: 1. The accuracy and precision of the titration are 0.3 per cent with cysteine concentrations between 5×10^{-4} and 10^{-4} *M*.

2. No sharp end point is obtained when the cysteine concentration is considerably greater than 10^{-4} *M*.

3. Iodide ion interferes if its molarity is greater than one-fifth of that of cysteine. Metals like zinc, cadmium, cobalt, and copper interfere.

Determination of Cystine. *1. Reduction with Sodium Sulfite.* In a 150-ml. beaker, place 25–100 ml. of a solution which is 0.2–0.3 *M* in ammonium nitrate, add the cystine sample so that its concentration in the mixture is between 5×10^{-4} and 10^{-4} *M*. Pass nitrogen through the solution, make the solution 0.1 *M* in ammonia and 0.1 to 0.025 *M* in sodium sulfite, and titrate with 10^{-3} *M* silver nitrate at an applied potential of −0.3 v. *vs.* S.C.E.

Note: Care must be taken to have an ample excess of sodium sulfite in the solution (at least 0.025 *M*). When too little sulfite is added the end point is not sharp.

2. Reduction with Sodium Amalgam. Place of the sample containing 1 to 0.3 mg. of cystine (corresponding to 0.27 to 0.08 mg. of titratable sulfur) in a small test tube. Make it 0.05 *M* in sulfuric acid and add about one drop of 1 per cent sodium amalgam. Allow the solution to react for about one-half hour with frequent shaking. Transfer the solution quantitatively into a 150-ml. beaker containing about 25 ml. of air-free ammonia buffer (0.1 *M* in NH_3, and 0.2 *M* in NH_4NO_3) and proceed as described in the analysis of cysteine.

Ammino Cupric Copper as Reagent. Unpublished investigations by Stricks and Kolthoff (University of Minnesota) have shown that an ammoniacal cupric copper solution reacts quantitatively with cysteine (RSH) in the presence of an excess of sulfite according to the equation:

$$2Cu^{++} + RS^{-} + SO_3^{--} \rightarrow CuRS + Cu^{+} + RSSO_3^{-}$$

This reaction can be made the basis of an accurate and rapid amperometric method for the determination of cysteine and cystine using the rotating platinum wire electrode as indicator electrode. The titration with cupric copper gives a sharper end point than that obtained in the argentometric titration, since no metal is deposited at the electrode after the end point in titrations with cupric copper. With strict exclusion of air from the titration mixture and titrating agent, it is possible to determine about 0.1 mg. of cystine in a 2×10^{-5} M solution with an accuracy of 2 per cent. This corresponds to the actual determination of about 0.014 mg. of sulfhydryl sulfur. The titration is carried out at an applied potential of -0.4 v. (S.C.E.) at which ammoniacal cupric copper gives a diffusion current corresponding to the reduction $Cu^{++} + e \rightarrow Cu^{+}$. Zinc and cadmium do not interfere; as a matter of fact, they accelerate the reaction between cupric copper and cysteine in the proper electrolyte. Cobalt has a strong retarding effect and must be absent.

Titrations of or with Iodine

Iodine can be titrated amperometrically with standard thiosulfate or arsenic trioxide solutions using a saturated calomel electrode as reference electrode without applying an external e.m.f. Solutions of iodine as dilute as 0.0001 N can be titrated with an accuracy greater than 1 per cent and even 0.00001 N solutions can be determined satisfactorily using a sensitive galvanometer. Wide use of the titration of microquantities of iodine or of substances which are readily oxidized by iodine, *e. g.*, trivalent arsenic and antimony, is anticipated.[86]

Normal and Tertiary Mercaptans. Primary mercaptans react with 0.5 mole of iodine, whereas tertiary mercaptans react with 1 mole of iodine per molecule of mercaptan:

$$2RSH + I_2 \rightarrow RSSR + 2HI \qquad \text{(primary)}$$

$$R'SH + I_2 \rightarrow R'SI + HI \qquad \text{(tertiary)}$$

Primary and tertiary mercaptans can be titrated amperometrically with iodine using a rotating platinum wire as indicator electrode and the saturated calomel electrode as reference electrode. When a mixture of a primary and tertiary mercaptan is titrated iodometrically, the amount of iodine used is much less than that calculated on the basis of the above equations because in the mixture part of the tertiary mercaptan reacts like a primary one with the formation of a mixed disulfide:

$$RSH + R'SH + I_2 \rightarrow RSSR' + 2HI$$

where R′SH is the tertiary, and RSH the primary, mercaptan. Transformation of the mercaptans into slightly dissociated mercaptides, such as lead mercaptide, prevent this reaction from taking place. Kolthoff and Harris[87] obtained good results in the iodometric titration of mixtures

[86] See, *e. g.*, D. P. Evans and N. T. Simmons, *J. Soc. Chem. Ind.*, **63**, 29 (1944).
[87] I. M. Kolthoff and W. E. Harris, *Anal. Chem.*, **21**, 963 (1949).

of primary and tertiary mercaptans in the presence of an excess of lead nitrate or perchlorate. The primary mercaptan yields the disulfide and the tertiary mercaptan in large dilutions yields the sulfenyl compound. Upon amperometric titration in an ammoniacal medium with silver nitrate (see p. 942) all mercaptans react in a molar ratio of 1/1.

Procedure A. Take an aliquot portion of a solution containing 1 to 40 mg. of mercaptan and titrate amperometrically with standard silver nitrate solution as described on page 942.

Procedure B. Dilute a second aliquot portion of the solution containing not more than about 5 mg. of mercaptan to about 100 ml. with 95 per cent ethanol. To the solution add 1 ml. of 1 *M* lead nitrate or perchlorate solution and enough 1 *M* perchloric acid to make the solution 0.01 *M* in perchloric acid. Titrate amperometrically in the cold with standard 0.005 *N* iodine, using a rotating platinum electrode short-circuited with the saturated calomel electrode through the microammeter.

If the total amount of mercaptan sulfur present from the silver nitrate titration (procedure A) is known, the amounts of primary and tertiary mercaptans present are readily calculated as follows:

Milliequivalents of silver nitrate (procedure A)

$$A = V_{AgNO_3} \times N_{AgNO_3} \times \frac{V(\text{total sample})}{V(\text{aliquot})}$$

Milliequivalents of iodine used (procedure B)

$$B = V_{\text{iodine}} \times N_{\text{iodine}} \times \frac{V(\text{total sample})}{V(\text{aliquot})}$$

Per cent of mercaptan present as tertiary mercaptan

$$\frac{B - A}{A} \times 100 = C$$

Per cent of mercaptan present as primary mercaptan $= 100 - C$

Titrations with Bromine (Potassium Bromate)

Bromine and bromate ion when added to an acid bromide solution yield diffusion currents at the rotating electrode which are proportional to the concentration of bromine. Chlorine behaves in a similar manner. On the other hand, hypobromite and hypochlorite ions were found (University of Minnesota) not to depolarize the dropping electrode. As might be expected, amperometric titrations with bromine or bromate–bromide in an acid medium can be made with the rotating electrode as indicator electrode; on the other hand, hypobromite is not suitable as a reagent since it does not yield a cathodic diffusion current.

Laitinen and Kolthoff[88] tested the general accuracy of the method by titrating arsenic trioxide solutions in 1 *N* hydrochloric acid and 0.05 *N* bromide with potassium bromate as the reagent. These authors applied an e.m.f. of 0.2–0.3 v. Later it was shown that the titration can be carried out without applying a voltage. The current is very small or zero with an excess of arsenic trioxide. After the end point has been reached the diffusion current of bromine is measured. The titration lines and the performance of the titration are similar to those described in the section on mercaptans. Oxygen does not interfere. With 0.001 *N* arsenic trioxide solutions an accuracy of 0.1 per cent was easily attained; with 0.0001 *N* solutions an accuracy and precision of 0.3 per cent were found. Even 10^{-5} *N* solutions can be titrated with satisfactory accuracy, taking due precautions to eliminate impurities which might affect the results. Trivalent *antimony* can be titrated like arsenic.

Undoubtedly many organic compounds which react rapidly by substitution (phenols and aniline) or addition (double bonds) with bromine can be titrated rapidly and accurately with bromine (or bromate–bromide) using the amperometric technique. Traces of phenol, aniline, and 8-hydroxyquinoline (oxine) have been titrated in this way in the senior author's laboratory. The titration of oxine should be particularly useful for the determination of traces of metals which yield insoluble 8-hydroxyquinolates.

A more detailed study has been made of the amperometric titration of traces of styrene.[89] It is necessary that the titration medium be a good solvent for styrene, as otherwise serious errors occur, due to volatilization of styrene during the course of the titration. This loss is promoted by the vigorous agitation of the solution by the rotating electrode. Methanol is a suitable solvent, since it does not react with romine during the titration. When a methanol–water mixture is used as a solvent it is recommended that the titration be carried out at 10° to 15° C. to minimize losses by volatilization. Styrene reacts with bromine according to the equation:

$$C_6H_5CH{=}CH_2 + Br_2 \rightarrow C_6H_5CHBrCH_2Br$$

Procedure. Into a 250-ml beaker pour about 75 ml. of methanol and 5 ml. of concentrated hydrochloric acid and cool in an ice bath to 5° to 10° C. Add the water solution of styrene from a pipet; the quantity of water added should not exceed 25 ml. (If a solution of styrene in some solvent other than water is used, the quantity of styrene added should not exceed 10 mg.) Cool for about 5 minutes more, until the temperature of the contents of the beaker is about 10° C.—the mixing of the water and methanol evolves considerable heat. Then insert the salt bridge and the rotating electrode, add about 1 g. of potassium bromide, and titrate with approxi-

[88] H. A. Laitinen and I. M. Kolthoff, *J. Phys. Chem.*, **45**, 1079 (1941).
[89] I. M. Kolthoff and F. A. Bovey, *Anal. Chem.*, **19**, 498 (1947).

mately 0.002 *M* potassium bromate, which has been made about 0.1 *M* in potassium bromide.

Before the end point a constant residual current of 0.2–0.5 microamp. is observed. After each addition of potassium bromate solution, the needle of the galvanometer deflects slightly to the right, but returns almost immediately to the residual current. As the end point is approached, the return of the needle becomes somewhat slower. Plot two or three points before the end point is reached, and three points after the end point, at intervals of 0.5 to 0.10 ml. As soon as the first excess of bromine is added, a diffusion current corresponding to the concentration of free bromine present is observed.

Quantities of the order of 5 mg. of styrene were titrated with an accuracy within 1 per cent. Even 0.5 mg. of styrene in a volume of 75 ml. could be titrated with an accuracy of 2 per cent.

Thiodiglycol. Sease *et al.*[90] made use of the oxidation of thiodiglycol in 50 per cent acetic acid by bromine to yield thiodiglycol sulfoxide for the determination of hydrolyzed bis(2-chlorethyl)sulfide. They used a pair of platinum electrodes having a potential difference impressed across them, and generated the bromine consumed in the titration electrolytically.

If two platinum electrodes are immersed in a solution containing bromide ion and a suitably small potential difference is applied between them, an appreciable current (more than 0.5 microamp.) flows only when the solution contains *bromine.* Preliminary experiments on the titration of thiodiglycol showed that the electrolytic generation of bromine in the solution was equivalent to the introduction of a corresponding amount of bromine in 50 per cent acetic acid. This made possible the elimination of unstable bromine solutions, which require frequent restandardization, and also permitted a permanent calibration in terms of electrical units.[91]

Chlorine, Chloramine, Chlorine Dioxide, and Chlorite in Water

Chlorine can be titrated accurately with arsenic trioxide solution, even at very large dilutions. Haller and Listek[92] use as indicator electrode a relatively large platinum wire wound around a porous cup. The reference electrode is a silver wire in saturated sodium chloride placed inside the cup. Instead of rotating the indicator electrode the solution is stirred thoroughly during the titration.

The amperometric titration technique is recommended for the determination of traces of residual chlorine, chloramine, chlorine dioxide, and

[90] J. W. Sease, C. Niemann, and E. H. Swift, *Anal. Chem.*, **19**, 197 (1947). See also R. J. Myers and E. H. Swift, *J. Am. Chem. Soc.*, **70**, 1047 (1948).

[91] Use of electrolytically generated hydroxyl ions from a sodium bromide solution in the titration of acids is made by J. Epstein, H. A. Sober, and L. D. Silver, *Anal. Chem.*, **19**, 675 (1947).

[92] J. F. Haller and S. S. Listek, *Anal. Chem.*, **20**, 639 (1948).

chlorite in water. In order to make chlorine dioxide harmless in the titration with arsenite at pH 7 the sample is first made alkaline with sodium hydroxide to a pH of at least 11 (at 25°) or 12 (at 0°):

$$2ClO_2 + 2OH^- \rightarrow ClO_2^- + ClO_3^- + H_2O$$

After 5 minutes' standing the pH is adjusted to 7 with a phosphate buffer and the titration is performed with a dilute standard arsenic trioxide solution. Under these conditions only the sum of *chlorine*, *hypochlorite*, and *hypochlorous acid* is determined.

Sum of Chlorine and Chloramine. The sample is treated with alkali as mentioned above and adjusted to pH 7. Then 1–2 g. of potassium iodide is added and the iodine titrated with arsenic trioxide solution.

Sum of Chlorine, Chloramine, and Chlorine Dioxide. The sample is not treated with alkali. The pH is adjusted to 7 with the phosphate buffer, iodide is added, and the solution is titrated. Under these conditions the chlorine dioxide is reduced only to chlorite.

Sum of Chlorine, Chloramine, Chlorine Dioxide, and Chlorite. Potassium iodide and sulfuric acid are added (to pH 2) and the mixture is allowed to stand for 5 minutes. The pH is adjusted to 7 and the iodine is titrated. Chlorine dioxide and chlorite are now reduced to chloride.

Residual Chlorine in Sewage. For the determination of residual chlorine in water Marks and Glass[93] add potassium iodide and titrate the iodine which is formed amperometrically with arsenic trioxide at a pH of around 7. This method cannot be applied to sewage, because the liberated iodine reacts with mild reducing agents always present in sewage. For this reason Marks *et al.*[94] add to the sewage an excess of a suitable reducing agent which can be back-titrated in a weakly acid medium with standard iodine solution by the amperometric technique. At a pH of 2.5 small amounts of iodine react so slowly with reducing substances in sewage that the amperometric end point is sufficiently stable. As a reducing agent phenylarsene oxide[95] was found suitable. This substance reacts quantitatively with iodine even at pH values below 2.5. To a 200-ml. sample they add 5 ml. of 0.00564 *N* phenylarsene oxide solution containing 7.5 g. of potassium iodide per liter. After thorough mixing the sample is treated with 2 ml. of 1 *N* hydrochloric acid. The excess reagent is back-titrated amperometrically with 0.0282 *N* iodine solution. The method gives results less subject to error than the 3-tolidine method.[96]

Cerous Cerium with Permanganate

Tripositive cerium behaves like Mn^{++} toward permanganate in the presence of sufficient pyrophosphate at pH 6 to 7, and is oxidized to Mn^{++++}

[93] H. C. Marks and J. R. Glass, *J. Am. Water Works Assoc.*, **34,** 1227 (1942).

[94] H. C. Marks, R. R. Joiner, and F. B. Strandskov, *Water & Sewage Works*, **95,** 175 (1948).

[95] See C. K. Banks and J. A. A. Sultzaberger, *J. Am. Chem. Soc.*, **69,** 1 (1947).

[96] H. C. Marks and R. J. Joiner, *Anal. Chem.*, **20,** 1197 (1948).

upon addition of permanganate. Goffart[97] has made this reaction the basis of a potentiometric and amperometric titration of cerium:

$$4Ce^{+++} + MnO_4^- + 15H_2P_2O_7^{--} + 8H^+ \rightarrow 4Ce(H_2P_2O_7)_3^{--} + Mn(H_2P_2O_7)_3^{---} + 4H_2O$$

Goffart uses a silver electrode coated with silver chloride as reference electrode. 200 ml. of a saturated solution of sodium pyrophosphate (approximately 0.3 *M*) adjusted to a pH of about 6 with the aid of hydrochloric acid is placed in a beaker. Then 25 ml. of cerous nitrate containing about 150 mg. of cerium is added and the mixture is titrated with 0.02 *N* permanganate. Before the end point a very slight positive current is observed due to the depolarizing effect of the pyrophosphatoceriate. After the end point the current increases rapidly with the excess of permanganate added. Near the end point it is necessary to wait for about 1 minute until a constant current reading is obtained. The current readings before and after the end point lie on two straight lines which intersect at the end point. The author does not state whether the method is suitable for the titration of microquantities of cerium.

Manganous Manganese with Permanganate

A similar method was developed by Goffart[98] for the determination of manganese. The solution is saturated with sodium pyrophosphate, the pH is adjusted to between 5 and 8, and the solution is titrated with permanganate solution. At the end point all the manganese is oxidized to Mn^{+++}. The end point is found in a similar manner as in the titration of cerium.

Ferrous Iron with Dichromate

Unfortunately, many of the common strong oxidizing agents like dichromate ion and ceric ion (and also hypobromite and hypoiodite) do not yield diffusion currents at the rotating platinum electrode even potentials that are so negative that the ions should be completely reduced. Evidently, these oxidizing agents give rise to complicated polarization phenomena at a rotating platinum cathode. When reducing agents are titrated at a relatively negative potential with the above-mentioned oxidizing agents no diffusion current is measured after the end point because of polarization of the indicator electrode.

In order to titrate amperometrically traces of ferrous iron with dichro-

---

[97] G. Goffart, *Anal. Chim. Acta*, **2,** 150 (1948).

[98] G. Goffart, *Anal. Chim. Acta*, **1,** 393 (1947).

mate, or traces of chromic acid with ferrous iron, Kolthoff and May[99] applied such a positive potential to the rotating electrode that ferrous iron yielded an anodic current, which was proportional to the iron concentration, while chromic acid, Cr^{+++}, and ferric iron did not cause any current at this potential. This condition is fulfilled when a potential of +1.0 v. (*vs.* S.C.E.) is applied to the rotating electrode. Thus, when an acid solution of chromate is titrated with a standard ferrous iron solution at this potential no current is observed until all the chromate has been reduced and an excess of ferrous iron is present. By measuring the current after adding successive amounts of ferrous iron and obtaining two or more values in the presence of an excess of ferrous solution the end point is found in the usual way by graphical extrapolation to zero current.

The standard iron solution used was 0.01 *N* in ferrous iron and 0.1 *N* in sulfuric acid. The chromate solution is made about 0.1 *N* in strong mineral acid (perchloric, hydrochloric, sulfuric, or nitric acid). The method was found to be accurate and precise to 0.5 per cent at concentrations of chromate as low as $1\text{–}2 \times 10^{-4}$ *M*. The method can also be used for the reverse titration of traces of ferrous iron with dichromate (or ceric sulfate). In such a titration the current decreases continuously and becomes zero at the end point, which is found graphically.

Kolthoff and Medalia[100] utilized the above method for the deterination of traces of organic peroxides (hydroperoxides) with an excess of ferrous iron, back-titrating this excess amperometrically with standard dichromate solution.

6. ADVANTAGES AND LIMITATIONS OF AMPEROMETRIC TITRATIONS

(a) Advantages

1. In general very dilute solutions can be titrated with a high degree of accuracy.

2. Precipitation titrations can be performed even when the solubility is relatively great, and under conditions where the potentiometric or indicator methods do not yield good results. However, solubility equilibrium must be attained relatively quickly for successful titration.

3. Foreign electrolytes, whose presence is harmful in conductometric titrations, do not interfere in amperometric titrations. Their presence is actually desirable in amperometric titrations to eliminate the migration current.

4. Substances which are not reduced or oxidized can be titrated if the

---

[99] I. M. Kolthoff and D. R. May, *Ind. Eng. Chem., Anal. Ed.*, **18,** 208 (1946).
[100] I. M. Kolthoff and A. I. Medalia, *Anal. Chem.*, **23,** 595 (1951).

reagent yields a diffusion current which is proportional to the concentration.

5. The titrations can be carried out rapidly because only a few measurements of current before and after the end point need be recorded.

6. It is immaterial whether the reaction which takes place during the titration is reversible or irreversible.

7. In routine work it is possible to simplify the polarographic equipment. The essential instrument is a sensitive galvanometer with a variable sensitivity.

8. Amperometric titrations, in general, yield more accurate results than direct polarographic measurements. It is of particular advantage that the results of the titration are independent of the characteristics of the capillary. Also, the temperature does not have to be known as long as it is kept constant during the titration.

9. In polarographic determinations the values of the diffusion current are more or less dependent upon the kind and concentration of supporting electrolyte. In amperometric titrations we are only interested in the relative change in the current and the supporting electrolytes have no effect upon the results as long as they do not interfere with the titration reaction.

10. In an amperometric titration the composition of the precipitate which is formed furing the titration is known. This is a special advantage when a precipitate decomposes on being washed or dried (organic reagents with metals).

11. Coprecipitation phenomena can be studied quantitatively.

(b) Disadvantages

Naturally, amperometric titrations are subject to the ordinary sources of error of volumetric determinations (instrumental errors, coprecipitations, etc.). In addition there is the serious limitation in the application of amperometric titrations that foreign substances which do not interefere in the titration reaction but which yield diffusion currents at the applied e.m.f. used for the titration must not be present in concentrations many times larger than that of the substance titrated. In their presence the relative changes in the current during the titration become smaller. If, for example, in the titration of a certain lead solution with oxalate the current decreases, say, from 60 to zero, a decrease from only about 660 to 600 would be found if the lead solution contained thallous ion in a concentration about ten times greater than that of lead. Under such circumstances one should try to find conditions under which the foreign constituent does not contribute to the current, or, if necessary, make a preliminary chemical separation.

Author Index, Volumes I and II

Volume I, pages 1–420; Volume II, pages 423–953

C

I

J

Subject Index, Volumes I and II

Volume I, pages 1–420; Volume II, pages 423–953

C

D

H

T